Lecture Notes in Computer Science 14146

The series Lecture Notes in Computer Science (LNCS), including its subseries Lecture Notes in Artificial Intelligence (LNAI) and Lecture Notes in Bioinformatics (LNBI), has established itself as a medium for the publication of new developments in computer science and information technology research, teaching, and education.

LNCS enjoys close cooperation with the computer science R & D community, the series counts many renowned academics among its volume editors and paper authors, and collaborates with prestigious societies. Its mission is to serve this international community by providing an invaluable service, mainly focused on the publication of conference and workshop proceedings and postproceedings. LNCS commenced publication in 1973.

Christine Strauss · Toshiyuki Amagasa ·
Gabriele Kotsis · A Min Tjoa · Ismail Khalil
Editors

Database and Expert Systems Applications

34th International Conference, DEXA 2023
Penang, Malaysia, August 28–30, 2023
Proceedings, Part I

 Springer

Editors
Christine Strauss
University of Vienna
Vienna, Austria

Gabriele Kotsis
Johannes Kepler University Linz
Linz, Austria

Ismail Khalil
Johannes Kepler University Linz
Linz, Austria

Toshiyuki Amagasa
University of Tsukuba
Ibaraki, Japan

A Min Tjoa 🆔
Vienna University of Technology
Vienna, Austria

ISSN 0302-9743   ISSN 1611-3349 (electronic)
Lecture Notes in Computer Science
ISBN 978-3-031-39846-9   ISBN 978-3-031-39847-6 (eBook)
https://doi.org/10.1007/978-3-031-39847-6

This Springer imprint is published by the registered company Springer Nature Switzerland AG
The registered company address is: Gewerbestrasse 11, 6330 Cham, Switzerland

Preface

Welcome to the proceedings of the 34th International Conference on Database and Expert Systems Applications (DEXA 2023). This gathering of brilliant minds from around the world serves as a testament to the tremendous progress made in the fields of data management, intelligent systems, and advanced algorithms. It is our great pleasure to present this compilation of papers, capturing the essence of groundbreaking research and innovative ideas that were shared during the conference.

The rapid advancements in technology have ushered in an era where data has become an invaluable asset, and its effective management and analysis have become critical for organizations across various domains. The integration of expert systems, artificial intelligence, and machine learning techniques has revolutionized the way we approach data, enabling us to extract insights, make informed decisions, and create intelligent systems that can adapt and learn from the vast amounts of information available.

This conference has provided a platform for researchers, academics, industry experts, and practitioners to come together and exchange their knowledge, experiences, and ideas. The proceedings reflect the diverse range of topics covered during the event, including but not limited to data modeling, database design, query optimization, knowledge representation, rule-based systems, natural language processing, deep learning, and neural networks.

The papers included in this volume represent the collective efforts of the authors who have dedicated their time and expertise to advancing the frontiers of database systems, expert systems, artificial intelligence, and machine learning. Each paper has undergone a rigorous review process by a panel of experts in the respective fields, ensuring the highest standards of quality and relevance.

As you delve into the pages of these proceedings, you will witness the fascinating discoveries, novel methodologies, and practical applications that are shaping the future of data-driven decision making. From the development of efficient algorithms for data processing to the creation of intelligent systems capable of autonomous reasoning and decision making, the papers within this volume illuminate the vast potential of interdisciplinary research.

We are proud to report that authors from more than 37 different countries submitted papers to DEXA this year. Our program committees have conducted close to five hundred single-blind reviews, with each submission receiving three reviews, on average. From 155 submitted papers the program committee decided to accept 49 full papers, 35 short papers with an acceptance rate of 31%, a rate lower than previous DEXA conferences.

We would like to express our heartfelt gratitude to the authors for their contributions and the dedication they have shown in presenting their work. We also extend our sincere appreciation to the members of the program committee, whose rigorous evaluation and insightful feedback have played a crucial role in shaping this collection.

Finally, we would like to thank the conference organizers, keynote speakers, and attendees for their invaluable support in making this event a resounding success. Their commitment to advancing the frontiers of knowledge and fostering collaboration has been instrumental in creating an environment conducive to intellectual growth and innovation.

It is our hope that these proceedings serve as a source of inspiration and knowledge for researchers, students, and professionals alike, as they embark on their own journeys to unravel the mysteries of data, expert systems, artificial intelligence, and machine learning.

August 2023

Christine Strauss
Toshiyuki Amagasa
Gabriele Kotsis
A Min Tjoa
Ismail Khalil

Organization

Program Committee Chairs

Christine Strauss	University of Vienna, Austria
Toshiyuki Amagasa	University of Tsukuba, Japan

Steering Committee

Gabriele Kotsis	Johannes Kepler University Linz, Austria
A Min Tjoa	Vienna University of Technology, Austria
Robert Wille	Software Competence Center Hagenberg, Austria
Bernhard Moser	Software Competence Center Hagenberg, Austria
Ismail Khalil	Johannes Kepler University Linz, Austria

Program Committee Members

Seth Adjei	Northern Kentucky University, USA
Riccardo Albertoni	CNR-IMATI, Italy
Sabri Allani	University of Tunis El Manar, Tunisia
Idir Amine Amarouche	USTHB, Algeria
Nidhi Arora	Amazon, India
Mustafa Atay	Winston-Salem State University, USA
Radim Bača	VSB - Technical University of Ostrava, Czech Republic
Ladjel Bellatreche	LIAS/ENSMA, France
Nadia Bennani	INSA de Lyon, France
Djamal Benslimane	Lyon 1 University, France
Karim Benouaret	Université Claude Bernard Lyon 1, France
Vasudha Bhatnagar	University of Delhi, India
Athman Bouguettaya	University of Sydney, Australia
Omar Boussaid	Université Lumière Lyon 2, France
Stephane Bressan	National University of Singapore, Singapore
Barbara Catania	University of Genoa, Italy
Brice Chardin	ENSMA, France
Asma Cherif	King Abdulaziz University, Saudi Arabia
Ruzanna Chitchyan	University of Bristol, UK
Soon Chun	City University of New York, USA

Camelia Constantin	Sorbonne University, France
Deborah Dahl	Conversational Technologies, USA
Matthew Damigos	Ionian University, Greece
Jérôme Darmont	Université Lyon 2, France
Soumyava Das	Teradata Labs, USA
José Gaviria de la Puerta	Universidad de Deusto, Spain
Vincenzo Deufemia	University of Salerno, Italy
Sabrina De Capitani Di Vimercati	Università degli Studi di Milano, Italy
Ivanna Dronyuk	Lviv Polytechnic National University, Ukraine
Cedric du Mouza	CNAM, France
Joyce El Haddad	Université Paris Dauphine - PSL, France
Markus Endres	University of Applied Sciences Munich, Germany
Nora Faci	Université Lyon 1, France
Bettina Fazzinga	University of Calabria, Italy
Jan Fell	Vienna University of Economics and Business, Austria
Flavio Ferrarotti	Software Competence Centre Hagenberg, Austria
Lukas Fischer	Software Competence Center Hagenberg, Austria
Flavius Frasincar	Erasmus University Rotterdam, The Netherlands
Bernhard Freudenthaler	Software Competence Center Hagenberg, Austria
Bouchra Frikh	Sidi Mohamed Ben Abdellah University, Morocco
Steven Furnell	University of Nottingham, UK
Pablo García Bringas	University of Deusto, Spain
Zoltan Geller	University of Novi Sad, Serbia
Manolis Gergatsoulis	Ionian University, Greece
Joseph Giovanelli	University of Bologna, Italy
Anna Gorawska	Silesian University of Technology, Poland
Sven Groppe	University of Lübeck, Germany
Wilfried Grossmann	University of Vienna, Austria
Giovanna Guerrini	University of Genoa, Italy
Allel Hadjali	ENSMA, France
Sana Hamdi	University of Tunis El Manar, Tunisia
Abdelkader Hameurlain	Paul Sabatier University, France
Hieu Hanh Le	Ochanomizu University, Japan
Sven Hartmann	Clausthal University of Technology, Germany
Manfred Hauswirth	TU Berlin, Germany
Ionut Iacob	Georgia Southern University, USA
Hamidah Ibrahim	Universiti Putra Malaysia, Malaysia
Sergio Ilarri	University of Zaragoza, Spain
Abdessamad Imine	Loria, France
Ivan Izonin	Lviv Polytechnic National University, Ukraine
Stéphane Jean	ISAE-ENSMA and University of Poitiers, France

Peiquan Jin	University of Science and Technology of China, China
Eleftherios Kalogeros	Ionian University, Greece
Anne Kayem	Hasso Plattner Institute, Germany
Carsten Kleiner	Hochschule Hannover - University of Applied Science and Arts, Germany
Michal Kratky	VSB-Technical University of Ostrava, Czech Republic
Petr Křemen	Czech Technical University in Prague and Cognizone, Czech Republic
Josef Küng	Johannes Kepler University Linz, Austria
Lynda Said Lhadj	Ecole nationale Supérieure d'Informatique, Algeria
Lenka Lhotska	Czech Technical University in Prague, Czech Republic
Jorge Lloret	University of Zaragoza, Spain
Qiang Ma	Kyoto University, Japan
Hui Ma	Victoria University of Wellington, New Zealand
Elio Masciari	Federico II University, Italy
Massimo Mecella	Sapienza Università di Roma, Italy
Sajib Mistry	Curtin University, Australia
Jun Miyazaki	Tokyo Institute of Technology, Japan
Lars Moench	University of Hagen, Germany
Riad Mokadem	Paul Sabatier University, France
Yang-Sae Moon	Kangwon National University, South Korea
Franck Morvan	IRIT and Université Paul Sabatier, France
Amira Mouakher	University of Perpignan, France
Philippe Mulhem	LIG-CNRS, France
Emir Muñoz	Genesys Telecommunications, Ireland
Francesc D. Muñoz-Escoí	Universitat Politècnica de València, Spain
Ismael Navas-Delgado	University of Málaga, Spain
Javier Nieves	Azterlan, Spain
Makoto Onizuka	Osaka University, Japan
Brahim Ouhbi	ENSAM, Morocco
Marcin Paprzycki	Systems Research Institute, Polish Academy of Sciences, Poland
Chihyun Park	Kangwon National University, South Korea
Louise Parkin	LIAS, France
Dhaval Patel	IBM, USA
Nikolai Podlesny	Hasso Plattner Institute and University of Potsdam, Germany
Simone Raponi	NATO STO CMRE, Italy
Tarmo Robal	Tallinn University of Technology, Estonia

Claudia Roncancio Ensimag, France
Massimo Ruffolo ICAR-CNR, Italy
Marinette Savonnet University of Burgundy, France
Rakhi Saxena University of Delhi, India
Florence Sedes IRIT and University of Toulouse III Paul Sabatier, France
Michael Sheng Macquarie University, Australia
Patrick Siarry Université Paris-Est Creteil, France
Gheorghe Cosmin Silaghi Babes-Bolyai University, Romania
Jiefu Song IRIT, France
Srinath Srinivasa Int. Institute of Information Technology, Bangalore, India
Bala Srinivasan Monash University, Australia
Christian Stummer Bielefeld University, Germany
Panagiotis Tampakis University of Southern Denmark, Denmark
Olivier Teste IRIT, France
A Min Tjoa Vienna University of Technology, Austria
Hiroyuki Toda Yokohama City University, Japan
Vicenc Torra Umeå University, Sweden
Nicolas Travers Pôle Universitaire Léonard de Vinci, France
Traian Marius Truta Northern Kentucky University, USA
Borja Sanz Urquijo Universidad de Deusto, Spain
Zheni Utici Georgia Southern University, USA
Yousuke Watanabe Nagoya University, Japan
Piotr Wisniewski Nicolaus Copernicus University, Poland
Vitaliy Yakovyna Lviv Polytechnic National University, Ukraine
Ming Hour Yang Chung Yuan Christian University, Taiwan
El Moukhtar Zemmouri Moulay Ismail University, Morocco
Qiang Zhu University of Michigan - Dearborn, USA
Yan Zhu Southwest Jiaotong University, China
Ester Zumpano University of Calabria, Italy

External Reviewers

Anys Bacha University of Michigan - Dearborn, USA
Zhengyan Bai JAIST, Japan
Wissal Benjira Pôle Universitaire Léonard de Vinci, France
Bernardo Breve University of Salerno, Italy
Simone Cammarasana IMATI-CNR, Italy
Renukswamy Chikkamath University of Applied Sciences Munich, Germany
Hsiu-Min Chuang Chung Yuan Christian University, Taiwan (R.O.C.)

Gaetano Cimino	University of Salerno, Italy
Kaushik Das Sharma	India
Chaitali Diwan	International Institute of Information Technology, Bangalore, India
Daniel Dorfmeister	Software Competence Center Hagenberg GmbH, Austria
Myeong-Seon Gil	Kangwon National University, South Korea
Ramón Hermoso	University of Zaragoza, Spain
A. K. M. Tauhidul Islam	Informatica, USA
Andrii Kashliev	Eastern Michigan University, USA
Khalid Kattan	University of Michigan - Dearborn, USA
Yuntao Kong	JAIST, Japan
Mohit Kumar	Software Competence Center Hagenberg GmbH, Austria
Yudi Li	Southwest Jiaotong University, China
Junjie Liu	Southwest Jiaotong University, China
Junhao Luo	Southwest Jiaotong University, China
Jorge Martinez-Gil	Software Competence Center Hagenberg GmbH, Austria
Moulay Driss Mechaoui	University of Abdelhamid-Ibn-Badis (UMAB), Algeria
Sankita Patel	Sardar Vallabhbhai National Institute of Technology, India
Maxime Prieur	CNAM & Airbus Defense and Space, France
Gang Qian	University of Central Oklahoma, USA
María del Carmen Rodríguez-Hernández	Technological Institute of Aragon, Spain
Hannes Sochor	Software Competence Center Hagenberg GmbH, Austria
Tung Son Tran	University of Applied Sciences Munich, Germany
Óscar Urra	Technological Institute of Aragon, Spain
Alexander Völz	University of Vienna, Austria
Haihan Wang	Southwest Jiaotong University, China
Yi-Hung Wu	Chung Yuan Christian University, Taiwan (R.O.C.)
Qin Yang	Southwest Jiaotong University, China
Chengyang Ye	Kyoto University, Japan
Kun Yi	Kyoto University, Japan
Chih-Chang Yu	Chung Yuan Christian University, Taiwan (R.O.C.)
Foutse Yuehgoh	Pôle Universitaire Léonard de Vinci, France
Shilong Zhu	Southwest Jiaotong University, China

Organizers

Abstracts of Keynote Talks

Physics-Informed Machine Learning

Stéphane Bressan

National University of Singapore, Singapore

Abstract. In 1687, Isaac Newton published his groundbreaking work, "Philosophiæ Naturalis Principia Mathematica". Newton's remarkable discoveries unveiled the laws of motion and the law of universal gravitation, propelling humanity's understanding of the physical world to new heights. In a letter to Robert Hooke in 1675, in response to an invitation to collaborate, Newton humbly remarked, "If I have seen further, it is by standing on the shoulders of giants." This metaphor swiftly became a powerful symbol of intellectual and scientific progress, signifying the idea that knowledge is built upon foundations laid by brilliant minds that came before us.

Fast-forwarding to the present, we find ourselves amidst a triumphant statistical machine learning revolution. In 2016, Google's AlphaGo, a deep reinforcement learning algorithm, astounded the world by outperforming a professional Go player. The following year, CheXNet, a deep convolutional neural network developed at Stanford University, surpassed radiologists in accurately detecting pneumonia from chest X-ray images. And in 2020, AlphaFold, a neural network model created by DeepMind, revolutionised protein structure prediction, surpassing other existing methods.

These advancements stand on the shoulders of giants. They owe their existence to the work of logicians, mathematicians, physicists, neurobiologists, computer scientists, and cyberneticists who have paved the way for the birth of modern machine learning models and algorithms. They also owe their existence to the work of material, electrical, electronics and other engineers, whose ingenuity has birthed the computer hardware and technology enabling such performance.

However, the remarkable ascent of machine learning is not solely reliant on these contributions. It thrives on the vast amounts of data permeating the global information infrastructure, enabling the construction of accurate representations of the world. What about knowledge?

In this context, we propose exploring and discussing how machine learning can both leverage and contribute to scientific knowledge. We explore how the training of a machine learning model can be informed by the fundamental principles of the very systems it seeks to comprehend and how it can create symbolic scientific knowledge. We explore applications in classical mechanics, fluid mechanics, quantum many-body systems, macroeconomics, chemistry, and astronomy. Along this journey, we cross the paths of such great minds as William Rowan Hamilton, Ernst Ising, Richard Feynman, and Johannes Kepler.

Data Integration Revitalized: from Data Warehouse through Data Lake to Data Mesh

Robert Wrembel

Faculty of Computing and Telecommunications,
Poznan University of Technology Poland

Abstract. For years, data integration (DI) architectures evolved from those supporting virtual integration (mediated, federated), through physical integration (data warehouse), to those supporting both virtual and physical integration (data lake, lakehouse, polystore, data mesh/fabric). Regardless of its type, all of the developed DI architectures include an integration layer. This layer is implemented by a sophisticated software, which runs the so-called DI processes. The integration layer is responsible for ingesting data from various sources (typically heterogeneous and distributed) and for homogenizing data into formats suitable for future processing and analysis. Nowadays, in all business domains, large volumes of highly heterogeneous data are produced, e.g., medical systems, smart cities, precision/smart agriculture, which require further advancements in the data integration technologies. In this paper, I present my subjective view on still-to-be developed data integration techniques, namely: (1) novel agile/flexible integration techniques, (2) cost-based and ML-based execution optimization of DI processes, and (3) quality assurance techniques in complex multi-modal data systems.

Contents – Part I

Query Optimization

Contents – Part II

Neural Networks

Keynote Paper

Data Integration Revitalized: From Data Warehouse Through Data Lake to Data Mesh

Robert Wrembel[1,2](✉) [iD]

[1] Poznan University of Technology, Poznań, Poland
robert.wrembel@cs.put.poznan.pl
[2] Artificial Intelligence and Cybersecurity Center, Poznań, Poland

Abstract. For years, data integration (DI) architectures evolved from those supporting virtual integration, through physical integration, to those supporting both virtual and physical integration. Regardless of its type, all of the developed DI architectures include an integration layer. This layer is implemented by a sophisticated software, which runs the so-called DI processes. The integration layer is responsible for ingesting data from various sources (typically heterogeneous and distributed) and for homogenizing data into formats suitable for future processing and analysis. Nowadays, in all business domains, large volumes of highly heterogeneous data are produced, e.g., medical systems, smart cities, smart agriculture, which require further advancements in the data integration technologies. In this keynote talk paper, I present my personal opinion on still-to-be developed data integration techniques - potential research directions, namely: (1) more flexible DI, (2) quality assurance in complex multi-modal systems, (3) execution optimization of DI processes.

Keywords: data integration architecture · data quality · performance optimization of integration process

1 Introduction

The *data integration* (DI) research area has been active for already six decades. A common goal of DI is to make heterogeneous and typically distributed data available for an end user in a unified format. Research and development resulted in a few standard DI architectures, namely: (1) federated and mediated, (2) data warehouse, (3) lambda, (4) data lake, (5) lake house, (6) polystore, and (7) data mesh/fabric. Each of these architectures has its advantages and disadvantages as well as different application fields.

In all of the aforementioned architectures data are moved from source systems into an integrated system by means of an integration layer. This layer is

C. Strauss et al. (Eds.): DEXA 2023, LNCS 14146, pp. 3–18, 2023.
https://doi.org/10.1007/978-3-031-39847-6_1

implemented by a sophisticated software, which runs the so-called DI processes (a.k.a. ETL - in data warehouse architectures, data processing pipeline - in data science, or data wrangling, or data processing workflows (DPW) [48,59]), each of which runs a sequence of tasks (steps). These tasks allow ingesting data from various sources and transforming data into formats suitable for analytical and machine learning applications.

For years it has been observed the widespread of complex, data-driven systems, e.g., medical systems, smart agriculture, smart cities. These systems produce huge volumes of highly heterogeneous (a.k.a. multi-modal) data that need to be integrated, to feed various descriptive analytics of prediction models. Thus, DI architectures re-gained their popularity but facing new challenges.

In this paper, which accompanies my keynote talk, I will present my subjective point of view on data integration challenges, in the context of data-driven systems. This point of view is based on my experience in being engaged in a few projects on data integration in: the financial sector, medical sector, agriculture, and software development. In this paper I will focus on the need for: (1) novel more flexible DI techniques, (2) advanced quality assurance techniques in complex systems, and (3) execution optimization of DI processes.

2   Data Integration Architectures: Overview

As mentioned in Sect. 1, a few data integration architectures have been developed so far. Figure 1 shows a generalized DI architecture, where data sources (DSs) are connected by a *data integration layer*, which exists in all the aforementioned architectures. In general, this layer is responsible for: (1) ingesting data from DSs, (2) transforming data into formats suitable for analytical and machine learning applications, (3) cleaning and homogenizing values, deduplicating data, and (4) making integrated data for applications. *Integrated data* delivered by the DI layer can be made available either as virtual or as materialized. In the first case, such integration (thus architectures) are called *virtual*. Integration techniques (architectures) that persistently store integrated data in a repository are called *materialized*.

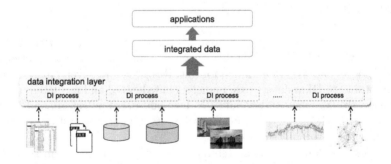

Fig. 1. A generalized data integration architecture

Virtual DI architectures include **federated databases** [17,24,60] and **mediator**-based systems [18,71]. In these architectures, data are integrated on demand by a software that is responsible for: (1) transforming source data models into a common integration model (frequently the relational one), (2) decomposing user queries into sub-queries and routing them into appropriate DSs for execution, (3) transforming the routed sub-queries into programs understandable and executable in a given DS, (4) transforming and integrating results returned by the queried DSs. The main difference between the federated and mediated architecture is that the first one is used to integrate databases built on the same data model (relational) and it uses one access interface (query language). The mediated architecture is applied to integrate not only databases but also other types of DSs.

Particular types of virtual DI architectures include [67]: polyglot, multi-store, and polystore. **Polyglot** allows to access multiple DSs built on the same data model by means of multiple access interfaces, e.g., SQL-like, procedural. **Multistore** allows to integrate DSs built on various data models and to access data via a single interface, e.g., a query language. **Polystore** allows to integrate DSs built on multiple data models by means of multiple access interfaces.

The first representative of a physical integration is a **data warehouse** (DW) architecture [43,70], where the integration is implemented by means of the socalled extract-transform-load (ETL) processes [63,68]. They are responsible for: (1) ingesting data from data sources and often storing them in an intermediate staging area, (2) transforming heterogeneous data into a common data model and schema used by an end user, (3) cleaning (e.g., removing errors, inconsistencies), normalizing, and eliminating duplicates, (4) loading data into a persistent repository, which is a data warehouse. This architecture is unrivalled in application domains like insurance, finances, trading, sales, which process large volumes of simple data, e.g., strings, numbers, and dates.

The standard DW architecture extended with capabilities of collecting data that arrive in data streams is called **lambda** [28,52]. It includes two data processing lanes - the standard batch one and the real-time one. The architecture was developed in order to be able to analyze batch-arriving data with stream-arriving data in the same system. Both lanes are integrated using the serving layer, which is typically implemented by means of virtual views (virtual integration) and/or materialized views (physical integration).

The proliferation of social media, IoT, robotics, and other industrial devices imposed a need for physically integrating data of complex, heterogeneous structures. To this end, the so-called data lake was proposed. A **data lake** is a repository that stores heterogeneous data ingested from DSs in their original formats [32,53]. Such data have to be further homogenized by DI processes, to produce data available for applications, e.g., [44]. In a pure data lake architecture, data are unified on-the-fly, like in a mediated system. This class of DI architectures is **hybrid**, i.e., it combines physical with virtual integration.

In a **data lakehouse** [25,34,62,73] data coming from a data lake are first unified by DI processes and then physically stored in one or more data warehouses, which are part of the whole architecture. Each data warehouse provides data prepared for specific analytical applications.

Recently, a technological concept called data mesh has gained popularity in DI. A **data mesh** defines a data architecture and data governance approach, where data ownership is decentralized [9,22]. Each component in a mesh is a DS having a dedicated owner. The owner is responsible for maintaining its data clean and up-to-date. Such a DS is made available via a standardized interface to other DS in the mesh. A data mesh architecture is implemented by a set of technologies, which are called **data fabric** [66]. A data fabric includes among others: data storage and data management systems (e.g., databases, distributed file systems), DI architectures, queuing systems and message brokers, data security and governance frameworks, as well as data analytics and visualization.

3   Data Integration Use-Cases

In this section three use-cases are outlined, which challenge DI processes.

3.1   Data Integration in Medicine

Medical systems are composed of multiple devices (e.g., imaging) and various information systems. Structured patient records are managed by the so-called Health Information System (HIS), whereas medical images are managed by the so-called Picture Archiving and Communication System (PACS). Medical images are transmitted via network and exchanged between systems by means of a standardized format called *DICOM* [23]. Medical data include various formats, like: (1) structured electronic patient medical record, (2) partially structured or unstructured short and long texts (medical interviews), (3) multiple types of medical images, e.g., X-ray, CT, MRI, PET, ultrasound.

Recent trends in medical data analytics tend to combine phenotype data (i.e., observable characteristics of a living organism, like physical, physiological, behavioral) with genotype data (encoded in DNA), in order to build a complex model of a given patient. The goal of this model is to predict future illnesses and to design a course of the most efficient treatment. Phenotype data combined with genotype data requires integrating all multi-modal data collected in the course of a patient history.

Since the volumes of data to be integrated and analyzed may be extremely large (especially multi-modal images), the most promising DI architecture for this domain would be a data lakehouse. In this architecture, all multi-modal data would be ingested into a data lake and then based on this repository, a few specialized data warehouses could be constructed, each for a given analytical purpose.

3.2   Data Integration in Smart Agriculture

Smart agriculture (a.k.a. precision agriculture, smart farming, digital farming, sustainable agronomy) refers to the application of advanced engineering technologies and sophisticated data analysis software to optimize agricultural practices. These technologies and software are used to monitor, analyze, and manage various aspects of crop production, livestock management, and resource utilization. Some key technologies used in smart agriculture include: (1) IoT sensors and other devices to collect in real-time data on soil parameters, crop growth, meteorological conditions, and livestock health, (2) imaging devices to collect images on crop illnesses, vegetation progress, and field conditions, (3) autonomous airborne and ground robots to perform field works (e.g., monitoring, planting, spraying, watering, harvesting).

Smart agriculture is a data driven business. Machinery used there produces highly heterogeneous and massive data. Running smart farming needs assistance from IT technologies such as: IoT, big data, analytics, computer vision, cloud computing, and artificial intelligence (AI) [2]. In this context, providing technologies for efficiently storing large volumes of multi-modal data, integrating, and making them suitable for further analysis is again a great challenge.

The DI architecture that seems to be adequate for this type of application scenario is again a data lakehouse - a data lake with all field data collected and a few specialized data warehouses, each of which serving data for a specific analytical task, e.g., scheduling robots tasks and trajectories, crop disease correlation with whether conditions, scheduling of field watering and spraying. On top of it, fast arriving data, in a from of streams, need to be analyzed in a real time, to handle unexpected events on robots, e.g., obstacles, equipment failures.

Assuring acceptable quality of field data may be difficult in smart farming. The fact that agrirobots and IoT devices operate in a harsh environment make the data error-prone, which result among others from: (1) areas without network coverage, (2) malfunctioning of IoT devices, (3) bad weather conditions (e.g., snowing) and dirt, which may distort generated signals.

3.3   Data Integration in Smart Cities

The concept of a *smart city* has been researched and developed already for years. This concept is supported by technologies, including among others: decision support and urban planning software, simulation sofware (digital twin), networks (wired and wireless), IoT, robots (airborne and ground). Some of the typical applications used within a smart city include: traffic management, energy management, waste management, water management, environment monitoring, public safety and security, urban planning, incident detection, predictive maintenance. All of these activities and applications are data-driven.

Similarly as in the two aforementioned use-cases, data produced within a smart city are highly multi-modal, which again makes their integration and analysis challenging. Some of data are made open, but they are not always well described by metadata. Therefore, understanding the structure of a given data set, how data were created and what observed phenomena do they represent, may not always be possible. Another problem is to locate a data set of interest. To this end, well organized open data sets are needed, augmented with rich metadata. Moreover, without rich metadata one is probably not able to fully assess the quality of such data.

In a smart city, data are produced and collected by many independent entities that make their data available via multiple separate services. For this reason, its seems that a data mesh architecture could be suitable for making these data repositories (and systems) interoperable. Moreover, recent explosion of AI algorithms applications in smart cities [1] requires DI integration techniques that support specific data pre-processing, suitable for AI algorithms.

4   Subjective View on Challenges in Data Integration

In the architectures outlined in Sect. 2, DI processes execute complex data tasks on data and they move large volumes of data from DSs into a target system. Therefore, their execution is typically time costly and reducing it (performance optimization) is of high importance.

In my subjective opinion, I strongly believe that the medical domain, smart farming, and smart cities generate the biggest challenges for DI. It is due to: (1) multi-modal data that the domains produce, (2) high volumes of data to be ingested and integrated, (3) fast changing data flows from machinery and IoT devices.

Multi-modality of data needs advanced storage, fast data access methods, data models/representations suitable for data analytics and AI. High data volumes need powerful computers to ingest and process data, often in real time. Fast changing data flows means that the throughput may change instantly from low to very high and a DI system must be able to handle such throughput peaks. In complex systems it is likely that new data producers appear frequently and they need to be integrated into the system. Such an integration, ideally should be automatic. This results in the need to develop new flexible or agile ways of integrating new data producers. Standard, static, ETL-like architectures seem not to be suitable for such applications.

In the following sections I outline challenges in DI that, in my opinion, are of high importance to be able to build efficient applications in data-driven business domains. These challenges include: (1) more flexible DI techniques, (2) quality assurance techniques, and (3) performance optimization of DI processes.

5   Flexible Data Integration

Designing and implementing methods for DI processes, used so far, seem not to be well suited for fast changing systems that feed with data an analytical system (e.g., a data lake or a lakehouse). Static connections to DSs and static mapping between DS schema objects and a destination schema objects are not sufficient. In the aforementioned complex application domains, new machinery and IoT devices may appear and disappear rapidly. Therefore, new means of connecting such kind of DSs are recommendable.

The first technology that comes into play are DS connectors - pieces of software that allow to access a given data source and ingest data from it. Typically, a connector is available as a library that has to be installed in an integration system. If multiple DS are to be connected, multiple separate dedicated connectors have to be installed. Connectors used as libraries have multiple disadvantages, e.g., difficult maintainability due to complex dependencies between versions of software installed in the system, limited support for new DSs, non-optimal performance, security leaks [14].

Nowadays, there exist a plethora of data management and data storage systems, which include also non-relational ones (e.g., NoSQL, graph, distributed file systems of row or columnar organization). For example, the number of different data sources reaches more than 350 [26]; IBM Cloud Pak for Data [40] includes over 100 built-in connectors to different data sources.

For a flexible data integration, the *library of connectors as a service* (LCS) may come into play [14]. Its basic idea is visualized in Fig. 2. Connectors to various DSs located under the *library of connectors* map native interfaces of DSs into a common access interface. This interface is made available to *applications* willing to access the DSs by means of the *connection server*. The *dispatcher* is responsible for instantiating a connector to a given DS and forwarding request to this connector, similarly as a mediator in the mediated DI architecture, see Sect. 2. Notice that the advantage of the LCS architecture is in its integrated additional services, available out of the box, for example: metadata management, data governance, data vault/credentials management, data access policy management, data access monitoring and collecting runtime statistics.

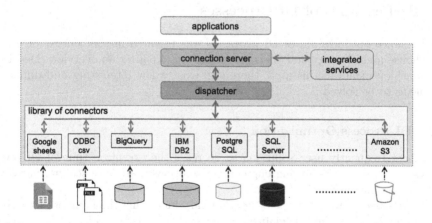

Fig. 2. The architecture of the *library of connectors as a service*

6 Data Quality

Data quality is challenging in traditional information systems, which store simple relational data (strings, numbers, and dates), see for example [16,62]. For relational data, multiple cleaning and homogenizing solutions have been successfully developed [20,21]. Assuring high data quality becomes more difficult in complex systems, like the aforementioned medical, smart agriculture, and smart cities. In such systems, there are basic sources of data errors, namely: (1) human factor, when a user enters data into a system using devices like keyboards, OCRs, and technologies like speech to text, (2) devices that may be malfunctioning, e.g., sensors, (3) a network that may be (periodically) malfunctioning or be unavailable, and (4) in some application areas - severe environmental conditions.

For checking and assuring the quality of multi-modal data, e.g., time series, graphs, large texts, as well as images, sounds, video sequences of various formats, dedicated quality metrics and dedicated data cleaning techniques are needed for each specific data format. Intensive research in this direction has already started - the https://dblp.org/ service lists over 6000 publications including keyword 'data quality' and the growth of publications on this topic in time is much faster than linear. Moreover, assuring data quality should be deployed not only at an integration layer, but primarily, if possible, at the devices (edge computing). Furthermore, ubiquitous AI algorithms need data of high quality as erroneous data used in the learning phase will produce erroneous models. Similarly, any bias in training data will result in AI models reflecting this bias. Moreover, the quality of ML models may depend on data pre-processing [8], therefore the development of a DI system that would assist a user in designing a DI process for a given application scenario is another topic to be researched - works in this direction have already started [29].

7   Performance of DI Processes

Performance optimization of DI processes was research in the context of data warehouse architectures and ETL processes, see [4,63] for an overview. Despite some achievements in this area, there still exist a few interesting and difficult problems to be solved.

7.1   DI Process Optimization

The most frequently used solutions to decrease the execution time of DI processes comprises the following engineering solutions: hardware scaling, parallel processing, and task orchestration.

Hardware scaling can be either vertical - increasing the number of CPUs, the size of RAM, adding specialized hardware like FPGAs [50,54] or horizontal - adding new processing nodes into a DI architecture. On such a hardware, DI processes or parts of them can be run in *parallel*, with data parallelism and/or task parallelism. In data parallelism, data are partitioned and the same task is run concurrently on all partitions. In task parallelism different pieces of code (tasks) are run concurrently on different CPUs (nodes).

Research still needs to be done to select those DI processes or particular tasks within a given process that could profit from parallelization. The next issue is to decide what would be the most efficient parallelization schemes for them. Some results in this direction have already been published [5,6], but they need an in-depth analysis and improvements.

Task orchestration consists in reordering tasks in a DI process, such that the final order is more efficient than the original one. This technique was originally researched for optimizing ETL processes in a relational data warehouse architecture. Finding an optimal sequence of tasks in a DI process requires finding and evaluating all possible valid orders of tasks. The evaluation is based on execution costs of the evaluated task orders. As the full search space is too large for real industrial DI processes, some search heuristics must be applied.

Pioneering works on this topic were presented in [64,65], which proposed some search space heuristics, followed by other techniques, like: (1) scheduling task execution [46] and (2) dividing complex task orders into linear orders and optimizing the linear orders [49]. Those solutions provide sub-optimal task sequences. In [33], the authors proposed a method for collecting execution statistics of an DI process for the purpose of using them in the optimization process. In [51], the idea from [49] was extended to apply parallel processing to each linear order of tasks (sub-process). [56] discussed a solution for constructing DI processes for data science problems, by modeling a DI process as a directed acyclic graph and applying AutoML techniques for finding a (sub-)optimal task order. [45] addressed the problem of integrating multiple DI processes to minimize the execution cost of a new integrated process.

Based on the aforementioned solutions, we can draw the following conclusions. First, defining the cost function for the whole DI process is difficult due to the complexity of such a process w.r.t. the number of tasks and the structure of the process, which may contain flow splits and conditional executions of tasks. Second, data characteristics to be used by a cost-based process optimizer are not known in advance (not like in databases), thus statistics on data must be computed either on-line or the execution of the process must be delayed until the statistics are computed. As a result, the optimization of such a process must be run before every execution. Third, in complex integration systems, new DSs to be plugged into the system appear and disappear frequently, thus, the structure of a DI process evolves, which results in execution cost evolution as well. Finally, in the existing DI tools (commercial and open) the aforementioned types of optimizations are not available.

Only two task orchestration techniques called push down and balanced optimization, are available in commercial DI tools. The *push down* optimization consists in moving some tasks into DSs to be executed there, to reduce a data volume (I/O) as soon as possible [42]. IBM extended this technique into the *balanced* optimization [41], where some tasks are pushed down into DSs, whereas other tasks are pushed-up into a destination storage system, to be executed there. Currently, these techniques work only on relational DSs. To the best of our knowledge, the push-down/up techniques for DSs other than relational ones have not been well researched yet, but some initial works in this direction has started [11,12]. For this type of optimization, there are needed techniques for:

(1) deciding what to push-down/up and (2) finding the most efficient implementations of a pushed-down/up task in source/destination systems, taking into account internals of these systems.

Another technique of reducing execution time of a DI process is to cache intermediate data produced by some integration tasks [35], in order to use them by other tasks within the same DI process or by different processes. A research problem here is to find out which results to cache to maximize performance. This problem is similar to the *materialized view selection* problem, researched for decades, e.g., [31,47].

7.2   UDFs in DI Processes Make Life More Difficult

Integrating multi-modal data frequently requires a custom data integration code, called a *user defined function* (UDF). Such a code may be implemented in any programming language and it is called from a DI engine as an external program. Such UDFs are frequently treated as *black-boxes*, since their internal logic and performance characteristics are not known to a DI process designer. It must be noted that even simple row data (text strings, dates, and numbers) in some cases have to be processed by UDFs. Advanced data cleaning and data deduplication in industrial projects may require non-standard algorithms, and such algorithms are called as external modules from a DI engine, e.g., [7,15]. As a consequence, optimization means for DI processes with UDFs are very limited, if not possible at all.

The approaches outlined in Sect. 7.1 do not support DI processes with UDFs. In order to optimize a DI processes with a black-box UDF, a DI engine must know performance characteristics of the UDF and (if possible) its semantics, in order to build its cost model.

Some approaches to learning such UDF characteristics have been proposed in the research literature. The most frequently proposed, and probably the easiest technique is to annotate a UDF with hints or rules that allow to figure out automatically some of its characteristics [30,38,39,58]. The annotations instruct DI engine how to re-order a workflow or instruct how to execute it in parallel. An alternative technique was proposed in [10]. It compares input and output attributes as well as input and output data, to figure out whether attribute projection and/or data filtering is executed by the UDF.

Complementary research results on optimizing the execution of UDFs (treated as white boxes) in databases were published in [57,61], where a UDF is translated either into relational algebraic expressions [57] or compiled into a low level executable code, merged with a compiled SQL code [61] for execution in a DBMS. The solution proposed in [27] allows to parallelize UDFs in the map-reduce framework, but UDFs must be implemented from scratch for this environment. In [3] the authors describe the so-called parallelization skeletons, which are code templates used for implementing UDFs for parallelized execution.

In recent years, a common trend is observed to support performance tuning of systems and software pieces by machine learning (ML) algorithms, e.g.,

[19,36,37,55], see also [69] for a brief overview. To this end, performance characteristics (typically, CPU, I/O, memory usage) must be collected, as part of program testing and/or during its normal execution. Based on the characteristics, various model building techniques are used, e.g., [72]. ML algorithms require large volumes of test data to learn reliable performance models. Thus, easy to configure and deploy architectures for running excessive experimental evaluations are needed, for the purpose of collecting massive performance data [13].

In the process of discovering characteristics of black-box UDFs, ML techniques may be profitable. Having collected large number of training performance data, during known UDF testing, one could apply ML techniques to discover patterns of resource usage of those known UDFs. They would serve as labeled performance patterns stored in a repository. For unknown black-box UDFs, the content of the repository could be used to find patterns that were the most similar to the ones exposed by a black-box UDF at hand. This way, one could get some insights (with a certain level of probability) on the semantics and performance of the black-box UDF.

8   Summary

This paper outlined challenges in DI integration, which in my personal opinion, are important to be investigated in the context of complex data-driven systems. These challenges include:

- novel techniques for flexible data integration;
- novel techniques for data quality assessment and assurance;
- cost-based performance optimization of DI processes with the support of ML;
- enhancement to performance optimization of DI processes by means of parallelization (methods for deciding which DI processes or particular tasks profit from parallelization and what are their efficient parallelization schemes);
- methods for discovering semantics of black-box UDFs with the support of ML;
- enhancement of the push-down/up techniques for non-relational DSs (methods for deciding which tasks profit from these techniques and how to efficiently implement a pushed-down/up task in a source/destination system).

The research literature on DI processes, includes multiple, but separate approaches to DI process design, implementation, and optimization. However, an end-to-end approach to DI process design at a logical level, physical level, optimization, and deployment (in the spirit of [59]) is still to be researched. Finally, designing DI processes for preparing data for ML algorithms is another topic to be researched.

References

1. Ahle, U., Hemetsberger, L., Lakomski, M., Wrembel, R.: AI and data: how cities of the future will use data in their development (2023)
2. Akkem, Y., Biswas, S.K., Varanasi, A.: Smart farming using artificial intelligence: a review. Eng. Appl. Artif. Intell. **120**, 105899 (2023)
3. Ali, S.M.F., Mey, J., Thiele, M.: Parallelizing user-defined functions in the ETL workflow using orchestration style sheets. Int. J. Appl. Math. Comput. Sci. **29**(1), 69–79 (2019)
4. Ali, S.M.F., Wrembel, R.: From conceptual design to performance optimization of ETL workflows: current state of research and open problems. VLDB J. **26**(6), 777–801 (2017). https://doi.org/10.1007/s00778-017-0477-2
5. Ali, S.M.F., Wrembel, R.: Towards a cost model to optimize user-defined functions in an ETL workflow based on user-defined performance metrics. In: Welzer, T., Eder, J., Podgorelec, V., Kamišalić Latifić, A. (eds.) ADBIS 2019. LNCS, vol. 11695, pp. 441–456. Springer, Cham (2019). https://doi.org/10.1007/978-3-030-28730-6_27
6. Ali, S.M.F., Wrembel, R.: Framework to optimize data processing pipelines using performance metrics. In: Song, M., Song, I.-Y., Kotsis, G., Tjoa, A.M., Khalil, I. (eds.) DaWaK 2020. LNCS, vol. 12393, pp. 131–140. Springer, Cham (2020). https://doi.org/10.1007/978-3-030-59065-9_11
7. Andrzejewski, W., Bebel, B., Boiński, P., Sienkiewicz, M., Wrembel, R.: Text similarity measures in a data deduplication pipeline for customers records. In: International Workshop on Design, Optimization, Languages and Analytical Processing of Big Data (DOLAP), volume 3369 of CEUR Workshop Proceedings, pp. 33–42. CEUR-WS.org (2023)
8. Bilalli, B., Abelló, A., Aluja-Banet, T., Wrembel, R.: PRESISTANT: learning based assistant for data pre-processing. Data Knowl. Eng. **123**, 101727 (2019)
9. Bode, J., Kühl, N., Kreuzberger, D., Hirschl, S., Holtmann, C.: Data mesh: best practices to avoid the data mess. CoRR, abs/2302.01713 (2023)
10. Bodziony, M., Krzyzanowski, H., Pieta, L., Wrembel, R.: On discovering semantics of user-defined functions in data processing workflows. In: International Workshop on Big Data in Emergent Distributed Environments (BiDEDE) @ SIGMOD/PODS, pp. 7:1–7:6. ACM (2021)
11. Bodziony, M., Morawski, R., Wrembel, R.: Evaluating push-down on nosql data sources: experiments and analysis paper. In: International Workshop on Big Data in Emergent Distributed Environments (BiDEDE) @ SIGMOD/PODS, pp. 4:1–4:6 (2022)
12. Bodziony, M., Roszyk, S., Wrembel, R.: On evaluating performance of balanced optimization of ETL processes for streaming data sources. In: International Workshop on Design, Optimization, Languages and Analytical Processing of Big Data (DOLAP), volume 2572 of CEUR Workshop Proceedings, pp. 74–78 (2020)
13. Bodziony, M., Wrembel, R.: Reference architecture for running large scale data integration experiments. In: Strauss, C., Kotsis, G., Tjoa, A.M., Khalil, I. (eds.) DEXA 2021. LNCS, vol. 12923, pp. 3–9. Springer, Cham (2021). https://doi.org/10.1007/978-3-030-86472-9_1
14. Bodziony, M., Wrembel, R.: Data source connectors layer as a service - design patterns. In: International Workshop on Design, Optimization, Languages and Analytical Processing of Big Data (DOLAP), volume 3369 of CEUR Workshop Proceedings, pp. 76–80. CEUR-WS.org (2023)

15. Boiński, P., Andrzejewski, W., Bębel, B., Wrembel, R.: On tuning the sorted neighborhood method for record comparisons in a data deduplication pipeline. In: International Conference on Database and Expert Systems Applications (DEXA). Springer, Cham (2023). Volume to appear of LNCS

16. Boinski, P., Sienkiewicz, M., Bebel, B., Wrembel, R., Galezowski, D., Graniszewski, W.: On customer data deduplication: lessons learned from a R&D project in the financial sector. In Workshops of the EDBT/ICDT Joint Conference, volume 3135 of CEUR Workshop Proceedings (2022)

17. Bouguettaya, A., Benatallah, B., Elmargamid, A.: Interconnecting Heterogeneous Information Systems. Kluwer Academic Publishers, Alphen aan den Rijn (1998). ISBN: 0792382161

18. Brezany, P., Tjoa, A.M., Wanek, H., Wöhrer, A.: Mediators in the architecture of grid information systems. In: Wyrzykowski, R., Dongarra, J., Paprzycki, M., Waśniewski, J. (eds.) PPAM 2003. LNCS, vol. 3019, pp. 788–795. Springer, Heidelberg (2004). https://doi.org/10.1007/978-3-540-24669-5_103

19. Chen, X., et al.: Leon: a new framework for ml-aided query optimization. VLDB Endowment **16**(9), 2261–2273 (2023)

20. Christophides, V., Efthymiou, V., Palpanas, T., Papadakis, G., Stefanidis, K.: An overview of end-to-end entity resolution for big data. ACM Comput. Surv. **53**(6), 127:1-127:42 (2021)

21. Chu, X., Ilyas, I.F., Krishnan, S., Wang, J.: Data cleaning: overview and emerging challenges. In: International Conference on Management of Data (SIGMOD), pp. 2201–2206. ACM (2016)

22. Dehghani, Z.: Data Mesh: Delivering Data-Driven Value at Scale. O'Reilly, Newton (2022). ISBN: 1492092398

23. DICOM. Dicom - digital imaging and communications in medicine. https://www.dicomstandard.org/

24. Elmagarmid, A., Rusinkiewicz, M., Sheth, A.: Management of Heterogeneous and Autonomous Database Systems. Morgan Kaufmann Publishers, Burlington (1999). ISBN: 1-55860-216-X

25. Errami, S.A., Hajji, H., Kadi, K.A.E., Badir, H.: Spatial big data architecture: from data warehouses and data lakes to the Lakehouse. J. Parallel Distrib. Comput. **176**, 70–79 (2023)

26. Fivetrain. Connectors for every data source. Accessed June 2023

27. Friedman, E., Pawlowski, P., Cieslewicz, J.: SQL/MapReduce: a practical approach to self-describing, polymorphic, and parallelizable user-defined functions. VLDB Endowment **2**(2), 1402–1413 (2009)

28. Gillet, A., Leclercq, É., Cullot, N.: Lambda+, the renewal of the lambda architecture: category theory to the rescue. In: La Rosa, M., Sadiq, S., Teniente, E. (eds.) CAiSE 2021. LNCS, vol. 12751, pp. 381–396. Springer, Cham (2021). https://doi.org/10.1007/978-3-030-79382-1_23

29. Giovanelli, J., Bilalli, B., Abelló, A.: Data pre-processing pipeline generation for AutoETL. Inf. Syst. **108**, 101957 (2022)

30. Große, P., May, N., Lehner, W.: A study of partitioning and parallel UDF execution with the SAP HANA database. In; Conference on Scientific and Statistical Database Management (SSDBM), p. 36 (2014)

31. Gupta, A., Mumick, I.S.: Materialized Views: Techniques, Implementations, and Applications. The MIT Press, Cambridge (1999)

32. Hai, R., Koutras, C., Quix, C., Jarke, M.: Data lakes: a survey of functions and systems (2023)

33. Halasipuram, R., Deshpande, P.M., Padmanabhan, S.: Determining essential statistics for cost based optimization of an ETL workflow. In: International Conference on Extending Database Technology (EDBT), pp. 307–318 (2014)
34. Harby, A.A., Zulkernine, F.: From data warehouse to Lakehouse: a comparative review. In: IEEE International Conference on Big Data, pp. 389–395 (2022)
35. Heidsieck, G., de Oliveira, D., Pacitti, E., Pradal, C., Tardieu, F., Valduriez, P.: Distributed caching of scientific workflows in multisite cloud. In: Hartmann, S., Küng, J., Kotsis, G., Tjoa, A.M., Khalil, I. (eds.) DEXA 2020. LNCS, vol. 12392, pp. 51–65. Springer, Cham (2020). https://doi.org/10.1007/978-3-030-59051-2_4
36. Hernández, Á.B., Pérez, M.S., Gupta, S., Muntés-Mulero, V.: Using machine learning to optimize parallelism in big data applications. Future Gener. Comput. Syst. **86**, 1076–1092 (2018)
37. Herodotou, H., et al.: Starfish: a self-tuning system for big data analytics. In: Conference on Innovative Data Systems Research CIDR, pp. 261–272 (2011)
38. Hueske, F., et al.: Peeking into the optimization of data flow programs with mapreduce-style UDFs. In: International Conference on Data Engineering (ICDE), pp. 1292–1295 (2013)
39. Hueske, F., et al.: Opening the black boxes in data flow optimization. VLDB Endowment **5**(11), 1256–1267 (2012)
40. IBM. IBM Cloud Pak for Data: Supported data sources. Accessed June 2023
41. IBM: Introduction to InfoSphere DataStage balanced optimization. Documentation. Accessed June 2023
42. Informatica: Pushdown optimization overview. Documentation. Accessed June 2023
43. Jarke, M., Lenzerini, M., Vassiliou, Y., Vassiliadis, P.: Fundamentals of Data Warehouses. Springer, Cham (2003). https://doi.org/10.1007/978-3-662-05153-5
44. Jemmali, R., Abdelhédi, F., Zurfluh, G.: Dltodw: transferring relational and NoSQL databases from a data lake. SN Comput. Sci. **3**(5), 381 (2022)
45. Jovanovic, P., Romero, O., Simitsis, A., Abelló, A.: Incremental consolidation of data-intensive multi-flows. IEEE Trans. Knowl. Data Eng. **28**(5), 1203–1216 (2016)
46. Karagiannis, A., Vassiliadis, P., Simitsis, A.: Scheduling strategies for efficient ETL execution. Inf. Syst. **38**(6), 927–945 (2013)
47. Kechar, M., Bellatreche, L.: Safeness: suffix arrays driven materialized view selection framework for large-scale workloads. In: Wrembel, R., Gamper, J., Kotsis, G., Tjoa, A.M., Khalil, I. (eds.) DaWaK 2022. Lecture Notes in Computer Science, vol. 13428, pp. 74–86. Springer, Cham (2022). https://doi.org/10.1007/978-3-031-12670-3_7
48. Konstantinou, N., Paton, N.W.: Feedback driven improvement of data preparation pipelines. Inf. Syst. **92**, 101480 (2020)
49. Kumar, N., Kumar, P.S.: An efficient heuristic for logical optimization of ETL workflows. In: Castellanos, M., Dayal, U., Markl, V. (eds.) BIRTE 2010. LNBIP, vol. 84, pp. 68–83. Springer, Heidelberg (2011). https://doi.org/10.1007/978-3-642-22970-1_6
50. Lerner, A., Hussein, R., Ryser, A., Lee, S., Cudré-Mauroux, P.: Networking and storage: the next computing elements in exascale systems? IEEE Data Eng. Bull. **43**(1), 60–71 (2020)
51. Liu, X., Iftikhar, N.: An ETL optimization framework using partitioning and parallelization. In: ACM Symposium on Applied Computing, pp. 1015–1022 (2015)
52. Munshi, A.A., Mohamed, Y.A.I.: Data lake lambda architecture for smart grids big data analytics. IEEE Access **6**, 40463–40471 (2018)

53. Nargesian, F., Zhu, E., Miller, R.J., Pu, K.Q., Arocena, P.C.: Data lake management: challenges and opportunities. VLDB Endowment **12**(12), 1986–1989 (2019)
54. Owaida, M., Alonso, G., Fogliarini, L., Hock-Koon, A., Melet, P.: Lowering the latency of data processing pipelines through FPGA based hardware acceleration. VLDB Endowment **13**(1), 71–85 (2019)
55. Popescu, A.D., Ercegovac, V., Balmin, A., Branco, M., Ailamaki, A.: Same queries, different data: can we predict runtime performance? In: Workshops @ International Conference on Data Engineering (ICDE), pp. 275–280. IEEE Computer Society (2012)
56. Quemy, A.: Binary classification in unstructured space with hypergraph case-based reasoning. Inf. Syst. **85**, 92–113 (2019)
57. Ramachandra, K., Park, K., Emani, K.V., Halverson, A., Galindo-Legaria, C.A., Cunningham, C.: Froid: optimization of imperative programs in a relational database. VLDB Endowment **11**(4), 432–444 (2017)
58. Rheinländer, A., Heise, A., Hueske, F., Leser, U., Naumann, F.: SOFA: an extensible logical optimizer for UDF-heavy data flows. Inf. Syst. **52**, 96–125 (2015)
59. Romero, O., Wrembel, R.: Data engineering for data science: two sides of the same coin. In: Song, M., Song, I.-Y., Kotsis, G., Tjoa, A.M., Khalil, I. (eds.) DaWaK 2020. LNCS, vol. 12393, pp. 157–166. Springer, Cham (2020). https://doi.org/10.1007/978-3-030-59065-9_13
60. Rusinkiewicz, M., Czejdo, B., Embley, D.W.: An implementation model for muldidatabase queries. In: Karagiannis, D. (ed.) Database and Expert Systems Applications, pp. 309–314. Springer-Verlag, Vienna (1991). https://doi.org/10.1007/978-3-7091-7555-2_52
61. Sichert, M., Neumann, T.: User-defined operators: efficiently integrating custom algorithms into modern databases. VLDB Endowment **15**(5), 1119–1131 (2022)
62. Sienkiewicz, M., Wrembel, R.: Managing data in a big financial institution: conclusions from a R&D project. In: Workshops of the EDBT/ICDT Joint Conference, vol. 2841 (2021)
63. Simitsis, A., Skiadopoulos, S., Vassiliadis, P.: The history, present, and future of ETL technology (invited). In: International Workshop on Design, Optimization, Languages and Analytical Processing of Big Data (DOLAP), volume 3369 of CEUR Workshop Proceedings, pp. 3–12. CEUR-WS.org (2023)
64. Simitsis, A., Vassiliadis, P., Sellis, T.K.: Optimizing ETL processes in data warehouses. In: International Conference on Data Engineering (ICDE), pp. 564–575. IEEE Computer Society (2005)
65. Simitsis, A., Vassiliadis, P., Sellis, T.K.: State-space optimization of ETL workflows. IEEE Trans. Knowl. Data Eng. **17**(10), 1404–1419 (2005)
66. Strengholt, P.: Data Management at Scale: Modern Data Architecture with Data Mesh and Data Fabric. O'Reilly, Newton (2023). ISBN: 1098138864
67. Tan, R., Chirkova, R., Gadepally, V., Mattson, T.G.: Enabling query processing across heterogeneous data models: a survey. In: IEEE International Conference on Big Data, pp. 3211–3220 (2017)
68. Thomsen, C.: ETL. In Encyclopedia of Big Data Technologies, Springer, Cham (2019). https://doi.org/10.1007/978-3-319-77525-8
69. Tsesmelis, D., Simitsis, A.: Database optimizers in the era of learning. In: International Conference on Data Engineering (ICDE), pp. 3213–3216 (2022)
70. Vaisman, A.A., Zimányi, E.: Data Warehouse Systems - Design and Implementation. Data-Centric Systems and Applications, 2nd edn. Springer (2022). https://doi.org/10.1007/978-3-662-65167-4

71. Wiederhold, G.: Mediators in the architecture of future information systems. Computer **25**(3), 38–49 (1992)
72. Witt, C., Bux, M., Gusew, W., Leser, U.: Predictive performance modeling for distributed batch processing using black box monitoring and machine learning. Inf. Syst. **82**, 33–52 (2019)
73. Zaharia, M., Ghodsi, A., Xin, R., Armbrust, M.: Lakehouse: a new generation of open platforms that unify data warehousing and advanced analytics. In: Conference on Innovative Data Systems Research (CIDR) (2021)

Data Modeling

Scalable Summarization for Knowledge Graphs with Controlled Utility Loss

Yi Wang[✉] [iD], Ying Wang, and Qia Wang

Southwest University, Chongqing 400715, China
echowang@swu.edu.cn

Abstract. Due to the explosive growth of semantic data over recent years, extracting critical and representative information from knowledge graphs (KGs), i.e., KG summarization is a challenging task under real-world resource constraints such as memories and response time. We present scalable utility-driven algorithms for summarizing large-scale KGs based on the features of entities. Specifically, we propose the notion of utility for KG summaries and develop the SiFS algorithm for efficient lossless KG summarization. Second, we present the highly scalable KG summarization algorithm ScFS for generating lossy summaries by user defined utility threshold. The experiments over real-world and synthetic datasets show that SiFS achieves more compact summary sizes and significant reductions in execution time. Moreover, ScFS outperforms state-of-the-art summarization methods significantly in summary size and is about two orders of magnitude better in terms of speed. Third, we present a query evaluation algorithm over KGs based on SiFS and ScFS. The experiments results show that the proposed summaries facilitate efficient and high-quality KG queries.

Keywords: Knowledge Graph compression · utility-driven summarization · RDF Graph summary · graph summarization

1 Introduction

Knowledge Graph (KG) summarization is a challenging task because not only the volume of real-world KGs are growing explosively, but also KGs have complex structures and lack of common schemas [1, 2]. The objective of KG summarization is to create a compact and meaningful representation of the original KG called the summary in order to solve problems such as exploration [2, 3], querying [4, 5] and error-detecting [6] of large-scale KGs.

The problem of KG summarization has received a great deal of attention in recent years. Graph structure-based summarization methods, which merge entities/nodes as supernodes based on the structure of the directed graphs formed by RDF data, constitute the most popular type of approaches [4, 5, 7, 8]. A typical structure-based summarization method is proposed in [4], where equivalence relations between nodes reflecting the similarity between nodes were defined. Nodes were merged by the equivalence relations and thus quotient graphs were formed as the graph summary. Partial order relations are also used to create KG summaries. Differently, González and Hogan [8] presented a lattice-based summary to dynamically model KGs. Entities having the same maximum set of properties (called characteristic set) were grouped together. Then the *partial order relations* between the sets of entities were identified based on the characteristic sets and the corresponding lattice was created as the summary.

Structure-based summarization has the advantage of helping users understand the KGs with a compact and meaningful summary graph, reflecting some kind of schema of the original KG. Due to the massive volume of current KGs which may contain hundreds of millions of triples, the overhead for calculating summaries is expensive. A major problem with current summarization methods is that the scalability of algorithms needs to be improved for very large KGs. For example, the time complexity for generating the lattice-based summary in [8] is $O(|P|\cdot|C|^2)$, where $|P|$ is the cardinality of properties and $|C|$ is the number of characteristic sets with $|C|\leq 2^{|P|}$. Moreover, real-world KGs are usually changed over time. Therefore, the overhead for computing summaries for large datasets is expensive.

Another major problem of the current KG summarization is the lack of utility control for lossy summaries. Utility of a summary refers to the "useful information" contained from the original KG [9]. For example, Song et al. [10] proposed a *d*-summary which describes node linking patterns within *d* hops based on frequent pattern mining technique. The *d*-summary includes top-*k* *d*-hop patterns, which is a lossy summary. However, how much loss of utility for the summary is not under controlled during the summary creation.

To address the above issues, we present two KG summarization algorithms: the lossless **Si**mple node **F**eature based **S**ummary (SiFS) and the lossy **Sc**alable node **F**eature based **S**ummary (ScFS) which can generate KG summaries efficiently with controlled utility loss. To the best of our knowledge, this is the first work to develop utility–driven KG summaries. Specifically, our contributions include:

- We propose SiFS for efficient lossless KG summarization. SiFS partitions nodes by their features with a hash function and generates the Hasse Diagram (HD) as the summary. By experiments over large KGs we show that SiFS outperforms similar lossless KG summarization algorithms in execution time and achieve great reduction in summary size.
- We define the utility metric for KG summary and propose a utility-driven KG summarization algorithm ScFS for lossy summarization. ScFS achieves high scalability and it outperforms state-of-the-art KG algorithms in both execution time and summary size.
- We present a query evaluation algorithm over KGs based on SiFS and ScFS and evaluated the algorithm over real-world and synthetic KGs. The results show that the two summaries improve the execution of KG queries. Especially for ScFS, with 80% of the utility requirements the coverage rate for entity search reaches 95% on average, and greatly reduces the query execution time.

2 Preliminaries

A KG is defined as $G = (V, P, R, L_V, \phi)$, where V is the set of nodes or entities, P is a set of relation types, i.e., properties, $R \subseteq V \times P \times V$ is a set of relations or triples between nodes, and L_V is a set of node labels or types, and $\phi: V \rightarrow \wp(L_V)$ is a function that maps a node to the set of types.

Definition 1 (*Node feature*). Given $G = (V, P, R, L_V, \phi)$, the feature of a node $v \in V$ is defined as a finite set of elements that characterizing v, denoted as $F(v) = \{f_1, f_2, ..., f_s\}$.

Definition 2 (*Feature Pattern, FP*). Given $G = (V, P, R, L_V, \phi)$, a feature pattern (FP) is defined as a tuple: $c = (W, T)$, where W is a subset of V and T is a subset of P and satisfies: (i) $\forall v \in W$, $F(v) = T$, and (ii) for any subset of V that include W, i.e., $W' \subseteq V$ and $W \subseteq W'$, the condition (i) does not hold for every node in W'.

Definition 2 describes that an FP is a maximum set of common features for a set of nodes. T in an FP c is called the *feature set*.

Definition 3 (KG summary based on FP). Given $G = (V, P, R, L_V, \phi)$, let C be the set of FPs formed by all the nodes of G, the summary of G is the Hasse Diagram (HD) formed by $(C, \subseteq)$, where $\subseteq$ is the subset relation between the feature sets in C. We denote the summary of G as $L = (C, E)$, where E describes the cover relations between the elements of C.

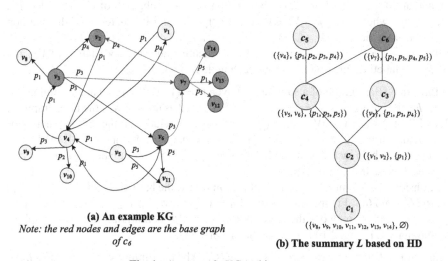

(a) An example KG
Note: the red nodes and edges are the base graph of c_6

(b) The summary L based on HD

Fig. 1. An example KG and its summary

Example 1. Figure 1 (a) shows a KG, where $V = \{v_1, v_2, v_3, v_4, v_5, v_6, v_7, v_8, v_9, v_{10}, v_{11}, v_{12}, v_{13}, v_{14}\}$ and $P = \{p_1, p_2, p_3, p_4, p_5\}$. We assume that the feature of each node is described by its out-going properties. For example, $F(v_1) = \{p_1\}$ and $F(v_7) = \{p_1, p_3, p_4, p_5\}$. Table 1 lists the node features and FPs. Figure 1 (b) is the KG summary based on the node features. Each FP is at a certain layer, determined by the cardinality of its feature set. For example, c_1 is at layer 1 because its feature set has one property and c_5

is at layer 4 because its feature set has four properties. We call the largest layer number the *Height* of the summary.

Table 1. Node features and FPs

Node feature	FP
$F(v_1) = \{p_1\}$	$c_1 = (\{v_8, v_9, v_{10}, v_{11}, v_{12}, v_{13}, v_{14}\}, \emptyset)$
$F(v_2) = \{p_1\}$	$c_2 = (\{v_1, v_2\}, \{p_1\})$
$F(v_3) = \{p_1, p_3, p_4\}$	$c_3 = (\{v_3\}, \{p_1, p_3, p_4\})$
$F(v_4) = \{p_1, p_2, p_3, p_4\}$	$c_4 = (\{v_5, v_6\}, \{p_1, p_3, p_5\})$
$F(v_5) = \{p_1, p_3, p_5\}$	$c_5 = (\{v_4\}, \{p_1, p_2, p_3, p_4\})$
$F(v_6) = \{p_1, p_3, p_5\}$	$c_6 = (\{v_7\}, \{p_1, p_3, p_4, p_5\})$
$F(v_7) = \{p_1, p_3, p_4, p_5\}$	
$F(v_8) = F(v_9) = F(v_{10}) = F(v_{11}) = F(v_{12}) = F(v_{13}) = F(v_{14}) = \emptyset$	

Definition 4 (Base graph of an FP). Given $G = (V, P, R, L_V, \phi)$, a summary $L = (C, E)$, and an FP $c = (W, T) \in C$, the base graph of c is a subgraph of G: $g_b = (V_b, P_b, R_b, L_V^b, \phi_b)$, where: (1) $V_b = V_\sigma \cup V_N$, $V_\sigma = \bigcup_{W \in c} W$ and V_N includes all the one-hop neighbor of the nodes in V_σ; (2) $R_b = \{(u, p, v) | u \in V_\sigma$ or $v \in V_\sigma\}$; (3) $P_b = \{p | p \in P$ and $(u, p, v) \in R_b\}$; (4) L_V^b is a subset of L_V which includes the node labels of V_b; and (5) ϕ_b is a labeling function that maps each node in V_b to its types.

Definition 5 (Base graph of a summary). Given $G = (V, P, R, L_V, \phi)$ and a summary $L = (C, E)$, the base graph $G_L = (V_s, P_s, R_s, L_V^s, \phi_s)$ of L is the union of the base graphs of all its FPs: (1) $V_s = V_\sigma \cup V_N$, $V_\sigma = \bigcup_{W \in c} W$ and V_N includes all the one-hop neighbor of the nodes in V_σ; (2) $P_s = \bigcup_{T \in c} T$; ; (3) $R_s = \{(u, p, v) | u \in V_\sigma$ or $v \in V_\sigma\}$; (4) L_V^s is a subset of L_V which includes the node labels of V_s; and (5) ϕ_s is a labeling function that maps each node in V_s to its types.

Consider the KG in example 1, by the definition 4, the base graph of c_6 is $g_b = (V_b, P_b, R_b, L_V^b, \phi_b)$ (c.f. Figure 1(a)), where $V_b = \{v_2, v_3, v_6, v_7, v_{12}, v_{13}, v_{14}\}$, $P_b = \{p_1, p_3, p_4, p_5\}$, and R_b includes the edges related to v_7. By definition 5, the base graph of the summary L in Fig. 2(b) is exactly the original KG.

Definition 6 (*Utility of a summary*). Let $L = (C, E)$ be the summary for a given KG, the utility of L is defined as $utility(L) = \sum_{(u,p,v) \in R_b} I(u, v)$, where $I(u, v)$ is the importance measure for the relation (u, p, v) in the based graph of the summary.

We normalize the utility of a summary by the maximum utility, i.e., the total utility of G, denoting as u_0:

$$u(L) = \sum_{(u,p,v) \in R_b} I(u, v)/u_0 \qquad (1)$$

Intuitively, the utility of a summary describes how much critical information is summarized by the summary. The utility value u is normalized between 0 and 1 and a greater value indicates a better summary [9]. Given a user-specified utility requirement τ, the objective of the KG summarization is to:

$$\text{minimize } |L| \text{ subject to } u(L) \geq T.$$

Table 2 lists the major notations used in this paper.

Table 2. Notations

Symbol	Definition
FP	feature pattern $c = (W, T)$, W is the set of nodes, T is the feature set
$L = (C, E)$	KG summary, C is the set of FPs, E is the links between FPs
G_L, g_b	the base graph of the summary L, the base graph of an FP
$u(L)$, u_0, τ	the normalized utility of the summary L, the total utility of the KG, the threshold of utility
Height	the largest layer of the HD-based summary
FS_i	the set of FPs at layer i
γ	the ratio for selecting the top-k FPs in each layer
$Cr(v)$, q_v, k_v	the weighted centrality value of v, the importance of its features, and the common node centrality metric

3   Lossless Simple Node Feature Based Summary (SiFS)

In this section, we present the algorithm: lossless **Si**mple node **F**eature based **S**ummary (SiFS) for constructing the lossless KG summary ($\tau = 1$).

Algorithm 1 relies on the hash function h to compute the node feature. The dictionary *mapF* stores the nodes having same the hash value with the hash value as the key. This process efficiently retrieves the common feature sets. However, there may be false positive, i.e., nodes having different features are hashed to a same value. Lines 5–11 of Algorithm 1 check the correctness of FPs and store the FPs by the length of node feature set. In practice, we use a very simple hash function: to mod a node feature by very large prime number and the no collision occurs. Algorithm 1 has $O(|R|)$ time complexity. Algorithm 2 **SiFS** calculates the lossless summary ($\tau = 1$) for a given KG. Line 2 of Algorithm 2 invokes Algorithm 1 to compute FPs and the height of the HD-based summary. Lines 3–9 build the summary of G by checking the inclusion relation of FPs at different layers. This task takes $O(Height^2 \cdot M)$ time complexity, where M is the maximum number of FPs in each layer. Since $Height \cdot M \leq |C|$ and $Height \leq |P|$, the time complexity for Algorithm 2 is $O(|C| \cdot |P|)$. Therefore, the total time complexity for the lossless KG summary SiFS is $O(|R|+|C| \cdot |P|)$.

Algorithm 1 Compute node feature sets

    Input G, h // h is the hash function to map node feature sets to numbers
1. $FS \leftarrow [\ ]$, $W \leftarrow [\ ]$, $mapF \leftarrow \varnothing$, $Height \leftarrow 1$ // FS is the set of FS_i, storing FPs by layer
2. for $v \in V$ do
3. $h_F \leftarrow h(F(v))$ // $F(v)$ is the set of node features for v
4. $mapF[h_F] \leftarrow mapF[h_F] \cup \{v\}$
5. for X in $mapF$ do
6.   while $X \neq \varnothing$ do
7. $u \leftarrow$ select-random-node(X)
8. $W(u) \leftarrow \{v \in X | F(u) = F(v)\}$ *//W(u) stores the nodes having same features*
                                                                       with u
9. $FS[|F(u)|]$.append($\{u\} \cup F(u)$)) // $|F(u)|$ is the cardinality of $F(u)$
10. $Height \leftarrow \max\{Height, |F(u)|\}$
11. $X \leftarrow X - W(u)$
12. **Return** W, FS, $Height$

Algorithm 2 Finding the lossless KG summary (SiFS)

    Input G, h
1. $L \leftarrow \varnothing$, $FS \leftarrow \varnothing$, $S \leftarrow \varnothing$
2. W, FS, $Height \leftarrow$ ComputeNodeFeatureSets(G, h)
3. for $i = 2$ to $Height$ do
3.   for $j = i - 1$ to 1 do
4.   if $j == i - 1$:
5. $L \leftarrow L \cup$ DirectCover(W, FS_i, FS_j) *//calculate relations between FPs at successive*
layers
6.   else if $j < i - 1$:
7. $L \leftarrow L \cup$ JumpLayerCover(W, FS_i, FS_j) // *calculate relations between FPs not at successive layers*
8. $j - -$;
9. $i + +$;
10. **Return** L

---

4 Lossy Scalable Node Feature Based Summary (ScFS)

Our strategy to build lossy **S**calable node **F**eature based **S**ummary (ScFS) for KGs is to construct the summary in an incremental way with controlled utility loss. Specifically, in each round of the utility computation, top-k FPs are selected to the summary. If the utility does not reach the requirement τ, another round continues.

In order to select top-k FPs from each layer in a round and maximize the utility, we need to rank the FPs. For each FP, we compute its score by Eq. (2).

$$Sr(c) = \max_{v \in W} Cr(v), \ c = (W, T) \in C \tag{2}$$

where $Cr(v)$ is the centrality score of v. In this paper we use a weighted centrality measure (3) combining both common node centrality and the importance of its features [21]. In Eq. (3), k_v is the centrality score for v which can be common centrality metric, e.g., degree or page rank centrality, q_v is the importance of its features (Eq. (4)), and α is a tuning parameter between 0 and 1. When $\alpha = 0$, $Cr(v)$ is decided by k_v, and when $\alpha = 1$, $Cr(v)$ is decided by q_v. The function *freq* in Eq. (4) is the frequency of a given feature and the formula is motivated by the similarity measure in graphs [22]. The intuition behind this centrality metric is that the nodes having large node centrality,.e.g., degree, and connected with low-frequency properties are important nodes. In fact, the function *freq* can be replaced by other functions related to p that characterizing the required importance of different properties.

$$Cr(v) = k_v^{1-\alpha} \times q_v^\alpha \tag{3}$$

$$q_v = \begin{cases} 1 & F(v) = \varnothing \\ \sum_{p \in F(v)} \frac{1}{\log(1+freq(p))} & F(v) \neq \varnothing \end{cases} \tag{4}$$

As the value of k is related to the number of FPs in each layer, we use the ratio γ to compute k as:

$$k = \lceil avg_{|FS|} \cdot \gamma \rceil \tag{5}$$

where $avg_{|FS|} = \frac{|C|}{Height}$ is the average number of FS in each layer.

Example 2 The base graph of the summary L for Fig. 1(a) is the original KG. Thus, $|R|=20$, and suppose k_v is the degree centrality of v, i.e., $k_v = d_v/2|R|=d_v/40$. q_v is calculated by the frequency of its features, here, the out-going properties. For example, v_3 connects to p_1, p_3, p_4, thus by Eq. (4) we have $q_{v_3}=14.30$. By Eq. (3) and with $\alpha = 0.5$, we obtain $Cr(v_3)=1.34$. Similarly, we have the weighted centrality scores for the rest of the nodes and the most central node is v_4 (2.64) and v_1 (0.41) is least central. By Eq. (2), we can calculate the scores for the six FPs and c_5 (2.64) is ranked the highest.

Algorithm 3 Building Lossy Summary by Layers (ScFS)
Input $G, FS, h, \tau, \gamma, u_0$
1. $FS \leftarrow \varnothing, S \leftarrow \varnothing, mapF \leftarrow \varnothing, mapI \leftarrow \varnothing, L \leftarrow \varnothing, utility \leftarrow 0$
2. $S, FS, Height \leftarrow$ ComputeNodeFeatureSets(G, h)
3. *Rank_FPs_by Layers*(S, FS)
4. $k \leftarrow \lceil avg_{|FS|} \cdot \gamma \rceil$
5. while $utility < \tau$ do
6. $\Delta L \leftarrow$ BuildConnection(T, FS, k) //Selecting top-k FPs in each layer and establish links
7. $\Delta u \leftarrow$ ComputeUtility$(\Delta L, mapI, g_b, Cr, u_0)$
8. $L \leftarrow$ Merge$(L, \Delta L)$ //Combine L with ΔL
9. $utility \leftarrow utility + \Delta u$
10. **Return** L

Algorithm 4 ComputingUtility

Input ΔL, $mapI$, g_b, Cr, u_0 //u_0 is the total utility of G, G_B is the set of base graphs of FPs
1. $\Delta u \leftarrow 0$
2. **for** v in W **do** //W is the set of nodes summarized by ΔL
3. $R_b \leftarrow$ fetchEdge(v, g_b) // fetchEdge retrieves relations related to v
4.   **for** e in R_b **do**
5.   **if** not hashedCheck(e, $mapI$) // checks if the utility is computed by the hashmap
6. $I(e) \leftarrow Cr(v)/d(v)+ Cr(u)/d(u)$, $u \in N(v)$
7. $\Delta u \leftarrow \Delta u + I(e)$
8. $\Delta u \leftarrow \Delta u / u_0$
9. **Return** utility

We present Algorithm 3 (**ScFS**) to create the lossy summary for a given KG by incrementally increasing the utility of the summary. Line 2 of Algorithm 3 invokes Algorithm 1 to compute node features and the height of the summary. Line 3 ranks the FPs by the scores of FPs calculated by Eq. (2). Line 4 calculates the value of k by the given ratio γ. Lines $5 - 9$ check the inclusion relations between the selected FPs in each layer and compute the utility of L incrementally by invoking Algorithm 4: ComputingUtility, until *utility* reaches the thresholdτ.

Algorithm 4 computes the utility of a given summary in an incremental fashion. For each node v in the summary, we fetch the related relations R_b from the corresponding base graph g_{bi} (Line 3). The function hashedCheck can hash the relation e and check if the hash value exits in *mapI* (Line 5). If it exists, the function returns true. If the hash value doesn't exist, the function adds the hash value to *mapI* and return false. This step is to avoid repeatedly computing utility of a relation. Line 6 of Algorithm 3 computes the utility of the relation e by Eq. (6) and Line 8 normalizes the utility of the summary. Algorithm 4 takes $O(|C|)$ time complexity. Therefore, the total time complexity for the lossy KG summary by layers is $O(|R|+|C|)$. Since $|C|$ is normally less than $|R|$, the time complexity for the algorithm is $O(|R|)$.

$$I(u, v) = \frac{C_r(u)}{d_u} + \frac{C_r(v)}{d_v} \tag{6}$$

Theorem 1 (Non-decreasing Utility Theorem) Given a KG G, if L_{t+1} and L_t are the lossy summaries generated at the $t + 1$ round and t round by Algorithm 4, then $u(L_{t+1}) \geq u(L_t)$ $(t \geq 0)$.

Proof. Since Algorithm 4 (ScFS) selects top-k FPs from each layer j $(1 \leq j \leq Height)$ in each round of iteration, L_{t+1} contains at most $Height \cdot k$ $(k \geq 1)$ FPs than L_t. By Eqs.(1) and (6), each FP $c = (W, T)$ can contribute utility $u(c) = \frac{1}{u_0}\sum_{(u,p,v) \in R_b} \frac{C_r(u)}{d_u} + \frac{C_r(v)}{d_v}$ to the utility of the summary. The larger the score of the FP $Sr(c)$ is, the more central nodes it contains and thus contributes more utility to the summary. Therefore, $u(L_{t+1}) \geq u(L_t)$ and the top-k strategy ensured by the ranking measure of FPs fastens the speed of $u(L)$ of reaching the threshold τ.

5 KG Search with Summaries

SPARQL is the standard W3C query language for accessing KG. A SPARQL query Q is a generalization of a KG, called Basic Graph Patterns (BGP), in which variables may appear as subject, property and object of triples. These graph patterns are matched against the KG and the matched graph is retrieved and manipulated according to the conditions given in the query. Algorithm 5 shows the search scheme for the proposed summaries SiFS and ScFS, which faciliate fast KG search under resource bounds. This is desirable in large-scale KG search with limited computational resources.

Algorithm 5 SearchwithHDS
    Input: $Q=\{t_1, t_2, ..., t_q\}$, L, G_B, N // G_B is the set of base graphs of FPs, N is the bound of
                                         query results
    1. Select_FP←{}, QR←{}
    2. for each t_i:
    3.   select_FP←MatchFP(t_i, L) //MatchFP find matching FPs from L
    4. for each X_i in Select_FP:
    5.   QR←Eval_L(X_i, G_B, N) // Fetch matching elements in the base graphs of FPs
    6. **Return** QR

Algorithm 5 includes two major tasks: FP matching (Lines 1–3) and base graph searching (Lines 4–5). In the FP matching step, the algorithm selects FPs from the summary L which contains the potential answers of Q as much as possible. A query Q is covered by a set of FPs if and only if the base graphs induced by the FPs: $\bigcup_{FP_i} Q_{FP_i}$ cover the query Q. The base graph searching step involves accessing the base graphs (can be stored as materialized views) of the selected FPs and fetching bounded numbers of elements as query results. The total complexity of the Algorithm 5 is $O(|C|q)$, where q is the number of triple patterns in the query.

6 Experiments

6.1 Goals and Datasets of the Experiments

Our goal is to validate the proposed approach: the two types of summaries: SiFS and ScFS for KG summarization. We evaluated the proposed methods from the two aspects: the performance of the algorithms in terms of execution time and summary size compared to similar summary methods and the query performance based on the proposed summaries. We developed the summary applications using Python version 3.8.5 and Neo4j version 3.5.26 (configured with 16GB working memory).We ran the summary algorithms on a machine with an Intel Core i7-3740QM CPU (2.70GHz, 4 Cores) and 32.0 GB memory. We carried out the experiments over the four datasets: YAGO[1], LinkedGeoData[2],

---

[1] https://www.mpi-inf.mpg.de/departments/databases-and-information-systems/research/yago-naga/yago/downloads. We used core datasets including yagoSimpleTypes, CORE including yagoFacts and yagoDateFacts, and yagoWikipediaInfo_en.
[2] http://linkedgeodata.org/.

Berlin SPARQL Benchmark[3] (BSBM) [23] and the Open Data Service of University of Southampton[4] (ODUS) shown in Table 3.

Table 3. Datasets

Datasets	Triples	Nodes	Properties	Relations
YAGO	20,450,000	4,104,072	15	8,776,501
LinkedGeoData	3,689,990	1,384,181	2022	1,132,444
BSBM	21,198,599	3,119,001	39	8,100,521
ODUS	599,987	121,330	420	229,197

6.2 Performance of the Summary Algorithms

Execution Time and Summary Size. We compared the proposed algorithms: SiFS and ScFS with the lattice-based summary (LS) of [8] and the quotient graph summary (QS) proposed in [4]. LS summarizes KGs as lattices. The elements in an LS are CSCs extracted from RDF triples. Differently, QS is a typical summarization method based on equivalence relations between nodes in a KG.

Figure 2 shows the results of the execution time and summary sizes for the four types of summaries over the datasets BSBM and LinkedGeoData. From Fig. 2 (a) we observe that the lossy summary ScFS ($\tau = 0.8$ and $\gamma = 0.2$) outperformed the other three summary algorithms while LS took the longest time. Although the scale of LinkedGeoData is smaller than that of BSBM, the former contains 2022 properties, almost 52 times than that of the latter. The large number of properties caused high computation cost for calculating CSCs of LS and FSs of SiFS. The lossy summary ScFS has advantage in this situation because only FSs with large scores were processed. Figure 2(b) and (c) show the sizes of the summaries for BSBM and LinkedGeoData. The sizes of the ScFS and SiFS refer to the numbers of FSs and the links between them. The size of LS refers to the numbers of CSCs and links between them. For QS, the size of the summary denotes the numbers of supernodes and superedges. It is obvious that the summaries generated by SiFS, ScFS, and LS were more compact than QS. Furthermore, ScFS obtained the most compact summary for the two datasets. QS, which is based on equivalence relations between nodes, generated the largest summaries compared to other three summary methods.

Performance and Parameters for ScFS. We carried out experiments for ScFS to find the impact of the parameter γ: the ratio of top elements selected in each layer, on the convergence of the utility. Figure 3 shows the relations between γ and the utility u of the summaries for the four datasets generated by ScFS. We observe that when γ was set to 0.2, the ScFS iterated at most 2 rounds to reach 80% of the total utility for BSBM, LinkedGeoData, and YAGO. ODUS took more rounds to reach the utility requirement.

---

[3] http://wifo5-03.informatik.uni-mannheim.de/bizer/berlinsparqlbenchmark/.
[4] http://data.southampton.ac.uk/.

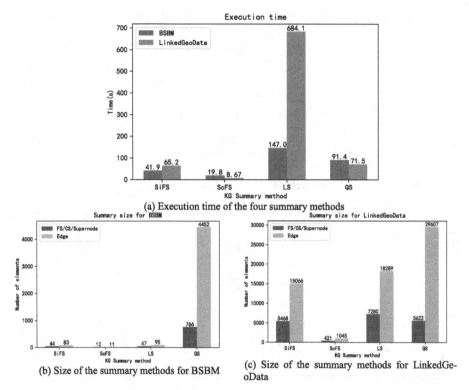

(a) Execution time of the four summary methods

(b) Size of the summary methods for BSBM

(c) Size of the summary methods for LinkedGe-oData

Fig. 2. Execution time and summary sizes for SiFS, ScFS, LS, and QS

ODUS has 392 FPs. The maximum length of the feature sets is 85, i.e., the height of the summary *Height* is 85. Thus, the average number of FPs in each layer $avg_{|FS|}$ is 4.6. If γ was set to 0.2, then top-1 FP was selected at each layer and the utility reached 0.81 after 5 rounds. If γ was set to 0.4, then top-2 FPs were selected and the utility reached 0.84 after 2 rounds.

Therefore, the value of γ impacts the convergence rate of the algorithm. There is a tradeoff between the computation cost and the convergence rate. Although we calculate the utility incrementally in each round, iterating more times increase the overhead. On the other hand, if γ is large, which means in each round a large portion of the FSs are selected to generate the summary. The results of Fig. 3 suggest the value of γ between 0.2 and 0.4.

6.3 Querying Evaluation

We evaluate how the proposed summarization method facilitates fast and high quality KG queries. Specifically, we compared the SearchwithHDS algorithm based on ScFS and SiFS with direct query. We performed two types of KG query over the four datasets, search-by-entity (s,?p,?o) and search-by-property (?s,p,?o). We randomly select 500 entities for search-by-entity and 20% of properties for search-by-property over each

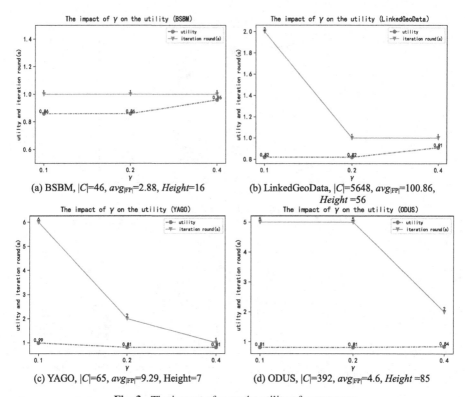

(a) BSBM, $|C|$=46, $avg_{|FP|}$=2.88, Height=16

(b) LinkedGeoData, $|C|$=5648, $avg_{|FP|}$=100.86, Height =56

(c) YAGO, $|C|$=65, $avg_{|FP|}$=9.29, Height=7

(d) ODUS, $|C|$=392, $avg_{|FP|}$=4.6, Height =85

Fig. 3. The impact of γ on the utility of a summary

dataset, and conducted 10 times for each type of query with a returned record for 100 triples. ScFS was set with parameters τ as 0.8 and γ as 0.2.

We define three metrics Eqs. (7)–(9) to evaluate the search results for the lossy summary generated by ScFS. The coverage rate of entities is defined in (7) as a ratio of the No. of entities in the base graph G_L of the summary L to the No. of selected entities in search by the search-by-entity queries. This ratio indicates how many entities are summarized by ScFS.

$$cov\_ent = \frac{No.\ of\ entities\ in\ the\ G_L\ of\ L}{No.\ of\ entities\ in\ search} \qquad (7)$$

Similarly, the coverage rate of properties is defined in (8) as a ratio of the No. of properties in the based graph G_L of the summary L to the No. of selected properties in search by the search-by-property queries. This ratio indicates how many properties are summarized by ScFS.

$$cov_{pr} = \frac{No.\ of\ properties\ in\ the\ G_L\ of\ L}{No.\ of\ properties\ in\ search} \qquad (8)$$

From the definition of base graph we know that if an entity is summarized in the summary L by ScFS, its base graph covers the associated relations and one-hop neighbors.

Similarly, if a property is summarized in the summary L by ScFS, its base graph covers the associated nodes. Therefore, the above definitions for coverage rates for entities and properties reflect the coverage of search triple patterns.

The compression rate of the lossy summary by ScFS is defined in (9) as the No. of FPs in the lossy summary $\hat{L}$ generated by ScFS to the No. of FPs in the lossless summary L generated by SiFS. The metric *comp_lossy* reflects the ability of ScFS for retrieving important of FPs while retaining the required utility.

$$comp\_lossy = \frac{No.\ of\ FPs\ in\ \hat{L}}{No.\ of\ FPs\ in\ L} \qquad (9)$$

Figure 4 (a) shows the results of the metrics *cov-ent*, *cov_pr*, and *comp_lossy* over the four datasets. The coverage rates of entities were all above 90% for the four datasets and the compression rates were at least 43.4% (ODUS) and at most 92.8% for Linked-GeoData. This indicates that SiFS generates concise summaries and at the same time retains key information about entities. The coverage rates for properties were at about 70% except 50% for LinkedGeoData. The result is reasonable because SiFS builds upon the FPs formed by properties. The summary for LinkedGeoData contains only 406 FSs compared to the lossless 5632 and half of the properties were filtered out. Figure 4(b) shows the time for querying the KGs by ScFS, SiFS and direct search without summary. As can be observed that ScFS performed best and remarkablly reduced the query time. Query with ScFS is more efficiently than direct search.

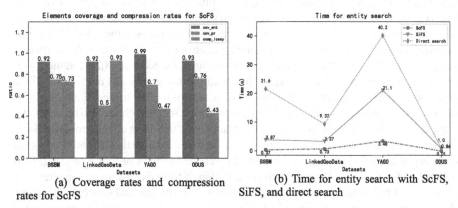

(a) Coverage rates and compression rates for ScFS

(b) Time for entity search with ScFS, SiFS, and direct search

Fig. 4. (a) Coverage rates and compression rates for ScFS (b) Time for entity search with ScFS, SiFS, and direct search

7 Related Work

Statistic-Based Summarization. Statistic-based summarization usually defines metrics to rank resources in KGs and select the most relevant resources as summaries. Pappas et al. [11] proposed six centrality metrics to select central nodes in KGs. Pires et al. [12] proposed using the concepts of centrality and frequency to select the most relevant resources as the summaries for ontologies. Similarly, Troullinou et al. [13] also proposed a summary method, called RDF Digest, to extract the most relevant paths in ontologies. The metrics such as relative cardinality of edge and relevance of node/edge were defined for evaluating resources in ontologies. Safavi et al. [11] studied personalized summaries of KGs. The personalized summarization was defined as a set of triples that maximized a user's defined metric over a given KG, subject to a user- and device-specific constraint on the summary's size. In [14], user preference was considered when creating ontology summaries. Relevance metrics were used to rank and select ontology resources as summaries. Presutti et al. [15] defined the concept knowledge pattern for resources in KGs as the properties by which instances of this type related to other individuals and the types of such individuals for each property. Then the key knowledge pattern paths were extracted based on the metrics of type betweenness and property betweenness.

Structure-Based KG Summarization. Since KGs are directed graphs in nature, a popular idea to solve the problem of KG summarization is from the aspect of graph structure. The structure-based summarization merges graph nodes or edges as supernodes or edges based on the structure of the directed graph formed by RDF data [16]. The set of supernodes and edges are called the summary of a KG. A typical structure-based summarization method is proposed in [4], where equivalence relations between nodes reflecting the similarity between nodes were defined. Nodes were merged by the equivalence relations and thus the quotient graphs were formed as the graph summary. Riondato [7] proposed a summary called k-summary, which partitions graph nodes into k sets (as supernodes) and superedges between supernodes with weights denoting numbers of original edges. Stefanoni et al. [5] presented a typed summary for KGs called SumRDF by merging resources of the same type into a partition. Some works generated partial order relations between elements as summaries for KGs. In [8], the authors presented a KG summary called, characteristic set lattice, to dynamically model KGs. First, Characteristic Set Concepts (CSCs) were extracted from RDF triples. Each CSC is a pair $c = (T, S)$, where T is the set of properties, which is called characteristic set, and S is the set of entities. Each CSC satisfies: all the entities in S have all the properties in T. The lattice of the CSCs were calculated based on the characteristic sets and used as the summary of the KG. Ferré [17] proposed an extended Formal Concept Analysis (FCA) [18] model for KGs which enriched the description of formal concepts. The context is a knowledge graph, concept intents are projected graph patterns, and concept extents are object relations. Alam and Napoli [19] transformed RDF triples into pattern structures and the resulting pattern lattice was the summary. In [20], pattern structure was used to classy RDF triples. The pattern concept lattice was built for discovering significant knowledge units from the of RDF triples.

8   Conclusion and Future Work

We presented an efficient utility-driven summarization method for large-scale KGs. Our method can be categorized into the structure-based summarization which establishes partial orders between node feature patterns. Specifically, we proposed utility-driven summarization algorithms: SiFS and ScFS which can generate KG summaries efficiently with controlled utility loss. We also developed the query evaluation algorithm based on the KG summaries generated by SiFS and ScFS. The experimental results verified that proposed summarization method can efficiently create compact summaries for KGs. Our approach shows significantly improved performance and generates more concise summaries. For the future work, we shall focus on: (i) Summarizing dynamic KGs incrementally without computing the summary from scratch. Since real-world datasets are usually updated periodically, developing algorithms that incrementally compute summaries based on the past versions is necessary. (ii) Generating summaries for streaming KGs. Compared to methods that calculate summaries based on the whole graphs, summaries of streaming graphs need to be generated on the fly based on partial knowledge of the graphs.

Acknowledgement. This research is sponsored by the Educational Reform Research Project of Southwest University (2022JY085) and the Fundamental Research Funds for the Central Universities-Doctoral Fund (SWU222001).

References

1. Čebirić, S., et al.: Summarizing semantic graphs: a survey. VLDB J. **28**, 295–327 (2019). https://doi.org/10.1007/s00778-018-0528-3
2. Liu, Q., Cheng, G., Gunaratna, K., Qu, Y.: Entity summarization: state of the art and future challenges. J. Web Semant. **69**, 100647 (2021)
3. Lissandrini, M., Pedersen, T.B., Hose, K., Mottin, D.: Knowledge graph exploration: where are we and where are we going? Dl.Acm.Org. 1–8 (2020)
4. Čebirić, Š, Goasdoué, F., Manolescu, I.: Query-oriented summarization of RDF graphs. In: Maneth, S. (ed.) BICOD 2015. LNCS, vol. 9147, pp. 87–91. Springer, Cham (2015). https://doi.org/10.1007/978-3-319-20424-6_9
5. Stefanoni, G., Motik, B., Kostylev, E. V.: Estimating the cardinality of conjunctive queries over RDF data using graph summarisation. In: Proceedings of the World Wide Web Conference, WWW 2018, pp. 1043–1052 (2018)
6. Belth, C., Zheng, X., Vreeken, J., Koutra, D.: What is normal, what is strange, and what is missing in a knowledge graph: unified characterization via inductive summarization. In: Proceedings of the World Wide Web Conference, WWW 2020, pp. 1115–1126 (2020)
7. Riondato, M., Garcia-Soriano, D., Bonchi, F.: Graph summarization with quality guarantees. In: Proceedings - IEEE International Conference on Data Mining, ICDM, pp. 947–952 (2014)
8. González, L., Hogan, A.: Modelling dynamics in semantic web knowledge graphs with formal concept analysis. In: Proceedings of the World Wide Web Conference, WWW 2018, pp. 1175–1184 (2018)
9. Hajiabadi, M., Singh, J., Srinivasan, V., Thomo, A.: Graph summarization with controlled utility loss. Association for Computing Machinery (2021)
10. Song, Q., Wu, Y., Dong, X.L.: Mining summaries for knowledge graph search. In: Proceedings of IEEE International Conference on Data Mining, ICDM, pp. 1215–1220 (2017)
11. Safavi, T., Belth, C., Faber, L., Mottin, D., Muller, E., Koutra, D.: Personalized knowledge graph summarization: From the cloud to your pocket. In: Proceedings of IEEE International Conference on Data Mining, ICDM, pp. 528–537 (2019)
12. Pires, C.E., Sousa, P., Kedad, Z., Salgado, A.C.: Summarizing ontology-based schemas in PDMS. In: Proceedings of International Conference on Data Engineering, pp. 239–244 (2010)

13. Troullinou, G., Kondylakis, H., Daskalaki, E., Plexousakis, D.: Ontology understanding without tears: the summarization approach. Semant. Web J. **8**, 797–815 (2017)
14. Queiroz-Sousa, P.O., Salgado, A.C., Pires, C.E.: A method for building personalized ontology summaries. J. Inf. Data Manag. **4**, 236–250 (2013)
15. Presutti, V., Aroyo, L., Adamou, A., Schopman, B., Gangemi, A., Schreiber, G.: Extracting core knowledge from linked data. In: CEUR Workshop (2011)
16. LeFevre, K., Terzi, E.: GraSS: graph structure summarization. In: Proceedings of the 10th SIAM International Conference on Data Mining, SDM 2010, pp. 454–465 (2010)
17. Ferré, Sébastien.: A proposal for extending formal concept analysis to knowledge graphs. In: Baixeries, J., Sacarea, C., Ojeda-Aciego, M. (eds.) ICFCA 2015. LNCS (LNAI), vol. 9113, pp. 271–286. Springer, Cham (2015). https://doi.org/10.1007/978-3-319-19545-2_17
18. Wille, R.: Restructuring lattice theory: an approach based on hierarchies of concepts. In: Ferré, S., Rudolph, S. (eds.) ICFCA 2009. LNCS (LNAI), vol. 5548, pp. 314–339. Springer, Heidelberg (2009). https://doi.org/10.1007/978-3-642-01815-2_23
19. Alam, M., Napoli, A.: An approach towards classifying and navigating RDF data based on pattern structures. In: CEUR Workshop Proceedings, pp. 33–48 (2015)
20. Reynaud, J., Alam, M., Toussaint, Y., Napoli, A.: A proposal for classifying the content of the web of data based on FCA and pattern structures. In: Kryszkiewicz, M., et al. (eds.) ISMIS 2017. LNCS (LNAI), vol. 10352, pp. 684–694. Springer, Cham (2017). https://doi.org/10.1007/978-3-319-60438-1_67
21. Opsahl, T., Agneessens, F., Skvoretz, J.: Node centrality in weighted networks: generalizing degree and shortest paths. Soc. Netw. **32**, 245–251 (2010)
22. Rehyani Hamedani, M., Kim, S.W.: AdaSim: a recursive similarity measure in graphs. In: International Conference on Information and Knowledge Management, Proceedings, pp. 1528–1537 (2021)
23. Bizer, C., Schultz, A.: The Berlin SPARQL benchmark. Int. J. Semant. Web Inf. Syst. **5**, 1–24 (2009)

Commonsense-Aware Attentive Modeling for Humor Recognition

Yuta Sasaki[1], Jianwei Zhang[1(✉)], and Yuhki Shiraishi[2]

[1] Iwate University, Morioka, Iwate, Japan
zhang@iwate-u.ac.jp
[2] Tsukuba University of Technology, Tsukuba, Ibaraki, Japan
yuhkis@a.tsukuba-tech.ac.jp

Abstract. Laughter has positive effects on health. Humor is an important component of daily communication and usually causes laughter. Hence, we can expect that the infusion of humor in a dialogue system will improve the user's physical and mental satisfaction. Since even humans have difficulty comprehending humor, appropriate knowledge is essential for humor understanding. In this paper, commonsense-aware modules are extrapolated to **P**re-trained **L**anguage **M**odels (PLMs) to provide external knowledge. We specifically extract keywords from a text and use COMET to obtain embeddings that represent the commonsense associated with the keywords. We attempt to detect humor that is not detectable by PLM alone. Our approach enables the model to access commonsense knowledge. Compared to the baseline, the number of humor detections increases, and recall is improved without a significant decrease in precision. Our best model significantly improves recall by 4.4% for a 0.4% reduction in precision in the HaHackathon dataset and by 20.3% for an 8.4% reduction in precision in the Humicroedit dataset compared to the baseline. We also observe the changes in prediction and processing speed so as to analyze the characteristics of the proposed method and the issues for its social implementation.

Keywords: Humor recognition · Commonsense-aware attention · Knowledge-intensive NLP · Commonsense knowledge · Pre-trained Language Model (PLM) · Keyword extraction

1 Introduction

Humor, which can usually cause laughter, is an important component of daily communication. Tamada et al. [22] indicate that laughing with others reduces the risk of functional disability in old age, and infer that humor that brings laughter is important from a healthcare perspective. Yamagoe et al. [27] show that laughter reduces the stress response and improves cognitive function by watching a Japanese comedy, which is called "manzai". As we can see, many medical and

This work was supported by JSPS KAKENHI Grant Numbers 19K12230, 22K12271.

psychological studies have shown the positive effects of laughter. By automating humor understanding and generation, not only the implementation of human-interactive chatbots or AI assistants but also the social implementation of inter-active robots as counselors or comedians can be expected. Conversations with such systems may physically and psychologically satisfy the users. Therefore, we focus on automatic humor recognition as the first step in introducing humor into dialogue systems.

In order to integrate laughter and humor into a system, it is important to understand humor automatically. In general, a joke, which is a kind of humor, consists of two parts: a setup, which makes up a story, and a punchline, which concludes the story. A joke includes the elements of wordplay, which hints at stereotypes, misunderstandings, and irony. This provides funniness but also cre-ates difficulties in the recognition of humor. Understanding humor is not an easy task for humans as well as AIs. To understand the punchline that completes the humor, it is necessary to understand the scene from the setup and to under-stand the commonsense that the words indicate. A rhetorical structure can be a trigger for humor understanding, but precise and deep commonsense and world knowledge are necessary to reliably understand humor. Stewart [21] shows that cultural differences can lead to misunderstandings of laughter through conver-sations between Spanish speakers and English speakers. This shows that correct knowledge is necessary to understand laughter and humor.

We illustrate the process of understanding humor with commonsense using the following example sentence:

"How did the astronaut break up with his girlfriend?"
"I just need some space."

The word "space" has multiple meanings including "distance" and "universe". If "space" is simply interpreted as "distance" between the astronaut and his girlfriend, this example cannot be recognized as humor. However, if people have commonsense that "space" associated with astronauts can also be interpreted as the "universe" where astronauts work, this example may probably be perceived as humor. Without sufficient knowledge to notice that "space" has a different interpretation, the humorous part of the sentence cannot be understood. Deep and sufficient knowledge is absolutely crucial to comprehend humor.

Various models for automatic humor recognition have been proposed [1,5,15]. However, few methods specifically focus on commonsense and world knowledge. Badri et al. [16] use multi-modality to predict laughter tracks. They demonstrate the error analysis of the predictions showing that humor, which requires an appropriate knowledge base, is a typical case of an incorrect prediction. Thus, it is conceivable that automatic humor recognition methods should have access to an appropriate knowledge base as humans do.

To address the above problem, we propose a method for humor recogni-tion by extrapolating a knowledge base to PLMs, focusing on commonsense. To enable the model to consider commonsense, we use COMET [3], a GPT-based [17] decoder trained to reconstruct commonsense from natural language. The resulting representation is used as a knowledge base and integrated with the representation of the text from the pre-trained Transformer-based [24] model using Multi-Head Dot Product Attention (MHA). Thus, we attempt to improve the performance of humor understanding by providing knowledge that cannot be considered only in PLMs. Experiments show that the proposed method can increase the number of humor detections and increase the humor sensitivity of the model.

Our contribution is as follows:

1. We propose an architecture for humor recognition that considers various types of commonsense by extrapolating COMET that reconstructs commonsense. We softly integrate commonsense into the context of a text by treating the output of COMET as knowledge embedding to perform humor recognition.
2. The experiments on two datasets show that commonsense increases the number of humor predictions while sharing many of the predictions made by PLM alone. It is also shown that recall can be improved without a significant decrease in precision.

2   Related Works

Mihalcea and Strapparava [15] use textual style features and content-based features for humor recognition employing classical methods. Chen et al. [5] propose a deep learning-based detection method by introducing CNN and Highway Networks. Weller and Seppi [25,26] contribute to the community by constructing a large dataset for humor recognition from Reddit[1]. They perform the recognition task by fine-tuning pre-trained Transformer-based models on that dataset. Annamoradnejad and Zoghi [1] notice that many jokes consist of setup and punchline and propose ColBERT, which considers the relationship between sentences by dividing the input text into sentences. In SemEval, competitions for humor recognition tasks have also been conducted, in which various systems have been proposed and many pre-trained Transformer-based models have been employed [9,14].

Although many methods for humor recognition and generation have been proposed, most of them use pre-trained Transformer-based models in recent years, and not many of them explicitly deal with commonsense and world knowledge. Zhang et al. [29] attempt to automatically generate a more consistent punchline by encoding the knowledge triplets of words extracted from Wikipedia as a graph structure.

In other tasks, several studies attempt to improve performance by accessing commonsense. Li et al. [11] investigate the effect of commonsense on sarcasm detection performance by extracting relevant commonsense from the input text and comparing two different knowledge selection strategies. Chowdhury and Chaturvedi [2] consider commonsense as a graph structure and apply GCN for irony detection. Yang et al. [28] treat commonsense as mental state knowledge and explicitly model the mental state of speakers. In other studies, commonsense

---

[1] https://www.reddit.com/.

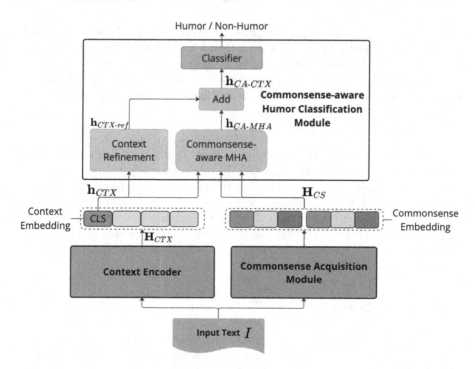

Fig. 1. Our proposed model architecture. It mainly contains three parts: (1) context encoder, (2) commonsense acquisition module, and (3) commonsense-aware humor classification module.

is also used in empathetic dialogue to attempt empathetic response generation [18, 23].

In this study, we attempt to augment the knowledge of the model for humor recognition by using commonsense as external knowledge and softly integrating it into the model.

3   Proposed Method

Our proposed model is illustrated in Fig. 1. Our model consists of three main parts: (1) context encoder, (2) commonsense acquisition module, and (3) commonsense-aware humor classification module.

3.1   Context Encoder

This encoder obtains the embedding that indicates the context of the input text. In this study, Transformer-based PLM is used as the context encoder. Given the input text I, the tokenizer of the PLM returns the token sequence X of I,

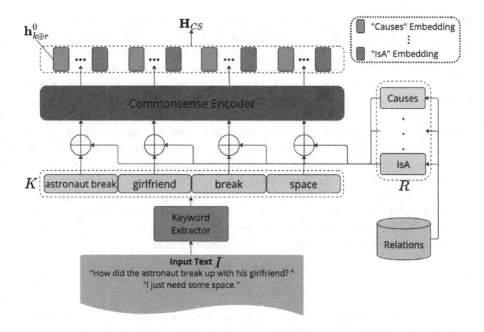

Fig. 2. Commonsense acquisition module.

prepending a special token [CLS] as x_0 to it:

$$X = tokenizer(I)$$
$$= \{x_0, x_1, ..., x_{L-1}\} \tag{1}$$

The PLM receives X and outputs the embeddings for each token. Especially, the embedding marked as h_{CTX}, corresponding to the first token x_0 which is [CLS], can represent the meaning of the entire input text:

$$\mathbf{H}_{CTX} = PLM(X) \tag{2}$$

$$\mathbf{h}_{CTX} = \mathbf{H}_{CTX}[0] \tag{3}$$

where $\mathbf{H}_{CTX} \in \mathbb{R}^{L \times d}$, $\mathbf{h}_{CTX} \in \mathbb{R}^d$, L is the length of the input text and d is the hidden size of the PLM output. Although various PLMs that can encode the meaning of sentences have achieved great success in recent years [7,10,12], in this experiment we used BERT [6], which is widely used in general and can achieve high performance.

3.2 Commonsense Acquisition Module

The commonsense acquisition module, illustrated in Fig. 2, is composed of a keyword extractor and a commonsense encoder.

Keyword Extractor. In order to capture the commonsense associated with an input text, it is necessary to capture the concepts of the words and phrases that constitute the text. To solve this problem, we have tried three types of input to the subsequent commonsense encoder: the raw input text, the summary of the input, and the keywords of the input. The commonsense encoder has a limitation on the input to generate knowledge. The length of tokens of the input to the encoder is short while fine-tuning. Since both the raw input text and the summary of the input are long, they may be unsuitable for the input to the encoder. Through the preliminary experiments, we find that the best method is to use the keywords of the input. Therefore, this component extracts important words and phrases from the input using a keyword extractor.

We use YAKE! as a keyword extractor [4]. With hyperparameters, it is possible to set the threshold of acceptable word duplicates and the maximum $n-$gram of keywords, and thus the output of keywords can be easily adjusted. The keyword extractor using YAKE! can be formulated as follows:

$$
\begin{aligned}
K &= YAKE!(I) \\
&= \{k_1, k_2, ..., k_m \mid m \leq M_{kwd}\}
\end{aligned}
\tag{4}
$$

where K is the resulting set of keywords or keyphrases and M_{kwd} is the maximum number of keywords to be input to the following commonsense encoder. m depends on the input text. In this study, top$-m$ keywords are extracted based on the scores of candidate keywords that YAKE! outputs.

Commonsense Encoder. The commonsense encoder obtains embeddings representing commonsense from the keywords extracted by the keyword extractor. We use COMET as this encoder. COMET is a pre-trained GPT-based model to generate commonsense in the form of natural language from input words and phrases. COMET has two models, which are trained on each of the two datasets, ConceptNet [20] and ATOMIC [19]. ConceptNet is a word-level commonsense knowledge base consisting of a graph structure of descriptions of word concepts. The edges of the graph are represented by relations, and there are 34 relations such as "IsA" and "Causes". ATOMIC is a large if-then knowledge base about daily human interaction. In this experiment, we target conceptual commonsense about the extracted keywords and keyphrases using COMET trained on ConceptNet. Concatenated texts of keywords and relations are input to COMET. For example, if the keyword is "astronaut break" and the relation is "IsA", "astronaut break IsA" is the input into COMET. Thus, the input $K' = \{k_1 \oplus r_1, k_1 \oplus r_2, ..., k_1 \oplus r_n, k_2 \oplus r_1, ..., k_m \oplus r_n\}$ to COMET is constructed from the keyword set K and the relation set $R = \{r_1, r_2, ..., r_n\}$. The relations that compose R can be chosen arbitrarily from 34 relations according to the task. $\oplus$ denotes the text concatenation operation. Instead of using COMET to explicitly generate the commonsense as a sentence, we let COMET act as an encoder by using the hidden state in the final layer corresponding to the last token as the commonsense embedding. We expect this to improve the affinity between the embedding obtained from the context encoder and the knowledge generated from COMET by softly integrating the knowledge base into the model. The commonsense encoder can be formulated as follows:

$$\mathbf{H}^i_{k\oplus r} = COMET(K'[i]) \tag{5}$$

$$\mathbf{h}^i_{k\oplus r} = \mathbf{H}^i_{k\oplus r}[l-1] \tag{6}$$

$$\mathbf{H}_{CS} = [\mathbf{h}^0_{k\oplus r}, \mathbf{h}^1_{k\oplus r}, ..., \mathbf{h}^{m \cdot n-1}_{k\oplus r}] \tag{7}$$

where $\mathbf{H}^i_{k\oplus r} \in \mathbb{R}^{l \times d}$ is the output embeddings of COMET for one pair of a keyword and a relation, $\mathbf{h}^i_{k\oplus r} \in \mathbb{R}^d$ is the embedding corresponding to the last token of the input, $\mathbf{H}_{CS} \in \mathbb{R}^{(m \cdot n) \times d}$ represents the keyword-based commonsense embeddings, $i \in \{0, 1, ..., m \cdot n - 1\}$, l are the maximum length of tokens that can be fed into COMET, and d is the hidden size of COMET.

3.3 Commonsense-Aware Humor Classification Module

This module is further composed of three components: Context Refinement, Commonsense-aware MHA & Add, and Humor Classifier.

Context Refinement. This module refines the context embedding $\mathbf{h}_{CTX}$ obtained in Eq. 3 for humor understanding and then outputs it as $\mathbf{h}_{CTX-ref}$. This process is formulated as follows:

$$\mathbf{h}_{CTX-ref} = \sigma(\mathbf{W}_{ref}\mathbf{h}_{CTX} + \mathbf{b}_{ref}) \tag{8}$$

where $\mathbf{h}_{CTX\text{-}ref} \in \mathbb{R}^d$, $\mathbf{W}_{ref} \in \mathbb{R}^{d \times d}$, $\mathbf{b}_{ref} \in \mathbb{R}^d$ and σ is the hyperbolic tangent.

Commonsense-Aware MHA & Add. This component integrates $\mathbf{h}_{CTX}$, context embedding, and $\mathbf{H}_{CS}$, commonsense embeddings. We expect the context to be used as the basis for humor understanding, and the conceptual commonsense of the keywords to be considered as background knowledge. Hence, we employ $\underline{C}$ommonsense-$\underline{A}$ware $\underline{M}$ulti-$\underline{H}$ead $\underline{A}$ttention (CA-MHA). The commonsense representation associated with or required by the context is obtained and added to the refined context embedding to realize the commonsense integration. Formulating Multi-Head Attention as $MHA(query, key, value)$, the module is as follows:

$$\mathbf{h}_{CA-MHA} = MHA(\mathbf{h}_{CTX}, \mathbf{H}_{CS}, \mathbf{H}_{CS}) \tag{9}$$

$$\mathbf{h}_{CA-CTX} = \mathbf{h}_{CTX-ref} + \mathbf{h}_{CA-MHA} \tag{10}$$

where $\mathbf{h}_{CA-MHA} \in \mathbb{R}^d$ represents the commonsense information selected by context and $\mathbf{h}_{CA-CTX} \in \mathbb{R}^d$ represents the commonsense-aware context.

Humor Classifier. The commonsense-aware context $\mathbf{h}_{CA-CTX}$ enhanced by commonsense-aware MHA & Add is input to the classifier for humor classification.

$$\mathbf{P} = softmax(\mathbf{W}_c\mathbf{h}_{CA-CTX} + \mathbf{b}_c) \tag{11}$$

where $\mathbf{W}_c \in \mathbb{R}^{2 \times d}$ and $\mathbf{b}_c \in \mathbb{R}^2$ are the weights and biases of the classifier, respectively. $\mathbf{P}$ represents the probability for each label.

Table 1. Summary of the datasets. The training, validation, and test sets are publicly provided.

Dataset	Train			Valid			Test		
	Pos	Neg	Total	Pos	Neg	Total	Pos	Neg	Total
HaHackathon	4,932	3,068	8,000	632	368	1,000	615	385	1,000
Humicroedit	4,731	12,573	19,304	1,173	3,665	4,838	1,466	4,582	6,048

3.4 Objective Function

The cross-entropy loss is used as the objective function for $\mathbf{P}(y)$, which is the probability corresponding to the label y, to optimize the model. This calculation is as follows:

$$\mathcal{L} = -(1 - y) \log \mathbf{P}(0) - y \log \mathbf{P}(1) \tag{12}$$

where $y \in \{0, 1\}$ is a discrete value and $\mathbf{P}(\cdot)$ is a continuous value.

4   Experimental Setup

4.1   Datasets

In this study, two public datasets are used for training and evaluation. The statistics of the datasets are shown in Table 1.

HaHackathon. This dataset is used in SemEval 2021 task 7[2]. 80% of the data comes from Twitter and the rest comes from Kaggle Short Jokes[3]. For humor detection, the annotators answer the question, "Is the intention of this text to be humor?" and thus each text is evaluated by 20 annotators. The humor label of a text is determined based on the majority of its votes.

Humicroedit. This dataset is used in SemEval 2020 task 7[4]. The dataset published in this competition is derived from a previous study by Hossain et al. [8]. This dataset is constructed by a task that rewrites one word from news headlines collected from Reddit to create a humorous sentence. The text consists of headings of 4–20 words. The degree of humor is rated by several annotators, with integer values from 0 to 3, and the average value is labeled as the score of the humor. Scores are set as follows:

0 - Not funny   1 - Slightly funny   2 - Moderately funny   3 - Funny

Since our task is binary classification, we set the data whose score is greater than 1 as Humor, and all the others as Non-Humor.

---

[2] https://competitions.codalab.org/competitions/27446.
[3] https://www.kaggle.com/datasets/abhinavmoudgil95/short-jokes.
[4] https://competitions.codalab.org/competitions/20970.

Notice the difference between the two datasets. Humicroedit contains data where the majority of words are the same but only one word differs. As the difference of this one word alone determines humor, it is even a difficult task for humans to recognize the change in humor that accompanies a change in one word. The classification on Humicroedit is a more difficult task than that on HaHackathon. Moreover, the two datasets differ in data size. HaHackathon has 8,000 training data, while Humicroedit has 19,304 training data.

4.2 Experimental Configuration

We use BERT (110M parameters) without introducing commonsense as a baseline, and bert-base-uncased[5] published in HuggingFace[6] as the initial values of the weights. The proposed method employs the same weights for the context encoder. We implement COMET (117M parameters) to generate commonsense using comet-commonsense and adopted a pre-trained model trained on Concept-Net[7].

AdamW [13] is adopted as the optimizer, the learning rate of BERT is set to 2e-5 and the learning rate of other modules is set to $5e-5$, while the weights of COMET are fixed. The batch size is 32. We attempt two patterns of inputting the relations to COMET. The first approach uses five relations, "IsA", "HasA", "Causes", "Desires" and "UsedFor". We adopt these five relations for commonsense acquisition because they are intuitively comprehensible. All 34 relations are used in the second approach. As the hyperparameters of YAKE!, dedupLim and $n-$gram are set to 0.3 and 3, respectively, and the maximum number of keywords M_{kwd} is 6. Due to the small size of data in the HaHackathon dataset, we set the dropout rate to 0.4 to prevent over-fitting. Since the Humicroedit dataset has sufficient data, we set the dropout rate to the default value of 0.1 for Humicroedit. The number of heads of CA-MHA is 8 or 12 in the experiments.

All experiments are performed on a CPU with 12 GB of memory and a Tesla T4 with 16 GB of memory using Google Colaboratory.

5 Experimental Results

In this section, we evaluate the performance of our models on the two datasets. Besides, to clarify the situation where our best model can perform well, we compare the difference in the tendency of predictions between the baseline BERT and our best model. Moreover, we discuss training and inference speeds since our models need more computational complexity than the baseline.

---

[5] https://huggingface.co/bert-base-uncased.

[6] https://huggingface.co.

[7] https://github.com/atcbosselut/comet-commonsense.

5.1 Model Performance

The results of our proposed method and the baseline are shown in Table 2. As evaluation metrics, precision (Prec), recall (Rec), and f1-score (F1) are employed. The proposed method is compared concerning the performance variation due to the change in the number of heads in CA-MHA and the difference in the relation used for commonsense acquisition. CA-MHA8- and CA-MHA12- denote the models with 8 and 12 heads, respectively. -5rels indicates that the five selected relations are used as input to the commonsense encoder, and -All indicates that all the relations are used.

Table 2. Performance comparisons on HaHackathon and Humicroedit. The top-1 score is highlighted in bold.

Model	HaHackathon			Humicroedit		
	Prec	Rec	F1	Prec	Rec	F1
BERT	**0.928**	0.908	0.918	0.620	0.475	0.538
CA-MHA8-5rels	0.924	**0.952**	**0.938**	0.536	**0.678**	**0.599**
CA-MHA8-All	0.895	0.923	0.909	**0.642**	0.242	0.352
CA-MHA12-5rels	0.918	0.910	0.914	0.627	0.392	0.483
CA-MHA12-All	0.911	0.900	0.905	0.591	0.536	0.562

From Table 2, we can see that using commonsense tends to increase recall. The performance of CA-MHA8-5rels is the best in both HaHackathon and Humicroedit. Compared to the baseline BERT, the model improves Rec by 4.4% for a 0.4% reduction in Prec in HaHackathon and by 20.3% for an 8.4% reduction in Prec in Humicroedit. These results show that our best model can significantly improve Rec without a significant drop in Prec and recognize humor more sensitively by enhancing knowledge through commonsense.

However, we observe that coarsely integrating as much commonsense as possible does not always lead to good performance. In HaHackathon, the -All model reduces all the metrics in contrast with the respective -5rels model. In Humicroedit, CA-MHA8-All significantly decreases Rec and F1 against CA-MHA8-5rels. This result indicates that it may be better to refinedly introduce knowledge selection and the combination of relations.

Focusing on the number of heads in CA-MHA, the performance of CA-MHA12-5rels is lower than that of CA-MHA8-5rels on both datasets. Compared to CA-MHA8-5rels, CA-MHA12-5rels decreases F1 by 2.4% in HaHackathon and by 11.6% in Humicroedit. This result indicates that the increased number of heads of CA-MHA does not always yield better results.

5.2 Efficacy of Commonsense

In this section, we analyze the performance impact of commonsense infusion. We compare the best model in this experiment, CA-MHA8-5rels, with the baseline model BERT. We evaluate the overlap and changes in the predictions of the baseline and the proposed model to clarify the efficacy of integrating commonsense. We also compare the training time and inference time between the proposed method and the baseline, respectively, and clarify the issues for social implementation.

Table 3. The overlap and changes in the predictions on the test set. The overlap shows the ratio of data on which both our best model CA-MHA8-5rels and the baseline BERT predict the same label. "$\hat{y}_{BERT} \rightarrow \hat{y}_{ours} : y$" is a notation where $\hat{y}$ denotes the predicted label and y denotes the target label. The columns, where our best model correctly predicts the label but the baseline does not, are bold.

Dataset	Overlap	$0 \rightarrow 1 : 1$	$0 \rightarrow 1 : 0$	$1 \rightarrow 0 : 1$	$1 \rightarrow 0 : 0$
HaHackathon	95.2%	**29**	11	2	**6**
Humicroedit	86.7%	**312**	453	15	**21**

Table 4. The training and inference speeds of BERT and our best model CA-MHA8-5rels. The speed unit is one second per step and the batch size on a step is set to 32.

Phase	BERT	CA-MHA8-5rels	Relative speed
Training	0.829 s	2.255 s	×2.720
Inference	0.070 s	1.344 s	×19.200

Overlap and Changes in the Predictions. In Fig. 3 and Fig. 4, the numbers of Humor predictions of the proposed model CA-MHA8-5rels, which are 634 on HaHackathon and 1,852 on Humicroedit, are much larger than those of the baseline BERT, which are 602 and 1,123 respectively on the two datasets. In Table 3, however, the overlap between the predictions of BERT and CA-MHA8-5rels is quite large. These figures and table show that the proposed model transforms the prediction of PLM used as a context encoder by treating commonsense as background knowledge. The number of predictions where the proposed model converts Non-Humor from the baseline to Humor is quite large, 40 for HaHackathon and 765 for Humicroedit, but not many for the reverse. This indicates that the model can be more sensitive to humor by utilizing commonsense. It is shown that the simple architecture integrating COMET embeddings can improve the model's performance for humor detection. However, the proposed model incorrectly converts predictions from Non-Humor to Humor for 11 out of 40 cases in HaHackathon and 453 out of 765 cases in Humicroedit. In an environment where appropriate predictions are required for Non-Humor, some solutions, such as filtering the predictions, should be further discussed.

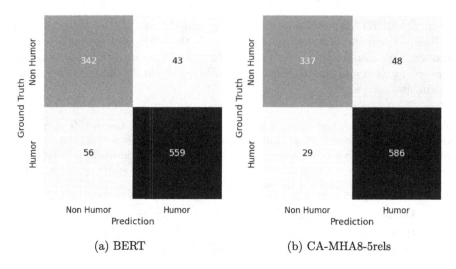

Fig. 3. Confusion matrix of BERT and CA-MHA8-5rels on HaHackathon.

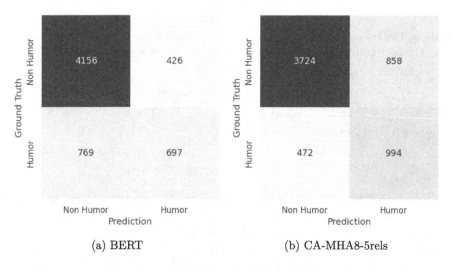

Fig. 4. Confusion matrix of BERT and CA-MHA8-5rels on Humicroedit.

Training and Inference Speeds. We compare the training and inference speeds of the baseline and the proposed method. According to Table 4, BERT is clearly faster in both training and inference speeds than our model. COMET accounts for most parameters of the additional modules associated with the use of commonsense. Even though the parameters of COMET are fixed while training, it is computationally expensive and a bottleneck in the overall processing system. In particular, a significant difference in speed is observed during inference, where the speed of the proposed method is 19.2 times slower than that of the baseline in one step, which takes 1.344 s. Although more humorous text can be detected with our proposed model, there are still challenges in applying this method to systems that require real-time processing, such as dialogue systems. In order to adapt to real-time processing, a module that integrates commonsense in a fast manner is crucial, which can be further future work.

6   Conclusion and Future Works

To improve the performance of humor recognition, we proposed a model that integrates the commonsense embeddings obtained from COMET with the context embedding of PLM by Commonsense-aware Multi-Head Attention. This model enables humor recognition with commonsense, shows an increase in the number of humor predictions compared to the number of humor predictions using PLM alone, and enables an increase in recall without a significant decrease in precision. Since the predictions from the proposed method share many of the predictions from the PLM used as a context encoder, the proposed model can make predictions in line with the knowledge without losing many of the PLM's prediction tendencies. We found that the use of keyword-based commonsense enhances the humor-sensitivity of the model. However, the adaptation of the proposed method to a PLM-only model results in a very large slowdown in both training and inference speeds. In order to deal with real-time systems, a fast-access method to commonsense is necessary to study.

In this study, BERT is used as a context encoder, and the proposed method is designed to extrapolate modules without dependence on PLMs. Using other PLMs, we will evaluate the generality of the proposed method and the trend of its impact on PLMs. In some cases, when all the input relations are used for COMET, the external knowledge becomes noise and the detection performance is not improved, but rather degraded. Thus, to promote more appropriate knowledge use, we will further study the module on knowledge selection and the combination of relations. In addition, the development of datasets for the humor recognition task is less sufficient than that for the other tasks. Therefore, constructing humor datasets is also one of our future works to develop the community of humor recognition.

References

1. Annamoradnejad, I., Zoghi, G.: ColBERT: using BERT sentence embedding for humor detection. arXiv preprint arXiv:2004.12765 (2020)
2. Basu Roy Chowdhury, S., Chaturvedi, S.: Does commonsense help in detecting sarcasm? In: Proceedings of the Second Workshop on Insights from Negative Results in NLP (2021)
3. Bosselut, A., Rashkin, H., Sap, M., Malaviya, C., Celikyilmaz, A., Choi, Y.: COMET: commonsense transformers for automatic knowledge graph construction. In: Proceedings of the 57th Annual Meeting of the Association for Computational Linguistics (2019)

4. Campos, R., Mangaravite, V., Pasquali, A., Jorge, A., Nunes, C., Jatowt, A.: YAKE! keyword extraction from single documents using multiple local features. Inf. Sci. **509**, 257–289 (2020)

5. Chen, P.Y., Soo, V.W.: Humor recognition using deep learning. In: Proceedings of the 2018 Conference of the North American Chapter of the Association for Computational Linguistics: Human Language Technologies (2018)

6. Devlin, J., Chang, M.W., Lee, K., Toutanova, K.: BERT: pre-training of deep bidirectional transformers for language understanding. In: Proceedings of the 2019 Conference of the North American Chapter of the Association for Computational Linguistics: Human Language Technologies (2019)

7. He, P., Liu, X., Gao, J., Chen, W.: DeBERTa: decoding-enhanced BERT with disentangled attention. In: Proceedings of the Ninth International Conference on Learning Representations (2021)

8. Hossain, N., Krumm, J., Gamon, M.: "President vows to cut <taxes> hair": Dataset and analysis of creative text editing for humorous headlines. In: Proceedings of the 2019 Conference of the North American Chapter of the Association for Computational Linguistics: Human Language Technologies (2019)

9. Hossain, N., Krumm, J., Gamon, M., Kautz, H.: SemEval-2020 task 7: assessing humor in edited news headlines. In: Proceedings of the Fourteenth Workshop on Semantic Evaluation (2020)

10. Lewis, M., et al.: BART: denoising sequence-to-sequence pre-training for natural language generation, translation, and comprehension. In: Proceedings of the 58th Annual Meeting of the Association for Computational Linguistics (2020)

11. Li, J., Pan, H., Lin, Z., Fu, P., Wang, W.: Sarcasm detection with commonsense knowledge. IEEE/ACM Trans. Audio Speech Lang. Process. **29**, 3192–3201 (2021)

12. Liu, Y., et al.: RoBERTa: a robustly optimized BERT pretraining approach. arXiv preprint arXiv:1907.11692 (2019)

13. Loshchilov, I., Hutter, F.: Decoupled weight decay regularization. In: Proceedings of the Seventh International Conference on Learning Representations (2019)

14. Meaney, J.A., Wilson, S., Chiruzzo, L., Lopez, A., Magdy, W.: SemEval 2021 task 7: HaHackathon, detecting and rating humor and offense. In: Proceedings of the 15th International Workshop on Semantic Evaluation (2021)

15. Mihalcea, R., Strapparava, C.: Making computers laugh: investigations in automatic humor recognition. In: Proceedings of Human Language Technology Conference and Conference on Empirical Methods in Natural Language Processing (2005)

16. Patro, B.N., Lunayach, M., Srivastava, D., Sarvesh, S., Singh, H., Namboodiri, V.P.: Multimodal humor dataset: Predicting laughter tracks for sitcoms. In: Proceedings of the 2021 IEEE Winter Conference on Applications of Computer Vision (2021)

17. Radford, A., Narasimhan, K., Salimans, T., Sutskever, I., et al.: Improving language understanding by generative pre-training (2018). https://cdn.openai.com/research-covers/language-unsupervised/language_understanding_paper.pdf

18. Sabour, S., Zheng, C., Huang, M.: CEM: commonsense-aware empathetic response generation. In: Proceedings of the 2022 AAAI Conference on Artificial Intelligence, vol. 36 (2022)

19. Sap, M., et al.: ATOMIC: an atlas of machine commonsense for if-then reasoning. In: Proceedings of the 2019 AAAI Conference on Artificial Intelligence (2019)

20. Speer, R., Chin, J., Havasi, C.: ConceptNet 5.5: an open multilingual graph of general knowledge. In: Proceedings of the 2017 AAAI Conference on Artificial Intelligence (2017)

21. Stewart, S.: The many faces of conversational laughter. ERIC (1997)
22. Tamada, Y., et al.: Does laughing with others lower the risk of functional disability among older Japanese adults? The JAGES prospective cohort study. Prevent. Med. **155**, 106945 (2022)
23. Tu, Q., Li, Y., Cui, J., Wang, B., Wen, J.R., Yan, R.: MISC: a mixed strategy-aware model integrating COMET for emotional support conversation. In: Proceedings of the 60th Annual Meeting of the Association for Computational Linguistics (2022)
24. Vaswani, A., et al.: Attention is all you need. In: Proceedings of the 31st Conference on Neural Information Processing Systems (2017)
25. Weller, O., Seppi, K.: Humor detection: a transformer gets the last laugh. In: Proceedings of the 2019 Conference on Empirical Methods in Natural Language Processing and the 9th International Joint Conference on Natural Language Processing (2019)
26. Weller, O., Seppi, K.: The rJokes dataset: a large scale humor collection. In: Proceedings of the 12th Language Resources and Evaluation Conference (2020)
27. Yamakoshi, T., et al.: 笑いによるストレス応答抑制と認知機能改善効果. In: Proceedings of the Annual convention of the Japanese Association of Health Psychology (2021)
28. Yang, K., Zhang, T., Ananiadou, S.: A mental state knowledge-aware and contrastive network for early stress and depression detection on social media. Inf. Process. Manage. **59**, 102961 (2022)
29. Zhang, H., Liu, D., Lv, J., Luo, C.: Let's be humorous: knowledge enhanced humor generation. arXiv preprint arXiv:2004.13317 (2020)

A Study on Vulnerability Code Labeling Method in Open-Source C Programs

Yaning Zheng⑩, Dongxia Wang, Huayang Cao, Cheng Qian, Xiaohui Kuang,
and Honglin Zhuang⁽✉⁾

National Key Laboratory of Science and Technology on Information System Security,
Beijing, China
qiancheng@nudt.edu.cn, zhlxsjl@163.com

Abstract. Various existing vulnerability databases and open-source
code platforms have accumulated a large amount of vulnerability information, and extracting vulnerability code samples from this information can help research the causes of vulnerabilities, develop vulnerability
detection technologies, and detect potential vulnerabilities. In this work,
we collected 13 vulnerability code datasets involving various applications and analyzed these datasets in seven aspects, such as data sources,
labeling methods, application scenarios, etc. We found several defects in
these datasets, including duplicated data, incomplete information, and
inaccurate labels. We also analyzed the extraction and labeling methods
of these datasets and proposed three labeling technology frameworks:
labeling based on text description, labeling based on patch analysis, and
labeling based on vulnerability scanning. The proposed frameworks can
be used to evaluate existing labeling methods and guide the future work
on labeling vulnerability code samples, which can help form a better
vulnerability code dataset.

Keywords: code label · patch analysis · vulnerability datasets

1 Introduction

Potential vulnerabilities rise rapidly as the number and complexity of software
and connected devices increases significantly, bringing serious challenges to information system security. To better manage and study vulnerability information,
various vulnerability datasets have been constructed based on publicly available
vulnerabilities [1–7]. Some datasets [3–5] mainly store text description information of vulnerabilities, which provide very limited support for vulnerability
research. In comparison, the vulnerability codes can provide direct support for
studying the causes of vulnerabilities, detecting potential vulnerabilities and
developing corresponding detection methods. In recent years, some work used
deep-learning models to mine vulnerability patterns, which also requires a large
number of high-quality vulnerability codes for training.

Studies [8,9] have shown that the lack of accurate and real-world datasets has
become an important obstacle in the field of vulnerability analysis. Jimenez [10]
further points out that unreliable code sample labeling information can greatly

C. Strauss et al. (Eds.): DEXA 2023, LNCS 14146, pp. 52–67, 2023.
https://doi.org/10.1007/978-3-031-39847-6_4

mislead the experimental conclusions. The existing literature [11–15] mentions some problems of the existing datasets, but no work systematically studies the code sample labeling problem in datasets to the best of our knowledge.

In this paper, we review three kinds of vulnerability code labeling methods on 13 datasets consisted of C codes. We first evaluated and analyzed these datasets from the aspects of data sources, application scenarios, etc. Then we studied the corresponding code sample labeling methods, and obtain some findings that we believe very helpful for the future work. The research in this paper attempts to answer the following three questions.

RQ1-Where are the data sources for vulnerability datasets and what are the application scenarios? By answering this research question, it will be helpful to understand the current state of generation and application of existing vulnerability datasets.

RQ2-What technologies are used to label the vulnerability samples? By answering this research question, we comprehensively surveyed the labeling methods of existing datasets.

RQ3-What is the defects of current vulnerability code labeling methods and how we may improve in the future work? By answering this research question, we help to discover new research points for code labeling methods.

The rest of the paper is organized as follows: Sect. 2 introduces the existing vulnerability source datasets and motivation; Section 3 discusses the existing vulnerability sample labeling methods. Section 4 describes the findings and future work. Section 5 concludes the paper.

2 Motivation

There are a very wide variety of source code vulnerability datasets available, depending on different application scenarios. We collected papers of source code

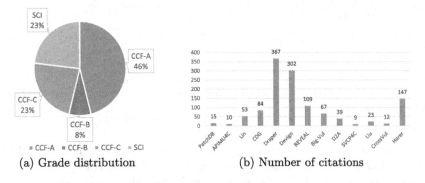

(a) Grade distribution   (b) Number of citations

Fig. 1. Grade distribution and citations of 13 selected papers

vulnerability detection for the C language and the code datasets they use as widely as possible within the last 5 years. We made a CCF grade distribution and citations given by Google Scholar for these 13 papers, as shown in Fig. 1.

As shown in Table 1, We analyzed the datasets in these papers. NVD, synthetic, GitHub denote the data source. The 'open' column indicates whether it is open source. Most of the above datasets are generated from information of CVE, NVD and GitHub.The 'type' column indicates whether the vulnerability type is indicated.

Table 1. Real-world C vulnerability datasets

Dataset	Application scenario	Collection method	NVD	Synthetic	GitHub	Type	Labeling	Granularity	Open
PatchDB [2]	Identify security-related patches	Automatic	yes	yes	yes	yes	NVD	Function	yes
APIMU4C [6]	API misuse detection	Artificial	no	yes	yes	yes	Artificial	Function	yes
Lin* [16]	Evaluating the performance of different deep neural networks on vulnerability source code detection	Automatic	yes	no	yes	no	NVD+commit diff	Function/File	part
CDG [17]	Deep learning based fine-grained source code vulnerability detection	Automatic	yes	yes	yes	yes	NVD+commit diff	Slice	yes
Draper [18]	Deep learning based source code vulnerability detection	Automatic	yes	no	yes	no	clang cppcheck flawfinder	Function	yes
Devign [1]	Detecting source code vulnerabilities using graph neural networks	Automatic	no	no	yes	no	Artificial+commit diff	Function	part
REVEAL [19]	Detecting source code vulnerabilities using graph neural networks	Automatic	no	no	yes	no	commit	Function	yes
Big-Vul [20]	Vulnerability Detection	Automatic	yes	no	yes	yes	CVE+commit diff	Function	yes
D2A [21]	Reducing false positives in vulnerability detection tools	Automatic	no	no	yes	yes	infer	Trace	yes
SVCP4C [22]	Generate buffer overflow type vulnerability dataset	Automatic	no	no	yes	yes	SonarCloud	File	yes
Liu* [23]	Study the distribution of vulnerabilities inside project	Automatic	yes	no	yes	no	commit+patch	Function	yes
CrossVul [24]	Generate vulnerability datasets for follow-up studies	Automatic	yes	no	yes	yes	NVD+commit diff	File	yes
Harer* [25]	Using machine learning to detect vulnerabilities	Automatic	no	no	yes	no	clang static analyzer-SA	Function	no

* The name of the dataset is not given in the representative article, we use the name of the author instead.

Vulnerability code datasets can help people learn patterns or features, which can help improve the capability of vulnerability detection. For example, deep learning-based vulnerability detections generally use vulnerability code datasets as training data to obtain a vulnerability discriminative model for vulnerability detection. Saikat [19] found that existing deep-learning vulnerability detection models that using pre-trained models directly to detect real-world vulnerabilities suffer an average performance degradation of about 73%. Even if these models are retrained using real data, the performance degrades by about 54% compared to the reported results in their papers.

By studying the above datasets, we summarize the reasons as follows.

(1) Existing datasets suffer from data loss or data duplication problems in varying degrees. For example, the CDG [17] dataset in Table 1 suffers from slicing data duplication problem and the D2A [21] dataset suffers from commit version duplication problem, etc. Grahn [13] explores seven C/C++ datasets and evaluates their applicability to machine learning-assisted vulnerability detection. That none of the datasets contained a complete C/C++ elements, about 11% code elements missing.

(2) Many source code datasets are limited to the function-level and do not capture some inter-procedural information which is critical to the complete analysis of the code. Models trained on such datasets do not really capture inter-procedural flows, resulting in a high rate of false positives.

(3) The inaccurate labeling results of the dataset lead the model to extract features that are not relevant to the vulnerability and increase the model noise.

Taking the dataset in [16] as an example, we selected the open-source LibPNG dataset to check the data labels using the static analysis tools cppcheck and flawfinder. The results are shown in the Table 2. We found that: it is inaccurate for labeling the unpatched part as benign as. The code that is not temporarily found to be vulnerable does not mean that there is no vulnerability in it. For the functions marked as non-vulnerable in LibPNG, we selected 10 samples among those detected as error by both cppcheck and flawfinder for manual verification and found that there exists vulnerabilities.

Table 2. LibPNG label accuracy verification.

LibPNG	Nums[a]	cppcheck	flawfinder	Common[b]
Non_vulnerable_function	577	286	83	50
vulnerable_ functions	45	Where are the data sources for33	11	10

[a] 'Nums' represents the number of functions in the original folder.
[b] 'Common' represents the number of common functions detected by cppcheck and flawfinder.

Based on our analysis of 13 high-level papers, it can be sure that there are still many problems in the existing datasets. We found that this is closely related to the labeling methods, and we implement a comprehensive review in the following section, so we researched the labeling methods.

3 Labeling Methods

Code labeling methods mark whether a code sample contains vulnerabilities or not. Based on the investigation of the vulnerability datasets in Section II, we classify code labeling methods into the following three categories: labeling based on text description, labeling based on patch analysis, and labeling based on vulnerability scanning. Labeling based on text description and labeling based on patch analysis refer to the imprecise vulnerability information that has been published, combined with various data and analysis to accurately identify the vulnerability. Labeling based on vulnerability scanning is to label the detected vulnerabilities by vulnerability scanning tools in unknown-source code. The labeling produced by these methods is imprecise. For example, specific location information is missing from NVD. It's impossible to judge if the commit on Github is functional or security-related.

Following the classification of the three labeling methods, we analyzed the vulnerability dataset in Table 1.

3.1 Labeling Based on Text Description

The labeling based on text description mainly starts from NVD, CVE, GitHub, etc., and extracts the specifically needed patches and vulnerability source code based on the provided text description information.

Gu [6] collected all the commit information, patch files, etc. and removed the commits that did not change the .c source file. After that, the vulnerability-related commits were filtered according to keywords. This method produces better results of relevance. However, this method significantly reduces the number of candidate functions we can flag and still requires a lot of manual checking, which is not suitable for large datasets. Moreover, some of the keywords are ambiguity, and may be just functional patches. The whole dataset labeling and construction process are highly manual involved, and the accuracy relies on human knowledge.

Wang [2] constructed a semi-artificial and semi-real-world dataset PatchDB. The patches given from the NVD hyperlinks are labeled as vulnerable. Out of all the commits of the 313 GitHub projects, excluding the 4076 vulnerable patches obtained from the NVD above, the remaining commits were marked as non-vulnerable patches. However, the non-vulnerable patches obtained in this way do not completely exclude vulnerable patches.

Lin [16] labels the samples based on the NVD and CVE page description. If a fragment of vulnerable code is located within a function, the corresponding version of the program source code will be downloaded and the source code function will be marked as vulnerable code. For the vulnerability location not explicitly given in NVD and CVE, the CVE-ID is searched in GitHub commit as a keyword, and the vulnerability fragment is finally found according to the diff file. To obtain the non-vulnerable functions, all functions from the non-vulnerable files are collected and marked as non-vulnerable. This labeling method ignores the contextual information and extracts only the functions where the fragment of vulnerability is located, rather than the data flow or control flow context of the vulnerability fragment. And the patched vulnerability fragments are not always the crash point of the function.

Devign [1] is a real-world vulnerability dataset. After filtering out non-security related commits based on security-related keywords, security-related commits are manually reviewed and security-related functions are labeled vulnerable. And the patched version of the function is labeled not-vulnerable. In the Devign dataset, if a commit was deemed to fix a bug, then all functions patched by that commit were labeled as vulnerable, which is incorrect in many cases.

REVEAL [19] is collected in a similar way to Devign [1]. For each vulnerability patch file, from k-1 to k versions, all modified functions in the file are labeled vulnerable, and the corresponding function in the next version of the file is labeled not-vulnerable, in addition, functions that have not been changed in the file are also labeled not-vulnerable. This labeling method ignores the fact that some of the code changes are functionally relevant.

Fan [20] collected the vulnerabilities descriptions from CVE, downloaded the code based on the CVE information and its published links to the relevant GitHub code, and extracted the patches related to the vulnerabilities. Based on the patches extracted from the commit, vulnerabilities/non-vulnerabilities are labeled after using the code change before and after. Unlike Devign and REVEAL, Big-Vul is constructed by leveraging and linking the CVE, project bug reports, and commits, which helps improve the accuracy of identifying vulnerability commits related to code changes compared to filtering commits by security keywords.

Nikitopoulos [24] constructed a cross-language real-world dataset: CrossVul. The authors corresponded the hyperlinks in the description information in the NVD to GitHub, and after filtering invalid links, used the git-diff command to identify the individual files modified in each commit, obtaining files that contain security patches and files that do not. The labeling method is similar to [16].

The text description relies heavily on information provided by the NVD, CVE, and GitHub. However the patch information for these disclosed vulnerabilities is incomplete in many cases. There are two types of vulnerability repository patch disclosures, one is attached to the reference link, and the other provides identification patches or patch tag content. For example, the NVD reference link is only a reference link and may not be completely correct, i.e. the diff pointed to by the reference link may not be the patch for that CVE. [26] evaluated more than 6,000 vulnerability patch disclosures and concluded, based on its links and labels, that about 40% of CVEs in the NVD do not provide the corresponding patch information, and part of the patch information is still wrong, especially as the patch label identifies this category, which is less than 50% for the utility dataset.

[1,6,19] used keyword filtering for security-related commissions. However, it is not accurate by only using keyword filtering. By keyword filtering in open source software openssl's commit, we found that some keywords have two sides, as shown in the following Table 3. Some experiments found that the false positive rate and false negative rate of keyword-based methods are as high as 36% and 11% [27], respectively.

Table 3. Disambiguation of keywords

Keyword	Positive	Negative
fix	Fix memory leak	fix broken tests, fix logger printout, fix CS violation, adds tests for fix
check	Check memory leak, Add missing check for OPENSSL_strndup, check the return value of CRYPTO_strdup	Check format, test check property, add check for
bug	Fix bug in scrypt KDF provider dup method	Log bug, test bug
error	Error code, Fixed typo in, inner_evp_generic_fetch, error handling	Clear incorrectly, reported errors in

Zafar [28] proposes a rule matching-based deep learning commit classifier. The authors defined five bug-fixing-commit rules and 11 not-a-bug-fixing-commit rules based on existing commits. The results show that the above method outperforms the keyword-based filtering in terms of accuracy.

For the combined use of NVD information and GitHub commit information, [29] proposed PatchScout, which extracts vulnerability information in NVD and code change information in code repository, calculates the association features in four dimensions. The results show that this method has significant improvement over existing methods (search with CVE-ID, checking commit-like URLs, checking patch-tagged URLs), up to 8 times.

3.2   Labeling Based on Patch Analysis

Security patches enhance the security of software by targeting specific security vulnerabilities. Non-security patches include bug fix patches and functional patches. Bug fix patches make the software run more robustly and reduce the possibility of crashes by correcting software errors. Functional patches add new features or updated existing features to the software. The purpose of patch analysis is to exclude modifications that are not related to the root cause of the vulnerability, such as, functional modifications, namespace modifications, formatting modifications, etc.

Wang [30] proposes PatchRNN, a deep learning-based system to automatically identify whether the patches in open-source software are security-relevant. The system uses diff codes and commits to capture more comprehensive features. PatchRNN represents the code, and text message token after serialization, such labeling will lose the contextual semantic information. And the recognition based on model learning is insensitive to small code changes.

Disco [31] detects code security iterations by comparing historical changes to code on the stack-overflow and determines whether code is secure by determining whether it has been patched or not. By comparing the features that have been proposed to detect code security updates, the authors finally selected three features to determine whether a code change is security-related, security-related API changes, security-related keyword changes, and code control flow changes. The changed part of the patch is matched with these 3 features, and the matched parts are considered as the security-related patches. This method lacks many features that do not belong to these three features, and does not locate the line of vulnerability.

Li [32] constructed the slice-level vulnerability dataset CDG. The code examples include rich inter-process context, but the examples are a subset of program slices and thus are not valid programs. Despite that this may eliminate noise, it limits the ability of the model to learn naturally valid code structures. The labeling method only considers the cases where the vulnerability patch file contains minus lines, and directly labels slices containing the minus lines in the vulnerability patch file as containing vulnerabilities and the matching lines as vulnerable. Slices that do not contain the minus lines in the vulnerability patch file are labeled as not vulnerable. It is not possible to handle the case where the vulnerability patch file contains only the plus line. Since the dataset comes from bugs already identified by NVD, the labeling quality may be good. However, the number of such examples is still limited and may not be sufficient for model training.

Tang [33] pointed out that the slice-level vulnerability dataset constructed in [32] lacks path sensitivity, and thus the slices generated by vulnerable code fragments and non-vulnerable code fragments will be consistent when slicing. There are two main reasons for this: first, control dependencies are rough descriptions of the relationships between two statements (i.e., the presence or absence of dependencies) and do not specify the paths of statements (i.e., dependencies on legitimate or illegitimate values); second, the process of reorganizing the order of statements and brute force overlay may lead to the direct adjacency of statements that are not in the same control range, thus generating path insensitivity. Based on this, the paper proposes path-sensitive code gadget.

Liu [23] built a real-world vulnerability dataset. In this work, in addition to the labeling of vulnerability functions by the same method as in the previous work, more fine-grained labeling of vulnerability rows is added. For the vulnerability line labeling method, 2 cases are considered: if the vulnerability patch only has the plus line, the place where the key variable is defined or referenced in the plus line is designated as the line where the vulnerability is located; if the vulnerability patch directly submits a new function, then the function is considered to be non-vulnerable. The labeling method does not consider the case of minus lines.

3.3 Labeling Based on Vulnerability Scanning

The labeling based on the vulnerability scanning is to label the test object after the tools finish vulnerability scanning.

Russell [18] complements the SATE IV Juliet manual dataset with real vulnerability data, obtained by using three different static analysis vulnerability detection tools on GitHub and Debian software. The results of each static analyzer were manually mapped to the corresponding CWEs, and it was determined which CWEs could lead to potential security vulnerabilities. The three static detection tools, Clang, Cppcheck, and Flawfinder, inherently have a high rate of false positives and misses. This work does not further examine this vulnerability dataset, but simply coarsely classifies the detection categories of the three static detection tools manually. The use of inaccurate detection results to label the sample functions directly affects the quality of the dataset.

Harer [25] built datasets from Debian packages and GitHub. After removing duplicate data, the authors use Clang static analyzer (SA) to remove warnings from Clang output that are not related to security vulnerabilities. Those without warning messages are marked as Good and those with messages are marked as Buggy. In this work, the authors use static analysis based labeling as the only labeling method without validation. The accuracy of the static analyzer affects the quality of the dataset. The samples were labeled by the static analysis tool, which generates many false positives and thus requires subsequent manual validation, but this can only be applied to small datasets.

Zheng [21] constructed a million samples real-world dataset, D2A. They proposes an approach based on diff analysis applied to automate real large programs. Using the static analysis tool infer to label the dataset, it can produce more information related to bugs, such as bug type, location, function path, etc. The data in the article shows that the accuracy of using the static analysis tool infer labeling is 53%.

Raducu [22] built the real-world dataset: SVCP4C, which focuses on buffer overflow type vulnerabilities. The annotation method is processing code in an open source project using the SonarCloud vulnerability detector. All files in the dataset provided by the authors contain vulnerabilities and comments detailing the vulnerable lines. SVCP4C has only 1,104 unique groups after deduplication, a reduction of 90.29%.

4 Findings and Future Work

By sorting through the existing labeling efforts, we broadly classify them into three categories: labeling based on text description, labeling based on patch analysis, and labeling based on vulnerability scanning. We propose a framework for each of labeling technology to summarize the current labeling methods completely, and list some defects along with possible future directions.

4.1   Summary of Papers Distribution

By summarizing the statistics of the dataset labeling methods mentioned in 16 papers in the last 5 years, we get the distribution of the three labeling methods as in Fig. 2.

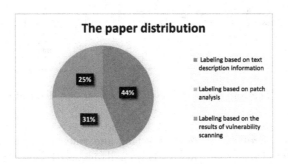

Fig. 2. Distribution of papers in the three categories of labeling methods.

Labeling based on text description is most often used because it is based on recognized vulnerability information, which is highly credible and easily accessible. Labeling based on path analysis is deeper than labeling based on text description. There are fewer labeling papers based on the results of vulnerability scanning because the accuracy of the scanning tools can not be guaranteed.

4.2   The Framework of Labeling Based on Text Description

The framework based on text description information is shown in the Fig. 3. One is to start from downloading the known vulnerability database NVD page description information, then locate the hyperlinks marked patches using some text mining technology. The other source is commits, some classification algorithms are used to filter the security related ones. Sometimes it is necessary to use a combination of these two sources to improve the accuracy. The code extraction stage (the following two frameworks are similar and will not be described again) is to locate the security patches for project backtracking, slicing and extracting by means of context analysis and path-sensitive analysis. This step is trying to make the obtained code fragments enough to extract vulnerability features. After that, the code fragments are de-duplicated and unified in format, and finally stored as a dataset.

Two key elements should be focused on in the above framework: hyperlink positioning and commit filtering.

Patching is the main means of blocking vulnerabilities and attacks, but the vulnerability attacks are evolving fast and patch management is chaotic. Patch intelligence is difficult to collect and patch effectiveness is difficult to evaluate. Existing methods propose improvements in pinpointing hyperlinks and security

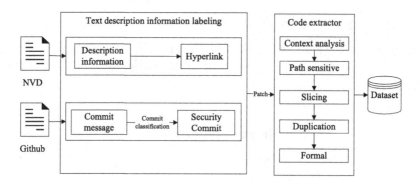

Fig. 3. The framework of labeling based on text description.

commit, but both involve feature engineering and rule matching, and selecting vulnerability-related effective features is a challenge, with inaccuracy problems at fine granularity and difficulty in pinpointing at coarse granularity.

4.3   The Framework of Labeling Based on Patch Analysis

By observing the collected diff data, it is found that different CVEs may refer to the same diff file. The accuracy of the diff files needs to be judged and labeled. False positive may exist when labeling only based on text description if labeled from text description information only. Some patches are of types not about source files but are classified as patch fixes by the author. Some patches fix configuration files, system builds, etc. Bugs not related to source code include incorrect or incomplete documentation, incorrect test or test input data, and incorrect build system configuration.

For the above problems, we propose a labeling framework based on patch analysis. As shown in the Fig. 4, there are variable definition substitution, variable assignment of initial value substitution or if/for, and other control statement substitution in the vulnerability patch. The key variables of the patches are found by different types of statement analysis based on + and − line comparison; meanwhile, the source code before and after patched are analyzed and the root cause of the vulnerability is found by using program analysis methods such as AST, CFG, DFG, etc. The latter analysis is more in-depth and can explore more information than the simple line of comparison.

Most of the patch databases are given in NVD hyperlinks, and the information is not accurate. The current labeling analysis for patch analysis is relatively shallow, and most of them stop at the comparison of + and − lines. The comparison based on + and − lines belongs to the text diff algorithm in source code change studies, such as the famous GNU diff, which compares source code at the plain text level. Text diff algorithms can detect inserted, deleted, and updated text lines, but such algorithms are difficult to infer the root cause of vulnerabilities because vulnerability is not always determined by a particular line. It is

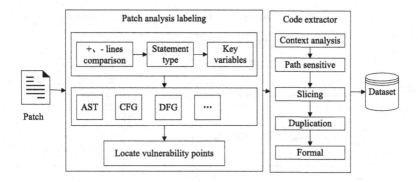

Fig. 4. The framework of labeling based on patch analysis.

contextually linked, and it is difficult to infer syntactic semantic changes in the cause of vulnerabilities without using the syntactic information contained in the source code.

A deeper patch analysis should go back to the source code for program analysis. As AST can express the structural information of the code very well, some AST analysis methods such as GumTree [34], MTDIFF [35], IJM [36] can be used. All these three tools are for JAVA language, but there is applicability to AST difference computation because the structure of AST is independent of the grammar of the specific language, and AST uses the context-independent method for syntax analysis of source code. RefactoringMiner [37], FixMiner [38], and ChangeDistiller [39] can be used to mine code change information. In addition to this, the code refactoring detection tools Ref-Finder and Ref-diff can be used.

4.4   The Framework of Labeling Based on Vulnerability Scanning

Based on the labeling of vulnerability scanning tool results, there is no restriction on the detection object, any source code items can be. However, the existing labeling results have a high false positive rate and are limited by the capability of the vulnerability scanning tool itself, and the labeled results are grain-coarse. Therefore, as shown in the Fig. 5, we propose a framework of labeling vulnerability scanning. The framework input is an arbitrary project, and the output is a dataset, and the main part is the vulnerability scanning tool labeling and code extraction. The input dataset is first tested by a variety of vulnerability scanners, here according to the vulnerability scanner capabilities, each focus, and infer can be analyzed across functions, cppcheck and flawfinder can be analyzed for code that can not be compiled. Each vulnerability scanner's report is different, and for each report, we add CVE verification on it to show that the label is valid.

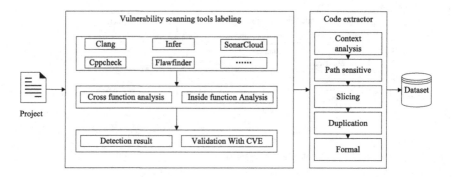

Fig. 5. The framework of labeling based on vulnerability scanning.

4.5   Summary

The choice of labeling method usually depends on the type of data source collected. For labeling based on text description and labeling based on patch analysis, the data source is publicly available vulnerability information, but non-vulnerability labeling is quite subjective because there is no relevant label source, so it can bring impact on labeling accuracy. The labeling based on the results of vulnerability scanning is heavily dependent on the accuracy of the scanning tools. We summarize the three labeling methods in the Table 4.

Table 4. Summary of labeling technology

Labeling method	Advantage	Disadvantage	Key Technologies	Future work
Labeling based on text description	High credibility	Incomplete information disclosure	Text mining analysis technology, Crawler	Patch information mining technology, Commit classification
Labeling based on patch analysis	Fine grained, Locate to the line of vulnerability, Analyze in depth	High computing cost	Slicing technology, Program analysis technology	Patch program analysis, Code change analysis, AST-difference
Labeling based on vulnerability scanning	Informative labeling	High rate of false positives	Vulnerability scanning tools, Slicing technology	Vulnerability scanning tools

5   Conclusion

In this paper, we systematically investigated the literature related to dataset generation and labeling, and find that there is a lack of common, widely-accepted vulnerability-related datasets. We also analyze 13 vulnerability code datasets and their labeling methods, and propose frameworks for labeling different input objects on the datasets. The research is analyzed and summarized for the intelligent trends presented by these annotation techniques and the problems faced.

High-quality real-world datasets need to have trusted labels and rich context associated with code examples. In future work, we plan to propose a new way of labeling vulnerability samples and generating high quality datasets to be applied to vulnerability detection. In addition, we plan to study the validation of dataset labeling quality.

References

1. Zhou, Y., Liu, S., Siow, J., Du, X., Liu, Y.: Devign: effective vulnerability identification by learning comprehensive program semantics via graph neural networks (2019)
2. Wang, X., Wang, S., Feng, P., Sun, K., Jajodia, S.: PatchDB: a large-scale security patch dataset. In: 2021 51st Annual IEEE/IFIP International Conference on Dependable Systems and Networks (DSN), pp. 149–160 (2021)
3. Ghadhab, L., Jenhani, I., Mkaouer, M.W., Messaoud, M.B.: Augmenting commit classification by using fine-grained source code changes and a pre-trained deep neural language model. Inf. Softw. Technol. **135**, 106566 (2021)
4. NVD. https://nvd.nist.gov/
5. CVE. https://cve.mitre.org/
6. Gu, Z., Wu, J., Liu, J., Zhou, M., Gu, M.: An empirical study on API-misuse bugs in open-source C programs. In: 2019 IEEE 43rd Annual Computer Software and Applications Conference (COMPSAC), vol. 1, pp. 11–20 (2019)
7. SARD. https://samate.nist.gov/SARD
8. Semasaba, A., Zheng, W., Wu, X., Agyemang, S.: Literature survey of deep learning-based vulnerability analysis on source code. IET Softw. **14**, 654–664 (2020)
9. Lin, G., Wen, S., Han, Q.-L., Zhang, J., Xiang, Y.: Software vulnerability detection using deep neural networks: a survey. Proc. IEEE **108**(10), 1825–1848 (2020)
10. Jimenez, M., Rwemalika, R., Papadakis, M., Sarro, F., Traon, Y.L., Harman, M.: The importance of accounting for real-world labelling when predicting software vulnerabilities. In: Proceedings of the 2019 27th ACM Joint Meeting on European Software Engineering Conference and Symposium on the Foundations of Software Engineering, ESEC/FSE 2019, New York, NY, USA, pp. 695–705. Association for Computing Machinery (2019)
11. Croft, R., Xie, Y., Babar, M.A.: Data preparation for software vulnerability prediction: a systematic literature review. IEEE Trans. Softw. Eng. 1 (2022)
12. Croft, R., Ali Babar, M., Chen, H.: Noisy label learning for security defects (2022)
13. Grahn, D., Zhang, J.: An analysis of C/C++ datasets for machine learning-assisted software vulnerability detection. In: Conference on Applied Machine Learning for Information Security, Arlington, VA (2021)

14. Lin, Y., et al.: Vulnerability dataset construction methods applied to vulnerability detection: a survey. In Undefined (2022)
15. Liu, L., Li, Z., Wen, Y., Chen, P.: Investigating the impact of vulnerability datasets on deep learning-based vulnerability detectors. PeerJ Comput. Sci. **8**, e975 (2022)
16. Lin, G., Xiao, W., Zhang, J., Xiang, Y.: Deep learning-based vulnerable function detection: a benchmark. In: Zhou, J., Luo, X., Shen, Q., Xu, Z. (eds.) ICICS 2019. LNCS, vol. 11999, pp. 219–232. Springer, Cham (2020). https://doi.org/10.1007/978-3-030-41579-2_13
17. Li, Z., Zou, D., Xu, S., Chen, Z., Zhu, Y., Jin, H.: VulDeeLocator: a deep learning-based fine-grained vulnerability detector. IEEE Trans. Dependable Secure Comput. 1 (2021)
18. Russell, R.L., et al.: Automated vulnerability detection in source code using deep representation learning. In: Automated Vulnerability Detection in Source Code Using Deep Representation Learning, pp. 757–762 (2018)
19. Chakraborty, S., Krishna, R., Ding, Y., Ray, B.: Deep learning based vulnerability detection: are we there yet? (2020)
20. Fan, J., Li, Y., Wang, S., Nguyen, T.N.: A C/C++ code vulnerability dataset with code changes and CVE summaries. In: Proceedings of the 17th International Conference on Mining Software Repositories, pp. 508–512. Association for Computing Machinery, New York (2020)
21. Zheng, Y., et al.: D2A: a dataset built for AI-based vulnerability detection methods using differential analysis. In: 2021 IEEE/ACM 43rd International Conference on Software Engineering: Software Engineering in Practice (ICSE-SEIP), pp. 111–120 (2021)
22. Raducu, R., Esteban, G., Lera, F.J.R., Fernández, C.: Collecting vulnerable source code from open-source repositories for dataset generation. Appl. Sci. **10**(4), 1270 (2020)
23. Liu, B., et al.: A large-scale empirical study on vulnerability distribution within projects and the lessons learned. In: 2020 IEEE/ACM 42nd International Conference on Software Engineering (ICSE), pp. 1547–1559 (2020)
24. Nikitopoulos, G., Dritsa, K., Louridas, P., Mitropoulos, D.: CrossVul: a cross-language vulnerability dataset with commit data. In: Proceedings of the 29th ACM Joint Meeting on European Software Engineering Conference and Symposium on the Foundations of Software Engineering, ESEC/FSE 2021, New York, NY, USA, pp. 1565–1569. Association for Computing Machinery (2021)
25. Harer, J.A., et al.: Automated software vulnerability detection with machine learning (2018)
26. Min, Y.: 2022 Beijing cyber security conference (BCS). https://bcs.qianxin.com/speaker/detail?id=63
27. Berger, E.D., Hollenbeck, C., Maj, P., Vitek, O., Vitek, J.: On the impact of programming languages on code quality: a reproduction study. ACM Trans. Program. Lang. Syst. **41**(4), 21:1–21:24 (2019)
28. Zafar, S., Malik, M.Z., Walia, G.S.: Towards standardizing and improving classification of bug-fix commits. In: 2019 ACM/IEEE International Symposium on Empirical Software Engineering and Measurement (ESEM), pp. 1–6 (2019)
29. Tan, X., et al.: Locating the security patches for disclosed OSS vulnerabilities with vulnerability-commit correlation ranking. In: Proceedings of the 2021 ACM SIGSAC Conference on Computer and Communications Security, CCS 2021, New York, NY, USA, pp. 3282–3299. Association for Computing Machinery (2021)

30. Wang, X., et al.: PatchRNN: a deep learning-based system for security patch identification. In: MILCOM 2021–2021 IEEE Military Communications Conference (MILCOM) (2021)
31. Hong, H., Woo, S., Lee, H.: Dicos: discovering insecure code snippets from stack overflow posts by leveraging user discussions. In: Annual Computer Security Applications Conference, ACSAC, New York, NY, USA, pp. 194–206. Association for Computing Machinery (2021)
32. Li, Z., et al.: VulDeePecker: a deep learning-based system for vulnerability detection. In: Proceedings 2018 Network and Distributed System Security Symposium (2018)
33. SEVulDet: A Semantics-Enhanced Learnable Vulnerability Detector (2022)
34. Falleri, J.-R., Morandat, F., Blanc, X., Martinez, M., Monperrus, M.: Fine-grained and accurate source code differencing. In: Proceedings of the 29th ACM/IEEE International Conference on Automated Software Engineering, ASE 2014, New York, NY, USA, pp. 313–324. Association for Computing Machinery (2014)
35. Dotzler, G., Philippsen, M.: Move-optimized source code tree differencing. In: Proceedings of the 31st IEEE/ACM International Conference on Automated Software Engineering, ASE 2016, New York, NY, USA, pp. 660–671. Association for Computing Machinery (2016)
36. Frick, V., Grassauer, T., Beck, F., Pinzger, M.: Generating accurate and compact edit scripts using tree differencing. In: 2018 IEEE International Conference on Software Maintenance and Evolution (ICSME), pp. 264–274 (2018)
37. Tsantalis, N., Mansouri, M., Eshkevari, L.M., Mazinanian, D., Dig, D.: Accurate and efficient refactoring detection in commit history. In Proceedings of the 40th International Conference on Software Engineering, ICSE 2018, New York, NY, USA, pp. 483–494. Association for Computing Machinery (2018)
38. FixMiner: Mining relevant fix patterns for automated program repair. Empirical Software Engineering
39. Fluri, B., Wuersch, M., Inzger, M.P., Gall, H.: Change distilling: tree differencing for fine-grained source code change extraction. IEEE Trans. Softw. Eng. 33(11), 725–743 (2007)

Adding Result Diversification to kNN-Based Joins in a Map-Reduce Framework

Vinícius Souza[1], Luiz Olmes Carvalho[2], Daniel de Oliveira[3], Marcos Bedo[3(✉)], and Lúcio F. D. Santos[1]

[1] Federal Institute of North of Minas Gerais, IFNMG, Montes Claros, MG, Brazil
{vinicius,lucio.santos}@ifnmg.edu.br
[2] Institute of Mathematics and Computing, UNIFEI, Itajubá, MG, Brazil
olmes@unifei.edu.br
[3] Institute of Computing, UFF, Niterói, RJ, Brazil
{danielcmo,marcosbedo}@ic.uff.br

Abstract. While the k-Nearest Neighbors (kNN) join fetches the k closest objects from a dataset for each element of a reference collection, a kNN join with result diversification aims at retrieving the k nearest objects to each reference entry that are dissimilar among themselves. Under the Metric Space Model, the distance-based ternary relationship between each search reference, the dataset, and the result set can be explicitly used to define a coverage-based criterion, namely *Influence*, that ensures diversification by dismissing nearby regions during the join operation. However, adding result diversification to kNN joins by means of *Influence* criteria in large-scale, big data frameworks is still an open issue since existing algorithms do not consider shared-nothing environments. To fulfill this gap, we extend the nested *Better Results with Influence Diversification* algorithm (BRID_k) to a Map-Reduce framework. In particular, this study introduces two new algorithms: the P-BRID_k and the SP-BRID_k. The P-BRID_k method relies on partitioning the objects by their proximity to a set of *pivots* so that the search space locality is preserved throughout the mapped distance-based comparisons. The SP-BRID_k method expands the P-BRID_k by using a data replication strategy where a window of the search space is copied across the partitions for enhancing the *Influence*-based pruning of the nearest objects. We performed an extensive evaluation of both methods over low and high-dimensional datasets on an Apache Hadoop cluster, and the results indicate that *(i)* P-BRID_k has consistently outperformed the nested BRID_k implementation, with gains up to 80% in terms of Recall (fraction of points among true diversified neighbors), *(ii)* fine-tuned SP-BRID_k has enhanced the P-BRID_k performance at a small overhead cost in data replication, and *(iii)* the SP-BRID_k elapsed time has scaled smoothly with the number of partitions, yielding high Recalls for kNN joins with result diversification for a controlled overhead ratio.

This study was supported by CAPES, CNPq and FAPERJ (grant numbers SEI-016517/2021 and E-26/202.806/2019).

Keywords: Metric Spaces · kNN · Result Diversification · Big data

1   Introduction

The massive amount of data produced by daily applications (*e.g.*, health monitoring, and social media) offers unlimited possibilities for value creation and societal changes. The analysis of such *big data* is typically carried out on parallel computing frameworks (*e.g.*, Hadoop and Spark), providing fast and scalable performance [3,12,15,16]. Another important aspect to benefit from *big data* is the underlying search model, which must deliver transparent answers and handle huge data variety [1,6]. The Metric Space Model offers a solid alternative for browsing through big data as its single requirement is pairing data with a distance function that complies with metric properties, including non-negativity and triangle inequality [8].

Under this paradigm, the higher the distance between two data objects, the most dissimilar they are. Accordingly, a similarity search operator is a systematic method for organizing the distances from dataset objects to any set of reference elements so that only objects satisfying a distance-based criterion are retrieved. A known similarity search operator is the k-Nearest Neighbors (kNN) join, which fetches the k closest objects from a dataset for each element of a reference collection [2]. An example of use for the operator is as follows. Suppose a film director running a remake of *"Once Upon a Time in the West"* wants to explore possibilities for the cast by considering the original starring actors. While the kNN join enables limiting the prospects for each role, it does not ensure each position is assorted enough to skip redundant possibilities, such as actors with the same lineaments and play style. The kNN join with result diversification enriches the kNN join operator by enforcing that the retrieved neighbors are also dissimilar according to distance coverage/optimization rules [7,13].

A number of approaches have discussed the implementation of kNN joins [3, 16]. For instance, the QuickJoin algorithm introduced in [9] was a seminal approach to solve kNN joins in centralized environments [9], while the proposal in [15] relies on a hyperplane-based window to devise a non-blocking strategy to handle kNN joins in a shared-nothing framework by using a Map-reduce implementation. The strategy in [5] uses space-filling curve mappings to enhance load balance and shows that good-quality *pivot* objects have an impact on the join operation [5]. This finding is also corroborated in [17], with the authors suggesting pivots chosen by approaches based on the dataset pairwise distance distribution have superior performance to randomly selected pivots [18,19].

On the other hand, just a handful of studies have addressed adding result diversification to the join operator to the best of our knowledge [7,13]. The study of [13] was the first to address the implementation of a kNN join operator by using incrementally-chosen coverage thresholds. They proposed a nested version of the BRID_k algorithm [14], whose thresholds are obtained dynamically after the ternary relationship among the search reference, the dataset, and the result set (the so-called *Influence* criteria) [10]. A more challenging aspect of kNN joins with result diversification is they require *synchronous* sorting, *i.e.*, greedily discarding objects, as non-diversified in a mapping phase may change the overall search outcome, which does not occur in kNN joins. Therefore, migrating the nested $BRID_k$ approach to efficiently and effectively execute *Influence*-based kNN joins in big data frameworks remains an open problem.

In this study, we introduce two Map-Reduced-based extensions to $BRID_k$: the $P\text{-}BRID_k$ and $SP\text{-}BRID_k$. They exploit the partitioning of data objects by their proximity to a set of pivot elements, chosen by the criterion of *maximal variance* in the pairwise distance distribution [18,19]. The $SP\text{-}BRID_k$ method also capitalizes on a data expansion strategy in which a pivot-based window of the search space is replicated across the partitions to enhance the *Influence*-based pruning of nearest objects. In summary, our main contributions are:

- $P\text{-}BRID_k$, a pivot-based method extending kNN joins with *Influence*-based result diversification to Map-Reduce frameworks,
- $SP\text{-}BRID_k$, a method that expands $P\text{-}BRID_k$ by using data replication within data partitions, and
- an extensive comparison of $P\text{-}BRID_k$ and $SP\text{-}BRID_k$ to nested $BRID_k$ over real-world datasets embedded in low and high-dimensional spaces, experimentally showing the advantages of our proposal.

The remainder of this paper is organized as follows. Section 2 presents the main concepts and summarizes related work. Section 3 introduces the $P\text{-}BRID_k$, and $SP\text{-}BRID_k$ methods, and Sect. 4 presents the experimental setup and evaluation. Section 5 discusses the conclusions and points to future directions.

2   Preliminaries

The Metric Space Model. A *Metric Space* is a pair $\mathcal{M} = \langle \mathbb{O}, \delta \rangle$, where $\mathbb{O}$ is a data domain and $\delta, \delta : \mathbb{O} \times \mathbb{O} \rightarrow \mathbb{R}_+$, is a *distance* function that complies with the following properties [8] for distinct objects $o_h, o_i, o_j \in \mathbb{O}$: *(i)* $\delta(o_i, o_j) = \delta(o_j, o_i)$ (Symmetry); *(ii)* $\delta(o_i, o_j) > 0$ (Positiveness); *(iii)* $\delta(o_i, o_i) = 0$ (Identity); and *(iv)* $\delta(o_i, o_j) \leq \delta(o_i, o_h) + \delta(o_h, o_j)$ (Triangle inequality).

Well-known distance functions include the Euclidean distance (L_2) and the entire Minkowski family (L_p) for d-dimensional spaces [4].

kNN Query. A k-Nearest Neighbor Query (kNN) query retrieves the set of the k closest elements from a dataset $\mathcal{O} \subseteq \mathbb{O}$ to a reference object $o_q \in \mathbb{O}$. Formally, an incremental kNN result set kNN $(o_q, \delta, k, \mathcal{O}) = \{o_1, o_2, ..., o_k\}$ is as follows

$$o_1 = o_i \in \mathcal{O}, \ \forall \ o_j \in \mathcal{O}, \ \delta(o_i, o_q) \leq \delta(o_j, o_q),$$

$$o_{m=2,...,k} = o_i \in \mathcal{O} \setminus \cup_{h=1}^{m-1} o_h, \ \forall \ o_j \in \mathcal{O} \setminus \cup_{h=1}^{m-1} o_h, \delta(o_i, o_q) \leq \delta(o_j, o_q)$$

kNN Join. A k-Nearest Neighbor join (kNN⋈) takes as input a reference collection set $\mathcal{Q} \subseteq \mathbb{O}$ and a dataset $\mathcal{O} \subseteq \mathbb{O}$, and retrieves a result set $\mathcal{R}$ with size at most $|\mathcal{R}| = k \times |\mathcal{Q}|$. The result set $\mathcal{R}$ is constructed by fetching every

pair $\mathcal{R} = \{\langle q, o_i \rangle \mid o_i \in k\text{NN}(q, \delta, k, \mathcal{O}), \; \forall \, q \in \mathcal{Q}\}$. In the case of ties at the kth position, any of the tied pairs can be arbitrarily chosen [8,10]. Figure 1(b) exemplifies a kNN join for a collection with two reference objects.

Although kNN is widely employed to recover data by proximity, it may *(i)* present non-determinism (with several elements qualifying for the kth position) or *(ii)* retrieve objects similar among themselves in the exploration of distance-dense datasets [10]. Result diversification extends kNN to recover objects dissimilar to each other [7]. In particular, *Influence* measures can create dynamic thresholds to prune the nearest candidates, ensuring diversification [13].

Influence Measures. Given two objects $o_i, o_j \in \mathbb{O}, o_i \neq o_j$, their mutual *Influence* (inverse dissimilarity) is calculated by $I(o_i, o_j) = 1/\delta(o_i, o_j)$. Given a query reference $o_q \in \mathbb{O}$, a diversified (*Influence*-free) neighbor $o_i \in \mathcal{O}$, and a dataset object $o_j \in \mathcal{O}$, then their *Influence* measures define a ternary relationship that indicates o_j is more influenced by o_i than o_q iff $I(o_i, o_j) > I(o_j, o_q)$. Thus, o_j should not be considered for the result set as o_i is already a diversified neighbor.

Influence Set. The *Influence Set* of a diversified neighbor o_i regarding a query reference $o_q \in \mathbb{O}$ encompasses every entry $o_j \in \mathcal{O} \setminus \{o_i \cup o_q\}$ that are *(i)* farther from o_q than o_i and *(ii)* more *Influenced* by o_i than o_q, i.e., $\ddot{I}_{o_i,o_q} = \{o_j \mid o_j \in \mathcal{O} \setminus \{o_i, o_q\}, I(o_i, o_j) > I(o_j, o_q) \wedge I(o_i, o_j) > I(o_j, o_q) \wedge I(o_i, o_q) \neq I(o_j, o_q)\}$. This concept is exploited by the *Better Results with Influence Diversification* algorithm (BRID$_k$) to define kNN queries with result diversification [10,14].

k_dNN Query. A kNN query with result diversification (k_dNN) fetches the k non-*Influenced* and nearest elements in $\mathcal{O} \subseteq \mathbb{O}$ to a reference object o_q so that $k_d\text{NN} \, (o_q, \delta, k, \mathcal{O}) = \mathcal{R} = \{o_1, o_2, ..., o_k\}$ is constructed as follows.

$$o_1 = o_i \in \mathcal{O}, \; \forall \, o_j \in \mathcal{O}, \delta(o_i, o_q) \leq \delta(o_j, o_q),$$

$$o_{m=2,...,k} = o_i \in \mathcal{O}, \; (\forall \, o_j \in \cup_{h=1}^{m-1} o_h \Rightarrow o_i \notin \ddot{I}_{o_j,o_q}) \wedge (\forall \, o_g \in \mathcal{O} \setminus \cup_{h=1}^{m-1} o_h \Rightarrow$$

$$(\delta(o_i, o_q) \leq \delta(o_g, o_q) \vee \exists \, o_j \in \cup_{h=1}^{m-1} o_h \Rightarrow o_g \in \ddot{I}_{o_j,o_q})).$$

kNN Join with Result Diversification. The result set $\mathcal{R}$ of a kNN join with result diversification (k_dNN$\bowtie$) is assembled from the input datasets $\mathcal{Q} \subseteq \mathbb{O}$ and $\mathcal{O} \subseteq \mathbb{O}$ where the retrieved set of joined pairs are as follows $\mathcal{R} = \{\langle q, o_i \rangle \mid o_i \in k_d NN(q, k, \mathcal{O}, \delta), \; \forall \, q \in \mathcal{Q}\}$. Figure 1(c) exemplifies a kNN join with result diversification for a collection with two reference objects.

Related Work. To optimize the join operation with MapReduce, the proposal in [11] implements prefix filtering to reduce the number of distance comparisons, while in [15], the authors use a hyperplane window to perform kNN joins in a Hadoop implementation. In [5], the authors discuss space-filling curve mappings to enhance load balance and show that the *pivot* choice for partitioned solutions affects the join operation. Such observation is also highlighted in [17], where authors suggest using the dataset pairwise distance distribution to find good pivots. The main difference between those methods and our study is that we focus on the challenge of including result diversification to kNN joins, which

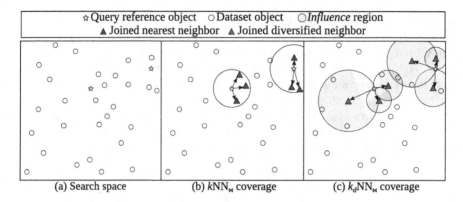

Fig. 1. (a) Example of a kNN⋈ for two random reference objects (stars) in an Euclidean space and $k = 3$ with (c) and without (b) result diversification. (Color figure online)

is usually carried by *synchronous* routines in centralized environments [10] (in opposition to the solutions for asynchronous kNN joins). Next, we discuss the nested kNN⋈ solution proposed in [13] and introduce two MapReduce solutions for the problem by considering pivot-based partitions in the search space.

3   Material and Methods

3.1   The Baseline Approach

The baseline strategy for executing an *Influence*-based k_dNN⋈ is the nested $BRID_k$ solution (N-$BRID_k$), summarized in Algorithm 1. Line 4 produces the candidate list of nearest neighbors for each query reference, whereas Lines 5–8 execute the inner k_dNN routine that prunes non-diversified neighbors. The object pairs are joined incrementally, and the *Influence* of each neighbor defines a monotonically increasing region of exclusion to the next neighbors (Line 6). In a centralized solution, the temporary list of pairs ($\mathcal{R}_{\text{temp}}$) is greedily constructed for each query reference, dynamically creating the regions of exclusions. However, every mapped data partition relies on its local, independent construction in a shared-nothing distributed environment. Thus, merging the individual result sets in the reduce phase implies that the result set includes only locally non-*Influenced* objects, *i.e.*, an element dismissed in a mapped partition may be globally not *Influenced* because the *Influence* sets were constructed after the data partition rather than the entire dataset. As a consequence, the N-$BRID_k$ translation to Map-Reduce produces locally diversified kNN joins, as follows.

Locally Diversified kNN Joins ($k_{\ell d}$NN⋈). Let a set of disjoint partitions $\mathcal{P}$ for the dataset $\mathcal{O} \subseteq \mathbb{O}$ (noted $\mathcal{P} \vdash \mathcal{O}$) be distributed across M workers with $|\mathcal{P}| = M$, then the kNN⋈ with result diversification retrieves the k locally non-*Influenced* neighbors in $\mathcal{O}$ to each entry in the reference set $\mathcal{Q} \subseteq \mathbb{O}$ so

---

Nested BRID$_k$ (Query set $\mathcal{Q}$, dataset $\mathcal{O}$, #neighbors k);

---

1 $\mathcal{R} \leftarrow \emptyset$;
2 **for** $q \in \mathcal{Q}$ **do**
3       $\mathcal{R}_{temp} \leftarrow \emptyset$;
4       $\mathcal{R}_{candidates} \leftarrow$ sortByDistance$(\mathcal{O}, q)$;
5   **for** $o_i \in \mathcal{R}_{candidates} \wedge |\mathcal{R}_{temp}| < k$ **do**
6   **if** $o_i \notin \cup_{o_j \in \mathcal{R}_{temp}} \ddot{I}_{o_j, q}$ **then** $\mathcal{R}_{temp} \leftarrow \mathcal{R}_{temp} \cup \{o_i\}$
7       $\mathcal{R} \leftarrow \mathcal{R} \cup (q \times \mathcal{R}_{temp})$;
8 **return** $\mathcal{R}$;

Algorithm 1: N-BRID$_k$: the nested-loop BRID$_k$ kNN join.

---

that $k_{\ell d}\text{NN}\bowtie(\mathcal{Q}, \delta, k, \mathcal{P} \vdash \mathcal{O}) = \mathcal{R} = \left\{ \begin{array}{c} \{\langle q_1, o_1 \rangle, \ldots, \langle q_1, o_k \rangle\}, \\ \ldots \\ \{\langle q_n, o_1 \rangle, \ldots, \langle q_n, o_k \rangle\}, \end{array} \right\}$, where for each

reference query $q \in \mathcal{Q}$ we have a *result pool* $\mathcal{R}_\mathcal{P} = \cup_{i=1}^M k_d\text{NN}(q, \delta, k, \mathcal{P}_i)$ that determines the pairs of joined elements $\{\langle q, o_1 \rangle, \ldots, \langle q, o_k \rangle\}$ as follows

$$o_1 = o_i \in \mathcal{R}_\mathcal{P}, \ \forall \ o_j \in \mathcal{R}_\mathcal{P}, \delta(o_i, o_q) \leq \delta(o_j, o_q),$$

$$o_{m=2,\ldots,k} = o_i \in \mathcal{R}_\mathcal{P}, \ (\forall \ o_j \in \cup_{h=1}^{m-1} o_h \Rightarrow o_i \notin \ddot{I}_{o_j, o_q}) \wedge (\forall \ o_g \in \mathcal{R}_\mathcal{P} \setminus \cup_{h=1}^{m-1} o_h \Rightarrow$$
$$(\delta(o_i, o_q) \leq \delta(o_g, o_q) \vee \exists \ o_j \in \cup_{h=1}^{m-1} o_h \Rightarrow o_g \in \ddot{I}_{o_j, o_q})).$$

3.2 The P-BRID$_k$ Method: A Novel Pivot-Based Approach

In contrast with the N-BRID$_k$, we propose the P-BRID$_k$ method by using *pivot* objects to cluster and partition the search space in a two-phase strategy, summarized in Fig. 2. We argue the adoption of good-quality pivots increases the clusters' cohesion and the possibilities for a $k_{\ell d}\text{NN}\bowtie$ to cover the first portion of a kNN$\bowtie$ result set. Since this first portion determines the growing pace for the *Influence* sets, it increases the overall chance of $k_{\ell d}\text{NN}\bowtie$ resembling the global $k_d\text{NN}\bowtie$ result set. The number of partitions determines the number of pivots, and we choose the data objects with the maximal variance in their pairwise distance distributions as pivots by using the sample-based heuristic in [18].

Map(Query set $\mathcal{Q}$, dataset $\mathcal{O}$, #neighbors k, #partitions M);

---

1 $\mathcal{P} \leftarrow$ maxVarPivots$(\mathcal{O}, M)$;/* M partitions with a pivot v each.   */
2 **for** $o \in \mathcal{O} \setminus \mathcal{P}$ **do**
3       $P_j \leftarrow P_j \cup \{o\} \mid P_j \in \mathcal{P}, \ \forall \ P_h \in \mathcal{P} \setminus \{P_j\}, \delta(P_j.v, o) \leq \delta(P_h.v, o)$;
4 **return** $\mathcal{R}_\mathcal{P} = \cup_{i=1}^M \text{N-BRID}_k(\mathcal{Q}, \mathcal{P}_i, k)$;/* Distribute to M workers.   */

Algorithm 2: Pivot-mapping phase of P-BRID$_k$.

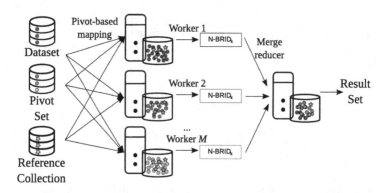

Fig. 2. P-BRID$_k$ Map-Reduce implementation overview.

Reducer(Query set $\mathcal{Q}$, results $\mathcal{R}_\mathcal{P} = \cup_{i=1}^{M}\mathcal{R}_{\mathcal{P}_i}$, #neighbors k);

---

1 $\mathcal{R} \leftarrow \emptyset$;
2 **for** $q \in \mathcal{Q}$ **do**
3       $\mathcal{R}'_\mathcal{P} \leftarrow \mathcal{R}_\mathcal{P}$;
4   **while** $|\{\langle q,o\rangle \in \mathcal{R}\}| < k \wedge |\mathcal{R}'_\mathcal{P}| > 0$ **do**
5           $w \leftarrow \langle q, o_i\rangle \in \mathcal{R}'_\mathcal{P}, \, \forall \, \langle q, o_j\rangle \in \mathcal{R}'_\mathcal{P}, \delta(q,o_i) \le \delta(q,o_j)$;
6           $\mathcal{R} \leftarrow \mathcal{R} \cup \{w\}$;/* Add next nearest and diversified pair   */
7           $\mathcal{R}'_\mathcal{P} \leftarrow \mathcal{R}'_\mathcal{P} \setminus \{w\} \setminus \{\langle q,o_j\rangle \in \mathcal{R}'_\mathcal{P}, o_j \in \cup_{\langle q,o_h\rangle \in \mathcal{R}}\ddot{I}_{o_h,q}\}$/* Pruning   */
8 **return** $\mathcal{R}$;

Algorithm 3: P-BRID$_k$ refinement in the reduce phase.

P-BRID$_k$ employs the pivots to create the data partitions in which every object is assigned to its closest pivot (ties are broken at random). The partitions are crossed against query objects and mapped to workers by the pivot, which executes N-BRID$_k$ to produce locally diversified sets. Next, the results are merged/refined in a reducer whose input is composed of $M \times k$ pairs[1]. Algorithm 2 summarizes the P-BRID$_k$ map phase, and Algorithm 3 the P-BRID$_k$ reducer. Those methods follow the *one-pass* premise, avoiding multiple data read/write operations during the distance comparisons within each partition.

Figures 3(a–b) present two examples of P-BRID$_k$ running for a single query reference $\mathcal{Q} = \{q\}$, $k = 5$, and two partitions $\mathcal{P} = \{P_1.v = v_1, P_2.v = v_2\}$. Trivially, if the partition covers the set of k globally diversified neighbors, then the result of the $k_{\ell d}$NN⋈ is equivalent to that of a k_dNN⋈ – Fig. 3(a). A globally diversified neighbor can also be retrieved even if it locates in a partition farther than previous neighbors – See object o_3 in P_2 in Fig. 3(c). In this case, the merger phase of P-BRID$_k$ discards the local candidate o_1 in P_2 (see the *Influence* regions in gray) because it is *Influenced* by o_2 in P_1, fetching the next candidate o_3 in P_2, which is also a globally diversified neighbor.

---

[1] If $M \times k$ exceeds the worker available memory, then the reduce task can be recursively split into two workers (and one merger) with $(M \times k/2)$ space requirement.

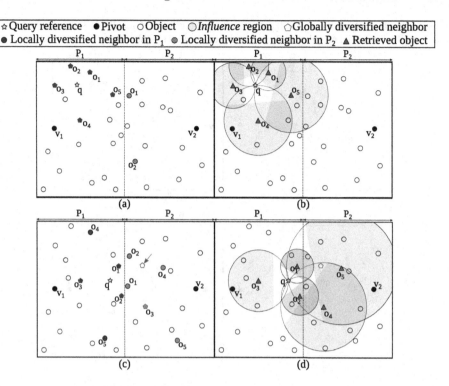

Fig. 3. Two scenarios (a–b) and (c–d) for P-BRID$_k$ joining a query reference q with $k = 5$ neighbors in an Euclidean space partitioned by two pivots $\{v_1, v_2\}$. Query locality determines the *global* coverage, *i.e.*, $|k_{\ell d}\text{NN}\bowtie \cap k_d\text{NN}\bowtie|$. (Color figure online)

However, the one-round processing may discard globally diversified neighbors in favor of locally diversified candidates. In the example of Fig. 3(c), a global neighbor (see the red arrow) is discarded because it is *Influenced* by o_2 in P_2 and never makes it to the reduction phase that would refine the answer. To soften this greedy aspect of P-BRID$_k$ and provide a *second-chance* for the merger to refine the search, we extend P-BRID$_k$ to consider the query locality context.

3.3   The SP-BRID$_k$: A Novel Context-Aware Approach

The proposed SP-BRID$_k$ method generalizes the P-BRID$_k$ proposal under the premise that preserving the query *spatial* locality context has the potential to increase the final number of globally diversified neighbors. Accordingly, we design SP-BRID$_k$ to replicate an ε-sized portion of border data (we call *context window*) to partitions whose frontiers overlap in at most ε, as follows.

Context Window. Let a partition $P_i \in \mathcal{P}, \mathcal{P} \setminus P_i \neq \emptyset$, with a pivot v be constructed over the dataset $\mathcal{O}, \mathcal{P} \vdash \mathcal{O}$. The context window $C_{i,\varepsilon}$ associated with P_i for a given threshold $\varepsilon \in \mathbb{R}_+$ is defined as follows $C_{i,\varepsilon} = \{o \in \mathcal{O} \setminus P_i \mid \forall P_j \in$

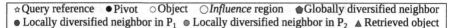

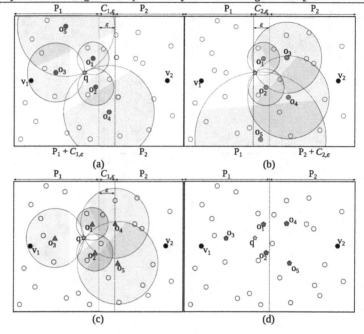

Fig. 4. SP-BRID$_k$ joining a query reference q in an Euclidean space partitioned by $\{v_1, v_2\}$ and window size ε with $k = 5$. (a–b) Locally diversified results. (c) SP-BRID$_k$ result set after reduction. (d) Globally diversified neighbors. (Color figure online)

$\mathcal{P} \setminus P_i, \delta(o, P_i.v) + \varepsilon \leq \delta(o, P_j.v)\}$. The set of context windows is $\mathcal{C} = \cup_{i=1}^{M} C_{i,\varepsilon}$, and $|\mathcal{C}|$ is the total data storage overhead distributed across M partitions.

The map stage of SP-BRID$_k$ collects the objects closer to the partition pivot and its related context window (see Algorithm 4), while the SP-BRID$_k$ reducer for locally diversified neighbors is carried out by the P-BRID$_k$ Algorithm 3. Thus, P-BRID$_k$ becomes a particular instance of SP-BRID$_k$ for $\varepsilon = 0.0$. Figures 4(a–c)

Map(Query set $\mathcal{Q}$, dataset $\mathcal{O}$, #neighbors k, #partitions M, window size ε);

---

1 $\mathcal{P} \leftarrow \texttt{maxVarPivots}(\mathcal{O}, M);$ /* M partitions with a pivot v each.   */
2 **for** $o \in \mathcal{O} \setminus \mathcal{P}$ **do**
3 $\quad$ | $\quad P_j \leftarrow P_j \cup \{o\} \mid P_j \in \mathcal{P}, \forall P_h \in \mathcal{P} \setminus \{P_j\}, \delta(P_j.v, o) \leq \delta(P_h.v, o);$
4 **for** $P_i \in \mathcal{P}$ **do**
5 $\quad$ | $\quad P_i \leftarrow P_i \cup C_{i,\varepsilon}$ /* Add context window elements   */
6 **return** $\mathcal{R}_\mathcal{P} = \cup_{i=1}^{M} \texttt{N-BRID}_k(\mathcal{Q}, \mathcal{P}_i, k);$ /* Distribute to M workers.   */

Algorithm 4: Pivot and context mapping of SP-BRID$_k$.

present a SP-BRID$_k$ running for the same query scenario in Fig. 3(c). The partitions are augmented by the ε-based window, which enables sharing the query context within the borders of two partitions. Unlike P-BRID$_k$ (see Fig. 3(d)), the local results in P_1 are now preserved throughout P_2, enabling the SP-BRID$_k$ reducer to find the k globally diversified neighbors.

4 Empirical Evaluation

For the experimental evaluation, we start describing the infrastructure, setup, and employed datasets. Then, we compare the performance of P-BRID$_k$ against N-BRID$_k$ by examining the joining quality via Recall values. Next, we discuss the tuning of SP-BRID$_k$ regarding context windows and evaluate the overhead ratio associated with the window size. Finally, we assess the P-BRID$_k$ and SP-BRID$_k$ horizontal scalability according to Recall, overhead ratio, and elapsed time. For all tests, we selected a batch of 100 random objects as the query set, i.e., $|\mathcal{Q}| = 100$. Then, we removed the query sets from the original datasets before the join. The L_2 distance was employed to perform the similarity comparisons.

4.1 Experimental Setup

P-BRID$_k$ and SP-BRID$_k$ Implementation and Infrastructure[2]. We implemented P-BRID$_k$, SP-BRID$_k$, and the baseline N-BRID$_k$ in Apache Hadoop 3.3.1 (JDK 8) by using a containerized cluster with 11 nodes (01 master, 10 workers) running on top of a Linux-based QLustar server with 48 AMD Opteron 2.2GhZ processors, 94 GB RAM, and 1 TB SATA disk. The nodes were defined through *docker containers* with equally distributed resources (CPU, memory, and disk). The map and reduce phases were implemented through the Hadoop *jobConf*, which distributes the search execution parameters (*e.g.*, $\mathcal{P}, k$, and ε) to every node. P-BRID$_k$, SP-BRID$_k$, and N-BRID$_k$ reduction phase benefits from the Hadoop Map-Reduce native *secondary sort* routine to orderly access the list of local candidates according to their distance to the query references. We overload the *secondary sort*, *grouping*, and *partition* Hadoop operations to support distance-based ordering and object-based joining.

Datasets (see Footnote 2). We experiment on synthetic and real-world datasets with varying cardinality ($|\mathcal{O}|$), dimensionality (d), and intrinsic dimensionality (ID)[3]. Table 1 summarizes the employed datasets and further separates the datasets in *low-dimensional* (LDG) and *high-dimensional* (HDG) groups. The execution of $k_{\ell d}$NN joins in HDG sets is expected to be costlier than in LDG datasets due to the distance concentration phenomenon [10], thus providing edge scenarios for the P-BRID$_k$ and SP-BRID$_k$ evaluation.

---

[2] Source-code and datasets in https://github.com/rviniciussouza/BRIDkD.
[3] $ID = \lceil \mu_{\mathcal{O}}^2/2 \cdot \sigma_{\mathcal{O}}^2 \rceil$, where $\mu_{\mathcal{O}}$ and $\sigma_{\mathcal{O}}$ are the mean and standard deviation of the pairwise distance distribution within $\mathcal{O}$, respectively [4].

Table 1. List of evaluated datasets

Dataset	$|\mathcal{O}|$	$|d|$	ID	Description	Group
CITIES	25,375	2	2	Coordinates of US cities	LDG
NASA	40,150	20	6	Low-level features from satellite images	
GAUSS	$2 \cdot 10^6$	2	2	Synthetic iid dimensions (Standard distribution)	
UNIFORM	$2 \cdot 10^6$	2	2	Synthetic iid dimensions (Uniform distribution)	
MNIST	70,000	784	26	Handwritten digits	HDG
ALOI	72,000	144	3	3D color model images	
COLORS	112,682	111	12	Low-level features from color photos	
SIFT	$1 \cdot 10^6$	128	15	SIFT features from images	

Performance Analysis. We juxtapose the performance of the proposed methods against the baseline approach N-BRID$_k$ by using measures of Recall (fraction of points among true diversified neighbors), data replication (overhead ratio), and elapsed time. The overhead required by context windows is expressed as percentile of the dataset cardinality. Finally, elapsed time is reported as an average of ten executions.

4.2   P-BRID$_k$ Performance

Figure 5 presents the comparison between P-BRID$_k$ and N-BRID$_k$. The average Recall for an increasing neighborhood range $k = \{5, 8, 13, 15, 20\}$. N-BRID$_k$ considers the dataset randomly partitioned, while P-BRID$_k$ uses the pivots selected by the maximal variance criterion. The results indicate the N-BRID$_k$ Recall quickly degrades, while P-BRID$_k$ maintained a high Recall ratio in all cases.

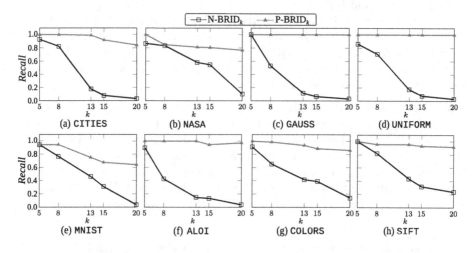

Fig. 5. P-BRID$_k$ Recall in low (a–d) and high-dimensional (e–h) datasets.

LDG Datasets. P-BRID$_k$ achieved an average Recall near 0.95 for the entire LDG, surpassing N-BRID$_k$ by up to 60% (NASA dataset ($k = 20$)) and 70% (CITIES dataset ($k = 20$)). In synthetic datasets GAUSS and UNIFORM, P-BRID$_k$ average Recall was 1.0 (exact matches between $k_{\ell d}$NN$\bowtie$ and k_dNN$\bowtie$) for all examined k, while N-BRID$_k$ Recall dropped with the range of neighbors. P-BRID$_k$ consistently outperformed N-BRID$_k$ by significant margins (even for $k = 5$), and every N-BRID$_k$ Recall (per query object) was tied or topped by P-BRID$_k$.

HDG Datasets. The evaluation for the high-dimensional group shows the N-BRID$_k$ performance deteriorates abruptly in comparison to the LDG datasets. The P-BRID$_k$ quality, however, was preserved, even for greater neighborhoods. In the cases of ALOI, COLORS, and SIFT datasets, P-BRID$_k$ average Recall constantly reached values above 0.87, and for the edge MNIST dataset, the P-BRID$_k$ Recall dropped only after $k = 8$, finishing with 0.67. The highest P-BRID$_k$ gain against N-BRID$_k$ was above 80% (ALOI), and we observed N-BRID$_k$ performances per query object were always equal to or below P-BRID$_k$.

Such a P-BRID$_k$ performance provides a *lower bound* for SP-BRID$_k$ behavior since P-BRID$_k$ can be seen as a SP-BRID$_k$ setup with no context window ($\epsilon = 0$). Thus, in the next experiment, we adjust the SP-BRID$_k$ parameters to enhance P-BRID$_k$ while minimizing the context window overhead.

4.3 SP-BRID$_k$ Tuning

We evaluated the impact of the context window size by setting ε as a percentage of the maximal pairwise distance within each data partition. In particular, we examined the range $\varepsilon = \{0\%, 10\%, 20\%, 30\%, 40\%, 50\%\}$ aiming at not reaching a total data overhead higher than the dataset cardinality itself. The overhead ratio was expressed as the proportion of dataset size copied throughout the partitions (the sum of context windows by data cardinality, *i.e.*, $|\mathcal{C}|/|\mathcal{O}|$). We also selected the fixed neighborhood size ($k = 15$) to evaluate as it showed to be an inflection point in the previous experiments in which P-BRID$_k$ achieved an average Recall below 0.8 for the first time regarding datasets MNIST, COLORS, and CITIES, *i.e.*, it is the first entry point for enhancements in P-BRID$_k$.

Figure 6 presents the SP-BRID$_k$ Recall and overhead ratio for LDG (a–b) and HDG (c–d) datasets. The Recall increased with the context window size in every dataset, reaching the maximum value for $\varepsilon = 50\%$ and exhibiting the same P-BRID$_k$ Recall for the window size $\varepsilon = 0\%$. The "elbow" of Recall curves was also observed with $\varepsilon = 20\%$.

LDG Datasets. The Recall for datasets CITIES and NASA increased after the adoption of a context window of 10%, reaching a close-to-top value for $\varepsilon = 20\%$. The results also reveal the Recall slowly increases (or peaked) after the $\varepsilon = 20\%$ threshold, but the overall overhead ratio continues to grow linearly with the window size, reaching up to 1.35 data duplication for $\varepsilon = 50\%$. UNIFORM and GAUSS datasets have required no context windows to reach a Recall of 1.0 for $k = 20$.

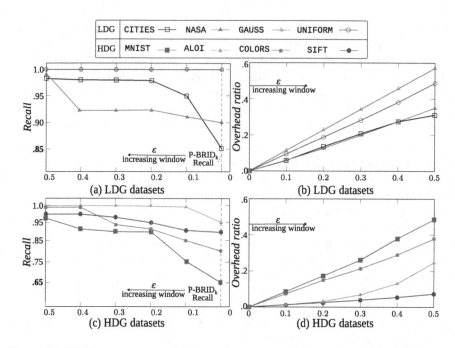

Fig. 6. SP-BRID$_k$ performance for increasing context window sizes. (a) and (c) Recall. (b) and (d) Data overhead.

HDG Datasets. Analogously to the LDG group, we observed the Recall increased for a window size $\varepsilon = 20\%$ at a 1.6 replication cost, on average. The highest Recall of 1.0 was achieved with this context window in the ALOI dataset, whereas for the edge MNIST case, a value higher than 0.9 was obtained. A similar outcome was observed for datasets COLORS and SIFT, which showed an average Recall above 0.95 for the same context window size. The data replication in HDG datasets was slightly sharper compared to LDG datasets, reaching up to 1.5 duplication for $\varepsilon = 50\%$ in the MNIST case.

Following those experimental findings for LDG and HDG datasets, we tune SP-BRID$_k$ by using a context window size $\varepsilon = 20\%$ in the next empirical evaluation of SP-BRID$_k$ scalability regarding Recall and processing performance.

4.4 Scalability

We examined the SP-BRID$_k$ scalability using a growing number of partitions $|\mathcal{P}| = \{10, 13, \ldots, 22\}$, each associated with a maximal variance pivot. In our evaluation, we considered the $\varepsilon = 20\%$ and $k = 15$. Figures 7(a–f) present the SP-BRID$_k$ Recall, overhead ratio, and CPU-only elapsed time. Results indicate Recall was above 0.8 for the range of partitions, whereas the average elapsed time smoothly reduced in the interval. They also reveal the overhead ratio was below data duplication ($2 \cdot |\mathcal{O}|$) even for the case with 22 partitions.

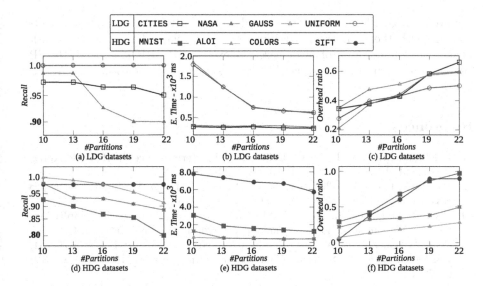

Fig. 7. SP-BRID$_k$ Recall (a–d), average elapsed time (b–e), and overhead ratio (c–f) for an increasing number of data partitions.

LDG Datasets. SP-BRID$_k$ showed a stable Recall regarding datasets GAUSS and UNIFORM for increasing partitions, with a Recall reduction (less than 10%) for 22 partitions in datasets CITIES and NASA. SP-BRID$_k$ elapsed time reduced proportionally with the partitions for all LDG datasets. Finally, the overall overhead ratio for LDG datasets was below 1.65 in all examined cases.

HDG Datasets. SP-BRID$_k$ has exhibited a higher Recall variance for HDG datasets than LDG datasets. The SP-BRID$_k$ Recall for MNIST was 0.8 with 22 partitions (the largest variation in the experiments), but the Recall differences for SIFT, COLORS and ALOI were less than 9% (in the worst case COLORS). The average elapsed time has decreased at a constant pace with the number of partitions, whereas the overhead ratio increased linearly, having an average slope greater than those of LDG datasets (nearly twice the replication).

Overall, SP-BRID$_k$ processing performance scaled smoothly with the number of partitions, with a *trade-off* between overhead and average elapsed time. Additionally, results suggest the number of partitions affects both the Recall and overhead of SP-BRID$_k$ in high-dimensional datasets.

5   Conclusion and Future Work

This study has introduced two Map-Reduce implementations for k_dNN join (see Footnote 2): P-BRID$_k$ and SP-BRID$_k$. Evaluations showed P-BRID$_k$ (and its generalized version SP-BRID$_k$) consistently outperformed the baseline approach N-BRID$_k$. We examined the SP-BRID$_k$ fine-tuning and found the context window $\varepsilon = 20\%$ as a balanced setting for Recall and overhead. The tuned SP-BRID$_k$

achieved a high Recall with controlled data replication and also showed to be scalable with the number of partitions regarding elapsed time. Future works include the *(i)* adaption of pivot-based indexes to SP-BRID$_k$, and *(ii)* the SP-BRID$_k$ implementation in a parallel shared-memory framework, *e.g.*, Apache Spark.

References

1. Armbrust, M., Das, T., et al.: Delta lake: high-performance ACID table storage over cloud object stores. VLDB **13**(12), 3411–3424 (2020)
2. Bohm, C., Krebs, F.: The k-nearest neighbour join: turbo charging the KDD process. Knowl. Inf. Syst. **6**(6), 728–749 (2004)
3. Čech, P., Maroušek, J., Lokoč, J., Silva, Y.N., Starks, J.: Comparing MapReduce-based k-NN similarity joins on Hadoop for high-dimensional data. In: Cong, G., Peng, W.-C., Zhang, W.E., Li, C., Sun, A. (eds.) ADMA 2017. LNCS (LNAI), vol. 10604, pp. 63–75. Springer, Cham (2017). https://doi.org/10.1007/978-3-319-69179-4_5
4. Chávez, E., Navarro, G., Baeza-Yates, R., Marroquín, J.: Searching in metric spaces. Comput. Surv. **33**(3), 273–321 (2001)
5. Chen, G., Yang, K., Chen, L., Gao, Y., Zheng, B., Chen, C.: Metric similarity joins using MapReduce. TKDE **29**(3), 656–669 (2016)
6. Chen, L., et al.: Indexing metric spaces for exact similarity search. Comput. Surv. **55**(6), 1–39 (2022)
7. Drosou, M., Jagadish, H., Pitoura, E., Stoyanovich, J.: Diversity in big data: a review. Big Data **5**, 73–84 (2017)
8. Hetland, M.L.: The basic principles of metric indexing. In: Coello, C.A.C., Dehuri, S., Ghosh, S. (eds.) Swarm Intelligence for Multi-objective Problems in Data Mining. SCI, vol. 242, pp. 199–232. Springer, Heidelberg (2009). https://doi.org/10.1007/978-3-642-03625-5_9
9. Jacox, E.H., Samet, H.: Metric space similarity joins. TODS **33**(2), 1–38 (2008)
10. Jasbick, D., Santos, L., Marques, P., Traina, A., Oliveira, D., Bedo, M.: Pushing diversity into higher dimensions: the LID effect on diversified similarity searching. Inf. Syst. **114**, 102–116 (2023)
11. Kim, C., Shim, K.: Supporting set-valued joins in NoSQL using MapReduce. Inf. Syst. **49**, 52–64 (2015)
12. Rong, C., Cheng, X., Chen, Z., Huo, N.: Similarity joins for high-dimensional data using Spark. Concurr. Comput.: Pract. Experience **31**(20), 1–17 (2019)
13. Santos, L.F.D., Carvalho, L.O., Oliveira, W.D., Traina, A.J.M., Traina, C.: Diversity in similarity joins. In: Amato, G., Connor, R., Falchi, F., Gennaro, C. (eds.) SISAP 2015. LNCS, vol. 9371, pp. 42–53. Springer, Cham (2015). https://doi.org/10.1007/978-3-319-25087-8_4
14. Santos, L., Oliveira, W., Ferreira, M., Traina, A., Traina, C., Jr.: Parameter-free and domain-independent similarity search with diversity. In: SSDBM (2013)
15. Silva, Y.N., Reed, J.M., Tsosie, L.M.: MapReduce-based similarity join for metric spaces. In: WCI, pp. 1–8 (2012)
16. Ukey, N., Yang, Z., Li, B., Zhang, G., Hu, Y., Zhang, W.: Survey on exact kNN queries over high-dimensional data space. Sensors **23**(2), 629 (2023)
17. Wu, J., Zhang, Y., Wang, J., Lin, C., Fu, Y., Xing, C.: Scalable metric similarity join using MapReduce. In: ICDE, pp. 1662–1665 (2019)

18. Yianilos, P.N.: Data structures and algorithms for nearest neighbor. In: ACM-SIAM Symposium on Discrete Algorithms, vol. 66, p. 311 (1993)
19. Zhu, Y., Chen, L., Gao, Y., Jensen, C.S.: Pivot selection algorithms in metric spaces: a survey and experimental study. VLDB J. **31**(1), 23–47 (2021). https://doi.org/10.1007/s00778-021-00691-4

Effective and Efficient Heuristic Algorithms for Supporting Optimal Location of Hubs over Networks with Demand Uncertainty

Alfredo Cuzzocrea[1,2]([✉]), Luigi Canadè[1], Giulia Fornari[3], Vittorio Gatto[3], and Abderraouf Hafsaoui[1]

[1] iDEA Lab, University of Calabria, Rende, Italy
alfredo.cuzzocrea@unical.it
[2] Department of Computer Science, University of Paris City, Paris, France
[3] ISIRES, Turin, Italy
{giulia.fornari,vittorio.gatto}@isires.org

Abstract. The problem faced in this paper concerns with finding the *optimal location for the hubs in a network, under demand uncertainty,* and where the allocation of the nodes is treated as *a second stage decision*. We proceed first with the definition of the mathematical model that we carved to fit the operational needs of *GUROBI Optimizer*. Afterwards, we propose *a collection of heuristic algorithms* able to solve the problem in a faster way with *sub-optimal solutions*. The heuristic algorithms proposed in our framework progressively reach a good approximation of the solution. Experimental results confirm the benefits of our approach.

Keywords: Network Algorithms · Demand Uncertainty · Heuristics

1 Introduction

In every kind of networks, transportation costs always represent an important issue, especially in large nets full of nodes and communication links and flows (e.g., [15]). This is why most of the time it is necessary to resort to intermediate nodes where it is possible to store the moving material or information, and to dispatch it to destination through the *shortest route* in terms of costs and/or time. These special intermediate stations are often called *hubs* and are fundamental in most of network engineering application fields, starting from telecommunications to postal and airlines services, and so forth. In most of the cases, hubs are considered required steps to reach the destination, whether considering a sms of a cellular network or a packet of a computer network. This is why the position of hubs within the network is of such importance.

This research has been made in the context of the Excellence Chair in Big Data Management and Analytics at University of Paris City, Paris, France.

The *hub location problem* consists of the decision of the number of hubs to use and in their location, as well as the assignment of other nodes to the hubs, aiming at minimizing the total cost that is composed by fixed and variable costs. Different constraints are possible here. Among these: (*i*) the number of selected hubs, which can be predetermined or left as a decision variable; (*ii*) whether the volume of traffic can concentrate in a hub, or single or multiple allocations of non-hub nodes to the hub is allowed. However, in all the variants, the objective is to find the location of the hub and the allocation of non-hub nodes so that the total cost is minimized.

In our research, we consider the *Stochastic Single-Allocation Hub Location Problem* (SAHLP) (e.g. [14]), which deals with the problem of positioning the hubs in the most optimal way for the users (depending on the specific application), by considering the allocation of the flows as *a second stage decision*. In this so-delineated scenario, *uncertainty* in flow demand is considered to act according to a *stochastic behaviour*, as the probability of having a greater number of requests for a hub towards specific directions rather than others is the most general way to deal with the real requirements of the target problem.

Following these considerations, in this paper we propose *three innovative heuristic algorithms that solve the optimal hub location problem with demand uncertainty*, whose main benefit consists in achieving good approximate solutions *while keeping acceptable computational time*. Our comprehensive experimental comparison confirm the benefits of our solutions.

2 Dataset Generation

In hub location problems, researchers are used to generate instances of some specific datasets that then exploited by the research community over the years, by also improving them where necessary. In this Section, we focus the attention on these classical datasets and their instances. In fact, the latter are used in our experimental evaluation.

The first one is the CAB dataset. CAB has been used frequently in the literature to test algorithms for solving P-hub problems and was introduced by O'Kelly in [19]. It is based on airline passenger flow between 25 US cities during 1970 and consists of 25 depots and 25 possible hub locations. Problems of size $n = 10, n = 15, n = 20, n = 25$ are extracted from this dataset by only considering a subset of nodes. Later, Abdinnour-Helm [1] adds fixed costs to CAB, such that fixed costs are equal for each possible hub location. Four different values for these costs are considered, namely 100, 150, 200 and 250. The CAB dataset is only tested for the basic model, because the capacity of the truck is set to 1. Therefore, the flow cannot be consolidated. The only reason why not all locations are chosen to act as a hub is the occurrence of a fixed cost for each established hub. This dataset was therefore unusable for our problem since we are considering an extension of the original problem where collection, transfer, and distribution costs are also taken into account.

The other dataset used in the hub location literature is the AP dataset. AP is based on the mail flow of the Australian post and was introduced by Ernst and Krishnamoorthy in [10]. It consists of 200 nodes, which represent postcode districts, along with their coordinates, flow volumes (mailflow), as well as the model parameters χ, α and δ (see Sect. 3 for a detailed description of these parameters). Unlike the CAB dataset, which stops at a maximum of $n = 25$, the AP dataset provides opportunities for researchers to tackle larger real-world problems, since problems of size up to $n = 200$ can be generated and solved. It is also possible to generate subset of the original dataset that still provide a good approximation of the original larger data. A feature of this dataset is that, since it is derived from a postal application, it contains non-uniform flows, particularly to and from the *Central Business District*. Moreover, the flow matrix W is not symmetrical, and, in addition to this, the diagonal elements W_{ii} need not be zero as a postcode district can send mail to itself. The AP dataset also includes capacities and fixed costs on the nodes. We consider two types of fixed costs: *Tight*, denoted as T, and *Loose*, denoted as L. Problems with fixed costs of type T have higher fixed costs for nodes with large flows. This makes it difficult for the model to nominate these high volume nodes as hubs (which would otherwise be "natural" candidates). Hence, these problems are more difficult to solve. By the contrary, problems with fixed costs of type L do not exhibit this trend.

Similarly to the work by Rostami *et al.* [21], we used the AP in our experimental work, by introducing it in our instance generator program. The AP dataset is available in the OR Library at [25]. In AP, we can find data useful to generate instances for up to 200 nodes. The parameters associated to the generation procedure are the following: (i) the number of nodes n; (ii) for each node, the coordinates x and y; (iii) the matrix W that represents the flow from a particular node to another node; (iv) the parameter p that represents the fixed number of hub; (v) for each node, the collection cost C_c; (vi) for each node, the transfer cost C_t; (vii) for each node, the distribution cost C_d; ($viii$) for each node, the fixed cost C_f.

As related to the instance generation, since our problem is stochastic in nature, we also consider different scenarios that characterize the generation of the target instances. In order to take into consideration the stochasticity of the problem, in the second stage we multiplied coefficient by p_s (see Eq. (3)) where, assuming ξ being a random parameter that follows a discrete distribution with finite support S_w modeled as follows:

$$S_w = s_1, \ldots, s_m \tag{1}$$

the corresponding probabilities are as follows:

$$p_{s_1}, \ldots, p_{s_m} \tag{2}$$

where:

$$p_s = P(\xi = s), s \in S_w \tag{3}$$

In order to translate the stochasticity of the problem in the code, we adopt a procedure similar to the classical Knapsack problem. Moreover, like in [21], in order to generate different flow scenarios for each source-sink pair, we make the assumption that the flow between two nodes can be modelled as a *Poisson distribution*, which is finally used to model demand uncertainty. As suggested in [21], then we multiply the items w_{ij} of the flow matrix W by a factor π_i that denotes the deviation from the base case. Further, we assume that π_i is uniformly distributed over the interval $[0.5, 1.5]$. Then, the demand value for a source-sink pair *(i,j)* in a stochastic scenario is chosen from a Poisson distribution with event rate as follows:

$$w_{ij}\pi_i\pi_j \tag{4}$$

with π_j also uniformly distributed between $[0.5, 1.5]$.

As far as the number of scenarios is concerned, in our experiments we used 5 different scenarios. This choice is due to the fact that, even in the reference paper [21], authors make use of 5 scenarios for their simulation, like also other similar experiences (e.g., [3]).

3 Mathematical Foundations

As introduced in Sect. 1, our problem consists in a single allocation hub location problem under demand uncertainty where the allocation of the spokes to the hubs is optimized as second stage decision after the uncertainty in the demand is addressed. This problem is different from fixed allocation case that is addressed in the literature, where the spokes are allocated to the chosen hubs *before* the uncertainty is addressed.

The stochastic SAHLP with variable allocation is formulated as a two-stage stochastic program with recursion. Here, first-stage decisions are the location of the hubs to be opened while second.stage decisions are the optimal allocation of the spoke nodes to the hub nodes as well as the routing of the flows.

In the following, we describe the proposed mathematical model and the related parameters. First of all, we consider a directed graph $G = (N, A)$ where $N = 1, 2, ..., n$ as representing the set of nodes that model origin, destination and possible hub location while A represents the set of arcs that report possible direct links between different nodes. The following ones are thus the useful parameters:

- *flow w_{ij}^s*: the amount of flow to be transported from node i to node j for each scenario $s \in S_w$;
- *distance d_{ij}^s*: the distance between node i and node j for each scenario $s \in S_w$;
- *outgoing flow O_i^s*: the sum of all the flow leaving from node i for each scenario $s \in S_w$ $O_i = \sum_{j \in N} w_{ji}^s$;
- *incoming flow D_i^s*: the sum of all the flow coming to node i for each scenario $s \in S_w$ $D_i = \sum_{j \in N} w_{ji}^s$;
- *fixed cost f_k*: for each $n \in N$, f_k, the fixed set-up cost for locating a hub at node k;

– *cost per unit of flow*: for each path $i - k - l - j$ from and origin node i to a destination node j passing through hubs k and l, the cost per unit flow is described as: $\chi\, d_{ik} + \alpha\, d_{kl} + \delta\, d_{lj}$, where χ, α and δ are the no-negative collection, transfer, and distribution costs, respectively, while d_{ik}, d_{kl} and d_{lj} are the distances between the pair of nodes, respectively – in particular, χ, α and δ are constant and their value is specified in the AP dataset as follows: $\alpha = 0.75$, $\chi = 3.0$ and $\delta = 2.0$.

In order to formulate the stochastic SHALP, the following allocation variables are introduced:

$$x_{ik}^s = \begin{cases} 1 & if\ a\ node\ i\ is\ allocated\ to\ a\ hub\ at\ node\ k,\ with\ s \in S_w \\ 0 & otherwise \end{cases} \tag{5}$$

and

$$z_k = \begin{cases} 1 & if\ a\ hub\ is\ located\ at\ node\ k \\ 0 & otherwise \end{cases} \tag{6}$$

The problem is then formulated as reported in Eq. (7).

$$\begin{aligned} min \quad & \sum_{k \in N} f_k z_k + \sum_{s \in S_w} p_s \sum_{\substack{i,k \in N \\ i \neq k}} c_{ik}^s x_{ik}^s + \\ & \sum_{s \in S_w} p_s \sum_{i,j \in N} \alpha w_{ij}^s \Big(d_{ij} z_i z_j + \sum_{\substack{l \in N \\ l \neq j}} d_{il} z_i x_{jl}^s + \\ & \sum_{\substack{k \in N \\ i \neq k}} d_{kj} x_{ik}^s z_j + \sum_{\substack{k,l \in N \\ i \neq k \\ j \neq l}} d_{kl} x_{ik}^s x_{jl}^s \Big) \end{aligned} \tag{7}$$

where:

$$c_{ij}^s = d_{ik}(\chi O_i^s + \delta D_i^s) \tag{8}$$

$$subject\ to \quad \sum_{\substack{k \in N \\ i \neq k}} x_{ij}^s = 1 - z_i \qquad i \in N, s \in S_w \tag{9}$$

$$x_{ik}^s \leq z_k \qquad i, k \in N, i \neq k, s \in S_w \tag{10}$$

$$z_i \in \{0,1\} \quad \forall i \in N \tag{11}$$

$$z_{ik}^s \in \{0,1\} \quad \forall i \in N, s \in S_w \tag{12}$$

where, in Eq. (7), we try to minimize the total cost of the network which includes the cost of setting up the hubs, the cost of collection and distribution of items between the spoke nodes and the hubs and the cost of transfer between the hubs. The fixed cost of setting up the hub is a first stage decision variable, meaning that we choose at first the nodes that have lower fixed cost without considering the demands. The other variables are instead second stage variables and they take into account the uncertainty of the demand.

In more detail, the first constraint (Eq. (8)) specifies that each node should be allocated to precisely one hub (i.e., single allocation), while the second constraint (Eq. (9)) enforces that a node i is allocated to a node k only if k is selected as a hub node.

Equation (10) and Eq. (11) are instead implicit constraints that impose a binary choice.

As it can be seen from the objective function (Eq. (7)), our problem is *quadratic* in nature, and it involves many variables and summations. Indeed, for its computation the complexity reaches a maximum formalized as follows:

$$O(s * n^4) \tag{13}$$

where s is the number of scenarios and n is the number of nodes.

This problem is considered NP-Hard from many other researchers, such as Ernst [11] and Silva-Cunha [22], while we find some arguments in the paper of Stanimirović [24] which considers the CSAHLP as a NP-Complete problem since its sub-problem, the USAHLP, is proven to be NP-Hard.

4   Baseline Solution

In this Section, we introduce the results obtained via using an *exact solver* with the AP dataset as input. In particular, we conducted our experiment using *GUROBI Optimizer 9.0* [13], and we implemented the mathematical model proposed in Sect. 3 in *Python 3.8*. All the experiments were conducted using an *Intel Core i5-5300U CPU @ 2.30* GHz × *4, 8* GB *RAM* with *OS Ubuntu 20.04.2*, with practical guidelines [12]. Then, in order to plot the results obtained we used the Python package *NetworkX* [18].

In Fig. 1 and Fig. 2, we report the results obtained using the AP dataset with 10 nodes (10L) and 25 nodes (25L), respectively. Here, we plot the position of the nodes (in respect to their x and y coordinates) while the size of the nodes is proportional to the outgoing flow O_i. Nodes that are chosen as hubs are then colored in yellow while the spokes nodes are colored in sky blue and each one of them is allocated to one hub.

From Fig. 1 and Fig. 2, we can see clearly the variable allocation decision. For example, in Fig. 1a, Fig. 1c and Fig. 1d the node number 4 is connected with hub number 6 while, instead, in Fig. 1b and Fig. 1e a different choice is made and the node 4 is connected with hub 2. Same thing happens in the case with 25 nodes where in Fig. 2a node 13 is allocated to hub 7 while in Fig. 2b the same node is connected to hub number 17.

In [21], i.e. the work that we use as reference, authors claimed that the usage of variable allocation strategy provides better results with respect to fixed allocation strategy. Indeed, variable allocation strategy results in an overall decrease of 2.0% for the simulation with 40 nodes, and of 8.7% for the simulation with 50 nodes.

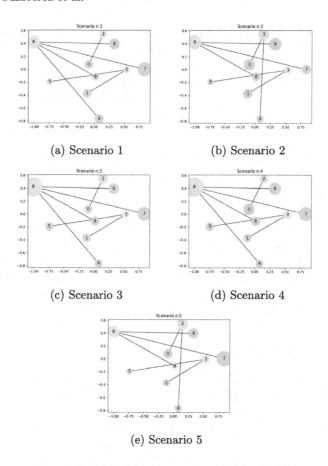

(a) Scenario 1   (b) Scenario 2

(c) Scenario 3   (d) Scenario 4

(e) Scenario 5

Fig. 1. Variable Allocation with 10 Nodes (10L)

5 Heuristic Algorithms

In the active literature, different kinds of heuristic approaches have been created according to the formulation of the problem. In particular, given the nature of the single allocation hub location problem, a large variety of heuristics have been proposed over the years. Alumur and Kara in [2] present the state of the art of the hub location problem and report on the different solution approaches used in the literature. Authors also claim that the most effective heuristics is the *Lagrangian relaxation-based heuristics* [20]. Here, authors make use of a previously proposed tight linear programming formulation and introduce a sub-gradient optimization based on the Lagrangian relaxation. However, to dramatically improve the performance of this approach, they specify a sub-problem of the Lagrangian relaxation model with a *cut constraint*.

According to [2], instead, the two best meta-heuristics are those proposed by Skorin-Kapov and Skorin-Kapov [23] and Ernst and Krishnamoorthy [11]. The

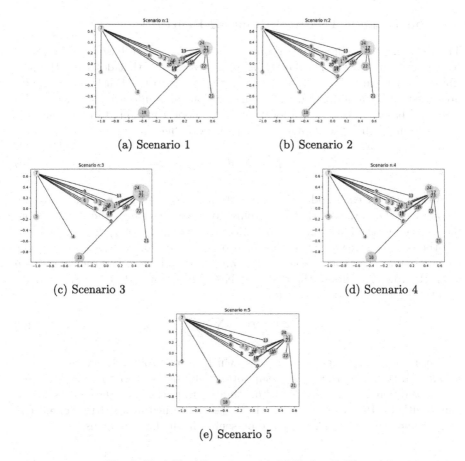

(a) Scenario 1   (b) Scenario 2

(c) Scenario 3   (d) Scenario 4

(e) Scenario 5

Fig. 2. Variable Allocation with 25 Nodes (25L)

first one is a heuristic method based on *Tabu Search* (TS) [27], where the problem of locating hub facilities and the problem of allocating the nodes to one and only one hub are treated in the same manner. The second one, instead, is a heuristic algorithm based on *Simulated Annealing* (SA) [26]. This approach, already used in other optimization problems like the *Traveling Salesman Problem*, consists, as translated to our case, in starting from a *random solution*, then defining a neighbourhood and the transitions that take a particular solution to one in its neighbourhood such that where only *feasible solutions* are considered. Ernst and Krishnamoorthy [11] present in detail the step of the algorithm, and, moreover, they use the previously-found upper bound to develop a *Linear Programming (LP)-based branch and bound solution* method.

The above-mentioned heuristic algorithms, even if widely used and well performing, do not work well in our case, since our main goal is to come up with a heuristic solution that is easy to implement and capable of reaching good results.

5.1   Fundamental Theory of Proposed Heuristics

The three heuristic algorithms proposed in this paper are based on the initial computation of a *penalty factor* pf, similar to a weight, that describes the possibility of a node to become a hub or not. Similar to the third starting point presented by Silva and Cunha [22], our solution takes into account the penalty factor pf based on the spatial location of the node i measured in terms of distances from all the other nodes in the network, defined as follows:

$$pf_i = \sum_j d_{ij} + d_{ji} \tag{14}$$

Moreover, for each node i, we compute a *decision factor* p_i that takes into account all the variables that contribute to the definition of the problem and the decision of the hubs, like, for instance: (i) the fixed cost f_i^s for each node i, (ii) the outgoing flow o_i^s for each scenario $s \in S$ and for each node i, (iii) the incoming flow S_i for each scenario $s \in S$ and for each node i, (iv) the summation of all the row elements of the cost matrix C_{ik}^s for each node i, (v) the penalty factor pf_i for each node i, as follows:

$$p_i = f_i + O_i^s + D_i^s + \sum_i \sum_s C_{ij}^s + pf_i \tag{15}$$

After computing these two factors, which, in our solution, act as a sort of *penalty function*, for the goal of supporting the decision of the hub location, our three different heuristic algorithms that best approximate the exact solution found with GUROBI are considered. These are named as: (i) SimpleHeu, (ii) HeuNew and (iii) HeuNew2. They are presented in the next Sections.

5.2   Algorithm SimpleHeu

The first heuristic algorithm SimpleHeu predicates that the decision of allocating a hub is simply due to the value obtained from the weights p_i. In more details, for each node i, we evaluate the value p_i and, if this value, is lower than the mean of p_i, denoted by P_{avg}, then the node is selected as a hub. Later, all the remaining nodes that were not chosen as hubs are connected to the nearest hub, following the O'Kelley nearest allocation heuristics [19]. As a result, it happens though that some nodes that are selected as hubs are not connected with any spoke. Therefore, as a final step, we check if there are hubs without links, and, if this happens, these nodes are converted to a non-hub nodes and they are reallocated to their nearest hub. SimpleHeu is reported by Algorithm 1.

5.3   Algorithm HeuNew

According to the main algorithmic framework, for the second heuristic algorithm, HeuNew, our first step is evaluating the p_i for each node i. Then, we choose as hub the nodes with the lowest p_i. The decision on the number of nodes that need

Algorithm 1. SimpleHeu

```
for (s in scenarios) do
    for (i in nodes) do
        generate penalty factors P[i]
    end for
end for
compute the mean of the penalty factors P_avg
for (i in nodes) do
    if (P[i] < P_avg) then
        node i becomes a hub
    end if
end for
for (s in scenarios) do
    evaluation of link matrix X
    for (i in nodes) do
        select chosen hubs i
        for (j in nodes) do
            if (hub without links) then
                cancel hub i
            end if
        end for
    end for
    recompute matrix X
end for
```

to be chosen as hubs is, in this case, the real unknown variable of the algorithm. Our idea is thus to try different number of hubs and choose the best option, by always considering the computational complexity of this decision step. For the case of HeuNew, our choice is to try for a number of hubs between 1 and 4. This is motivated by the results obtained with the exact solution, where we notice that, for each dataset, the number of hubs chosen is usually 2 or 3. Therefore, what is done in algorithm HeuNew is the evaluation of the objective function when the 1, 2, 3 or 4 nodes with the lowest p_i are chosen as hubs. Among these 4 objective functions, the lowest one is finally selected, and the solution is given by the corresponding hub chosen for that number. Even in this case, like for SimpleHeu, all the spokes are linked to the nearest hubs. HeuNew is reported by Algorithm 2.

5.4 Algorithm HeuNew2

In algorithm HeuNew, the choice of the number of hubs is dictated by the fact that the exact solution of the problem is known. This condition is of course inadmissible in a heuristic approach where it is not supposed the exact solution to be known. The main goal of the third heuristic algorithm HeuNew2 is therefore to choose the right number of hubs without having some a-priori knowledge on the exact solution.

Algorithm 2. HeuNew

---

   for (s in *scenarios*) **do**
      for (i in *nodes*) **do**
         generate penalty factors $P[i]$
      end for
   end for
   choose *Hubs* among *nodes* with lowest P
   for ($\#Hubs = 1, 2, 3, 4$) **do**
      for (i in $\#Hubs$) **do**
         all nodes i become hubs
      end for
      for (s in *scenarios*) **do**
         evaluation of link matrix X
      end for
      evaluation of obj function of_{new} for given X and given hubs
      if ($of_{new} <$ previously computed obj function of_{old}) **then**
         $of_{old} = of_{new}$
      end if
   end for

In HeuNew2, at first, as usual, we evaluate p_i for each node i. Then, like in HeuNew, we choose as hubs the nodes with the lowest p_i. The only difference is that, this time, we start with a number of hubs equal to the number of nodes. Then, we evaluate the objective function for this number of hubs and, as a subsequent step, the number of hubs chosen is halved and, for this new number, we compute the objective function. These two so-obtained objective functions are then compared, and the lowest one is chosen as *partial* solution. Then, the number of hubs is halved again (remember that we choose always the nodes with the lowest p_i), and a new objective function is evaluated and compared to the previous one. These steps are repeated until we end up with the best solution (i.e., the newest solution is worst than the previous one), and the latter is selected as final solution. HeuNew2 is reported by Algorithm 3.

6   Experimental Evaluation and Analysis

In this Section, we provide the experimental results retrieved for the three different heuristic algorithms SimpleHeu, HeuNew and HeuNew2, which are the main result og our research. Particularly, in Table 1, Table 2 and Table 3, we report the values obtained by SimpleHeu, HeuNew and HeuNew2, respectively. In particular, the last column *Gap %* reports the percentage difference between the solution obtained with the actual heuristics and the exact solution obtained with GUROBI. In more details, the gap g is computed as follows:

$$g = \left(1 - \frac{GUROBI\,objective\,function}{heuristic\,objective\,function}\right) * 100 \tag{16}$$

Algorithm 3. HeuNew2

---

for (s in *scenarios*) **do**
    for (i in *nodes*) **do**
        generate penalty factors $P[i]$
    end for
end for
choose *Hubs* among *nodes* with lowest P
$\#Hubs = \#nodes$
while ($\#Hubs > 0$) **do**
    for (i in $\#Hubs$) **do**
        all nodes i become hubs
    end for
    for (s in *scenarios*) **do**
        evaluation of link matrix X
    end for
    evaluation of obj function of_{new} for given X and given hubs
    if (of_{new} < previously computed obj function of_{old}) **then**
        $of_{old} = of_{new}$
    end if
    $\#Hubs$ divided by 2
end while

Table 1. Experimental Results with `SimpleHeu`

	SimpleHeu		
Data set	Obj. Function Value	Comp. Time [s]	Gap %
10L	204641729.12	0.0045	0.25
10T	312061592.19	0.0043	13.84
20L	308847636.15	0.0171	28.41
20T	374757279.6	0.0097	36.08
25L	353994905.59	0.0245	32.43
25T	529451193.12	0.0223	43.18
		Minimum Gap:	0,25
		Average Gap:	25,6983
		Maximum Gap:	43,18

As we can see from the analysis of the experimental results, `SimpleHeu` requires a very low computational time to solve the SAHLP, but it behaves badly for network with a large number of nodes, so that obtaining a value of the objective function too different from the one computed with GUROBI. For `HeuNew` and `HeuNew2`, instead, we are able to generate a solution that well approximate the exact one, even for a large number of nodes. As far as the computational time is concerned, `HeuNew` is slightly faster than `HeuNew2`. Moreover, at the bottom of

Table 2. Experimental Results with HeuNew

HeuNew			
Data set	Obj. Function Value	Comp. Time [s]	Gap %
10L	217417690.19	0.1587	6.11
10T	287847096.25	0.1695	6.6
20L	238768933.41	2.2613	7.39
20T	255443235.21	2.3201	6.22
25L	274981761.09	5.3129	13.01
25T	300835479.39	5.4361	0.0
		Minimum Gap:	0,0
		Average Gap:	6,555
		Maximum Gap:	13,01

Table 3. Experimental Results with HeuNew2

HeuNew2			
Data set	Obj. Function Value	Comp. Time [s]	Gap %
10L	217417690.19	0.2071	6.11
10T	287847096.25	0.2052	6.6
20L	238768933.41	3.3987	7.39
20T	264207465.3	3.3861	9.33
25L	274981761.09	9.2601	13.01
25T	300835479.39	9.1225	0.0
		Minimum Gap:	0,0
		Average Gap:	7,0733
		Maximum Gap:	13,01

each table, we add some more detailed information about the gap, in particular the minimum and maximum gap obtained, and the average value of the gap over the six target datasets.

7   Conclusions and Future Work

Starting from open research challenges of the research community, in this paper we have proposed three innovative heuristic algorithms that solve the optimal hub location problem with demand uncertainty, whose main benefit consists in achieving good approximate solutions while keeping acceptable computational time. In addition to this conceptual contribution, we have also performed a comprehensive experimental evaluation and analysis, where we inspected the variation of several experimental parameters. Derived results have confirmed the benefits of our solutions. Future work is mainly oriented towards making our algorithms compliant with emerging *big data trends* (e.g., [4–9,16,17,28]).

Acknowledgement. This research is supported by the ICSC National Research Centre for High Performance Computing, Big Data and Quantum Computing within the NextGenerationEU program (Project Code: PNRR CN00000013).

References

1. Abdinnour-Helm, S.: A hybrid heuristic for the uncapacitated hub location problem. Eur. J. Oper. Res. **106**(2–3), 489–499 (1998)
2. Alumur, S.A., Kara, B.Y.: Network hub location problems: the state of the art. Eur. J. Oper. Res. **190**(1), 1–21 (2008)
3. Alumur, S.A., Nickel, S., da Gama, F.S.: Hub location under uncertainty. Transp. Res. Part B Methodol. **46**(4), 529–543 (2012)
4. Bellatreche, L., Cuzzocrea, A., Benkrid, S.: $\mathcal{F}\&\mathcal{A}$: a methodology for effectively and efficiently designing parallel relational data warehouses on heterogenous database clusters. In: Bach Pedersen, T., Mohania, M.K., Tjoa, A.M. (eds.) DaWaK 2010. LNCS, vol. 6263, pp. 89–104. Springer, Heidelberg (2010). https://doi.org/10.1007/978-3-642-15105-7_8
5. Campan, A., Cuzzocrea, A., Truta, T.M.: Fighting fake news spread in online social networks: actual trends and future research directions. In: 2017 IEEE International Conference on Big Data (IEEE BigData 2017), Boston, MA, USA, 11–14 December 2017, pp. 4453–4457. IEEE Computer Society (2017)
6. Coronato, A., Cuzzocrea, A.: An innovative risk assessment methodology for medical information systems. IEEE Trans. Knowl. Data Eng. **34**(7), 3095–3110 (2022)
7. Cuzzocrea, A.: Analytics over big data: exploring the convergence of datawarehousing, OLAP and data-intensive cloud infrastructures. In: 37th Annual IEEE Computer Software and Applications Conference, COMPSAC 2013, Kyoto, Japan, 22–26 July 2013, pp. 481–483. IEEE Computer Society (2013)
8. Cuzzocrea, A., Martinelli, F., Mercaldo, F., Vercelli, G.V.: Tor traffic analysis and detection via machine learning techniques. In: 2017 IEEE International Conference on Big Data (IEEE BigData 2017), Boston, MA, USA, 11–14 December 2017, pp. 4474–4480. IEEE Computer Society (2017)
9. Demchenko, Y., De Laat, C., Membrey, P.: Defining architecture components of the big data ecosystem. In: 2014 International Conference on Collaboration Technologies and Systems (CTS), pp. 104–112 (2014)
10. Ernst, A.T., Krishnamoorthy, M.: Efficient algorithms for the uncapacitated single allocation p-hub median problem. Locat. Sci. **4**(3), 139–154 (1996)
11. Ernst, A.T., Krishnamoorthy, M.: Solution algorithms for the capacitated single allocation hub location problem. Ann. Oper. Res. **86**, 141–159 (1999)
12. Fadda, E., Manerba, D., Cabodi, G., Camurati, P.E., Tadei, R.: Comparative analysis of models and performance indicators for optimal service facility location. Transp. Res. Part E Logist. Transp. Rev. **145**, 102174 (2021)
13. Gurobi: Gurobi - the fastest solver (2021). http://www.gurobi.com/. Accessed 1 Dec 2021

14. Hu, Q.M., Hu, S., Wang, J., Li, X.: Stochastic single allocation hub location problems with balanced utilization of hub capacities. Transp. Res. Part B Methodol. **153**, 204–227 (2021)
15. Klingman, D., Napier, A., Stutz, J.: NETGEN: a program for generating large scale capacitated assignment, transportation, and minimum cost flow network problems. Manag. Sci. **20**(5), 814–21 (1974)
16. Leung, C.K., Cuzzocrea, A., Mai, J.J., Deng, D., Jiang, F.: Personalized Deepinf: enhanced social influence prediction with deep learning and transfer learning. In: 2019 IEEE International Conference on Big Data (IEEE BigData), Los Angeles, CA, USA, 9–12 December 2019, pp. 2871–2880. IEEE (2019)
17. Malek, Y.N., Najib, M., Bakhouya, M., Essaaidi, M.: Multivariate deep learning approach for electric vehicle speed forecasting. Big Data Min. Anal. **4**(1), 56–64 (2021)
18. NetworkX: Networkx - network analysis in python (2021). http://networkx.org/. Accessed 1 Dec 2021
19. O'kelly, M.E.: A quadratic integer program for the location of interacting hub facilities. Eur. J. Oper. Res. **32**(3), 393–404 (1987)
20. Pirkul, H., Schilling, D.A.: An efficient procedure for designing single allocation hub and spoke systems. Manag. Sci. **44**(12), 235–242 (1998)
21. Rostami, B., Kämmerling, N., Naoum-Sawaya, J., Buchheim, C., Clausen, U.: Stochastic single-allocation hub location. Eur. J. Oper. Res. **289**(3), 1087–1106 (2021)
22. Silva, M.R., da Cunha, C.B.: New simple and efficient heuristics for the uncapacitated single allocation hub location problem. Comput. Oper. Res. **36**(12), 3152–3165 (2009)
23. Skorin-Kapov, D., Skorin-Kapov, J.: On tabu search for the location of interacting hub facilities. Eur. J. Oper. Res. **73**(3), 502–509 (1994)
24. Stanimirović, Z.: Solving the capacitated single allocation hub location problem using genetic algorithm. In: Recent Advances in Stochastic Modeling and Data Analysis, pp. 464–471. World Scientific (2007)
25. University, B.: Ap data set (2021). http://people.brunel.ac.uk/~mastjjb/jeb/orlib/files/phub1.txt. Accessed 1 Dec 2021
26. van Laarhoven, P.J.M., Aarts, E.H.L.: Simulated annealing. In: Simulated Annealing: Theory and Applications. Mathematics and Its Applications, vol. 37, pp. 7–15. Springer, Dordrecht (1987). https://doi.org/10.1007/978-94-015-7744-1_2
27. de Werra, D., Hertz, A.: Tabu search techniques. Oper.-Res.-Spektrum **11**(3), 131–141 (1989)
28. White, L., Burger, K., Yearworth, M.: Big data and behavior in operational research: towards a smart or. In: Behavioral Operational Research, pp. 177–193 (2016)

DMIS: Dual Model Index Structure for Enhanced Performance on Complexly Distributed Datasets

Lanzhong Liu, Xujian Zhao$^{(\boxtimes)}$, and Yin Long

School of Computer Science and Technology,
Southwest University of Science and Technology, Mianyang, Sichuan, China
liulanzhong@mails.swust.edu.cn, jasonzhaoxj@gmail.com

Abstract. Recently learned index was proposed to improve index performance, where error-driven methods are widely adopted. However, when applied to datasets with complex data distributions, the method may produce overdispersed models. Specifically, data sets with complex distributions are highly irregular and difficult to describe using parameterized distribution laws, which ultimately affects the performance of the learned index. Aiming to the issue, we propose a Dual Model Index Structure (DMIS) that combines the learned index and traditional index to better handle complex datasets. Meanwhile, we propose an evaluation model that measures the compatibility of the data distribution with the learned index and explore a classification method to categorize datasets as either learning-friendly or non-learning-friendly. Our evaluation results demonstrate that the DMIS architecture improves performance by about 1.6 times compared to the state-of-the-art learned index and performs better under various workloads. The DMIS model for partitioning learning-friendly data enhances the model's universality, efficiently improves the data index's efficiency, and reduces the number of bottom segments of the index. Our work effectively mitigates the impact of dataset differences on learned indexes.

Keywords: Learned index · Database · Data Management

1 Introduction

Index technology is the key to achieving efficient data access in database systems. Both industry and academia have long been committed to researching and developing various index technologies, and this field has always been a hot topic in the database domain [1,5,9,18]. Recently learned indexes was proposed, which have shown significant potential in improving index performance [12]. However, it may not perform well on datasets with complex data distributions. Complex data distribution refers to datasets where data points are unevenly and irregularly distributed, without clear statistical patterns or potentially following multiple different distributions. Describing this type of data distribution

© The Author(s), under exclusive license to Springer Nature Switzerland AG 2023
C. Strauss et al. (Eds.): DEXA 2023, LNCS 14146, pp. 99–113, 2023.
https://doi.org/10.1007/978-3-031-39847-6_7

using traditional parameterized distribution laws becomes challenging, which may result in poor model fitting or overdispersed models. So as the complexity of the data distribution increases, the number of models required to build the learned index also increases. Retrieving information from a large number of models can make the process more complex and time-consuming. Additionally, ensuring the accuracy of the learned model results becomes more complex in such scenarios. We refer to this as the over-fragmentation of the learned index, which can also result in redundant space overhead.

In Table 1, we present the results of an experiment that demonstrates the varying performance of the learned index under different error ranges. When the error range is large, the learned index no longer provides a significant advantage.

Table 1. Performance evaluation of B+-tree and learned index under different error-bounds.

Error-bound	5.87	15.71	19.52
B+tree	1.67 Mops	1.65 Mops	1.66 Mops
Learned index	3.43 Mops	1.44 Mops	1.32 Mops

In general, learned index can be divided into two categories: error-driven linear segmentation model and count-driven model. Meanwhile, it is found that complex datasets are more likely to affect the performance of error-driven learned index [4,6,7,14,21]. This is because error-driven linear segmentation models typically linearly divide data segments by setting a parameter (error-bound ε) to ensure that the error of all models is below this threshold. However, the count-driven model that cuts data by calculating other metrics can not produce a globally optimal model. Therefore, we exploit the error-driven linear partitioning model as the basis of the proposed hybrid index structure.

Error-driven design aims to use error-bound as the splitting criteria in model building. Consequently, setting the error-bound is critical to the performance of the learned index model. A lower error-bound design can facilitate faster range lookup. However, it also results in more segments in the learned index, which makes the retrieval more complex and time-consuming. At the same time, if the index supports insertion, we need to provide a buffer for each fragment. Obviously, oversplit fragments require the model to provide more buffers to accommodate new inserts. During the retraining phase of executing the model, merging consumes a lot of computational performance, which degrades the performance of the learned index. In this paper, we propose a small-scale retraining mechanism to support insertion operation for the learned index, which reduces the cost of retraining the learned index and improves index efficiency.

In general, this paper makes the following main contributions:

- We model the relationship between error-bound settings and the performance of the data in an error-driven linear segmentation model. Using the performance of traditional tree models as a comparison, we evaluated what type of data would provide superior performance in a linear segmentation model, we call such data learning-friendly data. We build a detection model to automatically classify learning-friendly data and non-learning-friendly data.

- We build a set of linear interpolation models to organize the learning-friendly data, and we use the traditional models for the non-learning-friendly parts of the data. These two models are organized in the superstructure, resulting in a hybrid index structure. We follow the currently popular approach of creating buffers after non-learning-friendly data segments to support insertion and we devise a retraining method for it.
- We conducted experiments on four datasets and obtained excellent experimental findings in space and time, confirming the validity of our work.

The rest of the paper is organized as follows: Sect. 2 describes the related work; Sect. 3 gives the rules for learning-friendly data evaluation; Sect. 4 presents the DMIS; Sect. 5 shows the evaluation results; Sect. 6 concludes this paper.

2   Related Work

Traditional index structures are mainly divided into the tree-based index, the hash-based index, and the log-merge-tree structure [19]. The traditional index structure normally used attempts to speed up data retrieval by continually narrowing the scope of the data search. The main idea of the learned index is to view the index structure as a model where the inputs are keys and the outputs are positions. The current learned index structure aims to solve the problems of insertion support and parallel control in the originally learned index.

Insertion support for the index is key to an index structure that can be used for social production, and recent research has proposed two directions for providing insertions, using buffers and using the structure of the index itself. Xindex [20] and Finedex [14] propose a design that adds buffers between data segments to support the insertion of data. This approach requires two steps of merging and retraining to complete the update operation of the model. This step-by-step update is able to achieve concurrency control using optimistic locking and is therefore considered to be an excellent solution for learned index. Learned index using buffers typically uses an error-driven linear segmentation-based model as the base model, with a smaller error-bound to provide better performance ($\varepsilon = 0$ means that the model is able to query the location of the data exactly once). Using the index's own structure to support insert requires more space to support leaving spaces between data, and the structure will be changed more frequently.

There is also a newer view that the advantages of the learned index can be used in conjunction with traditional index [6,17]. For example, Llaveshi et al. [15] suggested using linear regression models to speed up the search for B+ trees in conventional nodes, and Hadian et al. [11] propose a new index structure, the IFB tree, which improves the performance of B+tree by combining interpolation methods with B+ trees. However, this structure for speeding up the traditional index still uses the B+tree as the base structure and does not take full advantage of the learned index. We use the learned index as the base structure to classify the data into learning-friendly and non-learning-friendly data. A dual-model index structure is proposed using the learned index combined with the underlying structure of the B+tree index.

3 Learning-Friendly Data Evaluation Rules

The existing error-driven learned index method implicitly assumes that the local distribution patterns of the dataset are consistent. Therefore, the same ε value is set to partition the dataset. Setting ε lower can make local data distribution more linear, which improves prediction accuracy. However, this also leads to a decrease in the number of data points covered by each segment, as well as an increase in the total number of segments, and a more complex index, thereby increasing time and space costs. On the contrary, adjusting ε to a relatively large value can reduce the difficulty of building the index and partitioning some non-linear data into one segment. Therefore, increasing the value of ε in error-driven method has spatial advantages but reduces prediction accuracy, which is disadvantageous for the learned index. Obviously, the performance of a learned index is closely related to the data distribution and the error-bound settings, and we want to determine the relationship between the two through segmentation rules.

Given a dataset as $D = \{(x_1, y_1), (x_2, y_2), (x_3, y_3), ..., (x_n, y_n)\}$ with size n, where $x_i \in \mathbf{X}$ denotes the key and $y_i \in \mathbf{Y}$ refers to the value. $\mathbf{X}$ can be segmented under error-bound(ε), meanwhile the set of segments is denoted as $R = \{[x_1, x_2, x_3, ..., x_i], [x_{i+1}, ..., xj], ..., [..., x_{n-1}, x_n]\}$. Here we define each segment as $S_i = a_i x + b_i$, and the size of the $\mathbf{S}$ is m. We denote the learned index total performance by the throughput as F_{learn}. Meanwhile, the time of single calculation and disk read is donated by T_{cal} and T_{read} respectively.

The average error-bound of a segment is the condition that determines the performance of a segment, and we can find data within the error-bound range regardless of the amount of data within the segment. We use a metric to evaluate whether the data is learning-friendly as $F_{learn} = g(D, \varepsilon, T_{cal}, T_{read})$. For the same data set, in the same working environment, only ε is variable. In fact, ε becomes the only variable affecting the metric's variation F_{learn}. Moreover, ε is both a parameter of segmentation and a metric that affects performance. We take a segment S_i and establish a mathematical relationship between ε and F_{learn}.

Figure 1 shows how the base index model finds the position of a key. As shown in Fig. 1(left), if we need to find the key in a queue of keys from a linear model, the base model needs a single computation and a width of error-bound binary search. The upper model also requires a similar operation to the base models. By the way, the minimum key of each base model is also its identifier. While using the same ε to build the upper and base models, we get the computation time for both the upper and base models $T = T_{cal} + (T_{cal} + T_{read}) * \log_2 \varepsilon$.

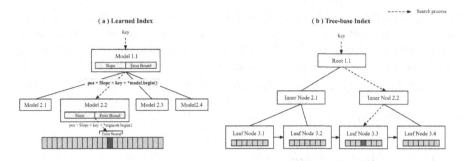

Fig. 1. The working process of a base model in the learned index.

For the same R_i, we use a linear segmentation model that exhibits the performance $F_{learn} = \frac{1}{(T_{cal}+(T_{cal}+T_{read})*\log_2 \varepsilon)*2}$. Similarly, for the model constructed using the dichotomous model, the theoretical performance is $F_{tree} = \frac{1}{(T_{cal}+T_{read})*\log_m n}$, m refers to the number of branches of the B+tree node.

We assume here that the data segment-friendly value function is $V(\varepsilon)$, and based on the known conditions above, we can have the following conclusion.

$$V(\varepsilon) = \frac{1}{(T_{cal} + (T_{cal} + T_{read}) * \log_2 \varepsilon) * 2} - \frac{1}{(T_{cal} + T_{read}) * \log_m n} \qquad (1)$$

We define the performance of segment establishment using traditional index F_{tree}. F_{learn} denotes the performance of an upper model constructed using the learned index. $F_{tree} = F_{learnS_i}$ is used as the performance threshold at which ε for that segment affects the change in F_{learnS_i}. Thus, if we evaluate a data segment formed by $F_{tree} < F_{learn}$, we refer to this segment as learning-friendly data, and while $F_{tree} > F_{learnS_i}$, we refer to this segment as non-learning-friendly data. This threshold does not always exist due to the computational power T_{cal}, the i/o power T_{read}, and the initial data set D.

Table 2. The evaluation criteria.

Range	Define
$V_t > 0$	Learning-friendly
$V_t = 0$	Balance
$V_t < 0$	Non-learning-friendly

To give a numerical indication of how good the data friendliness is, we provide evaluation criteria, as shown in Table 2.

We can define an equilibrium point V_t above the equilibrium point, i.e. in the positive range, indicating that the data is learning-friendly. A negative range indicates that the data is non-learning-friendly. t is the average number of lookups for the segment. $t = 0$ represents the optimal case for the learned index, where only one calculation is needed to arrive at the location, a situation similar to the hash index method. In what follows, we describe our rules for running the model based on the above definitions. Based on the existing evaluation rules (Equation.1 and Table 2), we can also deduce that $n < m^{\left(\frac{2*T_{cal}}{T_{cal}+T_{read}}+\log_2 \varepsilon\right)}$, which means that in some segments we can get better performance without building a learned index model.

4   Hybrid Index Structure

We describe the index model in two parts: index building and index operations. Figure 2 shows the framework of the index. The segmented data is initially organized into two models: the base linear model for the learning-friendly segments and the base tree model for the non-learning-friendly ones. The base linear model incorporates a linear model, along with a minimum key for identification purposes and a maximum error-bound to ensure boundary identification. In the base conventional tree model, a b+tree structure is utilized to store the data, and both the minimum and maximum keys are stored as identifiers. The base tree model offers a free partition to support data insertion, which we refer to as DMIS. Figure 2 demonstrates the process of data retrieval in the different base models, which will be explained in detail in Sect. 4.2.

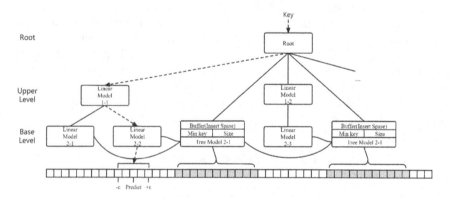

Fig. 2. The framework of the Dual-Model Index Structure (DMIS).

The part of model building is the process of building the index structure from the bottom to up on data segmented according to learning-friendly and non-learning-friendly data.

Index operations are required for persistence in order to keep the model effectively available. Update and rebuilding are the two main operations that change the structure of the model therefore they need to be optimized.

4.1   Index Building

Algorithm 1: Control Methods

---

input : A data stream D of size n
output: Segments sorted by control methods

1 $s \leftarrow 0$;
2 **for** $i \leftarrow 0$ **to** $n - 1$ **do**
3     $j \leftarrow i$;
4   **while** $j < n$ **do**
5   **if** Methods$(D\,[i,j])$ *is false* **then**
6   **if** $j - i > minVolume$ **then**
7              $Segment\,[s] \leftarrow$ NewSegment$(D\,[i, j - 1]\,, LearningFriendly)$;
8   **else**
9   **if** $Segment\,[s - 1]$ *is unLearningFriendly* **then**
10                $s \leftarrow s - 1$;
11             $i \leftarrow i - Segment\,[s]\,.size()$;
12             $Segment[s] \leftarrow$ NewSegment$(D\,[i, j - 1]\,, unLearningFriendly)$;
13          $s \leftarrow s + 1$;
14          $i \leftarrow j$;
15   *break* ;
16       $j \leftarrow j + 1$;
17   **if** $j = n$ **then**
18   **if** $n - i > minVolume$ **then**
19          $Segment\,[s] \leftarrow$ NewSegment$(D\,[i, j - 1]\,, LearningFriendly)$;
20   *break* ;
21   **else**
22   **if** $Segment\,[s - 1]$ *is unLearningFriendly* **then**
23             $s \leftarrow s - 1$;
24          $i \leftarrow i - Segment\,[s]\,.size()$;
25          $Segment[s] \leftarrow$ NewSegment$(D\,[i, n - 1]\,, unLearningFriendly)$;
26   *break* ;
27   **foreach** *element e of the line i* **do** Methods(e)

In order to build the DMIS, it is necessary to segment the data for the base model. Here it is possible to obtain a theoretically optimal classification result based on the analysis in the previous section. However, reproducing the algorithmic process can be difficult and overly complex, which can negatively impact the index cycle. Therefore, to achieve our goal of labeling learning-friendly data at this stage, we have filtered out some of the more complex algorithms and instead focused on simpler approaches. In the experimental section, we discuss five methods in detail.

At the end of the previous section, we mentioned that the amount of data is the criterion for evaluating whether the segments are conducive to learning. And the average error-bound constraint on data segments is also a criterion for evaluating which algorithm is better.

We divided the data segments into two parts, learning-friendly and non-learning-friendly data. After that, we will describe how to organize the segmented data in the models. Algorithm 1 shows the classification methods' work process.

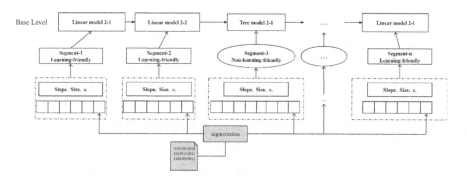

Fig. 3. Data classification and base level models construction process.

Figure 3 shows how we built the base model through segmentation. We use algorithms to segment all data and assign data volume, error, and linear model slope to each segment. We confirm whether the segment is learning-friendly by evaluating the segment model. For learning-friendly segments, we build linear models and assign parameters directly to the model. For non-learning-friendly segments, a tree model is established.

The upper model is constructed entirely using the linear model, following the idea of Recursive-Model Indexes (RMI) [12]. We collect the minimum identifiers from the base model as a new set of linear models, so that we can find the data location from the root node by looking twice through the upper and base models.

4.2   Index Operation

When the model has been initialized it will be depicted as the structure as shown in Fig. 3. If a lookup is carried out directly, we will get to the segment where the data is based on the linear prediction of the upper layer, where the previously designed identifier (the min key in the segment) is used. The lookup result of the upper model is a base model. For the upper model, we use a sliding window

---

Algorithm 2: Insert and Rebuild

---

input : A data x need to insert
output: A series of new base models

1 $posPtr \leftarrow$ Search(x);
2 **if** $posPtr$ *is null* **then** // Not exist and insert
3 | $baseModel \leftarrow$ SearchTreeBase(x) ;
4 | **if** Capacity$(baseModel) > treeCapacity$ **then** // Need to rebuild
5 | | $minModelptr \leftarrow$ SearchBase$(baseModel.minKey)$;
6 | | BuildBase$(minModelptr, baseModel)$;

7 UpperRebuild$(minModelptr, baseModel)$;

---

method to control the error. For the tree model results, we use the existing lookup methods in tree models.

The model needs to support data updates during the runtime. We divide the running process of the model into two stages: insertion and retraining.

To keep new insert data available for storage, we use an insert space to store the data. When there is data to be inserted, we first visit the location where this data is located, if it is in one of the base models, it means that this data already exists in our model and it is illegal to continue inserting. If the accessed data is not found, the model will be sequentially traversed to the next base tree model node. Then, the node will be placed in the insert space of that model. Finally, the minimum key stored in the model for this data will be updated based on this new insert data.

For the lookup after insertion, the base model already has data stored, so we need to ensure that all the data is looked up. Hence we use a similar strategy to Xindex. In addition, we continue to look up the next tree model's insert space if the key is not found, so we store a pointer to the adjacent partition in the base linear model to achieve a fast jump.

For the retraining, we always check the inserted data in the base tree model after insertion. If the amount of data is larger than the original amount of data in the model, the process of retraining will be triggered. We specify all models from the minimum key should be the linear model to the current tree model as the set of models to be retrained. We provide a space for this segment of the model to temporarily support the newly inserted data while building this segment of data following the process of building the base model. Once constructed, if the last few segments are base linear models, we link them to the inserted partitions in the next base conventional model. And we let the data in the temporary space continue to complete the insertion operation. The whole process is summarized in Algorithm 2.

5   Evaluation

5.1   Setup

In the paper, we implement DMIS in C++. Experiments are conducted with a single thread on a 2.1 GHz Intel Xeon Linux machine with 64 GB memory. We compare DMIS against four baselines: Masstree [16], Xindex [20], Learned index [12], and Finedex [14].

5.2   Datasets

Four datasets were used for the performance evaluation. The longitudes dataset consists of uses from Open Street Maps [2] for locations around the world. The longlat dataset consists of composite keys that combine longitude and latitude from Open Street Maps by applying the transformation $k = 180 \cdot floor(longitude) + latitude$ to each pair of longitude and latitude, and the resulting key k distribution is highly non-learning-friendly. The log-normal dataset is generated from a log-normal distribution with $\mu = 0$ and $\sigma = 2$, multiplied by 1e9 and rounded down to the nearest integer. The YCSB dataset is composed of user IDs generated according to the YCSB benchmark [3].

In Fig. 4, we plot the distribution of these four datasets using the Cumulative Distribution Function(CDF).

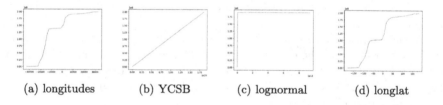

(a) longitudes   (b) YCSB   (c) lognormal   (d) longlat

Fig. 4. The CDF of datasets.

5.3   Design

We need to classify the data into two categories, and theoretically, all binary classification algorithms can be selected. Based on the practicality and applicability, we have selected the following five algorithms.

Top-Down (TD) [10]. This algorithm segments at the maximum error-bound position and tests whether the error-bound of the sub-segments is below the threshold. The algorithm will continue to recursively segment until the error-bounds of all segments are below this threshold.

Sliding Windows (SW) [8]. The left point is set at the first data point, and data is continuously added as the right point. If the error exceeds the threshold, the sub-segments between the two points are converted into a segment.

Piecewise Aggregate Approximation (PAA) [13]. This approach divides data into R segments and calculates the window length L by dividing the amount of data by the number of keys. Starting from the first point of the sequence, the first segment is formed by selecting the first $L - 1$ points.

Manhattan Distance Control (MDC). The segmentation of the data is based on the Manhattan distance formula, with a criterion of $|y_{i+1} - y_i| - |x_{i+1} - x_i| > k$. Here, x_i and y_i are the horizontal and vertical coordinates, respectively, of the i-th data point, and k is a threshold that controls the granularity of segmentation. If the formula is not satisfied, a new segment will be created.

Error-based Evaluation (EE) [17]. In this method, if the slope change of the data point is greater than forty percent, it will perform a new segmentation for the subsequent data points.

Table 3. Evaluation of classification algorithms.

Dataset	Method	Time(s)	Error-Bound	Total Segment	Non-learning-friendly
longitudes-100M	PAA	2.21789	29.4989	100000	0
	SW	62.4487	10.1354	698051	3545
	MDC	4.19642	1605.92	109157	2133
	EE	2.37225	1.46325e+07	4	1
	TD	17.6693	12.4312	234205	38521
lognormal-100M	PAA	2.21218	87.1038	100000	0
	SW	47.4864	10.3322	1022403	12243
	MDC	5.51938	12.2923	1953937	5652
	EE	3.6885	8.68162e+06	9	2
	TD	19.3573	11.3213	471614	8735

We used the five attractive algorithms above to perform segmentation on two representative datasets. We evaluated the time cost of the algorithms, as well as the segment average error, and the number of segments.

As shown in Table 3, the PAA algorithm has a low time overhead, but its classification results are heavily dependent on the dataset. Furthermore, the information provided by the algorithm varies significantly across different datasets. Additionally, this algorithm does not allow for defining a precise segment size in advance. Therefore, it is not suitable for our classification needs. The SW algorithm has good classification results but it is difficult to accept because of its long running time and high complexity, especially on the second dataset due to the complexity of the dataset itself, resulting in a large number of segments. MDC is also not considered for the time being due to its unstable performance.

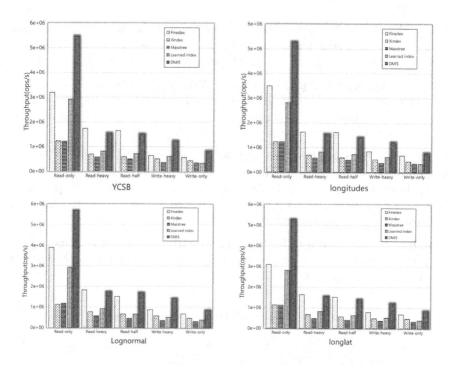

Fig. 5. Throughput evaluation under different workloads.

Because the EE algorithm is too simple, the classification result is poor. The TD algorithm shows stable results on different datasets while producing fewer segments. Based on the evaluation, we can get that the TD algorithm is the optimal algorithm to use as a learning-friendly data classification.

5.4   Result

As shown in Fig. 5, the dual-structure model can effectively improve the efficiency of the learned index. At the same time, DMIS shows relatively balanced performance on different datasets. Among them, the performance on the Read-heavy workloads is lower than other learned models. After insertion, in order to ensure the position of the key, the dual structure mode design is adopted, which reduces some performance. But in other workloads, the dual-model structure effectively improves the performance of the learned models, while ensuring that all index structures have more suitable model combinations, thereby improving the overall performance. Our structure also shows high performance under static workloads and stable efficiency under heavy insertion workloads.

Because our model is simple and efficient enough, it also has a significant advantage in construction time over complex structures, as shown in Table 4. To build a linear model faster, we use the method of linear interpolation. Since we

Table 4. Evaluation of index build time.

Index Methods	DMIS	Masstree	Finedex	Xindex
Build time(s)	27.12	32.41	39.76	60.33

limit the margin of error when segmenting, we ensure that the entire model is valid.

We perform statistics on the average error-bound of the learned models built on the longitude dataset. The training samples are the first 100 million pieces of data. In Fig. 6, results show that the classification algorithm we introduced leads to a significant reduction in the error-bound of the model, resulting in a significant increase in search efficiency. It is proved that TD is a useful tool for reducing model error-bound and improving index performance.

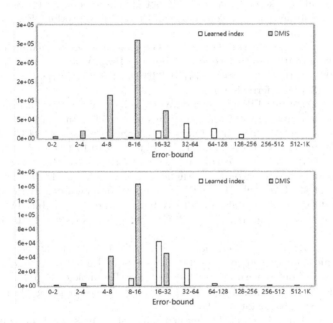

Fig. 6. Error-bounds under different sizes of segments.

6   Conclusion

We propose a novel index structure, namely DMIS, which combines traditional and learned indexes for efficient query processing with high accuracy and low error rates. To mitigate the performance degradation of learned indexes due to changes in data distribution or query patterns, we introduce a small-scale retraining mechanism for DMIS. Our experimental results demonstrate our proposal improves the efficiency and scalability of index structures in database systems. The retraining process ensures DMIS maintains long-term efficiency, resulting in efficient query processing with high accuracy and low error rates.

References

1. Alexiou, K., Kossmann, D., Larson, P.Å.: Adaptive range filters for cold data: avoiding trips to Siberia. Proc. VLDB Endow. **6**(14), 1714–1725 (2013)
2. Bennett, J.: OpenStreetMap. Packt Publishing Ltd. (2010)
3. Cooper, B.F., et al.: Benchmarking cloud serving systems with YCSB. In: Proceedings of the 1st ACM Symposium on Cloud Computing, pp. 143–154 (2010)
4. Ding, J., et al.: ALEX: an updatable adaptive learned index. In: Proceedings of the 2020 International Conference on Management of Data, SIGMOD Conference 2020, Online Conference, Portland, OR, USA, 14–19 June 2020, pp. 969–984. ACM (2020). https://doi.org/10.1145/3318464.3389711
5. Fan, B., et al.: Cuckoo filter: practically better than bloom. In: Proceedings of the 10th ACM International on Conference on Emerging Networking Experiments and Technologies, pp. 75–88 (2014)
6. Ferragina, P., Vinciguerra, G.: The PGM-index: a fully-dynamic compressed learned index with provable worst-case bounds. Proc. VLDB Endow. **13**(8), 1162–1175 (2020). https://doi.org/10.14778/3389133.3389135, https://www.vldb.org/pvldb/vol13/p1162-ferragina.pdf
7. Galakatos, A., et al.: Fiting-tree: a data-aware index structure. In: Proceedings of the 2019 International Conference on Management of Data, pp. 1189–1206 (2019)
8. Graefe, G.: Query evaluation techniques for large databases. ACM Comput. Surv. **25**(2), 73–169 (1993). https://doi.org/10.1145/152610.152611. ISSN 0360-0300
9. Graefe, G., Larson, P.-A.: B-tree indexes and CPU caches. In: Proceedings 17th International Conference on Data Engineering, pp. 349–358. IEEE (2001)
10. Gray, J., et al.: Data cube: a relational aggregation operator generalizing group-by, cross-tab, and sub-totals. In: Proceedings of the Twelfth International Conference on Data Engineering, pp. 152–159 (1996). https://doi.org/10.1109/ICDE.1996.492099
11. Hadian, A., Heinis, T.: Interpolation-friendly B-trees: bridging the gap between algorithmic and learned indexes. In: Advances in Database Technology - 22nd International Conference on Extending Database Technology, EDBT 2019, Lisbon, Portugal, 26–29 March 2019, pp. 710–713. OpenProceedings.org (2019). https://doi.org/10.5441/002/edbt.2019.93
12. Kraska, T., et al.: The case for learned index structures. In: Proceedings of the 2018 international conference on management of data, pp. 489–504 (2018)
13. Lambert, D., Pinheiro, J.C.: Mining a stream of transactions for customer patterns. In: Proceedings of the Seventh ACM SIGKDD International Conference on Knowledge Discovery and Data Mining, KDD 2001, New York, NY, USA, pp. 305–310. Association for Computing Machinery (2001). https://doi.org/10.1145/502512.502556. ISBN 158113391X
14. Li, P., et al.: FINEdex: a fine-grained learned index scheme for scalable and concurrent memory systems. Proc. VLDB Endow. **15**(2), 321–334 (2021). https://doi.org/10.14778/3489496.3489512, https://www.vldb.org/pvldb/vol15/p321-hua.pdf
15. Llaveshi, A., et al.: Accelerating B+ tree search by using simple machine learning techniques. In: Proceedings of the 1st International Workshop on Applied AI for Database Systems and Applications (2019)

16. Mao, Y., Kohler, E., Morris, R.T.: Cache craftiness for fast multicore key-value storage. In: Proceedings of the 7th ACM European Conference on Computer Systems, pp. 183–196 (2012)
17. Qu, W., Wang, X., Li, J., Li, X.: Hybrid indexes by exploring traditional B-tree and linear regression. In: Ni, W., Wang, X., Song, W., Li, Y. (eds.) WISA 2019. LNCS, vol. 11817, pp. 601–613. Springer, Cham (2019). https://doi.org/10.1007/978-3-030-30952-7_61
18. Richter, S., Alvarez, V., Dittrich, J.: A seven-dimensional analysis of hashing methods and its implications on query processing. PVLDB 9(3), 96–107 (2015)
19. Schraudolph, N.: Accelerated gradient descent by factor-centering decomposition. Technical report/IDSIA, 98 (1998)
20. Tang, C., et al.: XIndex: a scalable learned index for multicore data storage. In: Proceedings of the 25th ACM SIGPLAN Symposium on Principles and Practice of Parallel Programming, pp. 308–320 (2020)
21. Wu, J., et al.: Updatable learned index with precise positions. arXiv preprint arXiv:2104.05520 (2021)

Streaming Data Analytics for Feature Importance Measures in Concept Drift Detection and Adaptation

Ali Alizadeh Mansouri$^{(\boxtimes)}$ ⓘ, Abbas Javadtalab ⓘ, and Nematollaah Shiri

Department of Computer Science and Software Engineering, Concordia University,
Montreal, Canada
{aa.mansouri,abbas.javadtalab,nemat.shiri}@concordia.ca

Abstract. Numerous applications require the ability to detect and adapt to concept drifts in streaming data on the fly. This is challenged by limited computational resources and access to archival storage. In this paper, we study features that capture the evolving relationship between raw data features and target labels, and techniques to extract those features. In particular, we focus on the relationship between feature importance measures in streaming data and predictability performance of the main classifier. For this, we consider two groups of feature importance measures: impurity-based and permutation-based, both of which are computed over an auxiliary online gradient boosted decision trees ensemble that runs in parallel to the main classifier in processing the same data stream. We found strong evidence that feature importance measures follow the long-term trend of the performance metrics even if the data streams are non-stationary or deviate from the performance metrics in short-term. Our study also shows that classification models that process data with constant or monotonic rate of drift, are robust in terms of stationary nature of feature importance measures and learner's predictability performance. Moreover, we found evidence for more consistency and reliability of permutation feature importance measurements over impurity-based ones if data exhibits periodic or non-monotonic rates of drift, or if this knowledge is not known a priori. Our study and results indicate that the feature importance measures considered are viable sources of information for concept drift detection and adaptation problems. This has been established through a solution to these problems we developed based on vector error-correction analysis.

Keywords: Concept drift detection · Data stream analysis · Vector error-correction

1 Introduction

With continuous advances in computing technologies, mining techniques are gaining increased popularity for discovering hidden patterns in unbounded data

This work was partially supported by Concordia University.

C. Strauss et al. (Eds.): DEXA 2023, LNCS 14146, pp. 114–128, 2023.
https://doi.org/10.1007/978-3-031-39847-6_8

arriving at fast speeds for extended periods of time. Emerging streaming data analytics must detect pattern changes in the distribution of one or more variables that may affect applications' performance and adapt accordingly when such changes occur. This is referred to as *concept drift* [13, 22], which may result in deterioration of the learner's inference or prediction performance. Examples of real-life applications showing this behavior include healthcare, industrial sensor grids [30], environmental [7], smart cities and homes, [24], network infrastructure monitoring [27], business, e-commerce and insurance [1], and finance [15].

A major challenge in tackling the concept drift detection (CDD) problem is selection and analysis of the available detection information. There have chronologically been three major categories of techniques based on the kind of *information used* for CDD tasks. Early data feature change detection methods, such as [8], analyzed only independent features in raw data disregarding target labels y. These techniques are simple and light-weight to perform but are likely to overlook the association between the predictor X and response y variables. Next generation of CDD solutions, such as [25], have relied on the learner's predictability performance evaluation and feedback. They are more accurate than the raw data-only detectors, but still do not consider possible correlations between the features and response variables. Both of these categories of techniques suffer from using the same learner for the main classification and the CDD tasks. This potentially limits their effectiveness because the learner that gets adapted to CDD might perform poorly on its main classification task, and vice versa.

More recent CDD techniques use auxiliary models that run in parallel to the learner. These models have the advantages of being light-weight, decoupled, designed for the CDD task, and satisfy the stream processing requirements. Furthermore, they can analyze the evolving correlation of features and target labels more prominently than previous techniques. For example, Yang et al. [34] use an online sequential extreme learning machine (ELM) model [21] for CDD. In [2], we proposed Ensemble Learning Augmented Drift Detection (EnLAuDD), which employs an ensemble of a collection of simple (cheap) classifiers to model concept as a one-dimensional aggregation of changes in base classifiers. As noted in [31], these techniques suffer from a major shortcoming of ELM models, namely having high variance in each ELM model whose hidden layer does not reveal much information beneficial to the CDD tasks.

To alleviate the situation, we raise the following two questions:

1. What would be a slightly more expensive but yet affordable model than ELM to decrease the variance?
2. Would such a model provide better detection information beneficial to concept drift detection and adaptation?

The first question addresses data stream processing application's requirements. For this, we consider gradient boosted decision trees (GBDT) as suitable substituting models [11] than ELM to lower the bias and achieve higher accuracy.

The second question is concerned about effectiveness of a viable solution to the concept drift detection and adaptation (CDDA) problem, for which we consider feature importance measures (FIMs). Considering an ensemble of classification or regression trees trained on a stationary dataset, distribution of features whose scores are calculated and selected as split nodes do not change over time. However, few studies in the literature as well as our preliminary investigations showed that the importance of features evolves over time as data undergoes CD. Study of FIMs has been a subject of interest in offline machine learning, but not investigated thoroughly in data stream processing applications, particularly those undergoing CD. Therefore, these FIMs are a worthwhile source of information to learn, detect, or even predict CD in streaming data. Breiman [4,5] formalized impurity-based feature importance measurements for random forests. Impurity-based importance measurements are prone to high variance as they are calculated on training data only, and miscalculate on continuous and high-cardinality features. As a model-agnostic and more robust alternative, we also consider permutation feature importance measurements. Permutation importance measurement is the decrease in a model's performance when a single feature value is randomly shuffled [4].

Wang et al. [32] demonstrated the effectiveness of GBDT for concept drift adaptation (CDA). Barddal et al. [3] performed *feature selection* by training an Ada boosting ensemble of Hoeffding (online) tree stumps, but its performance can suffer from high bias. Cassidy and Deviney [6] concluded two online feature scoring metrics they applied to an ensemble of online random forests follow virtual CD. Gomes et al. [14] study CDD using two impurity-based feature scores from an incremental random forest and an ensemble of Hoeffding adaptive trees. However, both measures suffer due to the limitations of impurity-based scoring technique [33].

In this paper, we consider the constraints and requirements of data stream processing applications and study the relationship between FIMs analyzed from streaming data that exhibit different characteristics of CD and the predictability performance metric of the main classifier. The two groups of FIMs—impurity-based importance measurements and permutation importance measurements—are computed over an auxiliary GBDT ensemble model that runs in parallel to the main classifier but processes and analyzes the same streaming data. As such, the two models used are decoupled: the main classifier, whose remodeling can be potentially costly, has the task of processing streaming data with the goal of prediction on test instances. The auxiliary GBDT ensemble model is assigned the task of processing the same streaming data with the goal of CDD and possibly adaptation of the main classifier to changes in data. We specifically study the correlation of detection information, i.e., the two types of FIMs extracted from the auxiliary GBDT ensemble, with the performance of the main classifier. Therefore, no detection or adaptation are performed so that changes in FIMs and the main classifier in the face of CD can be monitored and analyzed with no intervention.

The main outcome of this study is providing evidence for strong correlation between FIMs computed from a decoupled, cost-effective model with the performance of a more accurate but costly model over time which acknowledges data

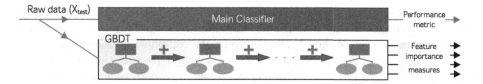

Fig. 1. Architecture of DSMS, the proposed data stream processing, to study the relationship of the main classifier's performance metrics with feature importance measures computed from a GBDT used as an auxiliary model.

stream processing requirements and encounters different types and rates of CD. Establishing the aforementioned correlation further provides:

- strong evidence to employ FIMs as a viable source of detection information for CDDA applications, that is, detection of and adaptation to changes reactively,
- better understanding of the behavior of CD in the underlying streaming data and processes, and,
- a way to investigate prediction of CD, that is, detection of and adaptation to changes proactively.

The rest of this paper is organized as follows. Technical background and the proposed methodology and statistical analysis are provided in Sect. 2. In Sect. 3, we present the results of our numerous experiments, results, followed by observations and findings of the aforementioned statistical analysis. The final section includes concluding remarks and directions for future work.

2   Methodology

We propose the following methodology to study the relationship between the main classifier's performance and the FIMs.

2.1   Variables

CD is defined as a change in the joint probability distribution of the dependent variable y and feature vector $\mathbf{x}$ between two points in time, as expressed in Eq. 1 [13,19,22].

$$p_{t_{i+1}}(y_j, \mathbf{x}) \neq p_{t_i}(y_j, \mathbf{x}) \tag{1}$$

We deploy a data stream management system (DSMS) where batches of streaming data that might exhibiting CD, denoted as $B^t = \{< X^t_{N \times D}, y^t_N >\}$, for $t \geq 1$, are provided as test data to the initially trained main classifier and GBDT model simultaneously, where D is the dimensionality of feature vector $\mathbf{x} \in X$. The two models are decoupled and do not interact with each other throughout the run of the streams. Figure 1 presents the architecture of our proposed solution DSMS for data stream processing.

The accuracy and F1 scores of the main classifier at each time step t are collected as predictability performance metrics, and eventually modeled as two univariate time series $\mathcal{A}_t$ and $\mathcal{B}_t$, respectively. We compute and model impurity-based feature importance measurements as $\mathcal{G}_{d,t}$ and permutation importance measurements $\mathcal{H}_{d,t}$ for each dimension $d \in D$ as univariate time series models. We represent impurity-based importance measurements of feature vector $\mathbf{x}$ as a multivariate time series $\mathcal{G}_t = \{\mathcal{G}_{d,t}\}$, for $1 \leq d \leq D$. Likewise, we denote permutation importance measurements as a multivariate time series $\mathcal{H}_t = \{\mathcal{H}_{d,t}\}$, for $1 \leq d \leq D$. Lastly, we form four multivariate time series models out of the two importance measurements types and the two predictability performance metrics, as denoted in Eq. 2.

$$\begin{aligned}
\mathcal{P}_t &= \{\mathcal{G}_t, \mathcal{A}_t\} \\
\mathcal{Q}_t &= \{\mathcal{G}_t, \mathcal{B}_t\} \\
\mathcal{R}_t &= \{\mathcal{H}_t, \mathcal{A}_t\} \\
\mathcal{S}_t &= \{\mathcal{H}_t, \mathcal{B}_t\}
\end{aligned} \tag{2}$$

2.2   Hypotheses

Based on the research questions presented in Sect. 1, we hypothesize that there exists meaningful relationship between FIMs computed from an auxiliary model and a main classifier's predictability performance as it deteriorates while undergoing CD. Specifically, we consider the following:

- an online, incremental GBDT as the auxiliary model
- a main classification model that is at least as costly as the auxiliary model in terms of computational resources and data stream processing requirements,
- an impurity-based feature importance measurement $\mathcal{G}_d$, which is the (normalized) total least squares improvement contributed by $\mathbf{x}_d$, as introduced in [11]
- a permutation-based feature importance measurement $\mathcal{H}_d$, which is the change in misclassification after noising feature $\mathbf{x}_d$ of test samples by random permutation [4,5]
- the main classifier's accuracy and F1 score as the predictability performance metrics, denoted as $\mathcal{A}_t$ and $\mathcal{B}_t$, respectively.

The null and alternative hypotheses for each type of FIMs and prediction performance metrics are stated in null hypotheses 1.–4. and hypotheses 1.–4..

Null Hypothesis (H_0) 1. *There is no relationship between impurity-based importance measurements computed on an online, incremental GBDT model and the main classifier's accuracy over time while each model analyzes streaming data exhibiting CD separately.*

Hypothesis (H_1) 1. *There exists statistically significant relationship between impurity-based importance measurements computed on an online, incremental GBDT model and the main classifier's accuracy over time while each model analyzes streaming data exhibiting CD separately.*

Null Hypothesis (H_0) **2.** *There is no relationship between permutation importance measurements computed an on online, incremental* GBDT *model and the main classifier's accuracy over time while each model analyzes streaming data exhibiting* CD *separately.*

Hypothesis (H_1) **2.** *There exists statistically significant relationship between permutation importance measurements computed on an online, incremental* GBDT*model and the main classifier's accuracy over time while each model analyzes streaming data exhibiting* CD *separately.*

Null Hypothesis (H_0) **3.** *There is no relationship between impurity-based importance measurements computed on an online, incremental* GBDT *model and the main classifier's F1 score over time while each model analyzes streaming data exhibiting* CD *separately.*

Hypothesis (H_1) **3.** *There exists statistically significant relationship between impurity-based importance measurements computed on an online, incremental* GBDT *model and the main classifier's F1 score over time while each model analyzes streaming data exhibiting* CD *separately.*

Null Hypothesis (H_0) **4.** *There is no relationship between permutation importance measurements computed an on online, incremental* GBDT *model and the main classifier's F1 score over time while each model analyzes streaming data exhibiting* CD *separately.*

Hypothesis (H_1) **4.** *There exists statistically significant relationship between permutation importance measurements computed on an online, incremental* GBDT*model and the main classifier's F1 score over time while each model analyzes streaming data exhibiting* CD *separately.*

2.3   Statistical Methods

The goal of this study in particular is to find out if the main classifier's predictability performance metrics as time series models share a common long-term stochastic drift with FIMs computed from a GBDT used as an auxiliary model as individual time series models constructed over the course of the stream. To this end, we have adopted multivariable regression analysis as the main statistical method in econometrics to establish relationships among the aforementioned time series models. The motivation for this choice follows.

Standard correlation statistics such as Pearson correlation coefficient (r), rank correlation coefficients such as Spearman's ρ and Kendall's τ, and Granger causality test can mislead to spurious relationships when data is non-stationary [17,23]. Moreover, if long-term information of the shared stochastic drift between studied variables (FIMs and performance metrics) appears in the levels of data, standard statistical practices such as vector autoregression (VAR) analysis become invalid as the common long-term information gets lost when differenced [26]. Engle and Granger [10] proposed to consider the presence of cointegration when testing for relationships between time series variables that are integrated of

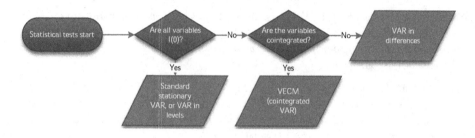

Fig. 2. Steps of the proposed multivariable regression analysis.

at least order one $I(1)$, which means non-stationary time series variables must be differenced at least once to become stationary. If two or more time series variables share a common stochastic trend and a linear combination of them is a stationary time series or one with a lower common order of integration, they are considered cointegrated. The cointegrating relationships among the variables can thus be modeled as a vector error-correction (VEC) model.

Since we deal with streaming data that exhibits CD, it is likely that any time series information produced by analyzing non-stationary streaming data is non-stationary per se as well. Therefore, we study the cointegration of the time series variables described earlier to avoid incorrect acceptance of spurious results. We adopt the Johansen method [18] for our study because it allows for multiple simultaneously cointegrating variables, requires no pretesting, and provides error correction features on the resulting VEC model.

3   Experiments, Results, and Analyses

3.1   Experimental Setup

The steps of our procedure for multivariable regression analysis are illustrated in Fig. 2. We perform this procedure on the four time series models of Eq. 2, using data gathered by running streams of all datasets in Sect. 3.2 with no CDD technique applied in order to analyze the behavior of the main classifier in the face of CD in the long run.

We start by testing each univariate time series model in $\mathcal{P}_t$, $\mathcal{Q}_t$, $\mathcal{R}_t$, and $\mathcal{S}_t$ for stationarity using Augmented Dickey–Fuller (ADF) tests. The null hypothesis of the ADF test is non-stationarity, and the alternative hypothesis is stationarity.

If we can reject the null, we can determine that all univariate variables in each of the multivariate series are stationary. This in turn enables us to form a VAR in levels from the FIMs and the performance metric of that multivariate time series model.

Otherwise, we use the Johansen method [18] to test if all univariate variables in each of the multivariate series are conintegrated. If the tests determine that the impact matrix C of the resulting VEC model has any rank $r > 0$, we conclude that there is statistically significant cointegration between FIMs of that multivariate series and the predictability metric with at least one cointegrating relation (a stationary linear combination) between them. However, if the tests determine that the impact matrix C of the resulting VEC model has rank $r = 0$, the error-correction term disappears, and we can form a VAR in differences out of the FIMs and the performance metric of that multivariate time series model.

3.2 Datasets

In our experiments, We used several synthetic and real-world datasets for a more thorough analysis and understanding of the relationship between FIMs and the main classifier's long-time performance using data with different characteristics. We have conducted experiments on the following synthetic datasets:

- *Rotating Checkerboard (RCB)* [20]. We have used the parameters in [9], and considered four rates of CD as constant (RCB-C), pulse (RCB-P), exponential (RCB-E), and sinusoidal (RCB-S) each with a batch size of 400 instances for a total of 1024 batches.
- *Streaming Ensemble Algorithm (SEA)* [28]. It has three continuous features, two of which affect the decision boundary while the third one is noise. We used the threshold values of θ considered in [9,28] for SEA1 and SEA2, and used $\theta = 8.0, 9.0, 7.5, 9.0$ for SEA3. This threshold changes three times suddenly throughout the dataset, resulting in three abrupt drifts. Each dataset consists of 200 batches of streams of size 250 instances each.

We also performed regression analysis on the following two real-world datasets, both of which exhibit gradual periodic drifts.

- *Bellevue NOAA weather dataset* [29]. This dataset consists of eight features as daily weather measurements, and two classes ("rain" and "no rain"). It has 605 batches, each batch containing 30 instances, with the first 36 batches used for training.
- *Electricity dataset (ELEC)* [12,16]. This dataset consists of five features affecting the change of electricity price and two classes ("up" and "down"). It contains a total of 944 batches with a batch size of 48 with the first 56 batches used for training.

3.3 Experimental Results

The results of the ADF tests are displayed in Table 1. The feature importance is extracted and tested for staionarity for each feature of synthetic and real-world datasets. Since importance measurements series $\mathcal{G}_t$ and $\mathcal{H}_t$ are multivariate, we rejected the null of ADF test for these series only if it could be rejected for all individual importance measurements series comprising $\mathcal{G}_t$ or $\mathcal{H}_t$. The highest significance level this could be achieved is noted in Table 1. Overall, the ADF stationarity test results indicate the necessity to test for cointegration of multivariate series $\mathcal{P}_t$, $\mathcal{Q}_t$, $\mathcal{R}_t$, and $\mathcal{S}_t$ on the next step except for the RCB-C dataset. We

Table 1. ADF stationarity test results.

Dataset	Series	ADF Test Result	Significance	Conclusion
RCB-C	$\mathcal{A}_t$	Stationary (I(0))	99%	Standard stationary VAR
	$\mathcal{B}_t$	Stationary (I(0))	99%	Standard stationary VAR
	$\mathcal{G}_t$	Stationary (I(0))	95%	
	$\mathcal{H}_t$	Stationary (I(0))	99%	
RCB-P	$\mathcal{A}_t$	Nonstationary	–	Test for cointegration
	$\mathcal{B}_t$	Nonstationary	–	Test for cointegration
	$\mathcal{G}_t$	Nonstationary	–	
	$\mathcal{H}_t$	Stationary (I(0))	95%	
RCB-E	$\mathcal{A}_t$	Stationary (I(0))	99%	Test for cointegration
	$\mathcal{B}_t$	Stationary (I(0))	99%	Test for cointegration
	$\mathcal{G}_t$	Stationary (I(0))	95%	
	$\mathcal{H}_t$	Inconclusive	–	
RCB-S	$\mathcal{A}_t$	Nonstationary	–	Test for cointegration
	$\mathcal{B}_t$	Nonstationary	–	Test for cointegration
	$\mathcal{G}_t$	Nonstationary	–	
	$\mathcal{H}_t$	Stationary (I(0))	99%	
NOAA	$\mathcal{A}_t$	Stationary (I(0))	99%	Test for cointegration
	$\mathcal{B}_t$	Stationary (I(0))	99%	Test for cointegration
	$\mathcal{G}_t$	Inconclusive	–	
	$\mathcal{H}_t$	Stationary (I(0))	90%	
ELEC	$\mathcal{A}_t$	Stationary (I(0))	95%	Test for cointegration
	$\mathcal{B}_t$	Nonstationary	–	Test for cointegration
	$\mathcal{G}_t$	Nonstationary	–	
	$\mathcal{H}_t$	Inconclusive	–	
SEA-1	$\mathcal{A}_t$	Nonstationary	–	Test for cointegration
	$\mathcal{B}_t$	Nonstationary	–	Test for cointegration
	$\mathcal{G}_t$	Nonstationary	–	
	$\mathcal{H}_t$	Inconclusive	–	
SEA-2	$\mathcal{A}_t$	Nonstationary	–	Test for cointegration
	$\mathcal{B}_t$	Nonstationary	–	Test for cointegration
	$\mathcal{G}_t$	Nonstationary	–	
	$\mathcal{H}_t$	Inconclusive	–	
SEA-3	$\mathcal{A}_t$	Nonstationary	–	Test for cointegration
	$\mathcal{B}_t$	Nonstationary	–	Test for cointegration
	$\mathcal{G}_t$	Nonstationary	–	
	$\mathcal{H}_t$	Inconclusive	–	

Table 2. The Johansen method test results for cointegration of feature importances with accuracy. EIG and TRC refer to the eigenvalue statistic and trace statistic in the test, respectively.

Dataset	Series	EIG	Sig. lvl.	TRC	Sig. lvl.	Conclusion
RCB-C	$\mathcal{P}_t$	$r(3)$—full rank	99%	$r(3)$—full rank	99%	VEC
	$\mathcal{Q}_t$	$r(3)$—full rank	99%	$r(3)$—full rank	99%	VEC
RCB-P	$\mathcal{P}_t$	$r(1)$	99%	$r(1)$	99%	VEC
	$\mathcal{Q}_t$	$r(3)$—full rank	99%	$r(3)$—full rank	99%	VEC
RCB-E	$\mathcal{P}_t$	$r(3)$—full rank	99%	$r(3)$—full rank	99%	VEC
	$\mathcal{Q}_t$	$r(3)$—full rank	99%	$r(3)$—full rank	99%	VEC
RCB-S	$\mathcal{P}_t$	$r(1)$	99%	$r(1)$	99%	VEC
	$\mathcal{Q}_t$	$r(3)$—full rank	99%	$r(3)$—full rank	99%	VEC
NOAA	$\mathcal{P}_t$	$r(2)$	99%	$r(4)$	99%	VEC
	$\mathcal{Q}_t$	$r(9)$—full rank	99%	$r(9)$—full rank	99%	VEC
ELEC	$\mathcal{P}_t$	$r(3)$	99%	$r(2)$	99%	VEC
	$\mathcal{Q}_t$	$r(7)$	99%	$r(7)$	99%	VEC
SEA-1	$\mathcal{P}_t$	$r(0)$	–	$r(0)$	–	VAR in diff
	$\mathcal{Q}_t$	$r(2)$	99%	$r(3)$	99%	VEC
SEA-2	$\mathcal{P}_t$	$r(1)$	99%	$r(1)$	90%	VEC
	$\mathcal{Q}_t$	$r(4)$—full rank	99%	$r(4)$—full rank	99%	VEC
SEA-3	$\mathcal{P}_t$	$r(0)$	–	$r(0)$	–	VAR in diff
	$\mathcal{Q}_t$	$r(2)$	99%	$r(3)$	99%	VEC

nonetheless tested this dataset's series alongside the others in order to verify the stationarity test results. Tables 2 and 3 demonstrate the results of the Johansen method to test for cointegration of impurity-based and permutation importance measurements with accuracy ($\mathcal{P}_t$ and $\mathcal{Q}_t$) and F1 score ($\mathcal{R}_t$ and $\mathcal{S}_t$), respectively.

Gradual vs. Abrupt Drifts. As it can be seen in Table 1, among datasets with gradual drifts (RCB-C, RCB-P, RCB-E, RCB-S, NOAA, ELEC), we can accept the alternative hypothesis of stationarity of all importance measurements and predictability performance series tested on the RCB-C dataset. The reason seems to be constant drift rate in this dataset. Therefore, $\mathcal{P}_t$, $\mathcal{Q}_t$, $\mathcal{R}_t$, and $\mathcal{S}_t$ can be modeled using a standard stationary VAR or VAR in levels. However, this would not be the case for the other RCB dataset variants where the drift rate is non-constant.

On datasets with abrupt drifts (SEA variants, for example), performance metrics were unanimously nonstationary, and impurity-based importance measurements were stationary. Yet, permutation importance measurements were stationary only for a subset of features. It appears that the reason lies in the fact that every abrupt change in these datasets is permanent until the next drift, hence resulting in a lasting deviation from the long-term trend.

Table 3. The Johansen method test results for cointegration of feature importances with F1 score. EIG and TRC refer to the eigenvalue statistic and trace statistic in the test, respectively.

Dataset	Series	EIG	Sig. lvl.	TRC	Sig. lvl.	Conclusion
RCB-C	$\mathcal{R}_t$	$r(3)$—full rank	99%	$r(3)$—full rank	99%	VEC
	$\mathcal{S}_t$	$r(3)$—full rank	99%	$r(3)$—full rank	99%	VEC
RCB-P	$\mathcal{R}_t$	$r(1)$	99%	$r(1)$	99%	VEC
	$\mathcal{S}_t$	$r(3)$—full rank	99%	$r(3)$—full rank	99%	VEC
RCB-E	$\mathcal{R}_t$	$r(1)$	99%	$r(1)$	99%	VEC
	$\mathcal{S}_t$	$r(3)$—full rank	99%	$r(3)$—full rank	99%	VEC
RCB-S	$\mathcal{R}_t$	$r(1)$	99%	$r(1)$	99%	VEC
	$\mathcal{S}_t$	$r(3)$—full rank	99%	$r(3)$—full rank	99%	VEC
NOAA	$\mathcal{R}_t$	$r(2)$	99%	$r(4)$	99%	VEC
	$\mathcal{S}_t$	$r(9)$—full rank	99%	$r(9)$—full rank	99%	VEC
ELEC	$\mathcal{R}_t$	$r(3)$	99%	$r(2)$	99%	VEC
	$\mathcal{S}_t$	$r(7)$	99%	$r(7)$	99%	VEC
SEA-1	$\mathcal{R}_t$	$r(1)$	99%	$r(1)$	99%	VEC
	$\mathcal{S}_t$	$r(2)$	99%	$r(2)$	99%	VEC
SEA-2	$\mathcal{R}_t$	$r(1)$	99%	$r(0)$	–	VEC or VAR in diff
	$\mathcal{S}_t$	$r(4)$—full rank	99%	$r(4)$—full rank	99%	VEC
SEA-3	$\mathcal{R}_t$	$r(1)$	99%	$r(1)$	99%	VEC
	$\mathcal{S}_t$	$r(3)$	99%	$r(3)$	99%	VEC

The conintegration results shown in Tables 2 and 3 show that in all datasets exhibiting gradual drift, the cointegration rank of the impact matrix C based on both eigenvalue and trace statistic is $r > 0$. The impact matrix C consolidates the long-term dynamics of each multivariate series: deviation from the stationary mean as *error* and adjustment speeds to *correct* or revert to the stationary mean over time. That means, there is highly significant evidence that there exists at least one cointegrating relation between any type of FIMs analyzed from the simpler GBDT auxiliary model and both predictability performance metrics of the main classifier in the face of CD.

Concerning our hypotheses stated in Sect. 2.2, we reject hypotheses 1.–4., and accept the alternative hypotheses 1.–4. with significant evidence provided by the results on datasets that exhibit gradual drifts.

It can be concluded that these FIMs follow the long-term trend of the performance metrics even if any of these series are non-stationary. Moreover, even if the FIMs deviate from the performance metrics in short-term (*error*), they revert (*correct*) to the long-term mean of the multivariate series.

The conintegration results shown in Tables 2 and 3 are consistent with the stationarity results of Table 1 on abrupt drift datasets. On these datasets, we can only reject hypotheses 2. and 4. and accept alternative hypotheses 2. and 4. for permutation importance measurements. We fail to reject hypotheses 1. and 3.

on abrupt drift datasets. Since the abrupt changes are sudden and persisting, we expect that the changes in importance measurements of features, except the third noise feature on SEA datasets, also persist for a longer time. That results in fewer cointegrating relations between FIMs and performance metrics. Indeed, in series with impurity-based importance measurements, that is $\mathcal{P}_t$ and $\mathcal{R}_t$, the experimental results confirm this expectation by showing hardly any meaningful relationship between these importances with either accuracy or F1 score.

Stable vs. Unstable Drift Rates. The *rate of drift* in RCB-C and RCB-E is constant and monotonic, respectively. All multivariate series are stationary in these two datasets, except for permutation importance measurements $(\mathcal{H}_t)$ on RCB-E. This is further supported by cointegration results where the impact matrix C is full rank based on both eigenvalue and trace statistic for both importance measurements types and accuracy.

On the other hand, the *rate of drift* is non-monotonic and periodic in RCB-P and RCB-S, respectively. The stationarity test results show that all series tested are non-stationary except permutation importance measurements. Conintegration tests results support this observation where C is restricted to reduced rank for impurity-based multivariate series $\mathcal{P}_t$ and $\mathcal{R}_t$.

Out of the two real-world datasets NOAA and ELEC, NOAA has a more consistent periodic behavior, that is, close to constant drift rate, and does not deviate drastically from its mean in long-term; therefore, it is not surprising that the stationarity of its performance metrics accuracy $(\mathcal{A}_t)$ and F1 score $(\mathcal{B}_t)$ are significant. Non-stationarity of this dataset's importance measurements is either failed to be rejected, or is inconclusive. That means while there is significant evidence for non-stationarity of some importance measurements, tests were insignificant for the others. The ELEC dataset, however, experiences a drastic change where data for two features become available part-way through the stream. The accuracy of the classifier in the face of these changes and possibly other phenomenon inducing CD is less significantly stationary, and its F1 score is non-stationary. Similar to NOAA, importance measurements computed on ELEC are either non-stationary or inconclusive.

Our cointegration results are also consistent here, as for NOAA most series tested are stationary with a significance level of 95%, whereas for ELEC only accuracy is stationary with a significance of 95%. Moreover, cointegration tests found more stationary linear combinations between FIMs and performance metrics for NOAA than ELEC, implying FIMs follow the long-term stochastic trend of performance metrics accuracy and F1 score more stably.

We can conclude that given data with constant or monotonic *rate of drift*, classification models perform more stably in the long run in terms of stationarity of both feature importance measurements analyzed from data, and their predictability performance. In contrast, when provided with data exhibiting non-monotonic or periodic *rate of drift*, both importance measurements and the predictability performance of classification models in long-term become nonstationary and unstable in levels.

Impurity-Based vs. Permutation Feature Importance Measurements.
We observe that the conintegration results indicate that permutation importance measurements are more stable in terms of following long-term trends of
performance metrics because the impact matrix C based on both eigenvalue and
trace statistic is full rank on all datasets with gradual drift except for ELEC. This
means we can form a stationary VAR in levels with an additional lag out of the
VEC model and its error-correction term. Even for ELEC, permutation importance
measurements had a higher rank, that is, a larger number of cointegrating relations with the performance metrics. Impurity-based importance measurements
achieved this full rank of cointegrating relations on more stable datasets with
constant or monotonic rates of drift.

Impurity-based importance measurements have the advantage of being computed as part of the auxiliary model construction; therefore, they are computationally less costly than permutation importance measurements. The former is
also more insightful as it abstracts *impurity* of predictor and response variables
according to some criterion. In our experiments, impurity-based importance measurements had also much less short-term variance than permutation importance
measurements.

We can conclude that impurity-based importance measurements are a viable
source of detection information on datasets with constant or monotonic rates
of drift with the advantage of less computational overhead, more insight on the
behavior of data, and less micro-variation. The test results provide evidence
for more consistency and reliability of permutation importance measurements
over impurity-based importance measurements if data exhibits periodic or non-
monotonic rates of drift, or if this prior information about data is unknown.

4   Conclusion and Future Work

In this work, we studied the relationship between two types of FIMs as detection
information analyzed from streaming data exhibiting different characteristics of
CD and the predictability performance metric of the main classifier considering
data stream processing application constraints and requirements.

As the main outcome of this study we found evidence for strong correlation between FIMs computed from a decoupled, cost-effective model with the
performance of a costly, though more accurate classification model.

A novel contribution of this study is a novel systematic approach on the
CDDA problem-solving methodology. More specifically, we investigated the direct
correlation of detection information with the main classifier's predictability performance rather than treating the problem as a black box, only comparing the
performance metrics of the main classifier *after* incorporating a CDDA technique.
This approach enabled us to analyze the long-term dynamics of the performance
of the classifier in the face of CD and its relationship to FIMs as a potential source
of detection information.

A major by-product of the cointegration analysis is the VEC model of the
FIMs and performance metrics which consolidates the long-term dynamics of

the system. We are currently investigating application of this model to forecast changes in the performance of the classifier in presence of concept drift.

References

1. Abdallah, A., Maarof, M.A., Zainal, A.: Fraud detection system: a survey. J. Netw. Comput. Appl. **68**, 90–113 (2016)
2. Alizadeh Mansouri, A., Javadtalab, A., Shiri, N.: An ensemble learning augmentation method for concept drift detection over data streams. In: Advances in Data Science and Information Engineering. Springer (2022)
3. Barddal, J.P., Enembreck, F., Gomes, H.M., Bifet, A., Pfahringer, B.: Boosting decision stumps for dynamic feature selection on data streams. Inf. Syst. **83**, 13–29 (2019)
4. Breiman, L.: Random forests. Mach. Learn. **45**(1), 5–32 (2001)
5. Breiman, L.: Manual on setting up, using, and understanding random forests v3.1. Stat. Dept. Univ. Calif. Berkeley CA, USA **1**(58), 3–42 (2002)
6. Cassidy, A.P., Deviney, F.A.: Calculating feature importance in data streams with concept drift using online random forest. In: 2014 IEEE International Conference on Big Data (Big Data), pp. 23–28 (2014)
7. Castro-Cabrera, P.A., Orozco-Alzate, M., Castellanos-Domínguez, C.G., Huenupán, F., Franco, L.E.: Supervised and unsupervised identification of concept drifts in data streams of seismic-volcanic signals. In: Simari, G.R., Fermé, E., Gutiérrez Segura, F., Rodríguez Melquiades, J.A. (eds.) IBERAMIA 2018. LNCS (LNAI), vol. 11238, pp. 193–205. Springer, Cham (2018). https://doi.org/10.1007/978-3-030-03928-8_16
8. Ditzler, G., Polikar, R.: Hellinger distance based drift detection for nonstationary environments. In: 2011 IEEE Symposium on Computational Intelligence in Dynamic and Uncertain Environments (CIDUE), pp. 41–48 (2011)
9. Elwell, R., Polikar, R.: Incremental learning of concept drift in nonstationary environments. IEEE Trans. Neural Netw. **22**(10), 1517–1531 (2011)
10. Engle, R.F., Granger, C.W.J.: Co-integration and error correction: representation, estimation, and testing. Econometrica **55**(2), 251–276 (1987)
11. Friedman, J.H.: Greedy function approximation: a gradient boosting machine. Ann. Stat. **29**(5), 1189–1232 (2001)
12. Gama, J., Medas, P., Castillo, G., Rodrigues, P.: Learning with drift detection. In: Bazzan, A.L.C., Labidi, S. (eds.) SBIA 2004. LNCS (LNAI), vol. 3171, pp. 286–295. Springer, Heidelberg (2004). https://doi.org/10.1007/978-3-540-28645-5_29
13. Gama, J., Žliobaitė, I., Bifet, A., Pechenizkiy, M., Bouchachia, A.: A survey on concept drift adaptation. ACM Comput. Surv. (CSUR) **46**(4), 44 (2014)
14. Gomes, H.M., de Mello, R.F., Pfahringer, B., Bifet, A.: Feature scoring using tree-based ensembles for evolving data streams. In: 2019 IEEE International Conference on Big Data (Big Data), pp. 761–769 (2019)
15. Hand, D.J., Adams, N.M.: Selection bias in credit scorecard evaluation. J. Oper. Res. Soc. **65**(3), 408–415 (2014)
16. Harries, M., Wales, N.S.: SPLICE-2 Comparative Evaluation: Electricity Pricing (1999)
17. He, Z., Maekawa, K.: On spurious Granger causality. Econ. Lett. **73**(3), 307–313 (2001)

18. Johansen, S.: Estimation and hypothesis testing of cointegration vectors in gaussian vector autoregressive models. Econometrica **59**(6), 1551–1580 (1991)
19. Khamassi, I., Sayed-Mouchaweh, M., Hammami, M., Ghédira, K.: Discussion and review on evolving data streams and concept drift adapting. Evol. Syst. **9**(1), 1–23 (2018)
20. Kuncheva, L.I.: Combining Pattern Classifiers: Methods and Algorithms, 2 edn. John Wiley & Sons, Hoboken (2014)
21. Liang, N.y., Huang, G.b., Saratchandran, P., Sundararajan, N.: A fast and accurate online sequential learning algorithm for feedforward networks. IEEE Trans. Neural Netw. **17**(6), 1411–1423 (2006)
22. Lu, J., Liu, A., Dong, F., Gu, F., Gama, J., Zhang, G.: Learning under concept drift: a review. IEEE Trans. Knowl. Data Eng. **31**(12), 2346–2363 (2019)
23. Maziarz, M.: A review of the Granger-causality fallacy. J. Philos. Econ. Reflect. Econ. Soc. Issues VIII **2**, 86–105 (2015)
24. Michaelides, M.P., Reppa, V., Panayiotou, C., Polycarpou, M.: Contaminant event monitoring in intelligent buildings using a multi-zone formulation. IFAC Proc. Vol. **45**(20), 492–497 (2012)
25. Sethi, T.S., Kantardzic, M.: On the reliable detection of concept drift from streaming unlabeled data. Expert Syst. Appl. **82**, 77–99 (2017)
26. Sims, C.A., Stock, J.H., Watson, M.W.: Inference in linear time series models with some unit roots. Econometrica **58**(1), 113–144 (1990)
27. Stolfo, S., Fan, W., Lee, W., Prodromidis, A., Chan, P.: Cost-based modeling for fraud and intrusion detection: results from the JAM project. In: Proceedings DARPA Information Survivability Conference and Exposition. DISCEX'00, vol. 2, pp. 130–144 (2000)
28. Street, W.N., Kim, Y.: A streaming ensemble algorithm (SEA) for large-scale classification. In: Proceedings of the Seventh ACM SIGKDD International Conference on Knowledge Discovery and Data Mining, pp. 377–382. KDD '01, Association for Computing Machinery (2001)
29. Unknown: Global Surface Summary of the Day - GSOD
30. Vergara, A., Vembu, S., Ayhan, T., Ryan, M.A., Homer, M.L., Huerta, R.: Chemical gas sensor drift compensation using classifier ensembles. Sens. Actuators B Chem. **166–167**, 320–329 (2012)
31. Wang, J., Lu, S., Wang, S.H., Zhang, Y.D.: A review on extreme learning machine. Multimed. Tools Appl. **81**(29), 41611–41660 (2022)
32. Wang, K., Lu, J., Liu, A., Zhang, G., Xiong, L.: Evolving gradient boost: a pruning scheme based on loss improvement ratio for learning under concept drift. IEEE Trans. Cybern. **53**(4), 2110–2123 (2023). https://doi.org/10.1109/TCYB.2021.3109796
33. White, A.P., Liu, W.Z.: Bias in information-based measures in decision tree induction. Mach. Learn. **15**(3), 321–329 (1994)
34. Yang, Z., Al-Dahidi, S., Baraldi, P., Zio, E., Montelatici, L.: A novel concept drift detection method for incremental learning in nonstationary environments. IEEE Trans. Neural Netw. Learn. Syst. **31**(1), 309–320 (2020)

An Approach for Efficient Processing
of Machine Operational Data

Ben Lenard[1,2], Eric Pershey[1], Zachary Nault[1], and Alexander Rasin[2(✉)]

[1] Argonne National Laboratory, Lemont, IL, USA
{blenard,pershey,znault}@anl.gov
[2] DePaul University, Chicago, IL, USA
blenard@depaul.edu, arasin@cdm.depaul.edu

Abstract. Supercomputers come in a variety of sizes and architectures with thousands of interconnected nodes. Most organizations are required to produce metrics for their funding sources to prove that these machines are being utilized and meeting the availability requirements. While tracking the state of an individual server is trivial, measuring uptime of a supercomputer with several thousand nodes spanning tens to hundreds of cabinets and rows with one or more mounted file systems is a complex task. Additionally, supercomputers have diverse architectures and System Logic (which includes unique characteristics of the machine itself such as networking topology, size, partitions, hardware layout, physical configuration and component hierarchy). These constraints complicate the computation of standardized metrics such as Mean Time To Failure (MTTI), Mean Time to Failure (MTTF), availability, and utilization.

At the Argonne Leadership Computing Facility (ALCF), we developed a tool that standardizes the analyses of these machines so that these metrics can be computed accurately and efficiently. We call this tool Operational Data Processing System (ODPS), and use it to process the data that Theta, a 4,392 node Cray XC40, generates. In addition to the XC40, this tool also works with Mira, a 49,152 node IBM BG/Q system that ALCF houses. This paper explores how ODPS processes the data from Theta and Mira, including the storage design decisions and architecture-independent approach to metric calculations. We quantitatively evaluate our approach, comparing it to alternative methods for storing and processing supercomputer machine state in the database.

Keywords: Uptime metrics · High Performance Computing · Supercomputer availability · Supercomputer utilization

1 Introduction

Complex problems require extraordinary computational resources. A typical supercomputer spans dozens of racks; it will contain compute nodes, memory DIMMs, power supplies, copper and optical cables, sensors, network switches, and much more. Additionally, these machines often mount large parallel filesystems and other storage devices such as Storage Area Network. Computing racks

C. Strauss et al. (Eds.): DEXA 2023, LNCS 14146, pp. 129–146, 2023.
https://doi.org/10.1007/978-3-031-39847-6_9

span multiple power circuits and may require water cooling. As such, these systems have many inter-dependencies and components which can fail at any time; failure scope can range from individual processors to whole machine sections.

In 2016, the ALCF acquired Theta [7], currently a 4,392-node, 24-rack Cray XC40 system with a 10PB Lustre file system. Each Theta node is equipped with a 64-core Intel Xeon Phi 7230 Knight's Landing processors with 16 GB of MCDRAM and 192 GB DDR4 RAM connected to a 10 PB Lustre filesystem. The ALCF also had a 49,152 node, 48-rack IBM BlueGene/Q system, Mira [3], accompanied by 26PB of GPFS file system. Mira has a 5D torus interconnect [3] whereas Theta has an Aries interconnect with Dragonfly configuration [7].

The ALCF reports to the United States Department of Energy on scheduling availability, overall availability, MTTI, MTTF, usage, and utilization for Mira and Theta [2]. Reporting these metrics becomes a challenge across thousands of compute nodes with complex networks, parallel filesystems, and hardware component hierarchies connecting them. For example, a node might be rebooted as part of a job; although it may appear unavailable, but as it was rebooted due to a user action, we consider that part of the job. Another example of complexity is a filesystem outage, when the reports may be showing that jobs are running – but the jobs are most likely dead and stuck in IOWait. This type of event must be detected and associated with an "unavailable" record, subtracting out jobs that did work and marking down the machine where jobs were not working. This can be a time intensive process that requires correlating data from many sources – we refer to this process as *Job Failure Analysis*. Similar to cloud computing, if a user process failed to execute, the user would request a refund due to a failure; thus, a process is needed to determine if a refund request is valid.

The ALCF needs a standardized method for deriving the uptime, utilization, and other metrics for the Operational Assessment Review Report [2,15,16]. While it is possible to create a unique set of tables for each machine's physical limitations and hierarchy, this greatly increases the calculation complexity. Instead, the ALCF abstracted the System Logic to compute standard metrics uniformly across all machines. We embed the System Logic into the data rather than tables, allowing us to track systems even as they change size over time.

In this paper, we describe ODPS, a standardized approach for converting the operational data generated by supercomputers and tracking the supercomputer's operational state regardless of the physical changes in the computer over time. We explore the database design and the efficiency of using a bitmask to track machine events over time, comparing it to other physical database designs and evaluating it with respect to the design objectives of this software. In sum, this paper describes the following significant contributions:

- Computing the metrics of Availability, Usage, Utilization, and MTBF as required by The United States Department of Energy
- Effectively finding target events and correlating them in time and space for use in job and hardware failure analysis
- Providing an API to access the events and machine structure on which many other tools can be built

2   Related Work

Related work for tracking the state of a supercomputer can be broken down into two areas: 1) large scale log processing and 2) efficiencies of using a bitmask to store and process data and how databases use them to index data.

Collecting and analyzing logs in a large-scale system is commonly used to perform diagnosis; a variety of papers cover this topic on Cray and IBM systems. We collect and process logs similar to prior work but with a few different caveats. Oliner and Stearly [17] focused on the log collection and log processing for modeling system and trying to predict failures. Lenard et al. [14] developed a system for generating and collecting large-scale distributed system events.

Bitmap indexes have been used for decades [5] to index and search large amounts of data efficiently in a database. They are often used in relational database management systems for scanning data quickly to retrieve rows or values in columnar data. Oracle allows for the direct creation of bitmap indexes instead of a B-Tree index [23], as bitmap index can be more efficient for scanning the data from the index. In some databases, such as Db2 LUW, the optimizer will automatically create and use bitmap indexes when multiple indexes on the same table are used in the predicate [10]. Furthermore, IBM Db2 BLU, a proprietary columnar store, utilizes bitmaps for processing SQL queries [22].

Bitmaps are also applied to other areas such as analyzing large data sets such as autoignition data [12]. The paper demonstrates that the use of bitmaps can facilitate efficient storage and processing speed of the dataset.

Sandia National Laboratories has developed an open-source tool tool called Lightweight Distributed Metric Service (LDMS) to gather metrics (memory, cpu, IO, etc.) about applications executing on supercomputers and stores them in a database or a flat file [1]. While these metrics are useful for identifying performance bottlenecks within the supercomputer, it does not correlate the data to applications and jobs running on the computer, which we need to determine utilization [4,6]. While LDMS can report on the the utilization of a node, we define utilization in terms of nodes being associated with jobs. Despite being a monitoring solution for supercomputers, LDMS provides data that is different from what we need. Furthermore, LDMS requires an infrastructure dedicated to monitoring, an excessive requirement for our reporting needs.

While processing the logs of a system is not a novel idea, and bitmaps have been in the database world for a while, the marriage of the two ideas within an application for the purposes of HPC metrics calculations and searching for faults over time and space presents a novel approach.

3   Paper Definitions

Event: For the purposes of this paper, an event occurs when the control system of the supercomputer changes the state of a given node. This state change can be the result of a failure either with the given node or the infrastructure, such as network, power, or cooling that supports the particular node.

RAS Events: Reliability, Availability, and Serviceability (RAS) event is when a component failed on the hardware or software of the system. The IBM Blue Gene [13] as well as the Cray XC [21] systems provide logging for capturing instances when something has failed.

MTTI: Mean Time to Interrupt is defined as the average outage time (scheduled or unscheduled). It is also known as Mean Time Between Interrupt:

$$\frac{TimePeriod-(ScheduledOutages+UnscheduledOutages)}{Count_{ScheduledOutages}+Count_{UnscheduledOutages}+1} \tag{1}$$

MTTF. Mean Time to Failure is defined as the time, on average, to an unscheduled outage on the system:

$$\frac{TimeInPeriod-UnscheduledOutages}{Count_{UnscheduledOoutages}+1} \tag{2}$$

Overall Availability: The overall availability is the percentage of time a system is available to users without interrupt. Outage time reflects both scheduled and unscheduled outages:

$$\frac{TimePeriod-UnavailableDueToOutages}{TimePeriod} * 100 \tag{3}$$

Scheduled Availability: Scheduled availability is the percentage of time a designated level of resource is available to users, excluding scheduled downtime for maintenance and upgrades:

$$\frac{TimePeriod-UnavailableDueToOutages}{TimePeriod-UnavailableDueToScheduledOutages} * 100 \tag{4}$$

Utilization: Utilization is the percentage of time the system was used out of the total number of available hours:

$$\frac{Core\ hours\ used\ in\ period}{Core\ hours\ available\ in\ period} * 100 \tag{5}$$

System Logic: System Logic is used to define the physical limitations and configuration of the supercomputer. These limitations were built into the machine and do not change unless by modifying the hardware. For example, configuration such as the memory mode of a node, or how the nodes are laid out and fiber connections within the supercomputer. System Logic is related to Business Logic and is generally applied before the Business Logic. These configurations can be very complex and are site-specific. For instance, Mira has 49,152 nodes, but we schedule resources in 512 node blocks because all the nodes in the block share the same 4 Input Output Nodes (IONs). We could go down to 128 node blocks, but they would share IONs for redundancy and one job could effect the IO characteristics of another job. Other machines (e.g., Cetus or Vesta) have different IONs to compute node ratios and we are able to run in smaller blocks with IO isolation. The decision to isolate IO traffic is Business Logic, but the way the ION are wired to provide IO isolation is System Logic.

Business Logic: Often overlooked, Business Logic includes site details such as the logic to rounding a timestamp. (e.g., having timestamp rounded to the nearest second or millisecond). For every second Mira runs, if we count all the cores, the total core hours per second is equal to $(49,152 * 16 \text{ cores})/3,600 = 218.453$ core hours per second. ODPS was designed to strictly account for every second of the machines use, per node/core. Similarly to timestamps, different sites could have different formulas for calculating Availability or Utilization.

Bitmap and Bitmask: Using bitmasks allows us to use simple operations such as 'xor', 'and', and 'or' when comparing two bitmasks against each other. The bitmask itself shows which components are marked as a dependent of the event. If two bitmasks are combined with an 'and' operation, and all bits are zero, there is no intersection. The order in which the bits refer to underlying hardware is encoded in the bitmap. It is beneficial to keep the index of the node close to the network links, because the node and its corresponding network links are likely to be used simultaneously. This idea enables us to compress the masks.

Location to Mask: Location to Mask (LTM) maps a node in the machine to the bitmask that represents the machine. For example, 4 nodes can be represented by the bitmask of '0000' with a bit per node; the first(MSB) node => first bit, second node => second bit, and so on. 1 means that the node is in use by the bitmask event. For example, bitmask '1010' represents an event that involves node #0 and #2. The bit positions also map to the node names, such as nid00000 and nid00002 being used in this event.

ETL: In data warehousing, Extract Transform and Load (ETL) commonly refers to the process of extracting the target data, transforming the data in the structure required by the organization, and loading the data into the database. In context of this paper, we parse the logs from the supercomputer control systems looking for error events, transform the data, and load the data into our system.

Time and Space correlation: On a given system, a job can consume the whole system or a partition of the machine. Time and space correlation help the ALCF identify which jobs are impacted by an interrupt since an interrupt can affect a portion of the machine. For example, if a key component which interconnected several nodes failed and was replaced, this lets us show the impact of that component during a given time frame.

Incarnation: We will use the term incarnation to refer to each instance of a supercomputer changes size. We are borrowing this term from Oracle as whenever you restore an Oracle database and open it with RESETLOGS [20], you have a new incarnation of the database and a list when the database changed incarnations. Similarly, whenever a supercomputer changes size, it is the same supercomputer but a different instance of it, and we track this in the LTM table.

4   ODPS

We developed ODPS to standardize different supercomputer systems, store the system state, and compute the system's metrics consistently. Within ODPS, we implemented a method of tracking machine state by using a bitmask to track the nodes, optimizing reporting calculations, as well as to correlate impacted jobs on the system when something does fail. Bitmask representation allows us to calculate the machines' metrics efficiently using bitwise operations such as 'xor', 'and', and 'or'. The bitmask also supports finding the intersection of a job and a component failure using bitwise 'and' operation. The goal of this paper is to illustrate efficiency of this approach to calculating supercomputer metrics in terms of both storage space, and SQL query execution runtime. We compare our proposed methods to storing the same data in a typical normalized row-based database. It is important to note that this system was developed to analyze and report historic metrics and is not designed for real-time monitoring.

ODPS is written in Python 3, allowing it to run on many platforms using a persistent Relational Database Management System (RDBMS). ODPS has been tested and used with Db2 and MySQL. ODPS provides an API via a REST interface for consumers and a Low-Level API for administrators to access the data without knowing the underlying data structures within the database.

4.1 Bitmask

The bitmasks are used in the database and within the ODPS API. The database can be used to search all the events for a given bit by applying an offset(+64) to the section that bit is set and executing database bitwise operations. Db2 and MySQL both support 64 integer bitwise operations. Since querying a row loads the entire database page and since we are doing the operation on a very small slice of data, we do not have to move that data across the network. After loading all the events, python code converts the bitmasks into Extended NumPy Arrays with a dtype of an unsigned integer 8 with overflow protection. Since NumPy Arrays [11] support incorporate bitwise operators, we leverage these operations on the full bitmask to calculate the metrics (see Sect. 6.2).

4.2 Location to Mask

The core of ODPS at an abstraction layer is a python class Location to Map (LTM). The LTM provides a common interface to process every machine's operational data using the bitmask representation, even as the machine changes size over time. The LTM layer consists of the Location To Mask Map (LTMM) and the Location to Map Base (LTMB) tables. As the system changes size, we need a mechanism to track it. The LTMB stores the metadata about the machine for a given period of time, type of nodes and the smallest machine unit that will be mapped to a bit in the mask. We store one LTMB per machine for a period of time. The LTMB stores the metadata about the machine as a whole whereas the LTMM stores information about each node as it relates to the LTMB.

For each LTMB record, or each time the machine changes size, the nodes for the machine are added into the LTMM table. In essence, the LTMB provides a context or incarnation of the machine. The LTMM stores metadata about each node; for Theta we store the NID, NID Name, IP address, as well as component name (CName). These are stored in machine independent structures with the true name mapping of NID and such within the LTMB. The LTMB is refreshed by updating the end timestamp every 2 min by an ETL on Theta (no other fields can be updated). LTMMs cannot be updated because we could lose history and we cannot be sure that the component remains in the same location in the bitmask. If something else changes, a new LTMB is created along with a new set of LTMMs. It is possible to keep one live copy and store a journal of changes to another table, but the system is meant for historic or range based queries, supporting queries across the boundary of a machine changing size or structure.

4.3   ETL

ODPS uses the operational data and metadata of the machine. Thus, we consider the ETL process for both of these areas. For example in Theta, the data is derived from Cray hardware supervisory system database. On Cray XC40 system, the Cray hardware supervisory system database is where the node inventory is kept by the Cray management stack; on the IBM BG/Q system this is similar to the control system. If the machine size has changed due to adding or removing nodes, a new LTMB is created followed by adding the nodes to the LTMM with the latest incarnation. Between Mira and Theta, the system control software is different based on vendor and series; therefore, information is extracted from different sources.

The operational data ETL is the second category of ETL for ODPS. The ODPS ETL process translates lines of a log file that contain the nodes into rows that contain the bitmask that represent the included nodes. These log sources, for Theta, include:

– Cobalt (scheduler/resource manager) Logs (Usage[1], Availability)
– ALPS (Usage, Availability[3], MTTI[3], MTTF[3])
– SMW messages (MTTI, MTTF, Availability)

5   Methodology

We used two years of operational data from Mira and two years of operational data from Theta in our experiments. We explore different methods to represent this data and consider storage efficiency and query execution performance for calculating our target metrics. We compare storage space required, execution runtime, and the complexity of deploying these tables. In the past, we have encountered scalability issues with naive implementations for tracking node state.

5.1   Static Table Approach

One approach to storing the state of the numerous nodes of a supercomputer is to deploy a single table for the entire supercomputer. If we took this approach,

---

[1] Denotes the primary source of the data; all other data is supplemental.

Theta would have 4,392 columns in addition to the metadata the describes the event such as timestamp, text about the event, duration (if known), and anything else. In Mira, if we represented each 512-node block as a column, the Mira table would have 100 columns plus the event's metadata.

Assuming we used this approach, each node, or smallest unit in the system, would be assigned to a column. The maximum number of columns for a table in Db2 11.1 is 1,012 for tablespaces of 8k or more [9]. MySQL 8 has a column limit of 4,096 with a max row size of 64k [19] and Oracle 18 has a limit of 1,000 columns per table [18]. That being said, a traditional table could encapsulate Mira but for a system of Theta's size, a single table would not be able to contain enough columns for each node in the system. Furthermore, as we move toward larger and larger systems as we head towards the exascale, and as processors are not becoming faster, the node count on these systems will increase over time. While MySQL might be able to have a column count of 4,096 columns, this would not be enough to handle Theta, let alone future systems.

In addition to having too many columns for a single table, static table approach is difficult to adjust with changing of the machine size. Supercomputers can change their size due to deployment stages, upgrades, or downgrades. While one could alter the table to add columns and account for upgrades, one would need to modify the application code to handle these changes.

Given the limitation of RDMSes, a column per node of the supercomputer is not a viable option since the size of the current supercomputer exceeds the limitations of the databases discussed above.

5.2 Materialized Views and Views

In order to improve performance, we might create a materialized view (MV) or a materialized query table (MQT). An MV is a pre-built query with data objects stored so that queries will run faster. If we were to assign one node per column and built the MV based on the rows in Sect. 5.3, the MV also has the limitation of the number of columns that the underlying RDMS will support. For Oracle 18, a MV can have a maximum of 1,000 columns [18]; Db2 11.1 can have a max of 1,012 columns in an MQT and 5,000 in a view [9]. Furthermore, we would need some method to adapt to changes to the supercomputer size.

5.3 Row Per Event and Node

Another approach to storing the nodes that are impacted by a System, Job, or Task Event is to have two tables for each event type. One table to describe the event and the metadata surrounding the event, and another table would contain the impacted nodes. For example, the System Event table would have begin and end times, a description, and a generated identification number. This identification number would be then inserted into the second table along with the identification number of the corresponding impacted node; there would be one row per node and event in the second table. MySQL supports up to a 64-terabyte table, no known limit on row count limit, and Db2 can store 128×10

to the 10 rows before partitioning, and Oracle does not have a row limit. This approach is feasible for how we store the operational data within the data.

5.4 Bitmask

Another approach to storing the nodes that are impacted by a System, Job, or Task Event is to have one row per event and encode the affected nodes within a bitmask. Each bit within the bitmask would represent the machine and the nodes impacted by a given event. For example, the System Event table store begin and end times, a description, an identification number generated at creation time, and a varchar to store the nodes description. We would still store the nodes in a table that would contain the metadata about each node but we would encode the effected nodes within the bitmask. For ODPS we use a bitmask of 28k and use a tablespace of 32k so that one row will fit into a single tablespace page.

6   Comparative Analysis

6.1   Raw Storage

We will now examine the database and the SQL operations to extract the data for our calculations. Regardless of how we store the supercomputer state, we need to store the metadata about the nodes, and the changes to its size over time. In our database, regardless of the method of storing state, we will be using the same two tables which track the incarnations of the supercomputer's over time. These two tables are used to convert from the bitmask to the nodes. We use LOCATION_TO_MASK_BASE and LOCATION_TO_MASK_MAP tables to track the incarnations. LOCATION_TO_MASK_BASE provides the base map for the machine's layout and other meta information such as map id, parent map, start and end timestamps, unit sizes and bits for the ODPS application. In the second table, LOCATION_TO_MASK_MAP we store meta data about each node such as map id and the corresponding LOCATION_TO_MASK_BASE id, and other attributes.

Whenever a supercomputer changes size, we generate a new base id as well as new entries for each node for that incarnation of the supercomputer. While this may seem excessive, it is important to historically determine the supercomputer's size and components at a given point in time. In order to make the mask easier to read and store in a database, we store the bitmask as a hex mask.

For a Cray system we store the CName, NID, IP address, and node's classification such as 'compute.' In terms of our IBM BG/Q, we also store each incarnation within LOCATION_TO_MASK_BASE table and then each midplane within the LOCATION_TO_MASK_MAP. Mira has never changed sizes, but we have the capability of storing the different incarnations. For the BG/Q's, we store the 'midplane' within the LOCATION_TO_MASK_MAP table since the System Logic determines that this is the smallest granularity that we can use for calculations.

Regardless of how we store the supercomputer's state, these two tables will remain the same for the comparisons since they provide functionality outside of the state of the 'compute' or 'midplane.'

Within the application two tables track the supercomputer state: SYSTEM_ EVENT and RESOURCE_MANAGER_EVENT. As their names suggest, SYSTEM_EVENT stores system events about the supercomputer and RESOURCE_MANAGER_EVENT stores information from our scheduler, resource manager and control system.

Table 1. SYSTEM_EVENT table as a single table with the bitmask

SYSTEM_EVENT:

Column_Name	Data_Type	Size	Null	Default
SYSTEM_EVENT_ID	INT		No	
PARENT_SYSTEM_EVENT_ID	INT		No	
PIT_START_ID	INT		No	
PIT_END_ID	INT		No	
EVENT_STATE_NAME	VAR	16	No	
TIME_START	TS	6	No	1970-01-01-00.00.00.000000
DIM_DATE_HOUR_START_KEY	INT		Yes	
TIME_END	TS	6	No	1970-01-01-00.00.00.000001
DIM_DATE_HOUR_END_KEY	INT		Yes	
HEXMASK	VAR	28672	No	
BIT_COUNT	INT		No	
EVENT_TYPE_NAME	VAR	256	No	
MACHINE_NAME	VAR	64	No	
DETAIL_JSON	VAR	256	Yes	
LOCATION_TO_MASK_BASE_ID	INT		No	
INSERTED_TIMESTAMP	TS	6	No	
UPDATED_TIMESTAMP	TS	6	No	
IS_DELETED	INT		No	0

SYSTEM_EVENT is a table that tracks the state of a node within the super-computer; every time a state change is detected a new row is inserted into this table. The bitmask formatted table is described in Table 1. Since we use a var-char 28,672 for the bitmask, which is encoded as a hex mask, we can store a mask large enough to represent 114,688 nodes. We used such a large field to future proof ODPS. The maximum row size for this table is 29,340 therefore in Db2 we used a 32k tablespace in order not to fragment the row. As of MySQL 5.7, support for 32k and 64k pages is available. Since Theta is only 4,392 nodes and we convert the bitmask into hex, the row size for Theta is far less than the 29,340 maximum row size. If we were to transform SYSTEM_EVENT into two tables, one that represents the event and another to represent the nodes, the row size will be much smaller but there will be a number of node rows per event Table 2 illustrates the schema that uses two tables.

Table 2. SYSTEM_EVENT as two tables without the bitmask

SYSTEM_EVENT:

Column_Name	Data_Type	Size	Null	Default
SYSTEM_EVENT_ID	INT		No	
PARENT_SYSTEM_EVENT_ID	INT		No	
PIT_START_ID	INT		No	
PIT_END_ID	INT		No	
EVENT_STATE_NAME	VAR	16	No	
TIME_START	TS	6	No	1970-01-01-00.00.00.000000
DIM_DATE_HOUR_START_KEY	INT		Yes	
TIME_END	TS	6	No	1970-01-01-00.00.00.000001
DIM_DATE_HOUR_END_KEY	INT		Yes	
EVENT_TYPE_NAME	VAR	256	No	
MACHINE_NAME	VAR	64	No	
DETAIL_JSON	VAR	256	Yes	
INSERTED_TS	TS	6	No	
UPDATED_TS	TS	6	No	
IS_DELETED	INT		No	0

SYSTEM_EVENT_NODE:

Column_Name	Data_Type	Size	Null	Default
SYSTEM_EVENT_ID	INT		No	
NODE_ID	INT		No	
INSERTED_TS	TS	6	No	
UPDATED_TS	TS	6	No	
IS_DELETED	INT		No	0

In our evaluation, we use data between 8/1/2016 and 11/1/2018, which contains 3,076,656 events generated from Theta. These events are loaded into the database by our ETL process that processes the Cray logs. While we hope that

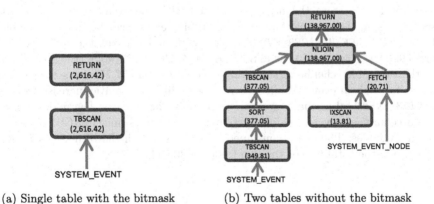

(a) Single table with the bitmask   (b) Two tables without the bitmask

Fig. 1. Db2 query plan comparison – single table versus two table design

the Cray software stack provides all of the events within the logs, we added the logic to ensure that if events were missing from the logs, for whatever reason, are software would handle such a condition. In other words, we validate the logs against the content of the of the system's databases. In a two-table version schema of our database SYSTEM_EVENT table also has 3,076,656 events but it also has 49,843,290 SYSTEM_EVENT_NODE table to describe all the impacted nodes. In terms of size, the table that houses the bitmask is slightly smaller then the table split into event and data, 3,390.01 GB and 3,474.66 GB respectively. While the sizes are similar, the cost of querying this data is dramatically different. The explain plans were generated on Db2 and are measured in timerons. A timeron is a unit of measure created by IBM for Db2 [8], and the lower the number of timerons the more efficient. For the simple query to show all the nodes that has system events between a time range, the SQL statement would be as follows for the table containing the bitmask:

```
SELECT * FROM SYSTEM_EVENT WHERE TIME_START >='2018-07-04'
                    AND TIME_START < '2018-07-14'
```

The SQL statement results in a cost of 2,616.42 timerons (see Fig. 1a). Alternatively, the query for selecting the nodes with the same predicate clause would be expressed as the following query. That SQL statement results in a cost of 138,967.00 timerons (see Fig. 1b).

```
SELECT * FROM SYSTEM_EVENT e, SYSTEM_EVENT_NODE n
WHERE e.SYSTEM_EVENT_ID = n.SYSTEM_EVENT_ID
AND TIME_START>='2018-07-04'AND TIME_START<'2018-07-14'
```

Similarly to the SYSTEM_EVENT table, we use date range from 8/1/2016 to 11/1/2018 from Theta where 5,990,941 resource managers events are stored in a table named RESOURCE_MANAGER_EVENT. Similar to the SYSTEM_EVENT table, RESOURCE_MANAGER_EVENT is also populated through ETL processes that parse the logs of our scheduler, resource manager, and control system. For comparison, we also split this table into a two table version, one for the metadata and another to store the nodes associated with the event. Table 3 shows the schema using the bitmask and the two table version can be found in Table 4. In the RESOURCE_MANAGER_EVENT tables, with and without the bitmask, there are 5,990,941 rows and in the RESOURCE_MANAGER_EVENT_NODE (the second table in the two-table schema version) table there are 155,657,751 rows that list the nodes for a given event. While this may seem like a dramatic increase from the SYSTEM_EVENT table, many system events only encompass a few nodes whereas resource manager events seldom encompass a few nodes since most HPC jobs run on many nodes. The space consumed by the table with the bitmask is 7,837.59 GB, while the total size of the two tables is 20,944.32 GB.

Table 3. RESOURCE_MANAGER_EVENT as a single table with the bitmask

RESOURCE_MANAGER_EVENT:				
Column_Name	Data_Type	Size	Null	Default
RESOURCE_MANAGER_EVENT_ID	INT		No	
PARENT_RESOURCE_MANAGER_EVENT_ID	INT		No	
TIME_START	TS	6	No	1970-01-01-00.00.00.000000
DIM_DATE_HOUR_START_KEY	INT		Yes	
TIME_END	TS	6	No	1970-01-01-00.00.00.000001
DIM_DATE_HOUR_END_KEY	INT		Yes	
HEXMASK	VAR	28672	No	
BIT_COUNT	INT		No	
EVENT_TYPE_NAME	VAR	256	No	
MACHINE_NAME	VAR	64	No	
DETAIL_JSON	VAR	256	Yes	
LOCATION_TO_MASK_BASE_ID	INT		No	
INSERTED_TS	TS	6	No	
UPDATED_TS	TS	6	No	
SOURCE_NAME	VAR	64	No	"
SOURCE_IDX	VAR	32	No	"
SOURCE_KEY	VAR	64	No	"
SOURCE_SUBKEY	VAR	32	No	"
ORPHAN	INT		No	0
SOURCE_PATH	VAR	512	Yes	
SOURCE_FILENAME	VAR	512	Yes	
SOURCE_HOSTNAME	VAR	128	Yes	
SOURCE_FILE_READ_TS	TS	6	Yes	

Table 4. RESOURCE_MANAGER_EVENT as two tables without the bitmask

RESOURCE_MANAGER_EVENT:				
Column_Name	Data_Type	Size	Null	Default
RESOURCE_MANAGER_EVENT_ID	INT		No	
PARENT_RESOURCE_MANAGER_EVENT_ID	INT		No	
TIME_START	TS	6	No	1970-01-01-00.00.00.000000
DIM_DATE_HOUR_START_KEY	INT		Yes	
TIME_END	TS	6	No	1970-01-01-00.00.00.000001
DIM_DATE_HOUR_END_KEY	INT		Yes	
EVENT_TYPE_NAME	VAR	256	No	
MACHINE_NAME	VAR	64	No	
DETAIL_JSON	VAR	256	Yes	
LOCATION_TO_MASK_BASE_ID	INT		No	
INSERTED_TS	TS	6	No	
UPDATED_TS	TS	6	No	
SOURCE_NAME	VAR	64	No	"
SOURCE_IDX	VAR	32	No	"
SOURCE_KEY	VAR	64	No	"
SOURCE_SUBKEY	VAR	32	No	"
ORPHAN	INT		No	0
SOURCE_PATH	VAR	512	Yes	
SOURCE_FILENAME	VAR	512	Yes	
SOURCE_HOSTNAME	VAR	128	Yes	
SOURCE_FILE_READ_TS	TS	6	Yes	
RESOURCE_MANAGER_EVENT_NODE:				
Column_Name	Data_Type	Size	Null	Default
RESOURCE_MANAGER_EVENT_ID	INT		No	
NODE_ID	INT		No	
INSERTED_TS	TS	6	No	
UPDATED_TS	TS	6	No	
IS_DELETED	INT		No	0

The following SQL statement is a simplified version of a SQL statement used within ODPS with arbitrary values in the where clause containing the bitmask:

```
SELECT * FROM RESOURCE_MANAGER_EVENT RME
WHERE RME.machine_name = 'theta' AND RME.time_end > '2018-07-04'
    AND RME.location_to_mask_base_id = 642
```

For the table with the bitmask, the Db2 cost of the query is 298,422.19 timerons, while the equivalent SQL statement spilt into two tables has a cost of 1,383,644.75 timerons.

6.2 Workflow of Calculated Availability

Calculating the overall availability is a critical metric used to gauge the effectiveness of a computing resource and a primary metric used the ALCF. It is also one of the more computationally heavy calculations as it takes into account a years worth of events. Furthermore, we account for every second of a machine, assuming nothing about its state without a record. This results in a large amount of historical data that is cumbersome to store in any other way besides as event with a bitmask describing effect nodes.

For this comparison we examined the bitmask method and the row per event per node (Sect. 5.3) method. We assumed that for both methods the underlying databases have been optimized for performance and any database queries will contribute little to the run time of the calculation. We have opted for a complexity analysis of the availability function within ODPS to show that a major factor affecting completion time is the total number of rows that are processed. It is important to note that this function can be easily parallelized to improve computation time. However, such analysis is beyond the scope of this paper. An algorithmic description of the availability calculation is detailed in Algorithm 1.

Algorithm 1. Availability algorithm

1: $TimeInPeriod \leftarrow$ Total time within range
2: $TimePossible \leftarrow$ TimeInPeriod * size of the Mask
3: $EventList \leftarrow$ list of all events in database over time range
4: $PITList \leftarrow$ list of all Points in Time
5: **for each** $Row \in Database$ **do**   ▷ Section 6.2
6:     $EventList.update(row)$
7: **for each** $Event \in EventList$ **do**   ▷ Section 6.2
8:     $PIT \leftarrow EventToPIT(Event)$
9:     $PITList.append(PIT)$
10: $Timeline \leftarrow GroupByTime(PITList)$ ▷ Section 6.2
11: $TimelineSorted \leftarrow SortByTime(Timeline)$ ▷ Section 6.2
12: $TimelineMask \leftarrow Normalize(TimelineSorted)$ ▷ Section 6.2
13: $TimelineMask \leftarrow Flatten(TimelineMask)$ ▷ Section 6.2
14: $TimelineMask \leftarrow Isolate(TimelineMask)$ ▷ Section 6.2
15: $TimelineMag \leftarrow TimelineMasktoMagnitude(TimelineMask)$ ▷ Section 6.2
16: $TimeConsumed \leftarrow calcArea(TimelineMag)$ ▷ Section 6.2
17: $Availability \leftarrow TimeConsumed/TimePossible * 100$ ▷ Section 6.2

Request Events from DB: First, we must get all the events from the database. For the bitmask approach this returns a row for every event over a given time range with the bitmask that describes the nodes. For the row per event per node method, this returns a row for every node that was involved in every event over a given time range. We are assuming that this query is not impacting the final time to completion and does not contribute to the complexity.

Points in Time (PIT): Each event is converted into two Points in Time (PIT): one marking the start, another marking the end. A PIT keeps track of the node(s) affected by the event, how the event has affected system state (up +1, down -1, or neutral 0), and what direction the event is facing in the timeline: forward (event start) or backward (event end). This operation has a complexity of $O(N)$.

Group PIT by Time: Loop over all PITs and return a dictionary of PITs grouped by time. This operation has a complexity of $O(N)$.

Sort PIT by Time into Timeline: The PITs now need to be sorted into a true timeline by organizing them by time. This operation uses an internal call to Python's sorted() which has a corresponding complexity of $O(N \log(N))$.

Normalize Timeline: Walk through the timeline and normalize all PITs at each time step by getting the mask sum. This combines masks of nodes affected and (+1, 0 −1) into a single PIT by summing all the +1 and the −1 masks. Any nodes affected in multiple ways are normalized to +1, 0, −1. For example, if a node was marked up (+1) by one PIT and marked down (−1) by another, the resulting state would be neutral (0). A node that is marked up (+1) in one PIT and marked up again (+1) by another PIT would result in a state of +2. However, this is normalized to +1. This operation has a complexity of $O(N)$.

Flatten Timeline: Walk the timeline and reduce each PIT into a single number representing the total system state. For example at some time T a 4 node system with a normalized PIT may look like this: +1, +1, +1, 0. Meaning at that time T the system as a magnitude of 3. These magnitudes are then integrated to generate a step plot showing units available in time. This operation has a complexity of $O(N)$.

Isolate Timeline: Walk though the timeline and check that all events are within the timeline range. If anything falls outside the range, truncate the event. This operation has a complexity of $O(N)$.

Mask to Magnitude Timeline: Walk though the timeline and sum each bitmask of ones. Then add edges by adding data to each change in size. If the magnitude goes from 0 nodes (@T0) to 10 nodes (@T1), a graph would show a slope up to 10, but those ten are not used until T1, so we add a magnitude of 0 at T1', taken from T0, right before T1, in the sorted list, to correctly form the area. After all this is complete the integration can be done. $O(N)$.

Calculate Area: Perform a second integration on the magnitude timeline and find time consumed by events. This will give us the area used by the events. This operation has a complexity of $O(N)$

144   B. Lenard et al.

Availability: Finally, knowing the total area(time*len(mask)) within the given period with the area consumed by events, availability can be calculated.

Net Complexity: Combining all of the operation complexities, the cost of the availability calculation is dominated by the timeline sort. This results in a net complexity $O(N \log(N))$ where N is twice the number of rows in the database. Before performing this complexity analysis, we expected for this function to be linear with the majority of speed-up coming from the bit operations.

Having an understanding of this function's complexity highlights the importance of reducing the number of database rows that need to be processed. The bitmask method achieves this by storing all node information with the event; reducing the total number of rows returned. For example, for the search of events generated from 8/1/2016 until 11/1/2018, 9.07×10^6 events (and therefore rows) were logged into ODPS. If the same events were logged in the row per event per node method there would be a total of 2.06×10^8 rows. Therefore, the availability calculation performed using the row per event per node method, it would take approximately 27 times longer to complete versus the bitmask method.

7   Results and Discussion

In our comparisons, we have shown that the disk storage and database query plan costs are significantly reduced by storing the data represented as a bitmask rather than storing a row per node per event. Storing the bitmask is similar to a materialized query within a column; when querying the event rows, the operation becomes free. In correlating events and jobs that are executing, we shown that it is also significantly faster to utilize the bitmask. ODPS with the bitmask has allowed operation of the system calculate our metrics from 2016 for Theta in seconds rather than hours using a naive method. This method is much more efficient by converting the common structure, the bitmask, at database load. The common structure has been used in every supercomputer we have, no matter the complexity or hierarchy of nodes and allows for a common representation, code reuse and deep performance tuning. Much of the time of generating usage or availability is attributed to looking up nodes, resolving overlapping events and setting states of nodes at a time while walking over all events, nodes and times. For more complex machines, given a node, one will have to figure out a hierarchy of what subset of nodes that node is contained within, referred to as blocks. This translation is done one time from block to nodes to bitmask. The bitmask also allows for parallel computation of availability or usage by breaking up in either chunks of time or nodes or both. This is done to speed up the computation and is embarrassingly parallel. Even further, if using a true bitmask structure of 1's and 0's, we can leverage bitwise operations of the processor by breaking the bitmask into 64bit integers.

8   Future Work

As machines grow in size, we are looking toward the future and exploring options for when the node counts exceed current database limits for tablespace pages.

Db2, Oracle, and Postgres impose a page size limit of 32k which might not be large enough for future systems. While RDBMSes do support extended or chained rows, accessing a row that spans multiple pages can cause inefficiencies incurring multiple IOPS per retrieved row. For example, if a system had 1,048,576 nodes, the bitmask would be 256k, and that would consume roughly 8 32k tablespace pages (including the overhead within the page) per row. We would need to modify the bitmask approach so it will scale to larger systems.

We are also looking at incorporating this method into our scheduler since the it also tracks the machine's state. Since ODPS is tracking the state of the system just as the scheduler does, we could directly track the state of the system with the resource manager and persist all changes to a data store. That data store could be ODPS as it's designed with the scale and complexities of our systems in mind. This also would allow a realtime component as we would have direct access to any state changes and we could leverage them to also remove the need for ETLs.

Another area of future development and research is to encapsulate the location to bitmask within the database itself with support for Db2, Postgres, and possibly MySQL. This would eliminate the network transfer latency for searching. Right now, we have Python code to convert the bitmask to the location, whereas we would like to see this built into the database itself.

Since we utilize bitmasks for locations of the nodes and their corresponding state, we also need to explore creating bitwise operations that are larger than 64 bits. Several database platforms support bitwise operations but only support a small number of bits since they internally use a BIG Int. We also plan to opensource ODPS following the Argonne National Laboratory process for publicly releasing software.

Acknowledgement. This research used resources of the Argonne Leadership Computing Facility, which is a DOE Office of Science User Facility supported under Contract DE-AC02-06CH1135. The authors would also like to acknowledge the review and editing help by Nick Scope.

References

1. https://www.sandia.gov/sandia-computing/high-performance-computing/lightweight-distributed-metric-service-ldms/
2. ALCF: 2016 operational assessment report argonne leadership computing facility. https://www.alcf.anl.gov/files/CY2016_OAR_ALCF_3_3_2017.pdf
3. ALCF: Mira. https://www.alcf.anl.gov/mira
4. Bhalachandra, S., Austin, B., Wright, N.J.: Understanding power variation and its implications on performance optimization on the Cori supercomputer. In: 2021 International Workshop on Performance Modeling, Benchmarking and Simulation of High Performance Computer Systems (PMBS), pp. 51–62 (2021). https://doi.org/10.1109/PMBS54543.2021.00011
5. Chan, C.Y., Ioannidis, Y.E.: An efficient bitmap encoding scheme for selection queries. SIGMOD Rec. **28**(2), 215–226 (1999). https://doi.acm.org/10.1145/304181.304201

6. Feldman, S., Zhang, D., Dechev, D., Brandt, J.: Extending LDMS to enable performance monitoring in multi-core applications. In: 2015 IEEE International Conference on Cluster Computing, pp. 717–720 (2015). https://doi.org/10.1109/CLUSTER.2015.125

7. Harms, K., et al.: Theta: rapid installation and acceptance of an XC40 KNL system. Concurr. Comput. Pract. Exp. **30**(1), e4336 (2018). e4336 cpe.4336, https://onlinelibrary.wiley.com/doi/abs/10.1002/cpe.4336

8. IBM: Explain information for data operators. https://www.ibm.com/support/knowledgecenter/en/SSEPGG_11.1.0/com.ibm.db2.luw.admin.perf.doc/doc/c0005140.html

9. IBM: Sql and xml limits. https://www.ibm.com/support/knowledgecenter/SSEPGG_11.1.0/com.ibm.db2.luw.sql.ref.doc/doc/r0001029.html

10. IBM: Types of index access. https://www.ibm.com/support/knowledgecenter/en/SSEPGG_11.1.0/com.ibm.db2.luw.admin.perf.doc/doc/c0005301.html

11. Idris, I.: NumPy Beginner's Guide. Packt Publishing Ltd., Birmingham (2013)

12. Koegler, W., Chen, J., Shoshani, A.: Using bitmap index for interactive exploration of large datasets. In: 15th International Conference on Scientific and Statistical Database Management, 2003, pp. 65–74, July 2003

13. Lakner, G., Knudson, B., et al.: IBM System Blue Gene solution: Blue Gene/Q System Administration. IBM Redbooks, Indianapolis (2013)

14. Lenard, B., Wagner, J., Rasin, A., Grier, J.: SysGen: system state corpus generator. In: Proceedings of the 15th International Conference on Availability, Reliability and Security, pp. 1–6 (2020)

15. McNally, S.T., et al.: High performance computing facility operational assessment 2016-oak ridge leadership computing facility. Technical report, Oak Ridge National Lab. (ORNL), Oak Ridge, TN (United States) (2017)

16. NERSC: Nersc operational assessment review highlights. https://www.nersc.gov/assets/NUG-2016-business-day/3-OAR-Highlights-NUG-Mar-2016.pdf

17. Oliner, A., Stearley, J.: What supercomputers say: a study of five system logs. In: 37th Annual IEEE/IFIP International Conference on Dependable Systems and Networks (DSN'07), pp. 575–584, June 2007. https://doi.org/10.1109/DSN.2007.103

18. Oracle: Logical database limits. https://docs.oracle.com/en/database/oracle/oracle-database/18/refrn/logical-database-limits.html#GUID-685230CF-63F5-4C5A-B8B0-037C566BDA76

19. Oracle: Mysql : Mysql 8.0 reference manual : C.10.4 limits on table column count and row size. https://dev.mysql.com/doc/refman/8.0/en/column-count-limit.html

20. Oracle: Rman data repair concepts. https://docs.oracle.com/cd/E11882_01/backup.112/e10642/rcmrvcon.htm#BRADV117

21. Pautsch, G., Roweth, D., Schroeder, S.: The cray® xc^{TM} supercomputer series: energy-efficient computing. Technical report (2013)

22. Raman, V., et al.: Db2 with BLU acceleration: so much more than just a column store. Proc. VLDB Endow. **6**(11), 1080–1091 (2013)

23. Sharma, V.: Bitmap index vs. b-tree index: which and when? https://www.oracle.com/technetwork/articles/sharma-indexes-093638.html

PrivSketch: A Private Sketch-Based Frequency Estimation Protocol for Data Streams

Ying Li[1,2], Xiaodong Lee[1,2(✉)], Botao Peng[1(✉)], Themis Palpanas[3], and Jingan Xue[4]

[1] Institute of Computing Technology, Chinese Academy of Sciences, Beijing, China
{XL,pengbotao}@ict.ac.cn
[2] University of Chinese Academy of Sciences, Beijing, China
[3] LIPADE, Université Paris Cité, French University Institute (IUF), Paris, France
[4] Huawei Technologies, Shenzhen, China

Abstract. Local differential privacy (LDP) has recently become a popular privacy-preserving data collection technique protecting users' privacy. The main problem of data stream collection under LDP is the poor utility due to multi-item collection from a very large domain. This paper proposes PrivSketch, a high-utility frequency estimation protocol taking advantage of sketches, suitable for private data stream collection. Combining the proposed background information and a decode-first collection-side workflow, PrivSketch improves the utility by reducing the errors introduced by the sketching algorithm and the privacy budget utilization when collecting multiple items. We analytically prove the superior accuracy and privacy characteristics of PrivSketch, and also evaluate them experimentally. Our evaluation, with several diverse synthetic and real datasets, demonstrates that PrivSketch is 1–3 orders of magnitude better than the competitors in terms of utility in both frequency estimation and frequent item estimation , while being up to ∼100× faster.

1 Introduction

Motivation. Collecting user data, often in the form of a data stream, in order to analyze them and provide some services has become a common practice. However, data collection may expose user information, which is a major concern. Local Differential Privacy (LDP) is popular to protect individual privacy during data collection and has been widely used in technology companies (such as Apple, Google, Microsoft). It perturbs data locally before sending them to the collector and enables the collector to obtain approximate statistics on the perturbed data, to avoid the risk of disclosing user privacy. A parameter ϵ is used to quantify the amount of perturbation, which determines the degree of privacy protection and the utility of the privacy-preserving algorithm.

C. Strauss et al. (Eds.): DEXA 2023, LNCS 14146, pp. 147–163, 2023.
https://doi.org/10.1007/978-3-031-39847-6_10

Utility Problem. Although several studies have focused on the frequency estimation problem under LDP, they do not perform well when used in a data stream context, due to following reasons. First, existing solutions consider a unified size for the data items generated by different users (i.e., data length) that is based on unrealistic assumptions (assume only one item in a collection interval) [10,25], or a predefined/estimated unified size L for each collection (padding and sampling) [19,22,23], in both cases hurting utility. Second, the large domains of several data streams (e.g., URLs, and IP) lead to excessive computation and communication costs, as well as significant perturbation errors (some of the existing literature on frequency estimation [22,23] is only applicable to small cardinality domains).

Sketching is widely used in streaming data processing for compressing sparse data from a large domain (e.g., when a tiny percentage of webpages are accessed by any individual user). The uniform size of sketches makes it possible to unify the data length of different users without extra padding and sampling [19], and leads to efficient storage. The sketching has been combined with LDP in the Private Count-Mean Sketch (PCMS) algorithm [20] proposed by Apple. However, it operates at the granularity of single items, which hurts performance. When considering extending it to multi-item collections, the following problems emerge. (i) The error introduced by sketching algorithms is not considered. Aggregating sketches from users directly is equivalent to encoding all data into one sketch, leading to increasing errors of collisions, i.e., data hashed in the same position. (ii) To maintain user-level privacy, allocating the privacy budget for each counter in the sketch is required, resulting in substantial inaccuracies and poor utility.

Our Solution. We propose PrivSketch, a high-utility privacy-preserving sketch-based frequency estimation protocol that leads to lower errors when compared to existing solutions. PrivSketch proposes an innovative LDP collector-side workflow that decodes the perturbed sketch before aggregating and calibrating it, which avoids the error introduced by the collisions when aggregating all perturbed sketches in the traditional decode-after workflow. In addition, PrivSketch utilizes the ordering matrix extracted from the original sketch, which enables the collector to obtain the minimum index information, while ensuring the privacy of each user's sketch (cf. proof in Sect. 3.4). This effectively reduces the minimum calculation error caused by disturbance and is the first attempt to improve utility using background information. Furthermore, PrivSketch uses the sampling technique to improve the utilization of the information encoded by the sketch, transmitting relatively accurate information with a limited privacy budget. Thus, it reduces the error caused by the uniform allocation of the privacy budget when encoding multiple items.

Contributions. Our contributions are summarized below.

- We propose a novel LDP protocol, PrivSketch, that is suitable for frequency estimation in data streams where multi-item encoding is needed. It is the first sketch-based privacy-preserving protocol that considers the errors introduced by the sketches with a novel decode-first workflow. It employs background information to reduce the minimum value calculation errors of sketches and utilizes a sampling technique to improve the privacy budget utilization.

- We prove (cf. Sect. 3.4) that the ordering matrix as background information does not expose the original value of each counter in the sketch, which meets the privacy needs of users. We introduce a new definition of the indistinguishable input set, where the collector cannot distinguish any two values. We observe that the utility of LDP algorithms can be improved with appropriate additional background information , but does not harm the users' privacy.

- We evaluate our approach on both synthetic and real datasets. We compare it with extensions and variants of existing algorithms, including the multi-item encoding extension and its min-variant algorithm. The utility of the protocol proposed in this paper is 1–3 orders of magnitude more accurate than existing algorithms, and up to ∼100× faster.

2   Background and Preliminaries

Local Differential Privacy (LDP). Differential privacy (DP) [13] is a technology with quantified privacy protection, but relies on a trustworthy third-party collector. To remove the trust in the collector, LDP [11] was proposed where original data are only accessible by users, and the collector only receives the perturbed data. A mechanism $\mathcal{M}$ satisfying LDP can be defined as follows.

Definition 1 (ϵ-Local Differential Privacy [11]). *A randomized algorithm $\mathcal{M}$ satisfies ϵ-local differential privacy ($\epsilon > 0$), if and only if for any two input tuples $x, x' \in \mathcal{D}$ and output y, then $\frac{\Pr[\mathcal{M}(x)=y]}{\Pr[\mathcal{M}(x')=y]} \le e^\epsilon$.*

Thus, a smaller ϵ means large perturbation and more indistinguishable, but lower utility. There is an important property of LDP:

Theorem 1 (Sequential Composition Mechanism [17]). *Assume a randomized algorithm $\mathcal{M}$ consists of a sequence of randomized algorithms $\mathcal{M}_i (1 \le i \le t)$. When for each $i, \mathcal{M}_i$ satisfies ϵ_i-LDP, $\mathcal{M}$ satisfies $\sum_{i=1}^{t} \epsilon_i$-LDP.*

Randomized Response Mechanism (RR) [15,28]. This fundamental LDP mechanism achieves plausible deniability by allowing users not to give the original value. Specifically, for binary values, users answer the original value with probability p, and the opposite value with probability $q = 1 - p$. To achieve ϵ-LDP, the worst case is $\frac{\max \Pr[\mathcal{M}(x)=y]}{\min \Pr[\mathcal{M}(x')=y]} = \frac{p}{1-p} = e^\epsilon$, therefore $p = \frac{e^\epsilon}{1+e^\epsilon}$. Denote $\Pr[x = 1]$ the percentage of $x = 1$. For the collector, $\Pr[y = 1] = p \Pr[x = 1] + (1 - p)(1 - \Pr[x = 1])$ and $\Pr[y = 0] = p(1 - \Pr[x = 1]) + (1 - p) \Pr[x = 1]$. $\Pr[y = 1]$ and $\Pr[y = 0]$ represent the probability of the output y taking the value of 1 and 0, respectively, which can used to obtain the unbiased estimation of $\Pr[x = 1]$ and $\Pr[x = 0]$.

Count-Min Sketch (CMS). A common approach to compress data from a large domain is the sketching algorithm. The Count-Min Sketch [9] is one of the most popular sketching algorithms due to its efficiency. The sketching uses a matrix X consisting of $K \times M$ counters, bound to K hash functions $H_1, H_2, \ldots, H_K : \{1, \ldots, d\} \mapsto \{1, \ldots, M\}$. It consists of two phases: (i) update, where K hash functions are used to hash the updated item x, and then the corresponding counters are updated, i.e. $X_{k,H_k(x)} = X_{k,H_k(x)} + 1, \forall 1 \leq k \leq K$; (ii) query, where item x's count $c(x)$ is estimated, denoted by $\tilde{c}(x)$, based on the corresponding counters in the sketch, i.e. $\min_{1 \leq k \leq K} X_{k,H_k(x)}$ [9].

Private Count-Mean Sketch (PCMS-Mean) [20]. PCMS-Mean estimates frequency under LDP, where the user perturbs data before sending them to the collector. Specifically, for item x, each user chooses a hash function H_k and updates $X_{k,H_k(x)} = 1$ (other positions keep as -1), then, perturbs X_k using RR and sends the perturbed result $\hat{X}_k$ to the collector. The collector constructs a matrix of size $K \times M$ where each row is the sum of the perturbed rows indexed by k, and estimates the frequency by averaging the sum of k counters corresponding to K hash functions. The algorithm assumes that each user generates only one item. Thus, for any two rows from different users X_k and $X'_{k'}$, at most two positions can be different. To protect these two positions under privacy budget ϵ, the parameter p in RR is set to $\frac{e^{\epsilon/2}}{1+e^{\epsilon/2}}$ (cf. Theorem 1).

When extending PCMS-Mean to encode multiple items, the number of different positions in any two rows from different users is up to M due to unlimited items of each user. Thus, to protect the privacy of each position, the parameter p is set to $\frac{e^{\epsilon/M}}{1+e^{\epsilon/M}}$. This naive solution works poorly when M is large. The irrational allocation of ϵ is one of the reasons. In addition, the error introduced by the sketching algorithm is also non-negligible. The estimation error of different sketching algorithms varies. The error of the Count-Min Sketch is smaller than that of the Count-Mean Sketch [8]; hence, we use the Count-Min Sketch.

Problem Definition. This paper studies the frequency estimation problem under LDP for data streams, where data are generated from a very large domain. There is an untrusted collector and a set of n users represented by $U = \{U_1, U_2, \ldots, U_n\}$. Each user, U_i, has a set of items of length $L^{(i)}(L^{(i)} \geq 0)$, which is denoted by $S^{(i)} = \{S_1^{(i)}, S_2^{(i)}, \ldots\}, |S^{(i)}| = L^{(i)}$. Each item $S_\ell^{(i)}(0 \leq \ell \leq L^{(i)})$ is discrete value and drawn from a large domain $\mathcal{D}$ of size $|\mathcal{D}| = d$, that is, $S_\ell^{(i)} \in \mathcal{D}$. In this paper, we focus on estimating the frequency of each item from $\mathcal{D}$, that represents the proportion of users who possess the item. Formally, the frequency for each value $x \in \mathcal{D}$ is defined as: $f(x) = \frac{|\{i|\exists \ell, 0 \leq \ell \leq L^{(i)}, S_\ell^{(i)} = x\}|}{n}$.

3   PrivSketch Solution

PrivSketch is a LDP protocol based on CMS to solve the frequency estimation problem in data stream collection. PrivSketch uses a novel collector-side workflow (cf. Sect. 3.2) and the ordering matrix (cf. Sect. 3.2) to reduce errors introduced by sketches. PrivSketch also uses a sampling technique to increase the information utilization in sketches under a limited privacy budget.

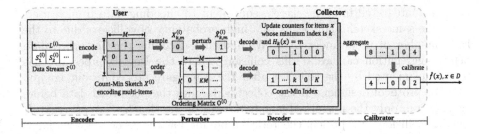

Fig. 1. Overview of PrivSketch.

Algorithm 1: PrivSketch

Input: $\{S^{(1)}, S^{(2)}, \ldots, S^{(n)}\}, \epsilon, K, M, D \subset \mathcal{D}$
1 select a set of hash functions $\mathcal{H} = \{H_1, H_2, \ldots, H_K\}$;
2 **for** *each* $i \in [1, n]$ **do**
3      $\hat{X}^{(i)}, O^{(i)} \leftarrow$ PrivSketch-User $(S^{(i)}, \epsilon, n, K, M, \mathcal{H})$;
4   send $\hat{X}^{(i)}, O^{(i)}$ to the collector;
5 **for** *each* $x \in \mathcal{D}$ **do**
6   set $\hat{\mathcal{X}} \leftarrow \{\hat{X}^{(1)}, \hat{X}^{(2)}, \ldots, \hat{X}^{(n)}\}$;
7   set $\mathcal{O} \leftarrow \{O^{(1)}, O^{(2)}, \ldots, O^{(n)}\}$;
8      $\hat{f}(x) \leftarrow$ PrivSketch-Collector $(x, \epsilon, n, M, \mathcal{H}, \hat{\mathcal{X}}, \mathcal{O})$;
9 **return** $\{\hat{f}(x)|x \in D\}$

Figure 1 provides a high-level overview of PrivSketch workflow. At the user end, the encoder encodes items using CMS and the perturber perturbs a sampled one counter in the sketch using RR. Then, the perturbed counter $\hat{X}_{k,m}^{(i)}$ is sent to the collector with an ordering matrix $O^{(i)}$ which reflects the order of all counters in the original sketch $X^{(i)}$. At the collector end, the decoder restores $\hat{X}_{k,m}^{(i)}$ to the original domain $\mathcal{D}$ by calculating each item's minimum index based on $O^{(i)}$ and updating counts of items x whose minimum index equal to the sampled k and $H_k(x) = m$. Then, the calibrator estimates items' frequency by aggregating restored counts from users and calibrating the perturbation error. The protocol is shown in Algorithm 1. We elaborate on its novel designs and details next.

3.1 Decoding-First Collector-Side Workflow

An important characteristic of PrivSketch is the decoding-first feature on the collector side, which is designed to reduce the collisions in the private sketching algorithm. The naive protocol, PCMS-Min as traditional LDP protocols, consists of three steps: Encode, Perturb, and Aggregate [24]. Collisions can occur in the Encode and Aggregate procedure. During encoding, the collision is caused by that different items are hashed into the same positions, which can be reduced by a good choice of the sketching parameters. During aggregation, sketches from n

users are integrated into one sketch, equivalent to encoding data from n users using the same sketch. This leads to a high probability of collisions due to the large number of users under LDP. We find that decoding the perturbed data before aggregation can avoid this collision, where the Decode procedure has been implemented by the collector after the Aggregate procedure but ignored by LDP protocol designers. If the collector decodes the perturbed data before aggregation, only the perturbed counts instead of the sketches are aggregated, thus, no collisions. We present theoretical proof for how the decode-first workflow reduces collision errors following. Note in our design, we use Calibration instead of Aggregate to describe the procedure where aggregating and calibrating errors caused by the perturbation.

Theorem 2. *For estimating the frequency of a value* $x \in \mathcal{D}$ *using Count-Min Sketch,* $\min_k \sum_{i=1}^{n} X_{k,H_k(x)}^{(i)}$ *represents the results of aggregating sketches before decoding,* $\sum_{i=1}^{n} \min_k X_{k,H_k(x)}^{(i)}$ *represents the results of decoding sketches before aggregating, the following formula holds:*

$$\min_k \sum_{i=1}^{n} X_{k,H_k(x)}^{(i)} \geq \sum_{i=1}^{n} \min_k X_{k,H_k(x)}^{(i)} \geq nf(x) \tag{1}$$

where $f(x)$ *represents the true frequency of* x.

Proof. For each user U_i and any $1 \leq k \leq K$, $X_{k,H_k(x)}^{(i)}$ reflects the occurrence of both x and $x'(x' \neq x)$, which are hashed into the same position with x.

$$X_{k,H_k(x)}^{(i)} = \mathbb{1}\{x \in S^{(i)}\} \vee \mathbb{1}\{x' \in S^{(i)}, H_k(x) = H_k(x')\}.$$

For the minimum index k where $X_{k,H_k(x)}^{(i)}$ is minimal, the equation above holds. As a result, $\sum_{i=1}^{n} \min_k X_{k,H_k(x)}^{(i)} = nf(x) + \sum_{i=1}^{n} \mathbb{1}\{x' \in S^{(i)}, x \notin S^{(i)}, H_{\min_k}(x) = H_{\min_k}(x')\} \geq nf(x)$. Moreover, $\sum_{i=1}^{n} X_{k,H_k(x)}^{(i)} \geq \sum_{i=1}^{n} \min_k X_{k,H_k(x)}^{(i)}$, $1 \leq k \leq K$. Considering $\min_k$ is one of the case that belongs to $[1, K]$, we can conclude that $\min_k \sum_{i=1}^{n} X_{k,H_k(x)}^{(i)} \geq \sum_{i=1}^{n} \min_k X_{k,H_k(x)}^{(i)}$.

Thus, when an unbiased estimation of the query result of the original Count-Min Sketch is achieved, the decode-first collector-side workflow brings fewer errors. Next, we introduce how to ensure an unbiased estimation in PrivSketch.

3.2   Ordering Matrix Generation

In PrivSketch, the minimum index of the perturbed count can be changed by the randomized response mechanism which hinders an unbiased estimation. As shown in Fig. 1, the collector queries the perturbed sketch $\hat{X}^{(i)}$ and estimates based on it. Assume the calibration for estimation of the frequency

$f(x)$ in $\mathcal{D}$, is based on sketches $\hat{X}^{(i)}$ with a linear function $h(x)$, i.e. $\hat{f}(x) = h(\sum_{i=1}^{n} \min_k \hat{X}^{(i)}_{k,H_k(x)})$. PrivSketch needs to satisfy the expectation of the variable after perturbation is an unbiased estimation of the result from querying the original sketch $\tilde{f}(x) = \frac{1}{n} \sum_{i=1}^{n} \min_k X^{(i)}_{k,H_k(x)}$. That is,

$$\mathbb{E}[\hat{f}(x)] = \mathbb{E}[h(\sum_{i=1}^{n} \min_k \hat{X}^{(i)}_{k,H_k(x)})] = \tilde{f}(x) = \frac{1}{n} \sum_{i=1}^{n} \min_k X^{(i)}_{k,H_k(x)}.$$

Assuming that the row indices of the minimum count for x in the perturbed and original sketches are k' and k, if $k \neq k'$,

$$\mathbb{E}[\hat{X}^{(i)}_{k',H'_k(x)}] = pX^{(i)}_{k',H'_k(x)} - qX^{(i)}_{k',H'_k(x)} = (p-q)X^{(i)}_{k',H'_k(x)} \geq (p-q)\min_k X^{(i)}_{k,H_k(x)},$$

where p and q represent the probability of keeping the original value and flipping to the opposite value, respectively. Due to the randomization, the minimum count in the perturbed sketch is not always in the same position as in the original sketch, i.e., $k \neq k'$. However, because the gap between different counts in sketches is diverse and related to the count of specific items, it is difficult to turn the above inequality into an equation by constructing a $h(x)$. To solve this problem, we propose the ordering matrix.

The ordering matrix $O^{(i)}$ is the background information provided by users, to assist the collector in getting the same row index of the minimum value as the original matrix, which takes advantage of the insensitivity of LDP to any background information to keep the privacy. The ordering matrix $O^{(i)}$ is a $K \times M$ matrix, where each position represents the serial number of the corresponding position in the original sketch $X^{(i)}$ ordered by count. Firstly, each counter $X^{(i)}_{k,m}$ is distributed into different groups $G^{(i)}_v$ according to its count v. As a result, $G^{(i)}_v$ includes a set of counters $\{(k,m)|X^{(i)}_{k,m} = v\}$ and its length is denoted by $|G^{(i)}_v| = g_v$. Secondly, each group G_v is bound with its order range $R^{(i)}_v = [\sum_{v' \leq v} g_{v'}, \sum_{v' \leq v} g_{v'} + g_v]$. Thirdly, we randomly sample an order without replacement from $R^{(i)}_v$ for each counter in $G^{(i)}_v$ where the order selected for each counter $X^{(i)}_{k,m}$ is denoted as $r^{(i)}_{k,m}$. Finally, we update the ordering matrix $O^{(i)}_{k,m} = r^{(i)}_{k,m}$. Thus, the collector can get the same minimum index by comparing the order of counters in $O^{(i)}$: this has the same result as calculating the minimum index on the original sketch $X^{(i)}$. An example is shown in Fig. 2.

In the following, we prove the estimation is unbiased in PrivSketch (Sect. 3.3), and analyze the impact of the ordering matrix on privacy (Sect. 3.4).

3.3 Utility Proof and Improvements

We present the protocol details on the user- and collector-side. We prove that the estimations are unbiased, and analyze the variances of errors, then employ sampling to achieve high utility.

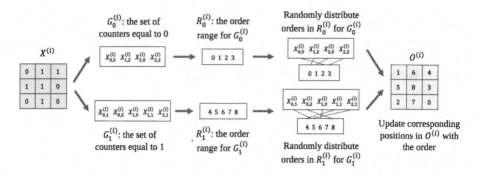

Fig. 2. The process of generating the ordering matrix.

Algorithm 2: PrivSketch-User

Input: $S^{(i)}, \epsilon, n, K, M, \mathcal{H}$

1 initialize a sketch $X^{(i)} \leftarrow \{0\}^{K \times M}$;

2 **for** *each* $\ell \in [1, L^{(i)}]$, *each* $k \in [1, K]$ **do**

3 $\quad \Big|\ X^{(i)}_{k, H_k(s^{(i)}_\ell)} = 1$;

4 generate the ordering matrix $O^{(i)}$;

5 **for** *each* $k \in [1, K]$, *each* $m \in [1, M]$ **do**

6 $\quad \Big|\ $ sample r from $[0, 1]$ uniformly;

7 $\quad \Big|\ $ **if** $r < \frac{1}{e^{\epsilon/KM}+1}$ **then**

8 $\quad \Big|\quad \Big|\ \hat{X}^{(i)}_{k, H_k(s^{(i)}_\ell)} = -2X^{(i)}_{k, H_k(s^{(i)}_\ell)} + 1$;

9 $\quad \Big|\ $ **else**

10 $\quad \Big|\quad \Big|\ \hat{X}^{(i)}_{k, H_k(s^{(i)}_\ell)} = 2X^{(i)}_{k, H_k(s^{(i)}_\ell)} - 1$;

11 **return** $\hat{X}^{(i)}, O^{(i)}$

User-Side Protocol (Algorithm 2). It consists of an encoder (lines 1–4) and a perturber (lines 5–10). In the encoder, each user records locally whether x appears, because our objective is to obtain the frequency of any value x in $\mathcal{D}$ (instead of counts). Consequently, an update in the encoder is a boolean disjunction, not an integer addition. Each position $X_{k,m}$ is initialized as *False* (i.e., 0). When x is hashed to $X_{k,m}$, the update is $X_{k,m} = X_{k,m} \vee True = True$ (line 3). After encoding, the ordering matrix is computed by the encoder. The perturber uses the randomized response mechanism (as in PCMS-Mean) to perturb each value to the opposite value with a probability of $\frac{1}{e^{\epsilon/KM}+1}$ due to at most $K \times M$ different positions.

Collector-Side Protocol (Algorithm 3). First, the decoder estimates the perturbed frequency of the value x using the perturbed minimum in $\hat{\mathcal{X}}$. The position of the minimum is provided by the background information $\mathcal{O}$ (line 4). Next, the calibrator removes the perturbation error to obtain the final estimation (line 6). The utility proof of the protocols follows.

Algorithm 3: PrivSketch-Collector

    Input: $x, \epsilon, n, M, \mathcal{H}, \hat{\mathcal{X}}, \mathcal{O}$

1 select a set of hash functions $\mathcal{H} = \{H_1, H_2, \ldots, H_K\}$;

2 $C(x) \leftarrow 0$;

3 **for** *each* $i \in [1, n]$ **do**

4      $k_{\min} \leftarrow \arg\min_k O^{(i)}_{k, H_k(x)}$;

5      $C(x) \leftarrow C(x) + \hat{X}^{(i)}_{k_{\min}, H_{k_{\min}}(x)}$;

6 $\hat{f}(x) \leftarrow \frac{1}{2}(\frac{e^{\epsilon/KM}+1}{e^{\epsilon/KM}-1}\frac{C(x)}{n} + 1)$;

7 **return** $\hat{f}(x)$

Theorem 3. *Let $C(x)$ denote the perturbed counters for each value in $\mathcal{D}$. $\hat{f}(x) = \frac{1}{2}(\frac{e^{\epsilon/KM}+1}{e^{\epsilon/KM}-1}\frac{C(x)}{n} + 1)$ is an unbiased estimation of $\widetilde{f}(x) = \frac{1}{n}\sum_{i=1}^{n}\min_k X^{(i)}_{k, H_k(x)}$ which is the frequency inferred from the original count-min sketch. Furthermore, the variance of $\hat{f}(x)$ is $\frac{e^{\epsilon/KM}}{n(e^{\epsilon/KM}-1)^2}$.*

Proof. For each user U_i, the counters for the item x in row k of perturbed sketch $\hat{X}$ is denoted by $\hat{X}^{(i)}_{k, H_k(x)}$, which value is determined by $X^{(i)}_{k, H_k(x)}$ (lines 6–10 in Algorithm 2). $C(x)$, which represents the result by aggregating the perturbed counters at the minimum position $\min_k \hat{X}^{(i)}_{k, H_k(x)}(x)$, equal to $\sum_{i=1}^{n}\min_k \hat{X}^{(i)}_{k, H_k(x)}(x)$, satisfies: $\mathbb{E}[C(x)] = 2(p-q)n\widetilde{f}(x)+(2q-1)n$, $\mathrm{Var}[C(x)] = 4n\{(p+q-1)(p-q)\widetilde{f}(x) + q(1-q)\}$, where $n\widetilde{f}(x)$ is the estimated number of users with x in their sequences using the set of original sketch $\mathcal{X}$. In our protocol, $p = \frac{e^{\epsilon/KM}}{e^{\epsilon/KM}+1}, q = \frac{1}{e^{\epsilon/KM}+1}$. Thus, the expectation of $\hat{f}(x)$, can be shown to be equal to $\widetilde{f}(x)$ as follows, which means the estimation is unbiased. And its variance of $\hat{f}(x)$ is satisfied:

$$\mathbb{E}[\hat{f}(x)] = \frac{1}{2}(\frac{e^{\epsilon/KM}+1}{e^{\epsilon/KM}-1}\frac{\mathbb{E}[C(x)]}{n} + 1) = \widetilde{f}(x) \qquad (2)$$

$$\mathrm{Var}[\hat{f}(x)] = \frac{1}{4n^2}\frac{(e^{\epsilon/KM}+1)^2}{(e^{\epsilon/KM}-1)^2}\mathrm{Var}[C(x)] = \frac{e^{\epsilon/KM}}{n(e^{\epsilon/KM}-1)^2}. \qquad (3)$$

Sample the Sketches. Following the above design, larger K and M make the perturbation probability closer to $\frac{1}{2}$ as random. And the variance also increases at the same time. The limited privacy budget ϵ/KM for each counter makes the collector receive scarcely useful information from the perturbed sketches, making it difficult to infer the true frequency. To solve the problem, the sampling technique is a common solution, i.e., randomly sampling one from $K \cdot M$ counters on the user end. Thus, for each counter chosen, the privacy budget becomes ϵ. The variance now is $\mathrm{Var}[\hat{f}(x)] = \frac{KMe^{\epsilon}}{n(e^{\epsilon}-1)^2}$, which is linearly related to $K \cdot M$ due to the sampling error, thus increasing more slowly than the exponential relation in Eq. (3). However, it is challenging to obtain the optimal sketching, because as K and M increase, the collision error introduced by Count-Min Sketch decreases, which is also related to the data domain size d and its distribution [9]. Though, we experimentally evaluate the effect of K and M on frequency estimation in Sect. 4.2. Besides, the utility of sampling in sketches is also verified by comparing with traditional PSFO [26] in Sect. 4.1.

3.4 Privacy Analysis

When the user sends only the perturbed counter $\hat{X}_{k,m}^{(i)}$ to the collector with the flipping probability $\frac{1}{e^\epsilon+1}$, ϵ-LDP is satisfied. However in PrivSketch, the user need also send the ordering matrix $O^{(i)}$ to the collector which may expose useful messages and indirectly damage the privacy. In the following, we analyze the influence of $O^{(i)}$ on privacy.

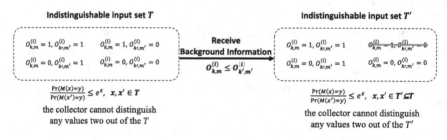

Fig. 3. Example of the effect of background information on indistinguishable input set.

The ordering matrix $O^{(i)}$ can be utilized to exclude some possible inputs for the collector, but the collector still cannot distinguish some inputs. As Fig. 3 shows, if $O_{k,m}^{(i)} \le O_{k',m'}^{(i)}$, $X_{k,m}^{(i)} = 1$ and $X_{k',m'}^{(i)} = 0$ will not hold at the same time. Thus, the cases of the possible sketches of users are reduced from 4 to 3 in the collector's view. To quantify the effect of the background information, we introduce *indistinguishable input set* to represent the possible inputs in the collector's view, denoted by T. According to the *LDP* definition, any two inputs are indistinguishable regardless of any background knowledge from the adversary. Therefore, we can deduce that any two inputs in the indistinguishable set still satisfy the *LDP* definition, even though the indistinguishable set becomes smaller than without the background information.

Theorem 4. *Consider a mechanism $\mathcal{M}$ that satisfies ϵ-LDP, its indistinguishable input set T, and any two inputs x, x'. When the collector receives any output y, along with the background information I, there exists an indistinguishable set $T' \subseteq T$ satisfying the following inequality: $\frac{\Pr[M(x)=y]}{\Pr[M(x')=y]} \le e^\epsilon$, $x, x' \in T'$.*

Table 1. Datasets characteristics

Dataset	n	d	max	min	P_{90}	Dataset	n	d	max	min	P_{90}
Kosarak	990002	41270	2498	1	15	AOL	521693	1632788	61932	1	62
Dataset1	10000	100000	123	1	80	Dataset2	100000	100000	117	1	78
Dataset3	100000	20000	112	1	73	Dataset4	100000	40000	107	1	72
Dataset5	100000	60000	110	1	74	Dataset6	100000	80000	109	1	75

Proof. For any I, T can be divided into two parts, T_+ and T_-. The former represents the inputs that are consistent with the information I, i.e., the possible inputs when I is true. The latter includes the inputs that contradict the information I, that is, the impossible inputs when I is true. Based on I, the collector can infer that the original input belongs to $T_+ (\subseteq T)$. For any two inputs $x, x' \in T_+$, x, x' is also in T. Therefore, following the definition of ϵ-LDP, $\frac{\Pr[M(x)=y]}{\Pr[M(x')=y]} \leq e^\epsilon$ is satisfied and any two input $x, x' \in T'$ is distinguishable.

The indistinguishable input set T' computed by the ordering matrix $\mathcal{O}$ is enough to protect the privacy of users in our problem. In PrivSketch, what each user needs to protect is its original sketch matrix $X^{(i)}$. Thus, the collector should not infer the value of any counter in $X^{(i)}$ is 1 or 0. In PrivSketch, counters can be divided into two groups, G_1 and G_0, and $g_1 + g_0 = KM$. Thus, when the collector receives $O^{(i)}$, the indistinguishable input set T' at most includes $KM + 1$ possible sketches with different sizes of each group. There are some constraints for sketches, e.g., it is impossible that $g_1 = 1, 2, 3$, because when there is an item occurred, for each $k \in [1, K]$, $\exists (k, m) \in G_1, m \in [1, M]$. Nevertheless, $\{0\}^{KM}, \{1\}^{KM} \in T'$ always holds. Thus, there is no counter with same value in different possible inputs, that is, its value is equal to 1 in some inputs and equal to 0 in the other inputs. The collector still cannot determine the value of each counter, which is sufficient to protect the privacy of users.

4 Experimental Evaluation

In this section, we evaluate the utility and running time of PrivSketch over synthetic and real datasets, and analyze how the main parameters affect its performance. For a comprehensive evaluation, we compare PrivSketch to the state-of-the-art PCMS-Mean [20], and PSFO [26] based on OLH [24] (denoted as PS-OLH in our experiments) for frequency estimation, and SVIM [26], a two-phase heavy hitter discovery protocol for discovery of frequent items.

Environment. We implement all LDP protocols in Python and conduct experiments on a server with 2 Intel Xeon 3206R Processors and 32G RAM running Centos. We repeat each experiment 10 times and report the average results.

Datasets. We use 6 synthetic datasets and 2 real datasets (see Table 1).

- **Synthetic Datasets:** These datasets follow Zipf distribution that real data stream often conforms to, with different number of users n and domain size d.

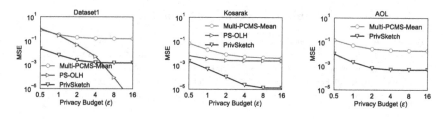

Fig. 4. Experimental results for frequency estimations.

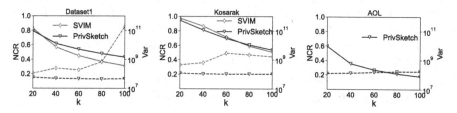

Fig. 5. VAR and NCR when varying parameter k.

- **Kosarak [4]:** This dataset contains the clicked items that anonymized users from a Hungarian online news portal, involving nearly 1M users and 40K items.
- **AOL [12]:** This dataset contains search queries of users on AOL between March 1 and May 31, 2016, with corresponding URLs clicked by them. The dataset includes more than 500K users with 1.6 million distinct URLs.

Parameters. The number of hash functions K is set to 4, and each hash function's hash domain size M is set to 128. The default privacy budget ϵ is 3, within the acceptable range in many works [8,19,27].

Evaluation Measures. We use the following measures, including running time.

- **Mean Squared Error (MSE).** We evaluate the frequency estimation accuracy by MSE: $\frac{1}{d}\sum_{x\in\mathcal{D}}(\hat{f}(x)-f(x))^2$, where $f(x)$ is x's true frequency.
- **Variance (Var).** We measure the error of estimating the top-k frequency terms using variance: $\frac{1}{|C_e \cap C_t|}\sum_{x\in C_e\cap C_t}(n\hat{f}(x)-nf(x))^2$.
- **Normalized Cumulative Rank (NCR).** To evaluate the estimation of frequent items, NCR measures how many top-k items are identified by the protocol with a quality function $q(.)$. It is calculated as follows: $\sum_{x\in C_e}q(x)/\sum_{x'\in C_t}q(x')$, where C_t and C_e represents the true top-k items and the estimated top-k items respectively. For $x\in C_t$ with a rank i, $q(x)=k+1-i$. For $x\notin C_t$, $q(x)=0$.

4.1 Comparing to Advanced Protocols

Experiments for Frequency Estimation: We compare our protocol to two advanced solutions: (i) a sketch-based solution, Multi-PCMS-Mean, which is an

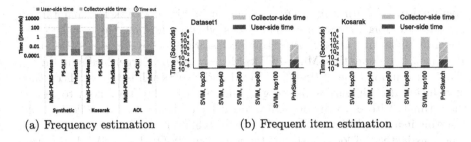

(a) Frequency estimation   (b) Frequent item estimation

Fig. 6. Comparison of running times.

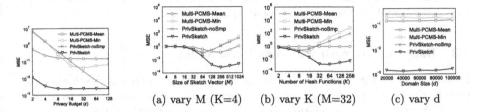

Fig. 7. MSE when   Fig. 8. MSE when varying parameters K, M, d.
varying ϵ, $n = 10^4$.

extended version of PCMS-Mean [20] for multi-item collection, and (ii) a non-sketch-based solution, PS-OLH, which is an advanced PSFO [26]. PSFO [26] combines the padding and sampling technique with a basic frequency estimation protocol to transform multiple-item into one-item problems. Because the optimal local hash (OLH) [26] performs best when $d \geq 3e^\epsilon + 2$ (i.e., for large domains), we choose the PSFO with OLH, i.e., PS-OLH, as our competitor. For a fair comparison, we assume the distribution of user input length is known and set the padding length l of PS-OLH to the 90th percentile of the user input [19] (avoiding to use the privacy budget to estimate l).

We evaluate the MSE of frequency estimation under different privacy budgets, varying from 0.5 to 16, on synthetic and real datasets. As shown in Fig. 4, PrivSketch performs best, especially for small privacy budgets, which indicates the high utility of PrivSketch and its strong privacy protection.

Experiments for Frequency Item Estimations: We also evaluate the performance of PrivSketch in frequent item mining (i.e., heavy hitter discovery), a popular application of frequency estimation. We compare it with the existing advanced multi-phase protocol, SVIM [26], which is the improved work after LDPMiner [19] and is also applicable in large domains. As shown in Fig. 5, PrivSketch performs better than SVIM, especially in frequency estimation for top-k items. It is expectable because PrivSketch has been designed for accurate frequency estimation, not frequent item identification.

Evaluation of Running Time. As shown in Fig. 6, Privsketch maintains a user-side running time smaller than 0.01 s while performing the calculation of the ordering matrix. Overall, PrivSketch is faster than PS-OLH and SVIM about 100 times, but slower than Multi-PCMS-Mean with much larger MSEs (in Fig. 4). The long running time of PS-OLH, SVIM and PrivSketch is the sacrificed time of reducing domain cardinality to gain high utility. Thus they need to restore the estimated items to the original domain for each user on the collector side, resulting in a complexity of $\mathcal{O}(nd)$. In PrivSketch, each user shares the same hash functions instead of local hash functions used in PS-OLH and SVIM, resulting in fewer hash function calculations on the collector side. We omit experimental results for PS-OLH and SVIM over AOL, because they need more than 10 days to compute, making them cumbersome to use in practice.

4.2 Experiments with Different Parameters

In this section, we compare PrivSketch with other sketch-based solutions to present the effect of our design under different parameters. In addition to Multi-PCMS-Mean, its min-estimation variant (denoted by Multi-PCMS-Min) and a middle version of PrivSketch without sampling (denoted by PrivSketch-noSmp) are also compared, to show the better utility of min estimation and the effect of our decode-first and sampling design.

Utility with Small Number of Users. We evaluate the MSE on Dataset1 with 10^4 users under a privacy budget range $[2, 128]$. Note the unrealistic privacy budget used here is to show the effect of our designs. In Fig. 7, PrivSketch always performs best especially under a small ϵ. We observe similar results (omitted for brevity) when n varies in $[10^4, 10^6]$. The result verifies that decode-first workflow with the ordering matrix effectively reduces the collision probability of sketches, and the min estimation has better accuracy than the mean estimation.

Impact of the Size of the Domain. We conduct this experiment on a group of synthetic datasets, which sets K, M and ϵ with default values, fixes $n = 10^5$, and varies the domain size d. As shown in Fig. 8(c), the errors of the four protocols only slightly increase with the increase of d. Theoretically, in a larger domain, when the sketch size is fixed, the collision probability increases, leading to an increase in error. However, since the items held by each user are sparse compared to the domain space, and the distribution of the number of items held by each user changes a little, the domain size change has a small impact. This confirms that sketching is an effective domain reduction and encoding method for data collection from a large domain.

Impact of the Parameters of the Sketch. In Fig. 8, we evaluate the effects of different K and M of the sketching using the datasets with parameters $n = 10^5$, $d = 10^5$ under $\epsilon = 3$. In Fig. 8(a), we can see that the utility of the PrivSketch is far better than the other three protocols under different M while fixing the hash vector size $K = 4$. As expected, increasing the size of the hash vector can reduce the estimation error. However, when M increases to a specific value, the error does not decrease but increases. This is because M affects two types of errors in these sketch-based LDP protocols. When M increases, the collision probability decreases, but the perturbation probability or the sampling errors increases. Varying the K brings a similar result to M, as shown in Fig. 8(b). However, the effects of K on Multi-PCMS-Mean protocol is different. Changes of K do not affect its MSE, because the effect of choosing one of the K hash functions when encoding is eliminated by the sum of K counters corresponding to K hash functions during the estimation process.

5   Related Work

Set-Valued Data Collection. The diverse set size is a challenge for set-value data collection under LDP. Padding and Sampling [19] is a common way to unify the set length, such as in PSFO [26], PrivSet [22]. Although Wang [23] proposes the wheel mechanism to reduce the computational overhead, these works do not aim at a large domain, where an efficient data structure is needed. Many works [3, 19,26,27], focus on frequent item mining, also known as heavy-hitters discovery in a huge domain. They utilize a multi-phase strategy to reduce the large domain size first, using a small part of the privacy budget to discover frequent candidates, and using the remaining part to obtain an accurate estimation. Nevertheless, this strategy is not suitable for estimating frequency.

Frequency Estimation with Hash-Encoding Technique. Under LDP settings, to reduce the data domain, RAPPOR [15] adopts Bloom filters to encode data, which requires expensive computations to use LASSO regression for the estimation. OLH [24] utilizes local hash functions to encode the user data, which requires a large number of hash calculations. With a simpler estimation solution, Count-Mean Sketch [20] was proposed to compute the populated emoji in IOS. [18,21] improve it by sending multiple sketches for each user, which also brings extra communication costs. [3] uses Count-Median Sketch with the Hadamard transform when computing the heavy hitters. [8] analyzes and compares LDP protocols with different sketching algorithms, including the Count-Min Sketch. However, these protocols, designed for the one-item collection, do not consider the error introduced by the sketching algorithm. Recently, [30] utilized hash functions to compute the frequency and the mean estimation of the k-sparse vector, with an assumption on the number of items each user generates.

Variants of LDP. Lots of works focus on optimizing the variants of LDP to improve its utility. Some works introduce extra trust in LDP, such as shuffling anonymized reports from users [7,14], and combining the centralized DP with the local version [2]. Some works introduce an extra parameter to relax the privacy constraint, such as [1,16] that use the distance metric of two inputs to improve the utility, which is inspired by the geo-indistinguishability concept [5]. Finally, some studies propose discriminative LDP based on different aspects, such as personalized privacy demand [6,29]. These works do not utilize the background information to enhance utility as we do in this paper.

6   Conclusions

This paper studies the frequency estimation problem under local differential privacy. We propose a privacy-preserving data collection protocol, PrivSketch, which does not expose the original value of any counter in the sketch. We experimentally verify the effectiveness of PrivSketch: it outperforms existing LDP protocols by 1–3 orders of magnitude and executes up to $\sim100\times$ faster.

Acknowledgments. We sincerely thank Dr. Zhenyu Liao for his insightful and constructive comments and suggestions on mathematical proof that help to improve the quality. This work is funded by NSFC Grant No. 62202450, Huawei New IP open identification resolution system project No. TC20201119008 and Postdoctoral Exchange Program No. YJ20210185.

References

1. Alvim, M.S., Chatzikokolakis, K., Palamidessi, C., Pazii, A.: Metric-based local differential privacy for statistical applications. arXiv preprint (2018)
2. Avent, B., Korolova, A., Zeber, D., Hovden, T., Livshits, B.: BLENDER: enabling local search with a hybrid differential privacy model. In: USENIX Security (2017)
3. Bassily, R., Nissim, K., Stemmer, U., Guha Thakurta, A.: Practical locally private heavy hitters. In: NIPS, vol. 30 (2017)
4. Bodon, F.: A fast apriori implementation. In: FIMI, vol. 3, p. 63 (2003)
5. Chatzikokolakis, K., Andrés, M.E., Bordenabe, N.E., Palamidessi, C.: Broadening the scope of differential privacy using metrics. In: De Cristofaro, E., Wright, M. (eds.) PETS 2013. LNCS, vol. 7981, pp. 82–102. Springer, Heidelberg (2013). https://doi.org/10.1007/978-3-642-39077-7_5
6. Chen, R., Li, H., Qin, A.K., Kasiviswanathan, S.P., Jin, H.: Private spatial data aggregation in the local setting. In: ICDE (2016)
7. Cheu, A., Smith, A., Ullman, J., Zeber, D., Zhilyaev, M.: Distributed differential privacy via shuffling. In: Ishai, Y., Rijmen, V. (eds.) EUROCRYPT 2019. LNCS, vol. 11476, pp. 375–403. Springer, Cham (2019). https://doi.org/10.1007/978-3-030-17653-2_13
8. Cormode, G., Maddock, S., Maple, C.: Frequency estimation under local differential privacy. PVLDB **14**(11), 2046–2058 (2021)
9. Cormode, G., Muthukrishnan, S.: An improved data stream summary: the count-min sketch and its applications. J. Algorithms **55**(1), 58–75 (2005)
10. Ding, B., Kulkarni, J., Yekhanin, S.: Collecting telemetry data privately. In: NIPS, vol. 30 (2017)
11. Duchi, J.C., Jordan, M.I., Wainwright, M.J.: Local privacy and statistical minimax rates. In: FOCS, pp. 429–438 (2013)
12. Dudek, G.: Aol search log (2007). http://www.cim.mcgill.ca/~dudek/206/Logs/AOL/
13. Dwork, C., McSherry, F., Nissim, K., Smith, A.: Calibrating noise to sensitivity in private data analysis. In: Halevi, S., Rabin, T. (eds.) TCC 2006. LNCS, vol. 3876, pp. 265–284. Springer, Heidelberg (2006). https://doi.org/10.1007/11681878_14
14. Erlingsson, Ú., Feldman, V., Mironov, I., Raghunathan, A., Talwar, K., Thakurta, A.: Amplification by shuffling: from local to central differential privacy via anonymity. In: SODA (2019)

15. Erlingsson, Ú., Pihur, V., Korolova, A.: RAPPOR: randomized aggregatable privacy-preserving ordinal response. In: CCS (2014)
16. Gursoy, M.E., Tamersoy, A., Truex, S., Wei, W., Liu, L.: Secure and utility-aware data collection with condensed local differential privacy. TDSC **18**, 2365–2378 (2019)
17. McSherry, F.D.: Privacy integrated queries: an extensible platform for privacy-preserving data analysis. In: SIGMOD, pp. 19–30 (2009)
18. Piao, C., Hao, Y., Yan, J., Jiang, X.: Privacy protection in government data sharing: an improved LDP-based approach. SOCA **15**, 309–322 (2021). https://doi.org/10.1007/s11761-021-00315-3
19. Qin, Z., Yang, Y., Yu, T., Khalil, I., Xiao, X., Ren, K.: Heavy hitter estimation over set-valued data with local differential privacy. In: CCS, pp. 192–203 (2016)
20. Team, D.P.: Learning with privacy at scale. Apple Mach. Learn. J. **1**(8) (2017)
21. Vepakomma, P., Pushpita, S.N., Raskar, R.: DAMS: meta-estimation of private sketch data structures for differentially private COVID-19 contact tracing. Technical report (2021)
22. Wang, S., Huang, L., Nie, Y., Wang, P., Xu, H., Yang, W.: PrivSet: set-valued data analyses with locale differential privacy. In: INFOCOM, pp. 1088–1096 (2018)
23. Wang, S., Qian, Y., Du, J., Yang, W., Huang, L., Xuy, H.: Set-valued data publication with local privacy: tight error bounds and efficient mechanisms. PVLDB **13**, 1234–1247 (2020)
24. Wang, T., Blocki, J., Li, N., Jha, S.: Locally differentially private protocols for frequency estimation. In: USENIX Security Symposium, pp. 729–745 (2017)
25. Wang, T., et al.: Continuous release of data streams under both centralized and local differential privacy. In: CCS (2021)
26. Wang, T., Li, N., Jha, S.: Locally differentially private frequent itemset mining. In: S&P, pp. 127–143 (2018)
27. Wang, T., Li, N., Jha, S.: Locally differentially private heavy hitter identification. TDSC **18**(2), 982–993 (2019)
28. Warner, S.L.: Randomized response: a survey technique for eliminating evasive answer bias. J. Am. Stat. Assoc. **60**(309), 63–69 (1965)
29. Yiwen, N., Yang, W., Huang, L., Xie, X., Zhao, Z., Wang, S.: A utility-optimized framework for personalized private histogram estimation. IEEE TKDE **31**, 655–669 (2018)
30. Zhou, M., Wang, T., Chan, T.H., Fanti, G., Shi, E.: Locally differentially private sparse vector aggregation. In: S&P (2022)

On Tuning the Sorted Neighborhood Method for Record Comparisons in a Data Deduplication Pipeline
Industrial Experience Report

Paweł Boiński[1] , Witold Andrzejewski[1] , Bartosz Bębel[1] ,
and Robert Wrembel[1,2](✉)

[1] Poznan University of Technology, Poznań, Poland
{pawel.boinski,witold.andrzejewski,bartosz.bebel,
robert.wrembel}@put.poznan.pl
[2] Artificial Intelligence and Cybersecurity Center, Poznań, Poland

Abstract. Assuring high quality of data stored in information systems (ISs) is challenging and it is one of concerns of companies. Typically, data stored in ISs are not free from errors, which include among others wrong and missing values as well as duplicates. Data deduplication has received a lot of attention from the research community. The research efforts have resulted in a state-of-the-art data deduplication pipeline, supported by software tools and algorithms. One of the tasks in the pipeline consists in reducing the complexity of records comparisons. This task is known as blocking. Multiple algorithms for blocking have been proposed and one of them is the *sorted neighborhood* method. In this paper, we focus on tuning and evaluating the method on a real data set composed of 5.5M of customer records. To the best of our knowledge, this is the largest real data set being used in research. The findings reported in this paper come from a R&D project run for a big company in a financial sector.

Keywords: data deduplication pipeline · customers' records deduplication · sorted neighborhood · moving window size

1 Introduction

Institutions and enterprises worldwide use data governance strategies to manage data collected by their day-to-day business applications. These strategies are supported by the most advanced state-of-the-art data management and data engineering solutions (typically commercial). Despite using these solutions, some of the collected data include errors, like typos, wrong values, outdated values, and duplicates. Faulty data mainly concern customers, both individuals and institutions, since such data are typically entered manually into a system, are imported from legacy systems, and change in time (e.g., last names, addresses).

A special case of faulty data are duplicated customer records. For example, in a financial institution duplicates may be created as a result of: (1) acquisition of another institution, with its proper customer repositories, (2) financial products that for each product require a separate customer record in a system, (3) the imperfection of a software and processes used in data governance [5].

A remedy for the problem of duplicated data is a process of deduplication, combined with data cleaning. In the research literature, a base-line data deduplication (DD) pipeline has been proposed [12–14,17,20,21], cf. Sect. 2. It has become a standard pipeline for multiple DD projects. A DD process needs to compare pairs of records and compute their similarities. In an ideal case, compared records were cleaned in advance, to correct typing errors and wrong values, substitute nulls with values as well as homogenize value formats (e.g., phone numbers, dates), abbreviations, and names (e.g., street names). However, in real DD projects it is impossible to fully clean all data [5,25].

A naive approach to discover duplicates is to compare records in pairs between all records in a data set, which results in a quadratic computational complexity. To reduce this complexity, multiple so-called *blocking* methods (algorithms) have been developed and reported in the research literature, e.g., [3,4,8,11]. One of the algorithms is *sorted neighborhood*. This algorithm runs a sliding window of a given size over a set of records.

The *sorted neighborhood* was proved to be adequate for efficient record comparison [27]. Moreover, it is intuitive and has lower computational complexity than a naive blocking. Therefore, *sorted neighborhood* has gained a popularity in various DD projects. It has been used to compare not only simple records but also XML objects [23], RDF objects [15], and images [18]. A non-trivial task in *sorted neighborhood* is to define an adequate size of the sliding window. A window that is too small prevents from discovering all potential duplicates, whereas a too big window results in unnecessary comparisons of records, which are not duplicates.

In this paper, we **contribute** our findings on tuning the size of the sliding window in the *sorted neighborhood* method (see Sect. 4). The tuning was assessed by a series of experiments, which were run on **real customers data**, within a **R&D project** in the financial sector. The experiments reported in this paper were run on a data set of over 5.5M records. To the best of our knowledge, it is **the largest real data set** used in research on data deduplication techniques and reported in the research literature.

2   Overview of Data Deduplication

In the simplest implementation of a duplicate discovery process, all records in a given data set are pairwise compared with each other, at a quadratic complexity. Such complexity is inadequate for large data sets. Therefore, in order to facilitate duplicate discovery, the state-of-the-art data deduplication pipeline was proposed. In this section we outline the pipeline and some aspects in the pipeline that are pertinent to the topic of this paper.

2.1 Basic Data Deduplication Pipeline

The basic state-of-the-art data deduplication pipeline was used in multiple deduplication projects, e.g., [12–14,17,20,21]. The DD pipeline includes four basic
tasks (see Fig. 1), namely:

- blocking (a.k.a. indexing) - it organizes records into groups (called block building), such that each group includes records that may include potential duplicates; next, records in blocks are being pairwise compared;
- block processing (a.k.a. filtering) - its goal is to eliminate records that do not
  need to be compared as they do not represent potential duplicates;
- entity matching (a.k.a. similarity computation) - it computes similarity values
  between records compared in pairs, i.e., a value of each attribute in one record
  is compared to a value of a corresponding attribute in the second record, in
  the same pair;
- entity clustering - it aims at creating groups of similar records, from pairs of
  records representing highly probable duplicates.

Fig. 1. The basic state-of-the-art data deduplication pipeline

Each of the four tasks in the DD pipeline is challenging and each task is
supported by dedicated algorithms, see [13] for a concise overview. In this paper,
we focus on a particular *challenge within the blocking task*. This challenge is in
optimizing record comparisons in groups.

2.2 Blocking

To reduce the quadratic complexity of record comparisons, the so-called *blocking*
method was developed [3,4,8,11]. The first step in blocking consists in clustering
similar records into a collection (called a block), such that similar records are
co-located in the same block. The goal of blocking is to: (1) maximize precision
and recall, and (2) reduce the number of unnecessary record comparisons [20].
Multiple blocking techniques (algorithms) have been developed and reported in
the research literature, see [20] for the most recent overview.

Once records are organized into blocks, block processing is run. It consists in
pairwise comparing records in a block. One of the algorithms used here is *sorted
neighborhood*, see Sect. 2.3.

Apart from the blocking algorithm that has to be carefully selected to serve
well a given problem at hand, another parameter of blocking is the so-called
blocking scheme [16,26]. It defines attributes that are used to organize records
into blocks. Notice that it is possible that only portions of attribute values (e.g.,

the year of a birthdate instead of the whole birthdate) can be used to organize records into blocks.

The deduplication task, which is based on *sorted neighborhood*, is guided by the following parameters: (1) a blocking scheme, (2) the size of a window in which records are compared, and (3) similarity measures to compute similarities between pairs of compared records. The values of these parameters impact precision, recall, and runtime of the task. Typically, their values are set by experts.

2.3   Sorted Neighborhood

As it has been already mentioned, the process of comparing records in pairs is often implemented by the *sorted neighborhood* (SN) method. It consists in comparing records with each other in a moving window of size w records. Thus, within a given move of the window, only records that are enclosed by the window are compared with each other. Next, the window moves one record further. This record is then compared with all records that are enclosed by the window. This process repeats until the last record in a data set is compared.

An example of three initial steps in the SN method is shown in Fig. 2, where a data file includes M records (denoted as *record1, record2, ..., recordM*) and the size of the moving window is equal to 3. In the first step, records 1, 2, and 3 are pairwise compared. In the second step, the window moves by 1 record to include *record4*, which is compared with records in the window, i.e., 2 and 3. This process repeats until reaching *recordM*.

In the simplest approach, the size of the window is fixed (further called a *fixed-size* window). One challenge of the SN method is to define size of the moving window, which would be adequate for a given problem at hand. A window of too small size will prevent discovering all potential duplicates, since not all potential duplicates will be enclosed by the window, thus will not be discovered. A window of too big size will cause runtime deterioration, as in the same window there will be compared records that are not duplicates, thus their comparisons will be useless.

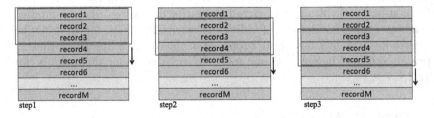

Fig. 2. Example steps in *sorted neighborhood*

A remedy for this problem is to use a window of a dynamic size [24,28]. The window will grow automatically until no potential duplicates are included in the window. This technique is known as a *dynamic* or *adaptive sorted neighborhood*. A few variants of the SN method with a dynamic window are possible. They define the moment until which the window is extended. The three intuitive approaches are the following:

- the window is extended until the values of selected compared attributes have exactly the same values; this approach is used in our implementation of the SN method;
- the window is extended until the similarity between the first and the last record in the window is greater than a given threshold value; tuning this value is nontrivial and it depends on the application domain and deduplication goals (discussing this topic is out of the scope of this paper);
- the window is extended if needed but until it reaches a predefined maximum size; this approach is also used in our implementation of the SN method.

2.4   Data Quality

In projects that process large amounts of data it is not possible to perfectly clean all data delivered to the DD pipeline. First, because not all data can be cleaned automatically - in such cases an expert knowledge and manual works are needed. Second, the amount of data that needs to be cleaned by a human may be too large to be done within a finite time and at reasonable monetary costs. Third, some data like last names cannot be changed (corrected) without an explicit permission of a customer, cf. [5,25]. As a consequence, data entering the DD pipeline may not be 100% clean, which is the case of our project. Erroneous data in our project include: transposed letters, missing letters, letters without diacritical Polish signs, in some cases abbreviated names of streets and cities, in some cases missing values of streets and/or cities.

For these reasons, data that are processed in the blocking task need to be compared based on an overall similarity value of the compared records. Such a similarity value, in turn, is computed based on similarities of values of attributes being compared. Typically the overall similarity of two records is computed as a weighted sum of similarities of individual attribute values. Multiple similarity measures for text data have been proposed in the research literature, their categories are outlined in Sect. 2.5, and more information about them can be found in [1,2,6,7,9,10,19].

2.5   Similarity Measures

Similarity measures represent one of the parameters to be set up in the *sorted neighborhood* method. Multiple similarity measures have been proposed and made available in various programming languages. The most popular similarity measures for text data are typically categorized as [2,10,19]:

- *edit distance* - a distance between character strings $s1$ and $s2$ is measured by the smallest number of edit operations that are required to convert $s1$ to $s2$, e.g., Levenshtein, Damerau-Levenshtein, Smith-Waterman;
- *n-grams* - a distance between $s1$ and $s2$ is measured by the number of n-grams common to both strings;
- *set similarity* - a distance between $s1$ and $s2$ is measured by the number of characters common to both strings, e.g., Overlap, Jaccard, Sorensen-dice;

Selecting the right similarity measure for a given DD problem is not straightforward. Challenges and solutions to this issue were analyzed in [1,2,6,7,9].

2.6 Computing Overall Record Similarity

In order to decide whether two records in a pair represent duplicates, an overall similarity of the whole records needs to be computed. To explain this, let us introduce the following notations:

- Let $R = \{A_1, A_2, \ldots, A_i\}$ be the schema of all records being compared, where A_i denotes an attribute.
- Let r_m be a record of schema R: $r_m = \{v_1^m, v_2^m, \ldots, v_i^m\}$, where v_i^m is the value of attribute A_i in record m.
- Let $simF(v_i^m, v_i^n)$ denote the similarity value between v_i^m and v_i^n, which is computed by means of similarity measure $simF$ (see Sect. 2.5).
- Let w_i denote the weight of attribute A_i.

Then, the similarity between r_m and r_n, denoted as $simF(r_m, r_n)$, is computed as a weighted sum of similarities of corresponding attribute values:

$$simF(r_m, r_n) = \sum_{i=1}^{n} w_i * simF(v_i^m, v_i^n).$$

3 Experimental Setup

In our project, three classes of record pairs are distinguished, which represent: (1) duplicates - denoted as T, (2) probable duplicates - P, and non-duplicates - N. Class T includes pairs such that $simF(r_m, r_n) \geq 0.92$, whereas class P includes pairs such that $0.85 \leq simF(r_m, r_n) < 0.92$.

3.1 Experimental Environment

The following test environment was used to perform the experiments. Customer data were stored in a relational database, and the deduplication process was performed using both SQL language commands and a program implemented in Python. The database served as the data source with sorting capability (necessary for *sorting neighborhood*), while the Python program performed all the computations for creating blocks and comparing records.

The experiments were run on a server with 64 GB RAM and Intel(R) Xeon(R) Gold 6226R CPU running at 2.9 GHz.

3.2 Data Sets

The experiments were run on a real data set including records describing customers of the financial institution. A customer record was composed of 23 typical
attributes describing a customer, all of them of text data types, which included
among others: (1) national ID, (2) first and last name, (3) living address, (4)
mailing address, and (5) contact numbers and emails. All these attributes were
used to compute similarities between record pairs. The total number of records
used in the experiments was equal to 5 557 224.

3.3 Blocking Scheme

In our DD pipeline, the *sorted neighborhood* algorithm was run on a customers
data set sorted by *birthdate* and *last name*. The sorting attributes had to fulfill
the following requirements: (1) sort records in such a way that potential duplicates were collocated close to each other, (2) not include nulls, (3) include low
number of erroneous values. The sorting keys were selected based on: data pro-
filing, a statistical method that we developed for this purpose, and on expert
knowledge.

3.4 Computing Record Similarities

In our approach, the similarity between two records was computed as a weighted
sum of similarities of corresponding attribute values, as described in Sect. 2.6.
The similarity measure ($simF$) that we applied was *Jaro-Winkler*, available in
the *textdistance* Python package. The selection of the measure was based on
the evaluation of similarity measures reported in [2]. Weights w_i of individual
attributes were found by means of: a mathematical programming algorithm [22],
experimental evaluation, manual tuning, and expert knowledge.

4 Results

The goal of the experimental evaluation was to find answers to the following
questions:

- How the number of discovered duplicates of classes T and P depends on the
  size of a moving window, for the *fixed-size* and *dynamic window* methods?
- How the percentage of discovered duplicates of classes T and P depends on
  the number of pair comparisons?
- How the percentage of discovered duplicates is impacted by a moving window
  size?

4.1 Number of Discovered Duplicates w.r.t. Window Size

These experiments assessed how the percentage of discovered duplicates of class T and class P depends on the size of a moving window, for the fixed-size and dynamic window methods.

The *fixed-size window* (denoted as FW) ranged from: (1) 2 to 10 records, increased by 1, (2) 10 to 100 records, increased by 10, and (3) 100 to 300 records increased by 50. For the dynamic window (denoted as DW), it was assumed that the minimum window size is analogous to that of a FW, but the maximum window size was always constrained to 300. This means that for the DW of size 2 we allowed its size to increase from 2 up to 300 records. Note that for the DW of maximum size 300, the content of generated blocks was exactly the same as for the FW of size 300.

In each execution of the SN algorithm, the number of pairs that were verified and the number of discovered duplicates T and probable duplicates P were measured. Figures 3 and 4 show the number of duplicates of class T and P discovered, respectively, depending on the type of a window used, i.e., fixed-size or dynamic.

In Fig. 3 we can observe how the number of discovered duplicates T changes depending on the type of a window and its size. For the smallest FW, i.e., the size of 2, slightly over 240 000 duplicates were discovered, and this number increases with the increasing size of the FW.

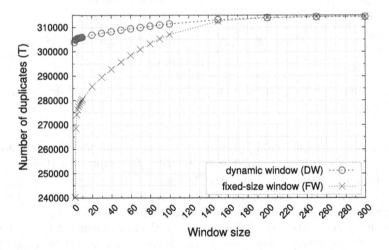

Fig. 3. The number of duplicates of class T discovered by the fixed-size and dynamic window methods

Looking at the results obtained using the DW, we can observe that for the smallest window, the number of discovered duplicates is already about 60 thousand more than for the FW. Of course, increasing the minimum size of the DW increases the number of discovered duplicates, but the growth rate decreases slowly with the increase of the window size. For the window of size of 150 records, the results of the DW and FW are very similar and differ only by 595 pairs (0.19%).

Similar results for duplicates of class P are shown in Fig. 4. In this case, a significant difference is observable only for the window with the smallest size, i.e., for the FW, 120 964 pairs were discovered, whereas for the DW, 136 037 pairs were discovered (that is, 15 073 more). For the window of size 3, the values are 134 788 (FW) and 137 554 (DW), respectively, i.e., the difference is equal to 2 766. Having performed a statistical test, it can be concluded that for windows of sizes greater than 5 there is no statistically significant difference in the results obtained by the FW and DW.

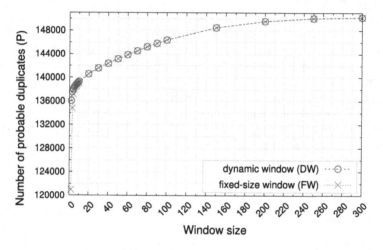

Fig. 4. The number of duplicates of class P discovered by the fixed-size and dynamic window methods

It is also worth mentioning that for duplicates of class T, the results for the DW were always at least as good as for the FW. For duplicates P we observed situations in which very few duplicates P were detected using the FW, while they were missing from the results obtained by the DW. The experiments revealed that the total number of such pairs was equaled to only 7.

The differences in the results for duplicates P and T are due to the properties of these different classes. Duplicates T usually are characterized by a significant number of repeated or very similar elements. Hence, using a dynamic window, neighbor groups of such similar values will be dynamically included in created blocks. This has the effect of producing very good results even for a very small window size. Duplicates P can be more diverse, including sorting key elements. Thus, the chance that they will be dynamically added to created blocks are much smaller.

4.2   Percentage of Discovered Duplicates w.r.t. Number of Pair Comparisons

Figures 5 and 6 show the percentage of discovered duplicates w.r.t. the number of compared pairs of records. The percentage is expressed in relation to the maximum number of duplicates discovered in the experiment (notice that this number was obtained for a window of size 300 records). Notice also that each pair comparison has a constant time cost. Thus, the number of comparisons can be directly translated into runtime.

The number of compared pairs depends on the window size, and for a FW it can be easily calculated. For a DW, this number depends also on the characteristics of the data being processed and it is limited by the maximum window size. In Fig. 5 and 6 we show the results (i.e., the percentage of discovered duplicates) w.r.t. the number of compared pairs (the minimum window size varied from 2 to 300).

Figure 5 shows the results for duplicates of class T. As we observe, the DW allows to discover approx. 97% of duplicates T, cf. to approx. 77% of duplicates P discovered by the FW (see the points close to value 0 on the X axis). By using the DW, 5 548 234 pairs were compared, whereas by using the FW, 5 405 830 pairs were compared, i.e., the DW compared only 2.6% more pairs than the FW, but it allowed to discover about 20% more pairs. The better performance of the DW is also well visible at $1 * 10^8$ comparisons, where the DW allowed to discover over 97.5% of duplicates, whereas the FW allowed to discover 92% of duplicates (having run the same number of comparisons).

Further, we can observe that when the number of comparisons increases, the percentage of discovered duplicates converges for the FW and the DW. At approximately $1.1 * 10^9$ comparisons, both types of windows produce the same number of discovered duplicates.

Figure 6 presents similar characteristics as the previous chart, but for discovered duplicates of class P. In this case, the characteristics of the FW and DW are very similar. The reason for such a behavior is that for duplicates of class P it is more likely that they do not share the same sorting key with each other, as compared to duplicates of class T. Thus, the possibility of discovering them is more affected by the minimum size of the window than by its dynamic expansion based on the common sorting key value. Consequently, the DW and FW allow to discover similar numbers of pairs of type P.

4.3   Percentage Change of Discovered Duplicates w.r.t. Window Size

Figures 7 and 8 show the changes in the number of duplicates discovered by *sorted neighborhood* by applying the DW of minimum size ws, compared to the discovered number of duplicates by using a previous window of minimum size, i.e., $ws - x$. To visualize this dependency, we defined a *discovery change ratio*

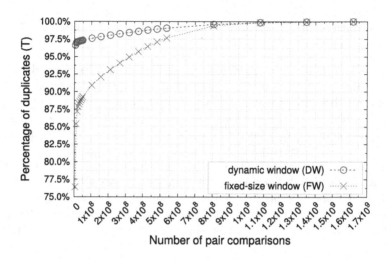

Fig. 5. The percent of duplicates of class T discovered w.r.t. the number of compared pairs

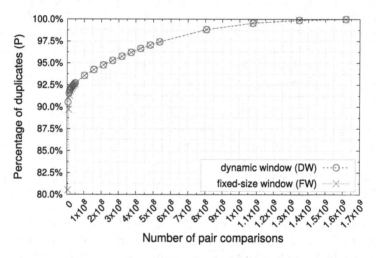

Fig. 6. The percent of duplicates of class P discovered w.r.t. the number of compared pairs

$dcr = (N_{ws} - N_{ws-x})/N_{ws-x}$, where N_{ws} and N_{ws-x} denote the number of duplicates discovered by using a window of size ws and $ws - x$, respectively; x represents a window size increase. Notice that x ranges from 1 to 50 depending on a window type (see Sect. 4.1). For example, value 0.5% on the Y axis means that increasing the window size by 1, results in discovering 0.5% more duplicates. Notice that in these figures a logarithmic scale is used.

The analysis of Fig. 7 reveals that when the minimum window size increases, the number of discovered duplicates decreases sharply. *dcr* for duplicates T is smaller than for duplicates P. This is due to the fact that for the dynamic window, a great number of duplicates of type T have been already discovered from the smallest window size (see Fig. 3).

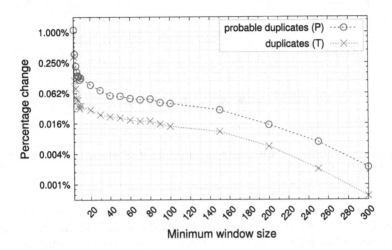

Fig. 7. The relative percentage change of the number of discovered duplicates w.r.t. the minimum size of a dynamic window

The possibility of discovering additional duplicates T when increasing the minimum size of the DW is limited as compared to the number of discovered duplicates of type P. Note that the increase of discovered duplicates T is about 3 times smaller than the increase of duplicates P. The smallest $dcr=3.67$ was measured for the largest window size, i.e., 300.

Even more interesting are the results for the FW, shown in Fig. 8. What primarily draws our attention is that the curves for the two types of duplicates intersect. Notice that the results for duplicates P are almost the same as in chart (Fig. 7). As we showed earlier (see Sects. 4.1 and 4.2), the number of discovered duplicates P depends mainly on the minimum window size, whereas the possibility of dynamically expanding the window does not significantly affect the number of discovered duplicates P.

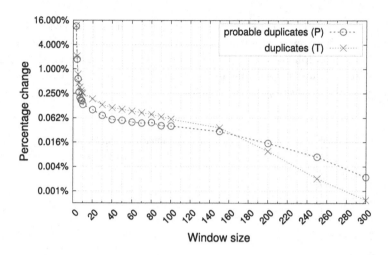

Fig. 8. The relative percentage change of the number of discovered duplicates w.r.t. the fixed-size window

The processing characteristics for duplicates T are completely different. For the smallest window size, the number of such duplicates discovered using the FW is relatively small as compared to the DW. Thus, increasing the size of the FW has a large effect on the number of discovered T duplicates. In the range of window size from 5 to approx. 170, the increments are larger than for duplicates P. The situation changes for larger window sizes.

One can notice that the results are resembling those obtained for the DW. This suggests that the maximum size of the DW that brings the greatest benefit from the use of dynamic window is approx. 170. Above this value, the values of dcr for the FW begin to be much less noticeable. This, of course, applies to processing where the minimum window size is less than 170.

5   Summary

In this paper we reported experimental evaluation of the *sorted neighborhood* method w.r.t.: the moving window size, for a dynamic window and a fixed-size window. The evaluation was performed on a real data set of over 5.5 million of records.

The most important findings from the evaluation are summarized as follows. First, for duplicates T and for the moving window of sizes < approx. 150 records, the dynamic window method allows to discover more pairs of duplicates T. For window sizes > approx. 150, the DW and FW methods allow to discover similar numbers of pairs. For duplicates P, the DW allows to discover more pairs than the FW only for window sizes < 4.

Second, the DW allows to discover more duplicates T than the FW, by running a lower number of pair comparisons. With limited computational resources, by applying the DW we can discover a set of duplicates T, which would require dozens of times more computations if using the FW. Both methods converge their results after exceeding $1 * 10^9$ comparisons. For duplicates P the characteristic of the DW is similar to the characteristics of the FW.

The findings from the experiments reported here were incorporated into the DD pipeline for customers data that we implemented in the R&D project (for details refer to [5]). The pipeline has already been deployed and run on over 20M records in the financial institution.

Acknowledgements. The project is supported by the grant from the National Center for Research and Development no. POIR.01.01.01-00-0287/19.

References

1. Alamuri, M., Surampudi, B.R., Negi, A.: A survey of distance/similarity measures for categorical data. In: International Joint Conference on Neural Networks (IJCNN), pp. 1907–1914. IEEE (2014)
2. Andrzejewski, W., Bębel, B., Boiński, P., Sienkiewicz, M., Wrembel, R.: Text similarity measures in a data deduplication pipeline for customers records. In: International Workshop on Design, Optimization, Languages and Analytical Processing of Big Data DOLAP, co-located with EDBT/ICDT. CEUR Workshop Proceedings, CEUR-WS.org (2023, to appear)
3. Baxter, R., Christen, P.: A comparison of fast blocking methods for record linkage. In: ACM SIGKDD Workshop on Data Cleaning, Record Linkage, and Object Consolidation (2003)
4. Bilenko, M., Kamath, B., Mooney, R.J.: Adaptive blocking: learning to scale up record linkage. In: The IEEE International Conference on Data Mining (ICDM), pp. 87–96. IEEE Computer Society (2006)
5. Boiński, P., Sienkiewicz, M., Bębel, B., Wrembel, R., Gałęzowski, D., Graniszewski, W.: On customer data deduplication: lessons learned from a R&D project in the financial sector. In: Workshops of the EDBT/ICDT 2022 Joint Conference. CEUR Workshop Proceedings, vol. 3135. CEUR-WS.org (2022)
6. Boiński, P., Sienkiewicz, M., Wrembel, R., Bębel, B., Andrzejewski, W.: Text similarity measures in a data deduplication pipeline for customers records. In: ACM/SIGAPP Symposium on Applied Computing SAC. ACM (2023, to appear)
7. Boriah, S., Chandola, V., Kumar, V.: Similarity measures for categorical data: a comparative evaluation. In: SIAM International Conference on Data Mining (SDM), pp. 243–254. SIAM (2008)
8. Cao, Y., Chen, Z., Zhu, J., Yue, P., Lin, C., Yu, Y.: Leveraging unlabeled data to scale blocking for record linkage. In: International Joint Conference on Artificial Intelligence IJCAI, pp. 2211–2217 (2011)
9. Christen, P.: A comparison of personal name matching: techniques and practical issues. In: International Conference on Data Mining (ICDM), pp. 290–294. IEEE Computer Society (2006)
10. Christen, P.: Data Matching - Concepts and Techniques for Record Linkage, Entity Resolution, and Duplicate Detection. DCSA, Springer (2012). https://doi.org/10.1007/978-3-642-31164-2
11. Christen, P.: A survey of indexing techniques for scalable record linkage and deduplication. IEEE Trans. Knowl. Data Eng. **24**(9), 1537–1555 (2012)

12. Christophides, V., Efthymiou, V., Palpanas, T., Papadakis, G., Stefanidis, K.: An overview of end-to-end entity resolution for big data. ACM Comput. Surv. **53**(6), 127:1–127:42 (2021)
13. Colyer, A.: The morning paper on An overview of end-to-end entity resolution for big data (2020). https://blog.acolyer.org/2020/12/14/entity-resolution/
14. Elmagarmid, A.K., Ipeirotis, P.G., Verykios, V.S.: Duplicate record detection: a survey. IEEE Trans. Knowl. Data Eng. **19**(1), 1–16 (2007)
15. Kejriwal, M.: Sorted neighborhood for the semantic web. In: AAAI Conference on Artificial Intelligence, pp. 4174–4175. AAAI Press (2015)
16. Kejriwal, M., Miranker, D.P.: An unsupervised algorithm for learning blocking schemes. In: IEEE International Conference on Data Mining, pp. 340–349. IEEE Computer Society (2013)
17. Köpcke, H., Rahm, E.: Frameworks for entity matching: a comparison. Data Knowl. Eng. **69**(2), 197–210 (2010)
18. Li, G., Wu, Q., Tu, D., Sun, S.: A sorted neighborhood approach for detecting duplicated regions in image forgeries based on DWT and SVD. In: IEEE International Conference on Multimedia and Expo ICME, pp. 1750–1753. IEEE Computer Society (2007)
19. Naumann, F.: Similarity Measures. Hasso Plattner Institute (2013)
20. Papadakis, G., Skoutas, D., Thanos, E., Palpanas, T.: Blocking and filtering techniques for entity resolution: a survey. ACM Comput. Surv. **53**(2), 31:1–31:42 (2020)
21. Papadakis, G., Tsekouras, L., Thanos, E., Giannakopoulos, G., Palpanas, T., Koubarakis, M.: Domain- and structure-agnostic end-to-end entity resolution with JedAI. SIGMOD Rec. **48**(4), 30–36 (2019)
22. Powell, M.J.D.: An efficient method for finding the minimum of a function of several variables without calculating derivatives. Comput. J. **7**(2), 155–162 (1964)
23. Puhlmann, S., Weis, M., Naumann, F.: XML duplicate detection using sorted neighborhoods. In: Ioannidis, Y., et al. (eds.) EDBT 2006. LNCS, vol. 3896, pp. 773–791. Springer, Heidelberg (2006). https://doi.org/10.1007/11687238_46
24. Ramadan, B., Christen, P., Liang, H., Gayler, R.W.: Dynamic sorted neighborhood indexing for real-time entity resolution. ACM J. Data Inf. Qual. **6**(4), 15:1–15:29 (2015)
25. Sienkiewicz, M., Wrembel, R.: Managing data in a big financial institution: conclusions from a R&D project. In: Workshops of the EDBT/ICDT 2021 Joint Conference. CEUR Workshop Proceedings, vol. 2841. CEUR-WS.org (2021)
26. de Souza Silva, L., Murai, F., da Silva, A.P.C., Moro, M.M.: Automatic identification of best attributes for indexing in data deduplication. In: Mendelzon, A. (ed.) International Workshop on Foundations of Data Management. CEUR Workshop Proceedings, vol. 2100. CEUR-WS.org (2018)
27. Vatsalan, D., Christen, P.: Sorted nearest neighborhood clustering for efficient private blocking. In: Pei, J., Tseng, V.S., Cao, L., Motoda, H., Xu, G. (eds.) PAKDD 2013. LNCS (LNAI), vol. 7819, pp. 341–352. Springer, Heidelberg (2013). https://doi.org/10.1007/978-3-642-37456-2_29
28. Yan, S., Lee, D., Kan, M., Giles, C.L.: Adaptive sorted neighborhood methods for efficient record linkage. In: ACM/IEEE Joint Conference on Digital Libraries JCDL, pp. 185–194. ACM (2007)

Managing Semantic Evolutions in Semi-Structured Data

Pedro Ivo Siqueira Nepomuceno[(✉)] and Kelly Rosa Braghetto

Department of Computer Science, University of Sao Paulo, Sao Paulo, Brazil
{pedro.siqueira,kellyrb}@ime.usp.br

Abstract. This paper introduces a model to store semi-structured data while documenting its semantic changes over time. The paper also presents algorithms for querying semantic evolved data, which conciliate the multiple versions the data may have. An implementation of the model and algorithms, MellowDB, was developed, and its performance was analyzed, showing the proposed algorithms and model are feasible.

Keywords: Databases · Semantic heterogeneity · Query Rewriting

1 Introduction

Several works have addressed database evolution in structured [3] and semi-structured databases [6]. Most, however, focus on schema evolution. Our work, on the other hand, focuses on operations over the attributes' values (semantic evolution), which change the data semantics over time. The Brazilian county of "Moji Mirim" for example, was renamed to "Mogi Mirim" in 2016 [5]. Official statistical data before 2016 refers to "Moji Mirim", while from 2016 and beyond, "Mogi Mirim" is referred to. In another example, "Laguna" was ungrouped in 2013 into "Laguna" and "Pescaria Brava". After ungrouping, numbers inform the population estimates for each new county. However, it is possible to group new estimates to make a grouped analysis using all previous registers.

Even when subtle, semantic heterogeneity can make old and new data incompatible so that they cannot be judiciously grouped or compared [9]. This paper presents a model to represent the semantic evolution of semi-structured data collections and algorithms for easily querying them. Both model and algorithms were implemented as a middle layer over MongoDB, and its performance was evaluated through extensive experiments.

This research is part of the INCT of the Future Internet for Smart Cities funded by CNPq proc. 465446/2014-0, Coordenação de Aperfeiçoamento de Pessoal de Nível Superior - Brasil (CAPES) - Finance Code 001, FAPESP proc. 14/50937-1, and FAPESP proc. 15/24485-9.

2   Related Work

Temporal Data Models (TDMs) preserve the complete history of data changes. This way, it is possible to retrieve current values and query states in specific moments of past time [8]. Most TDMs have been proposed or implemented using relational database management systems (RDBMS), although there are some implementations in semi-structured data, such as in JSON files [1].

TDMs do not directly tackle semantic evolution. Mainly because in semantic evolution, changes are generated following declared rules (the SEOs). But they do present a deep framework to deal with time, including timestamping, modeling, and querying techniques useful for dealing with semantic evolution.

Database Evolution demands special care to enable easy querying. The main strategy to support good querying interfaces for databases that suffered evolution is *query rewriting*. Moon et al. [7] and Möller et al. [6] developed systems capable of dealing with different schema versions using query rewriting as long as the evolution history is known. Another related technique (*delta code generation*) automatically generates views to mimic tables before and after the evolution. Herrmann et al. [4] presented a tool for generating delta code between schema versions.

It is important to notice that all the above-cited works deal with schema evolution. **Semantic evolution, which is the main target of this paper, is not dealt with**. In fact, semantic evolution is a less explored area in scientific literature. Ventrone [9] defined some types of semantic heterogeneity and evolution forms which result in operations similar to the ones considered in this work. However, no algorithms or models to deal with them were presented.

3   Framework to Handle Semantic Evolution Operations

This section formalizes a semantic evolution operation, the *translation*, to illustrate how to deal with semantic evolution in semi-structured collections. Other operations such as grouping and ungrouping can be defined similarly.

Definition 1. *A document $d = (t, V)$ contains a timestamp t and a set V of attribute-value pairs. The notation $V[a]$ will represent "the value of attribute a". In other words, $V = \{(a, v)|V[a] = v\}$.*

According to these definitions, documents may contain only simple values. The extension to consider complex values in nested structures is future work.

Definition 2. *The translation operation($T_{t_h,a,q,r}(d)$) transforms the value of attribute a of a document d from q to r starting at time t_h. This is defined as:*

$$T_{t_h,a,q,r}(d) = \begin{cases} (t, V \setminus \{(a, q)\} \cup \{(a, r)\}), & \text{if } t \leq t_h \text{ and } V[a] = q \\ d, & \text{otherwise} \end{cases} \quad (1)$$

```
{                          {                          {                          {                          {
  "s":1,                     "s":2,                     "o":"s23a",               "o":"s23a",               "o":"g567z",
  "time":"0001-01-01",       "time":"2016-01-01",       "V": {                     "V": {                     "V": {
  "next": {                  "prev": {                   "Country":"Brazil",        "Country":"Brazil",        "Country":"Brazil",
    "s":2,                     "s":1,                     "County":"Moji Mirim",     "County":"Mogi Mirim",     "County":"Rio de Janeiro",
    "type":"translation",      "type":"translation",      "Year":2015,               "Year":2015,               "Year":2016,
    "field":"County",          "field":"County",          "Population":91483         "Population":91483          "Population":6498837
    "from":"Moji Mirim",       "from":"Mogi Mirim",     },                         },                         },
    "to":"Mogi Mirim"          "to":"Mogi Mirim"        "s_min":1,                 "s_min":2,                 "s_min":1,
  }                          }                          "s_max":1,                 "s_max":2,                 "s_max":2
}                          }                          "evolved":[2]              "evolved":[1]              }
                                                      }                          }
```

     (a)   (b)

Fig. 1. (a) Versions collection C_s and (b) documents in the processed collection C_p. The first two of (b) are different semantic versions of the same document.

The example cited in Sect. 1 is a translation with q = $Moji\text{-}Mirim$, r = $Mogi\text{-}Mirim$ and t_h = 2016. The translation operation is reversible; it can be formalized similarly.

Definition 3. *A semantic evolution compatible collection C is composed of tuples (d, s), where d is a document and s is the semantic version of the document.*

 When a semantic evolution operation (SEO) takes place over a collection, first, a new semantic version is created. Then a new version of every document in the collection is created and associated with the new semantic version.

 Every version of a document is a copy of the original document after all changes from previous semantic operations are applied. Each tuple (d_1, s_i) is a *version* of the original document d_1. A *semantic version* s_i is as a subset of the semantic compatible collection, where all its associated documents have the same semantic interpretation for their attribute-value pairs.

4   Storage Model and Algorithms

The proposed model contains three collections. The *raw collection* stores the original documents. The *semantic versions collection* keeps metadata of the semantic versions. The *processed collection* stores documents in all semantic versions.

 The **Raw Collection** (C_r) contains the original document attribute-value pairs $(d_r.V)$ as well as its valid time $(d_r.time)$ and the *original version number* $(d_r.s)$ which is the version in effect using $d_r.time$ as reference.

 Each document (d_s) of the **Semantic Versions Collection** (C_s) contains: the *version number* $(d_s.s)$; the *valid time* $(d_s.time)$ of the version and the *next and previous version operation and arguments* $(d_s.next/d_s.prev)$ with the arguments of the SEO that needs to be applied to map the version into the next and the previous one (depending if it is a reversible operation or not). These two fields resemble a doubly linked list. Any arguments needed, such as the translation t_h, q, and r, are also included in these attributes. Figure 1a shows an example of two documents of C_s.

In the **Processed Collection** (C_p), storing one version of each document for each semantic version is impractical. A better approach is to associate documents with an interval of versions. Then, when there are no changes in a document, the version can only be extended. Each document d_p, besides the original attribute-value pairs set $(d_p.V)$ edited to fit its semantic version, includes the following metadata attribute-value pairs: the *original document ($d_p.o$)*, a reference to the original document in the raw collection; the *minimum ($d_p.s_{min}$)* and *maximum ($d_p.s_{max}$) version number* that define the limits of version range in which the copy of the document is valid; and the *evolution list ($d_p.evolved$)*, indicating every SEO that affected that document. Figure 1b shows an example of the processed collection for a document affected by a semantic evolution and one that has not. For documents that have not been affected, the full interval of semantic versions can be synthesized in only one processed document.

4.1   Semantic Operation Processing

The first semantic version document d_{s_1} is also created when the collection is created. Valid time of this version $(d_{s_1}.time)$ is set to zero $(d_{s_1}.t = 0)$.

When a SEO is executed, a new semantic version is created with a new version number. If the operation happened before another previously informed one, this number might be fractional to "fit" between two other pre-existing versions. The *prev* and *next* of neighboring versions must be reconnected correspondingly.

The next step is to process documents into the *processed collection* accordingly. For unaffected documents, limits of pre-existing processed documents are just extended. For each affected document, it is necessary to create another copy to represent it from that point in time. Figure 1b show an example of how documents stay when affected by a semantic operation (the Moji/Mogi Mirim case) and when not affected (the Rio de Janeiro case).

After all affected processed documents are copied or have their value extended, the SEO may occur. This step depends on the operation and will happen as stated in Definition 2 over the documents associated with the new semantic version s_j. Then, all posterior operations must be reapplied over these documents because their results might be different than before.

When new documents are inserted into the database, they must also be processed consistently, checking if it has been affected by any SEO.

4.2   Query Transformation Algorithm

When querying, it is necessary to consider semantic changes affecting queried attributes. This way, users may query an attribute by its old or new value seamlessly. To make an *attribute:value* filter query the procedure is:

1. Query the semantic versions collection (C_s) for any semantic "next" operation where the *attribute:value* combination has been transformed into another value (*attribute:new_value*). If there are any, add the *attribute:new_value :semantic_version_ number* of these semantic versions to a queue P.

While P is not empty, pop the first *attribute:value:semantic_version* tuple and make the same query again in C_s, to check if this attribute has been transformed into still another value. If it has, push the new *attribute:value :semantic_version* to P. If not, add to another list, L_2. This is to detect "new names" that could represent the queried value in the most recent version.

2. All of L_2 values will be used to compose the final query, using an OR operator in the selection criteria while filtering the semantic version in the *evolved* attribute. The original *attribute:value* is also added.

As an example, consider the collections shown in Figs. 1a to 1b and the query *"County":"Moji Mirim"*. The final version of the query (Q) will be:

$$Q = \{\text{"County":"Moji Mirim"}\}$$
$$\text{or } (\{\text{"county":"Mogi Mirim"}\} \text{ and } \{\text{"evolved\_contains":"1"}\}) \quad (2)$$

The final query should be executed in C_p, but only in the last semantic version subset. It considers both counties that were called "Moji Mirim" and were renamed to "Mogi Mirim" and counties that are still named "Moji Mirim" in order to consider homonyms also if they exist.

5   Implementation and Performance Analysis

To validate and evaluate the model and algorithms introduced in Sect. 4, we implemented MellowDB, a middle-layer library developed in Python to deal with semantic evolution in MongoDB. For now, it implements operations for insertion and querying. The developed code and all experiments scripts are publicly available on https://github.com/pisn/semantic_heterogeneous_database.

For the experiments, databases with 500K documents containing 20 fields, each with a domain of 20 possible values, were randomly generated. Five different scenarios were simulated in this phase: *Read-Only* (only queries), *Heavy Read* (95% of queries and 5% insertions), *Write Only* (only insertions), *Heavy Write* (95% insertions and 5% queries) and *50/50* (50% insertions and 50% queries). These scenarios were inspired by YCSB Framework workload scenarios [2]. Every experiment was repeated 5 times. Repetitions were executed over a newly created database in an environment with Debian 5.4.19-1 OS, Intel(R) Core(TM) i7-6700K CPU @ 4.00 GHz, and 30 GB RAM.

Figure 2a shows that queries suffer less overhead than inserts because documents are already pre-processed in C_p to be queried, while documents being inserted must pass through the evolution process. Figure 2b shows that the heterogeneity level of the database affects the insert operations, but not the queries, also because documents are already pre-processed to be queried.

MellowDB obviously added some overhead over the operations. However, querying without its aid would demand from users not only much more effort and time but also a deep knowledge of the database domain. Nevertheless, for the insertion of 500 documents, the worst average time was roughly 5 s.

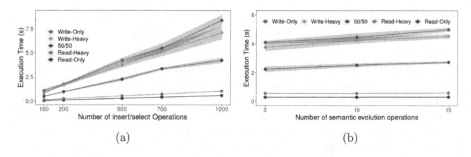

Fig. 2. Execution times (95% confidence interval) for all scenarios with different (a) quantities of select/insert operations and (b) levels of semantic heterogeneity.

6   Concluding Remarks

This work advances the state-of-art techniques in managing semi-structured data heterogeneity caused by database evolution. The formalization of the evolution operations and the storage model and algorithms to deal with them presented here are original contributions, there is no similar approach in the related work.

The theoretical framework, models, and algorithms provide tools to deal with semantic heterogeneity in semi-structured data. As long as the operations history is registered, users may query the database without being aware of details on the values' changes. Results show that the use of the proposed models is feasible, achieving desired results much faster and more conveniently than if the operations were manually treated.

References

1. Brahmia, S., Brahmia, Z., Grandi, F., Bouaziz, R.: τJSchema: a framework for managing temporal JSON-based NoSQL databases. In: Hartmann, S., Ma, H. (eds.) DEXA 2016. LNCS, vol. 9828, pp. 167–181. Springer, Cham (2016). https://doi.org/10.1007/978-3-319-44406-2_13
2. Cooper, B.F., Silberstein, A., Tam, E., Ramakrishnan, R., Sears, R.: Benchmarking cloud serving systems with YCSB. In: Proceedings of the 1st ACM symposium on Cloud computing, pp. 143–154 (2010)
3. Curino, C., Moon, H.J., Deutsch, A., Zaniolo, C.: Automating the database schema evolution process. VLDB J. 22(1), 73–98 (2013)
4. Herrmann, K., Voigt, H., Behrend, A., Rausch, J., Lehner, W.: Living in parallel realities: co-existing schema versions with a bidirectional database evolution language. In: Proceedings of the 2017 ACM International Conference on Management of Data, pp. 1101–1116. SIGMOD/PODS 2017 (2017)
5. Instituto Brasileiro de Geografia e Estatística - IBGE: Alterações topomínicas (2022). https://www.ibge.gov.br/geociencias/organizacao-do-territorio/estrutura-territorial/27336-alteracoes-toponimicas-municipais.html
6. Möller, M.L., Klettke, M., Hillenbrand, A., Störl, U.: Query rewriting for continuously evolving NoSQL databases. In: International Conference on Conceptual Modeling, pp. 213–221. ER 2019 (2019)

7. Moon, H.J., Curino, C.A., Deutsch, A., Hou, C.Y., Zaniolo, C.: Managing and querying transaction-time databases under schema evolution. Proc. VLDB Endowment **1**(1), 882–895 (2008)
8. Tansel, A.U., Clifford, J., Gadia, S., Jajodia, S., Segev, A., Snodgrass, R.: Temporal databases: theory, design, and implementation. Benjamin-Cummings Publishing Co., Inc. (1993)
9. Ventrone, V.: Semantic heterogeneity as a result of domain evolution. ACM SIGMOD Rec. **20**(4), 16–20 (1991)

Co-location Pattern Mining Under the Spatial Structure Constraint

Rodrigue Govan[1]([✉]) [iD], Nazha Selmaoui-Folcher[1] [iD], Aristotelis Giannakos[2],
and Philippe Fournier-Viger[3] [iD]

[1] Institute of Exact and Applied Sciences, University of New Caledonia,
98851 Nouméa Cedex, France
{rodrigue.govan,nazha.selmaoui}@unc.nc
[2] EPROAD, Université de Picardie Jules Verne, Amiens, France
[3] Big Data Institute, College of Computer Science and Software Engineering,
Shenzhen University, Shenzhen, China

Abstract. Most methods to find spatial co-location patterns (subsets of object features that are geographically close to one another) employ standard proximity measures (e.g. Euclidean distance). But for some applications, these measures do not work well since the spatial structure is not considered. This article proposes CSS-Miner, a co-location pattern mining approach under the spatial structure constraint. In this case, the street network of a city is used as a constraint. CSS-Miner has been applied to two real datasets with different points of interest.

Keywords: co-location · data mining · spatial data · spatial structure

1 Introduction

Discovering co-location patterns is a data mining task that aims at extracting knowledge and insights that integrate the spatial dimension to help decision-makers. A co-location (or *co-location pattern*) is a subset of spatial features that are frequently located in the same region. Despite numerous studies [8,9,14], most co-location pattern mining methods use standard distance functions (e.g. the Euclidean distance) to assess the proximity of spatial objects. For applications such as demographic analysis via points of interest (POIs), the Euclidean distance is not suitable since a path between two spatial objects can be significantly different from their Euclidean distance. Hence, other distance measures should be used.

In this paper, we propose CSS-Miner (CSS stands for **C**o-location under the **S**patial **S**tructure constraint), a co-location pattern mining approach for identifying interesting co-locations under the constraint of the spatial structure of a city's street network. CSS-Miner first constructs a graph under the spatial structure constraint using a shortest path algorithm, and then extracts maximal cliques to obtain spatial patterns. For evaluation, the proposed approach was

This work was supported by the ANR Grant SpiRAL ANR-19-CE35-0006-02.

applied on two datasets from the cities of Paris and Chicago, which allowed discovering relevant patterns.

The article is organized as follows. Section 2 reviews relevant work on spatial pattern mining, focusing on the event-based approach. Section 3 describes the proposed CSS-Miner approach to consider the spatial structure constraint. Then, Sect. 4 presents the data used for evaluation and the discovered patterns. Finally, a conclusion is drawn and perspectives are discussed.

2 Related Work

Huang et al. [6] described two main approaches for spatial pattern mining: the sequence-based approach and the event-based approach used in this paper.

The event-based approach (or join-less approach) focuses on the location of spatial objects and their proximity. Initially proposed by Shekhar et al. [9], this approach extracts subsets of objects that are spatially close together, and are called co-locations.

In this paper, we propose a method adopting the event-based approach to leverage the spatial dimension of objects and their proximity. To apply the event-based approach under the spatial structure constraint, maximal clique mining is used to extract co-location patterns. Therefore, the next Subsects. 2.1 and 2.2 respectively give an overview of approaches for maximal clique mining and key studies on co-location pattern mining and their interestingness measures.

2.1 Maximal Clique Mining

(**Complete graph**) Let $G = (V, E)$ be a graph with $V = \{v_1, v_2, \ldots, v_n\}$ the set of vertices and $E \subseteq \{(v_i, v_j) \in V^2 \mid \forall i, j \in \{1, \ldots, n\}$ and $i < j\}$ the set of edges (in this setting, all graphs considered are undirected.) If $(v_i, v_j) \in E$, then v_i and v_j are adjacent. A graph is complete if each pair of graph vertices is connected by an edge (adjacent).

(**Clique**) Let $G = (V, E)$ be a graph and $g = (V_g, E_g)$ be a subgraph such that $V_g \subseteq V$ and $E_g \subseteq \{(v_{g,i}, v_{g,j}) \in E \mid v_{g,i} \in V_g \land v_{g,j} \in V_g$ and $i \neq j\}$. A clique of G is a subgraph $g \subseteq G$ such that g is complete.

(**Maximal clique**) Given $G = (V, E)$ a graph and $g \subset G$ a clique, the clique g is said to be maximal if and only if there exists no clique g' such that $g \subset g' \subseteq G$.

Valiant [13] has shown that mining all maximal cliques is #P-complete. We can particularly mention the algorithm proposed by Tomita et al. [10] for its $O(3^{n/3})$ worst-case complexity in an n-vertex graph which is optimal as a function of n but also Cazals et al. [3] who consider a recursive approach to improve the mining performance.

Maximal clique mining methods are commonly used to mine co-location patterns [1,11]. By defining a graph network where vertices represent spatial objects and edges represent their neighborhood then by applying a maximal clique mining method, we can obtain subsets of objects that are all neighbors to each other. Therefore, in this paper, we will use the approach proposed by Tomita et al. [10] for its speed given the size of our datasets detailed in the Sect. 4.1.

2.2   Co-location Pattern Mining and Interestingness Measures

The event-based approach projects spatialized data with their coordinates and defines the proximity between each spatial object to extract patterns. In this section, we recall the co-location mining framework proposed in Shekhar and Huang [9], Huang et al. [6] and Yoo and Shekhar [14]. Let $\mathcal{F}$ be a set of features and $\mathcal{O} = \{o_1, o_2, \ldots, o_n\}$ be a database of spatial objects. Each object in $\mathcal{O}$ consists of a tuple <object_id, location, feature>, where feature $\in \mathcal{F}$. For example, in Fig. 1b, $\mathcal{F} = \{A, B, C\}$, $\mathcal{O} = \{A_1, B_2, \ldots, C_3\}$ with $A_1 = <1, (x_1, y_1), A>$, $B_2 = <2, (x_2, y_2), B>$, etc. A co-location $\mathcal{C}$ is a subset of features $\mathcal{F}$ associated to spatial objects $\mathcal{O}$. These co-location patterns represent pattern frequently located in neighbor objects. The neighborhood relationship is defined as a binary relation $\mathcal{R}(o, o')$ between two spatial objects o and o'. $\mathcal{R}$ can be based on a distance threshold between two objects, or based on their intersection. Several studies have been done in this vein [7,14]. Recently some researchers used a proximity measure that is not the Euclidean distance. For example, Yu [15] proposed the shortest path length as proximity measure. However, the author utilized a sequence-based approach with a limited number of neighbors, which can miss out some relevant information.

In the join-less approach, to determine if two objects are spatially close, the user sets a maximum distance threshold d. A graph is then constructed with vertices representing the spatial objects. Two vertices are adjacent if the associated spatial objects' distance falls within a threshold d (i.e., the spatial distance measure between these two vertices is less than d).

Interestingness measures have been developed to quantify interesting patterns. To measure whether a co-location pattern is interesting or not, the participation index (or prevalence), based on the participation ratio is used.

(**Participation ratio**) Let $\mathcal{C}$ be a co-location pattern. For an instance $f_i \in \mathcal{C}$, the participation ratio is given by:

$$Pr(f_i, \mathcal{C}) = \frac{|\{\text{ instances of } f_i \text{ participating in } \mathcal{C})\}|}{|\{\text{ instances of } f_i\}|} \tag{1}$$

Given the example of Fig. 1, let $\mathcal{C} = \{A, B\}$ be a co-location candidate and $I_C = \{(A_1, B_1), (A_1, B_2), (A_3, B_4)\}$ be the set of row-instances of $\mathcal{C}$. With A and B, two features having respectively, 3 and 4 instances, we have $Pr(A, \{A, B\}) = \frac{|\{A_1, A_3\}|}{|\{A_1, A_2, A_3\}|} = \frac{2}{3}$ and $Pr(B, \{A, B\}) = \frac{|\{B_1, B_2, B_4\}|}{|\{B_1, B_2, B_3, B_4\}|} = \frac{3}{4}$.

(**Participation index**) Let $\mathcal{C}$ be a co-location candidate, $I_C = \{I_1^C, \ldots, I_k^C\}$ be the set of row-instances of $\mathcal{C}$ and $\mathcal{F} = \{f_1, \ldots, f_n\}$ be the set of spatial features from the database $\mathcal{O}$. The participation index is defined by:

$$Pi(\mathcal{C}) = \min_{f_i \in \mathcal{C}} Pr(f_i, \mathcal{C}) \tag{2}$$

Using the previous example, we have as participation index:

$$Pi(\{A, B\}) = \min_{f_i \in \{A, B\}} Pr(f_i, \{A, B\}) = \min(\frac{2}{3}, \frac{3}{4}) = \frac{2}{3}$$

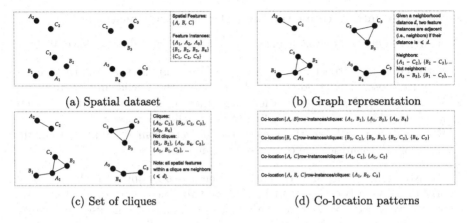

(a) Spatial dataset

(b) Graph representation

(c) Set of cliques

(d) Co-location patterns

Fig. 1. Example of co-location patterns based on a set of cliques from a spatial dataset.

In this paper, the prevalence measure will be used to determine whether co-location patterns in Sect. 4 are relevant or not.

As mentioned before, methods based on the join-less approach mostly used standard distance functions as proximity measure for spatial objects. By using standard distance measures, we may lose the spatial structure. For this reason, we will use the shortest path length as proximity measure.

2.3   Shortest Path Search

Over the last decades, the shortest path search has been a major problem in graph theory. The speed of search depends entirely on the number of vertices and edges in a graph. One of the first solutions was introduced by Dijkstra [4].

More recently, Varia and Kurasova [12] proposed an accelerated version of Dijkstra's algorithm, by adding two components: a bidirectional search and a parallelized process. To find the shortest path between two vertices v_i and v_j, authors applied Dijkstra's algorithm to find the shortest path from v_i to v_j and from v_j to v_i. Since Dijkstra's algorithm is based on a priority queue, parallel and bidirectional components use two priority queues. With these components, the two paths move forward simultaneously. According to their results, the improved approach is at least twice as fast as the standard algorithm.

To leverage the spatial structure constraint and accelerate the process, the parallel bidirectional Dijkstra's algorithm will be used.

3   Methods

Let consider a set of spatial objects $\mathcal{O}$ with a set of features $\mathcal{F}$. Let G_S be a graph representing the spatial structure as $G_S = (V_S, E_S)$ where V_S a set of vertices representing objects and E_S a set of edges.

3.1   Taking into Account the Spatial Structure Constraint

To analyze POIs, the spatial structure constraint is carried out in several steps:

1. For each spatial object $o_i \in \mathcal{O}$, we associate it in the spatial structure G_S with the closest object noted $o_S \in V_S$ (through the Euclidean distance);
2. We apply Dijkstra's algorithm for each object from V_S to the other objects located within a radius d according to the Euclidean distance;
3. If the shortest path length between two objects from V_S is lower than the threshold d, then they are considered as neighbors.

   To avoid unnecessary shortest path searches, we only apply the shortest path algorithm between two objects of V_S if these two objects are respectively associated to two objects of $\mathcal{O}$. Here, the Euclidean distance is only used in order to limit the number of shortest path search. Applying a distance radius threshold with the Euclidean distance will prevent computing irrelevant shortest paths. By triangular inequality, a spatial object located outside a distance radius d from another spatial object has a shortest path length greater than or equal to d.

3.2   Graph Construction

To extract our spatial patterns (co-locations) which are the maximal cliques, we chose to go on a graph construction $G = (\mathcal{O}, E_{\mathcal{O}})$ (under the spatial structure constraint) where $E_{\mathcal{O}} = \{(o_i, o_j) \mid \exists (o_{S,i}, o_{S,j}) \in E_S, D_{sp}(o_{S,i}, o_{S,j}) \leq d, \forall (i, j) \in [\![1, n]\!]^2, i \neq j\}$ with $o_{S,i}$ representing the object from the spatial structure associated to the spatial object $o_i \in \mathcal{O}$ and D_{sp} representing the distance obtained by Dijkstra's shortest path algorithm if it exists.

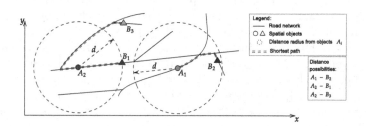

Fig. 2. Three possibilities of distance CSS-Miner can encounter

In the Fig. 2, A_i and B_i are objects from V_S explained in the Sect. 3.1. With d as the distance radius and the shortest path length threshold, we have:

- $d_2(A_2, B_3) > d$ so CSS-Miner will not compute $D_{sp}(A_2, B_3)$;
- $d_2(A_1, B_2) \leq d$ so CSS-Miner will compute D_{sp} and get $D_{sp}(A_1, B_2) > d$ so we will not consider A_1 and B_2 as neighbors;
- $d_2(A_2, B_1) \leq d$ so CSS-Miner will compute D_{sp} and get $D_{sp}(A_2, B_1) < d$ so we will consider A_2 and B_1 as neighbors.

In our approach, CSS-Miner processes two graphs: The first one representing the spatial structure and the second one representing the relationship of our spatial dataset created with the first graph.

4   Experimental Results

We apply CSS-Miner on two real datasets. Both have been created by collecting data from OpenData[1]. The first dataset is located in Paris city with High Schools, Movie theaters, Bicycle stations, Parks and Subway station variables having respectively 239, 85, 996, 722 and 326 spatial objects (2368 objects in total). The second dataset is located in Chicago city with High Schools, Bus station, Rail Lines station, Fast food chains, Bicycle stations and Parks variables having respectively 142, 5606, 124, 877, 1402 and 613 spatial objects (8764 objects in total). For each dataset, the entire process was carried out with a AMD Ryzen 7 3700X 8-core processor with 64GB of RAM. It took respectively, about 2 and 5 h to run the entire process on Paris and Chicago datasets.

Although we aim to analyze and understand the young population behavior, CSS-Miner is applicable to other demographic analysis, for instance: What are the daily habits of a manager compared to a student? Another POIs analysis can also be useful to develop a decision support tool to help developing the tourism of a city. Finally, the POIs analysis remain a very large subject to study.

4.1   Data Preprocessing

To integrate the spatial structure constraint, it is necessary to get access to that information. In this case, we used the road network as spatial structure. Here, we assume that the path is taken on foot because we wanted to integrate only data from OpenData platforms where the traffic noise is not always available. To get access to the road network of Paris and Chicago, we used OSMnx methods [2]. Once the street network is retrieved, it can be converted into a graph network with roads as edges and road intersections as vertices. At the end, the graph associated to Paris street network has 42,870 vertices and 241,016 edges and the graph associated to Chicago has 184,476 vertices and 1,217,928 edges.

4.2   Results

The Table 1 shows us the possible activities near High Schools in Paris, in particular Parks and Movie theaters. Due to limited page number, the Table 1 only displays few extracted patterns. We note through extracted co-location patterns, the ubiquity of High Schools and Bicycle variables, which also show us that the city of Paris helps young population to get around the city autonomously and practice a physical activity. It would be interesting to apply CSS-Miner to other french cities offering this service in order to confirm this trend.

---

[1] opendata.paris.fr/, data.iledefrance.fr/, data.cityofchicago.org/.

Since CSS-Miner integrates the road network as spatial structure constraint, we compared our co-location patterns with the ones without this constraint i.e., using only the Euclidean distance. The results show us that by taking into account the road network, co-location patterns not always have a prevalence greater than prevalence with the Euclidean distance as proximity measure.

Indeed, the extracted co-location patterns without constraint used a distance threshold equal to 500 (meters), just as CSS-Miner. By triangular inequality, a walking distance between two spatial objects is greater than or equal to their Euclidean distance. Therefore, without constraint, the co-location candidates contain more spatial objects, increasing the probability to have a high number of feature instances per variable, which can reduce their prevalence. This also explains why the {Parks, High Schools, Bicycle} co-location pattern has a decreasing prevalence from 0.89 to 0.56 by adding the Movie theaters variable.

Table 1. Extracted co-location pattern prevalence (Pi from Eq. 2)

City	Co-location pattern	Pi under constraint	Pi without constraint
Paris	{Parks, High Schools, Bicycle}	0.89	0.89
	{Parks, High Schools, Movie theaters, Bicycle} ...	**0.56**	0.44
Chicago	{Bus, Fast food chains, High Schools, Bicycle}	**0.58**	0.5
	{Bus, Fast food chains, High Schools} ...	**0.33**	0.17

Moreover, without constraint, the algorithm extracted patterns CSS-Miner did not extract: {High Schools, Subway} and {Parks, High Schools, Movie theaters, Subway} with a prevalence equal to 0.31 and 0.14 respectively without the constraint. It shows that even if the spatial objects are close to one another using the Euclidean distance, their shortest path length do not verify our proximity criterion, so they cannot be considered as close. At the end, CSS-Miner can extract more relevant patterns and filter not so relevant patterns.

The results show that most of High Schools in Chicago have a Fast food chains around it, so young population in Chicago will be more tempted to go eat in a Fast food at lunch or after school. The ubiquity of High Schools and Fast food chains variables can also be a sign of malnutrition in the US, at least in Chicago. To confirm this affirmation, it would be interesting to apply CSS-Miner in other US cities and verify the relevancy on a national scale. It would also be interesting to get a Fast food dataset in Paris (unavailable on the OpenData) to reveal if Fast food chains in Paris target young population as in Chicago.

5 Conclusion and Perspectives

In this paper, we introduced CSS-Miner, a co-location pattern mining approach integrating the spatial structure. We described how this constraint has been defined and taken into account, particularly with a road network and a shortest path search algorithm. To extract co-location patterns, we used the maximal clique mining approach with a restricted search radius and editable depending on the use case. Then, we applied the approach on two real datasets.

The next step of our work will be to integrate knowledge from experts [5], such as urban planners and geographers to verify the relevancy of the extracted patterns. Moreover, CSS-Miner will be applied on larger datasets to estimate the performance. Finally, future work will consider the altitude as spatial structure.

References

1. Bao, X., Wang, L.: A clique-based approach for co-location pattern mining. Inf. Sci. **490**, 244–264 (2019)
2. Boeing, G.: OSMnx: new methods for acquiring, constructing, analyzing, and visualizing complex street networks. Comput. Environ. Urban Syst. **65**, 126–139 (2017)
3. Cazals, F., Karande, C.: A note on the problem of reporting maximal cliques. Theor. Comput. Sci. **407**(1–3), 564–568 (2008)
4. Dijkstra, E.W., et al.: A note on two problems in connexion with graphs. Numer. Math. **1**(1), 269–271 (1959)
5. Flouvat, F., Van Soc, J.F.N., Desmier, E., Selmaoui-Folcher, N.: Domain-driven co-location mining: extraction, visualization and integration in a GIS. GeoInformatica **19**, 147–183 (2015)
6. Huang, Y., Shekhar, S., Xiong, H.: Discovering colocation patterns from spatial data sets: a general approach. IEEE Trans. Knowl. Data Eng. **16**(12), 1472–1485 (2004)
7. Kim, S.K., Lee, J.H., Ryu, K.H., Kim, U.: A framework of spatial co-location pattern mining for ubiquitous GIS. Multimed. Tools Appl. **71**(1), 199–218 (2014)
8. Koperski, K., Han, J.: Discovery of spatial association rules in geographic information databases. In: Egenhofer, M.J., Herring, J.R. (eds.) SSD 1995. LNCS, vol. 951, pp. 47–66. Springer, Heidelberg (1995). https://doi.org/10.1007/3-540-60159-7_4
9. Shekhar, S., Huang, Y.: Discovering spatial co-location patterns: a summary of results. In: Jensen, C.S., Schneider, M., Seeger, B., Tsotras, V.J. (eds.) SSTD 2001. LNCS, vol. 2121, pp. 236–256. Springer, Heidelberg (2001). https://doi.org/10.1007/3-540-47724-1_13
10. Tomita, E., Tanaka, A., Takahashi, H.: The worst-case time complexity for generating all maximal cliques and computational experiments. Theor. Comput. Sci. **363**, 28–42 (2006)
11. Tran, V., Wang, L., Chen, H., Xiao, Q.: MCHT: a maximal clique and hash table-based maximal prevalent co-location pattern mining algorithm. Expert Syst. Appl. **175**, 114830 (2021)
12. Vaira, G., Kurasova, O.: Parallel bidirectional Dijkstra's shortest path algorithm. Databases Inf. Syst. VI Front. Artif. Intell. Appl. **224**, 422–435 (2011)
13. Valiant, L.: The complexity of enumeration and reliability problems. SIAM J. Comput. **8**(3), 410–421 (1979)
14. Yoo, J.S., Shekhar, S.: A joinless approach for mining spatial colocation patterns. IEEE Trans. Knowl. Data Eng. **18**(10), 1323–1337 (2006)
15. Yu, W.: Spatial co-location pattern mining for location-based services in road networks. Expert Syst. Appl. **46**, 324–335 (2016)

Database Design

Enhancing Online Index Tuning
with a Learned Tuning Diagnostic

Haitian Hang⬩ and Jianling Sun$^{(\boxtimes)}$⬩

College of Computer Science and Technology, Zhejiang University, Hangzhou, China
{hanght,sunjl}@zju.edu.cn

Abstract. Indexes are vital for data retrieval performance. For online scenarios with dynamic workloads, index tuning is challenging. A commonly used strategy is to launch tuning requests periodically, yet resource-intensive tuning sessions can obstruct it, particularly when dealing with frequently varying workloads.

To tackle this challenge, we propose a learned tuning diagnostic that can be incorporated into the *Monitor-Diagnose-Tune* paradigm for online index tuning. Rather than invoking a comprehensive tuning tool every time a triggering condition occurs, the tuning diagnostic serves to determine whether a tuning session should be launched. By formulating the determination of sub-optimal index configurations as a classification task in machine learning, our approach can effectively identify whether the current index configuration is sub-optimal. To circumvent the need for costly data collection for each database instance, we propose a transferable representation of queries and indexes that allows for cross-database learning. Our comprehensive empirical results on the TPC-H and TPC-DS benchmarks demonstrate that our approach can reduce the total time by up to 13.3% and the number of optimizer what-if calls by up to 36% compared to the baselines, and validate the effectiveness of our transferable representation in cross-database learning.

Keywords: Index Tuning · Online Tuning · ML for DB

1 Introduction

Database management systems employ indexes to facilitate expedited data retrieval. The index tuning problem, which involves selecting an appropriate set of indexes, has been the subject of extensive research in recent decades [5,9–11,19,27,31]. These approaches aim to identify a set of indexes that minimize the estimated cost of a representative workload while complying with specific constraints, such as a limited storage budget. Although static analysis based on a representative workload can facilitate effective physical design, this approach can be computationally demanding and time-consuming, often requiring multiple *what-if* calls to the query optimizer and running for several minutes to hours [8].

In the context of modern dynamic workloads, the need for evolving index recommendations is essential, given that the present index configuration may become sub-optimal as the workload pattern changes. This is in contrast to static

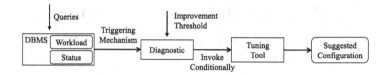

Fig. 1. Monitor-Diagnose-Tune Paradigm

cases, where tuning is a one-time effort. A *Monitor-Diagnose-Tune* paradigm, as illustrated in Fig. 1, is typically used to provide continuous recommendations. The database management system (DBMS) internally tracks workloads and their related information, which will later be utilized by the diagnostic. Upon the detection of a trigger condition, such as performance regression, the diagnostic conducts further analysis based on relevant information, such as index utilization, to determine whether any updates to the current indexes are necessary. If an update is required, a tuning tool is invoked to provide a fresh recommendation.

Motivation. While some online algorithms offer lightweight solutions for continuous tuning by identifying promising candidate configurations using statistics gathered from query execution, the reliability of their recommendations is not guaranteed [7,28]. In comparison, comprehensive analysis or modeling based on the workload characteristics available through existing static tools offers a better alternative in terms of recommendation quality. However, comprehensive analysis is resource-intensive and time-consuming, which can negate its benefits, especially when it does not recommend a better configuration or the improvement resulting from the recommended configuration is negligible. Even when comprehensive tuning is performed on a templated job with medium-sized memory constraints, it can still take tens of minutes, let alone the more complex and variable workloads in real-world scenarios. In some cases, commercial physical design tools require several hours for a single invocation. Determining whether there is a superior configuration requires executing a tuning session, which is problematic. While recently proposed learning-based index selection algorithms have demonstrated superior runtime performance, they still require extensive training duration and repetitive training costs for every new database instance, which limits their practical application [20,22,25,29,33–35].

Our Approach. To bridge the gap in the deployment of sophisticated static tuning tools in online scenarios, we propose investigating whether a comprehensive tuning session would produce a configuration that is noticeably better than the existing one. In this paper, we present a technique called *learned tuning diagnostic* to achieve this goal. Specifically, the problem of determining whether a tuning session would be worthwhile is formulated as a classification task in machine learning. Given the target workload, the current index configuration, the memory budget, and the expected improvement threshold, the learned tuning diagnostic determines whether a comprehensive tuning session would recommend a configuration that outperforms the previous one by more than the threshold. Our learned tuning diagnostic can help to prevent pointless tuning requests in a dynamic environment where the configuration needs to be updated frequently in response to workload changes.

Contribution. To summarize, we present the following contributions in this paper:

- We formulate the problem of determining whether a tuning session is worthwhile as a *classification* task in machine learning and propose a learned tuning diagnostic that fits into the Monitor-Diagnose-Tune paradigm for online index tuning.
- We propose a transferable representation of queries and indexes that enables cross-database learning and incremental training when the schema changes, thereby avoiding costly data collection and repetitive training efforts for each specific database instance.
- We implement a Monitor-Diagnose-Tune framework with our proposed learned tuning diagnostic and evaluate it on PostgreSQL for two standard benchmarks TPC-H and TPC-DS. The experimental results demonstrate the superiority of the learned diagnostic and the effectiveness of the proposed transferable representation of queries and indexes.

2 Preliminaries

2.1 Online Index Tuning

Let a workload $W = \{w_1, w_2, ..., w_n\}$ be a sequence of mini-workloads, where w_t is either a single query or a batch of queries at time interval t, and I be the set of available indexes. For an index configuration $s \subseteq I$, we denote the memory consumption to materialize s by $C_{storage}(s)$, and the set of index configuration feasible within a memory budget b is $\varsigma = \{s \subseteq I \mid C_{storage}(s) \leq b\}$. The goal of *online index tuning* is to find a sequence of index configuration $S = \{s_0, s_1, s_2, ..., s_n\}$ with w_t executed under the configuration s_t, which minimizes the total cost of W. The total cost of W under a configuration sequence S is defined as:

$$Cost\,(W, S) = \sum_{t=1}^{n} \left(c_{tune}\,(t)\,, c_{trans}\,(s_{t-1}, s_t)\,, c_{exec}\,(w_t, s_t) \right)$$

where $c_{tune}\,(t)$, $c_{trans}\,(s_{t-1}, s_t)$ and $c_{exec}\,(w_t, s_t)$ refer to the runtime of a tuning request, the transition time from index configuration s_{t-1} to s_t, and the execution time of workload w_t against the configuration s_t, respectively.

In the case where the entire workload sequence W is known or predictable in advance, online index tuning can be viewed as a *sequence tuning* problem [4] or an *action planning* problem [23]. These problems can be addressed using either heuristic or learning-based techniques to determine the optimal time to take action before executing the workload sequence. However, the uncertainty associated with forecasting long-term workloads limits the applicability of these algorithms in practice. Instead, other online tuning methods continuously monitor, diagnose, and adjust the configuration [6, 7, 28]. Given the impracticality of

---

in-depth modeling or analysis in online scenarios, most of these methods identify promising configuration changes based on statistics acquired from query execution. If high-quality configuration recommendations are desired, a static analysis based on either the most recent workload $\{w_i, w_2, ..., w_{t-1}\}$ or the upcoming workload w_t using offline tools is a more suitable alternative. Nevertheless, such comprehensive tuning requests come with high overhead, i.e., high $c_{tune}(t)$ at each time interval t, which limits its benefit. Additionally, the repeated what-if calls to the query optimizer cause significant CPU/memory resource consumption, placing a heavy burden on the production server.

To tackle this issue, an additional component, named the *Alerter*, has been introduced in the Monitor-Diagnose-Tune paradigm [6]. The Alerter analyzes whether a tuning session is worthwhile by presenting lower and upper bounds on the potential improvement that the tuning session could suggest. While the Alerter offers low-overhead diagnostics to prevent tuning requests that are unlikely to result in a significant improvement over the current configuration, it is coupled with the relaxation-based tuning approach [5], and thus, its effectiveness may not extend to other tuning algorithms.

While reinforcement learning-based approaches for offline tuning are faster in terms of tuning time (e.g., several seconds) [19], they require a lengthy training duration, which restricts their application. Furthermore, they are not suitable for online scenarios where unseen workloads are present due to their inability to generalize for unknown workloads. As a result, we have excluded them from the scope of this paper.

2.2 Workload and Index Configuration Featurization

As AI-enhanced database technology continues to evolve, a new area of research is emerging that aims to featurize queries or workloads. To enable data-driven learning-based techniques, queries are transformed into vectorized representations based on their raw text or logical (or physical) plans [12–17,21,24,30,32]. In certain tasks where query execution details are necessary, such as learned cost models [14,15,30], the operators in query plans are encoded along with other relevant information, while query text is used as input in other scenarios, such as SQL2Text [16]. Most workload-driven tasks, such as index tuning [25], knob tuning [21], and cost estimation [30], encode the columns and tables referenced in the query using one-hot representation. However, this representation, which is tied to a specific schema, limits cross-database learning. As a result, training data collection must be repeated, and a new model must be trained from scratch for each new database instance or when the schema is altered (e.g., new tables or columns are created). Additionally, NLP-based methods have been utilized to generate the encoding of SQL text [16,32]. Nonetheless, the same problem persists in the language model since database-specific vocabularies are used as input. Furthermore, in machine learning-based approaches for index tuning, not only the query but also the index configuration must be vectorized. One-hot encoding is commonly used, which encodes whether an index is present or not for each indexable attribute.

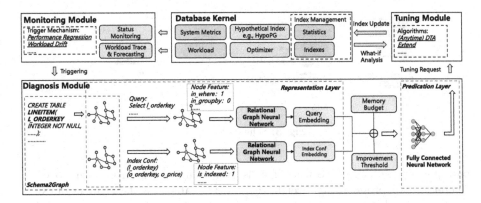

Fig. 2. Overview of a Monitor-Diagnose-Tune framework for online tuning with learned tuning diagnostic

In general, using one-hot encoding to featurize the workload and index configuration in machine learning-based approaches restricts the ability of cross-database learning and incremental updates when the schema is altered, impeding the deployment of learning-based approaches in practice. For example, enabling RL-based index recommendations for each database in a cloud platform is a significant undertaking, as each instance must train an agent independently, which typically takes several hours to several days. Although some transferable representations that enable generalization across databases have been proposed recently, they are based on query plans and are not applicable to our problem, as the query plan varies with different index configurations [13,15].

3   Overview

Architecture. Figure 2 provides an overview of the Monitor-Diagnose-Tune paradigm with our proposed learned tuning diagnostic. It consists of three major components: (1) a *monitoring* module that continuously monitors system metrics during workload execution and triggers the next procedure when necessary; (2) a *diagnosis* module that determines whether a tuning session would result in a worthwhile improvement beyond a specific threshold; and (3) a *tuning* module that invokes a resource-intensive tuning tool with a comprehensive analysis and deploys the appropriate configuration. When a trigger event occurs, such as performance regression, workload shift, or a fixed time interval, the diagnostic guides the launch of a tuning session. If the diagnostic decides that a tuning session is necessary, we run a comprehensive tuning tool based on the recent past workload or the upcoming workload, which can be known in advance or predicted depending on the situation. The focus of this work is the diagnosis module, and we do not delve into the trigger mechanism or index selection technique. It is our view that frequent workload changes in online scenarios make it impractical to launch sophisticated tuning sessions with long runtimes and

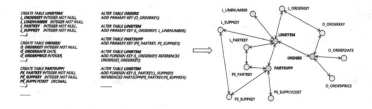

Fig. 3. Schema2Graph

multiple resource-intensive what-if calls every time a triggering event occurs. Therefore, our proposed lightweight diagnostic enables the deployment of existing tuning tools that provide better recommendations through comprehensive workload analysis, making them practical for use in online scenarios.

Workflow. The workflow of the learned tuning diagnostic is presented in the *Diagnosis* module in Fig. 2. It consists of three stages: (1) *schema2graph*, which converts the database schema to a directed graph; (2) *representation layer*, which uses relational graph convolutional networks (RGCN) to embed the query and index configuration into vectors based on the schema graph; and (3) *predication layer*, a fully connected neural network that determines whether the current configuration needs to be updated for the target workload. The first step is to generate a schema graph with vertices representing columns/tables and edges capturing the relationships between them. The representation layer uses features that capture the necessary information of the query, such as column/table references in different clauses, to encode the nodes in the graph. A graph convolutional neural network is used to generate the graph embedding, which represents the query embedding based on the schema graph with node features encoding the query information. The current index configuration is featurized in a similar vein. Finally, the graph embedding yields the vectorized representation of the current index configuration and the target workload. To diagnose tuning, we feed the representation of the workload and index configuration into a fully-connected neural network along with other parameters, such as the memory budget and the improvement threshold, to determine whether a tuning session is necessary for the target workload.

4   Learned Tuning Diagnostic Model

4.1   Model Design

Schema2Graph. We begin by converting the database schema to a directed graph, denoted by $G_{schema} = (\mathcal{V}, \mathcal{E}, \mathcal{R})$. Here, nodes (entities) are represented by $v_i \in \mathcal{V}$, and labeled edges (relations) are represented by $(v_i, r, v_j) \in \mathcal{E}$, where $r \in \mathcal{R}$ is a relation type. Each table and column in the schema is represented as a node in the graph. To capture how columns and tables relate to one another, we introduce four types of edges with 11 different labels, which are listed in Table 1. An example of the schema graph representation is illustrated in Fig. 3.

Table 1. Description of edge types used in the schema graph(* additional edge labels used in the index configuration representation).

Type of (v_i, v_j)	Label of r	Description
(Column, Column)	FK(Foreign Key)-Left	v_i is a foreign key for v_j
	FK-Right	v_j is a foreign key for v_i
	Two-Column-Index*	A two-column index (v_i, v_j) is built
(Column, Table)	PK(Primary Key)-Left	v_i is a primary key of v_j
	Belongs-to-Left	v_i is a column of v_j (not the primary key)
(Table, Column)	PK-Right	v_j is a primary key of v_i
	Belongs-to-Right	v_j is a column of v_i (not the primary key)
(Table, Table)	FK-Table-Left	Table v_i has a foreign key column in v_j
	FK-Table-Right	v_j has a foreign key column in v_i
	FK-Table-Both	v_i and v_j have foreign keys in both directions

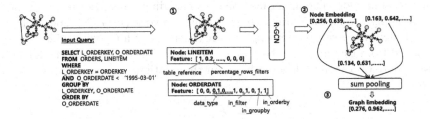

Fig. 4. Workload Featurization

Workload Representation. We now illustrate how to featurize the workload using our proposed graph-based representation (detailed in Fig. 4). Our methodology involves extracting relevant features from the query. This entails capturing the various ways in which columns and tables are referenced in different function calls and clauses, as well as taking into account the underlying data characteristics. Table 2 provides a list of the different features we extract for tables and columns. In addition to the semantic information, such as how columns are referenced in different clauses, we also incorporate some fundamental data features from DBMS optimizer. For example, the percentage of rows filtered by table condition, which enables us to account for the effective changes in data size and distribution when learning across databases.

In our methodology, distinct features are considered for the table and column nodes. However, we do not encode the node features in a node-wise manner. Instead, we unify the encoding for the table and column nodes by padding. For each column/table node, a node feature is defined, where each bit denotes a feature listed in the table above (except that the *data_type* feature adopts one-hot encoding and the *percentage_rows_filtered* is a decimal between 0 and 1). For the features that do not belong to this node type, the corresponding bit in the feature vector is set to 0. To illustrate, consider the column node O_ORDERDATE in Step ② in Fig. 4. The first two bits that represent the

Table 2. Query Feature Extraction

Type	Features	Description
Table	table_reference	is table referenced
	percentage_rows_filtered	the percentage of rows filtered by table conditions
Column	data_type	data type
	in_select	is column in a select clause
	in_functional_call	is column in a function call
	in_filter	is column in a filter condition
	in_join	is column in a join condition
	in_groupby	is column in a groupby clause
	in_orderby	is column in a orderby clause

data_type and *percentage_rows_filtered* features are set to 0, as O_ORDERDATE is not a table node. The underlined one-hot encoding indicates the data type.

We apply a relational graph convolutional network (R-GCN) [26] to embed each vertex into a vertex vector, given the schema graph G_q with node features extracted from the query q. The propagation model for calculating the forward-pass update of a vertex, denoted by v_i, in the relational graph is formulated as follows:

$$h_i^{(l+1)} = \sigma\left(\sum_{r \in \mathcal{R}} \sum_{j \in \mathcal{N}_i^r} \frac{1}{c_{i,r}} W_r^{(l)} h_j^{(l)} + W_0^{(l)} h_i^{(l)}\right)$$

Here, $h_i^{(l)}$ is the hidden state of vertex v_i in the i-th layer of the neural network. $\mathcal{N}_i^r$ denotes the set of neighbor vertices of v_i under relation r, and $\frac{1}{c_{i,r}}$ is a problem-specific normalization constant that can either be learned or chosen in advance (in our model, $c_{i,r} = |\mathcal{N}_i^r|$ is a normalized sum). Note that, in addition to the edges defined in Table 1, a self-connection edge is added for each vertex in the graph, so that the representation of a vertex at layer $l + 1$ can also be informed by the corresponding representation at layer l.

During the propagation of the graph neural network, the message-passing aggregates the information of both neighbor vertices and the vertex itself for each vertex, due to the self-connection edge. However, the original information of the vertex itself may still be diluted. To address this, we concatenate the hidden state of the last layer with the node features extracted before applying the graph neural network (i.e., $h_i^{(0)}$) to obtain the embedding of vertex v_i:

$$e_{v_i} = concat(h_i^{(l)}, h_i^{(0)})$$

We represent query q with the embedding of G_q, denoted as e_q. To compute a feature vector of the entire graph, we utilize a rule-based readout phase. In this phase, we collect all nodes whose corresponding columns/tables are referenced in the query q. We then use sum pooling of these nodes to form the graph representation, as follows:

$$e_q = e_{G_q} = sum_pool\left(\{e_{v_i} \,|\}\right)$$

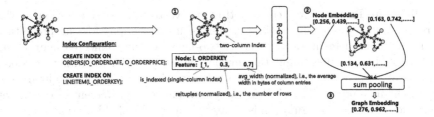

Fig. 5. Index Configuration Featurization.

Finally, we compute the embedding of the workload by calculating the weighted average of the query composition associated with query weights, as follows:

$$e_w = \sum_{q_i \in w} e_{q_i} \cdot f_{q_i}$$

Here, f_q represents the query weight, which is the proportion of query q arriving per time interval.

In comparison to one-hot encoding and language models, where input literals such as column and table names are used, our graph-based representation not only captures the schema definition but also the query details. Furthermore, our featurization approach is transferable across databases since it encodes node features without any literals. This makes cross-database learning feasible, allowing for leveraging data from millions of databases on a cloud platform.

Index Configuration Representation. Similarly, we featurize the index configuration based on the schema graph. The node features are constructed using the underlying data characteristics of each column and the index's build information. The difference between the index and the query featurization is that we do not extract any query-related features. As shown in Step ① in Fig. 5, the first bit of the node feature vector indicates that a single-column index is created on the corresponding column. The remaining bits describe the underlying data characteristics of that column, including the number of rows (*reltuples*) and the average width in bytes of column entries (*avg_width*).

In addition to the edge types introduced in workload representation, we add a new type of edge to indicate the construction of a two-column index. Specifically, when a two-column index $Index(v_{c1}, v_{c2})$ exists, we create an edge (v_{c1}, v_{c2}) with the label *Index-Left* and an edge (v_{c2}, v_{c1}) with the label *Index-Right*. As illustrated in Fig. 5, two directed edges, marked in red, are created to indicate the existence of $(O\_ORDERDATE, O\_ORDERPRICE)$.

It should be noted that we only consider indexes with widths of 1 and 2. As reported by Kossmann et al. [18], wider indexes with a width greater than 2 do not significantly improve the investigated benchmarks for most index tuning algorithms. Furthermore, the runtime of some algorithms would substantially increase when considering large index combinations. Therefore, our index configuration featurization is limited to a two-column index.

Predication Layer. After constructing the vectorized representation of the workload and index configuration, it can be used as input for the prediction layer. In our implementation, we use a two-layered fully-connected neural network to predict whether tuning the current configuration for the target workload is necessary. Specifically, it determines whether there is a better configuration within the memory limit, such that the performance improvement of the target workload under the configuration is greater than the threshold compared to that under the original configuration. Therefore, we concatenate the representation of the workload e_w, the representation of the index configuration e_i, the memory budget b, and the improvement threshold θ as input.

4.2   Training Model

Training Data Generation. Due to limited availability of real data, we need to generate training data by simulation. Specifically, considering n query templates from the selected benchmark, we randomly select m templates and assign random frequencies using a uniform distribution to generate a workload at each round. Next, we use an existing tuning tool to recommend an index configuration based on the generated workload. In the subsequent round, a new workload and corresponding configuration are generated in the same way. We then evaluate the performance of the new workload under the previous and new configurations, respectively. The label is assigned based on whether the performance improvement brought by the new configuration exceeds the set threshold. For example, at the beginning, we randomly generate a workload w_0 as described above and recommend an index configuration c_0 for w_0 using a tuning tool with a set memory budget b. At time t_1, a new workload w_1 and configuration c_1 for w_1 are generated. The performance of workload w_1 evaluated under configuration c_0 and c_1 is denoted as $cost(w_0, c_0)$ and $cost(w_1, c_1)$, respectively. If $\frac{cost(w_0,c_0)-cost(w_1,c_1)}{cost(w_0,c_0)} \geq \theta$, the label is set to to_tune; otherwise, it is set to not_to_tune. In this way, we can construct our data set using $< w_i, c_{i-1}, b, \theta >$ with an assigned label.

End-to-End Training. As discussed in Sect. 4.1, our diagnostic model comprises a representation layer, which is composed of two independent graph neural networks for featurizing the workload and index configuration, and a predication layer. Unlike the conventional approach of training these layers separately, we employ an end-to-end training strategy that involves joint training of the representation and predication layers. This approach is enabled by the differentiability of all steps in the training process.

5   Experimental Evaluation

5.1   Experimental Setup

Implementation and Environment. We have implemented a Monitor-Diagnosis-Tune framework that incorporates our learned tuning diagnostic. As

Table 3. Performance of learned tuning diagnostic.

Benchmark	$\theta = 5\%$			$\theta = 10\%$			$\theta = 15\%$		
	Precision	Recall	F1	Precision	Recall	F1	Precision	Recall	F1
TPC-H	0.884	0.869	0.876	0.919	0.907	0.907	0.923	0.929	0.926
TPC-DS	0.799	0.811	0.805	0.830	0.837	0.833	0.838	0.821	0.829

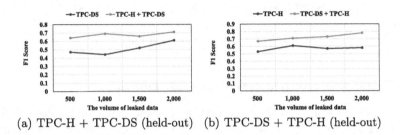

(a) TPC-H + TPC-DS (held-out)   (b) TPC-DS + TPC-H (held-out)

Fig. 6. Ability of cross-database learning.

the core of the tuning module, we have utilized the state-of-the-art index selection algorithm *Extend* [27]. For cost estimation, we rely on HypoPG [1], a PostgreSQL extension that allows the creation, deletion, and size estimation of hypothetical indexes.

Datasets and Workload. For our evaluation, we have selected two benchmarks: TPC-H [3] and TPC-DS [2]. We have excluded TPC-DS queries 4, 6, 9, 10, 11, 32, 35, 41, and 95, as well as TPC-H queries 2, 17, and 20, because their estimated costs are orders of magnitude higher than those of the other queries. This renders the index tuning problem less complex, as validated in [18]. We have generated the training data as detailed in Sect. 4.2 to train a learned tuning diagnostic for each benchmark. To simulate the online scenarios with dynamic workloads, we have randomly selected 10 templates from all query templates and generated 1000 queries for each time interval using the selected templates with uniform weight distribution.

Experimental Parameters. The parameters varied in our experiments include the storage space budget b and the improvement threshold θ. The space budget is expressed as a multiple of the raw data size. For instance, with 10 GB of data, a storage budget b of 0.5 represents 5 GB of storage space.

Evaluation Metrics. In our evaluation, we evaluate the performance of our learned tuning diagnostic using Precision, Recall, and F1 score. On one hand, we are interested in the ratio of true positives to total predicted positives, which is reflected by precision. The higher the precision, the less time is wasted on unnecessary tuning requests. On the other hand, we aim to identify situations that require tuning as much as possible. If all tuning opportunities are correctly identified, then the recall is 100%. Besides reporting the performance of the model, we also evaluate the overall time for online index tuning and report the performance of the Monitor-Diagnose-Tune framework with the learned diagnostic.

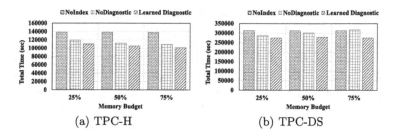

Fig. 7. Evaluation of total time

5.2 Performance of Learned Tuning Diagnostic

We first evaluate the classification quality of our learned tuning diagnostic by generating a dataset of size 10000 for each benchmark. We vary the improvement threshold θ in $[5\%, 10\%, 15\%]$. We do not examine our method with a higher threshold, such as 20%, as a new configuration is unlikely to bring an improvement beyond this threshold for the two benchmarks evaluated. The results for different improvement thresholds are presented in Table 3. Our proposed learned tuning diagnostic achieves precision and recall values above 0.8 for both benchmarks under most improvement threshold settings. These results demonstrate that our proposed diagnostic can effectively reduce unnecessary tuning requests while avoiding missing tuning opportunities. Note that we do not compare our approach with other database-specific query and index representations in this section. Such a comparison would be unfair as they do not consider the ability of cross-database learning.

5.3 Ability of Cross-Database Learning

To assess the model's ability to perform cross-database learning, we train it on one benchmark's complete dataset and a fraction of the dataset leaked from another benchmark, then test it on the held-out benchmark. This held-out setup simulates the scenario where there is insufficient training data to independently train a model for a new database deployment. Figure 6 shows the F1 score obtained from applying cross-database learning. The x-axis represents the volume of data leaked from the held-out database. For instance, Fig. 6(a) depicts the TPC-DS benchmark as the held-out database. The TPC-DS line corresponds to a model trained using only a fraction of the leaked data from the TPC-DS benchmark, while the TPCH + TPC-DS line illustrates the F1 score of a model trained using the entire training data from the TPC-H benchmark and the leaked data from TPC-DS. Due to the small size of the leaked data, the model trained using only the leaked data performs poorly. By combining the significant amount of training data generated by existing databases, we can train a more effective model.

Table 4. Total time breakdown for the TPC-H benchmark with storage budget $b = 50\%$ and improvement threshold $\theta = 15\%$

Methods	Runtime (1e5 sec)			
	Tuning requests	Index Update	Workload Execution	Total
NoIndex	0	0	1.384 (100%)	1.384
NoDiagnostic	0.057 (5.1%)	0.073 (6.5%)	0.993 (88.4%)	1.123
Learned Diagnostic	0.031 (2.9%)	0.032 (2.9%)	1.024 (94.2%)	1.087

5.4 Online Index Tuning with Learned Diagnostic

Since the primary goal of online index tuning is to minimize the total time, including runtime for the tuning session, we also evaluate the impact of our learned tuning diagnostic integrated into the Monitor-Diagnose-Tune framework on total runtime. We consider the online index tuning scenario described in Sect. 2.1, where we set the number of time intervals to 100. We compare our approach with two baselines: 1) *NoIndex*, a baseline without any secondary indices, and 2) a Monitor-Tune cycle without diagnosis, denoted by *NoDiagnostic*.

Based on the experimental results presented in Sect. 5.2, our learned tuning diagnostic model performs best when the improvement threshold is set to 15%. We argue that 15% is a suitable threshold since a lower threshold would incur unnecessary tuning costs that offset the benefits of a better configuration. To further evaluate the Monitor-Diagnose-Tune framework with our learned diagnostic, we set the improvement threshold to 15%.

Figure 7 shows the total time for different benchmarks under various memory budgets. Our method outperforms NoIndex and NoDiagnostic for both benchmarks. A monitor-tune cycle does not consistently perform better than NoIndex since the overhead of tuning sessions offsets the benefits of better configurations. Our learned diagnostic avoids unnecessary and time-consuming tuning requests, resulting in up to a 13.3% reduction in total time for processing the workload sequence, significantly enhancing the Monitor-Diagnose-Tune framework.

Moreover, we present a detailed breakdown of the total time for both benchmarks in Table 4 and Table 5. In comparison with NoDiagnostic, which tunes for each interval, our method is not superior in workload execution time since it cannot fully and accurately identify scenarios with better configurations. However, by utilizing our learned diagnostic, the runtime for tuning requests is reduced by up to 43%. Clearly, the benefits of such reduction outweigh the disadvantages in workload execution time.

Besides evaluating the total time for online index tuning, we also assess the benefits of the learned diagnostic in reducing resource consumption, specifically in terms of the number of optimizer What-if calls. The number of what-if calls directly impacts the burden on the database server. As illustrated in Fig. 8, our approach significantly reduces the number of what-if calls by avoiding unnecessary tuning requests. For instance, for the TPC-DS benchmark, the number of what-if calls is reduced by 37.4%, 34.5%, and 36.2% with the storage budgets of $b = 25\%$, 50%, and 75%, respectively.

Table 5. Total time breakdown for the TPC-DS benchmark with storage budget $b = 50\%$ and improvement threshold $\theta = 15\%$

Methods	Runtime (1e5 sec)			
	Tuning requests	Index Update	Workload Execution	Total
NoIndex	0	0	3.129 (100.0%)	3.129
NoDiagnostic	0.703 (23.3%)	0.077 (2.5%)	2.269 (75.2%)	3.016
Learned Diagnostic	0.401 (14.3%)	0.035(1.2%)	2.377 (84.5%)	2.813

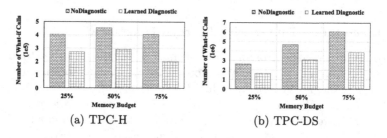

(a) TPC-H   (b) TPC-DS

Fig. 8. Evaluation of Optimizer What-if Calls

6   Conclusion

Although comprehensive index tuning algorithms are superior in terms of recommendation quality, they are time-consuming and resource-intensive, making them unsuitable for frequent tuning requests in online scenarios. This paper focuses on determining whether a sophisticated tuning session would result in a new configuration that exceeds a given performance threshold, without executing it. We introduce our key insight that this problem can be cast as a machine learning classification task, and propose a learned tuning diagnostic that enhances the Monitor-Diagnose-Tune paradigm for online index tuning. To avoid costly data collection and repeated training for each database instance, we propose a transferable representation of queries and indexes for cross-database learning. Our learned tuning diagnostic allows the Monitor-Diagnose-Tune framework to accurately avoid unnecessary tuning requests that cannot significantly improve performance.

References

1. HypoPG - Hypothetical Indexes for PostgreSQL. https://github.com/HypoPG/hypopg
2. TPC-DS benchmark. www.tpc.org/tpcds
3. TPC-H benchmark. www.tpc.org/tpch
4. Agrawal, S., Chu, E., Narasayya, V.R.: Automatic physical design tuning: workload as a sequence. In: Proceedings of the 2006 ACM SIGMOD International Conference on Management of Data (2006)

5. Bruno, N., Chaudhuri, S.: Automatic physical database tuning: a relaxation-based approach. In: ACM SIGMOD Conference (2005)
6. Bruno, N., Chaudhuri, S.: To tune or not to tune?: a lightweight physical design alerter. In: Very Large Data Bases Conference (2006)
7. Bruno, N., Chaudhuri, S.: An online approach to physical design tuning. In: 2007 IEEE 23rd International Conference on Data Engineering, pp. 826–835 (2007)
8. Chaudhuri, S., Narasayya, V.: AutoAdmin "what-if" index analysis utility. ACM SIGMOD Rec. **27**(2), 367–378 (1998)
9. Chaudhuri, S., Narasayya, V.R.: An efficient cost-driven index selection tool for Microsoft SQL server. In: Very Large Data Bases Conference (1997)
10. Choenni, S., Blanken, H.M., Chang, T.: Index selection in relational databases. In: Proceedings of ICCI 1993: 5th International Conference on Computing and Information, pp. 491–496 (1993)
11. Dash, D., Polyzotis, N., Ailamaki, A.: Cophy: a scalable, portable, and interactive index advisor for large workloads. arXiv abs/1104.3214 (2011)
12. Deep, S., Gruenheid, A., Koutris, P., Naughton, J.F., Viglas, S.: Comprehensive and efficient workload compression. arXiv abs/2011.05549 (2020)
13. Ding, B., Das, S., Marcus, R., Wu, W., Chaudhuri, S., Narasayya, V.R.: AI meets AI: leveraging query executions to improve index recommendations. In: Proceedings of the 2019 International Conference on Management of Data (2019)
14. Gao, J., Zhao, N., Wang, N., Hao, S., Wu, H.: Automatic index selection with learned cost estimator. Inf. Sci. **612**, 706–723 (2022)
15. Hilprecht, B., Binnig, C.: Zero-shot cost models for out-of-the-box learned cost prediction. Proc. VLDB Endow. **15**, 2361–2374 (2022)
16. Jain, S., Howe, B., Yan, J., Cruanes, T.: Query2vec: an evaluation of NLP techniques for generalized workload analytics. arXiv, Databases (2018)
17. Kipf, A., Kipf, T., Radke, B., Leis, V., Boncz, P.A., Kemper, A.: Learned cardinalities: estimating correlated joins with deep learning. arXiv abs/1809.00677 (2018)
18. Kossmann, J., Halfpap, S., Jankrift, M., Schlosser, R.: Magic mirror in my hand, which is the best in the land? Proc. VLDB Endow. **13**, 2382–2395 (2020)
19. Kossmann, J., Kastius, A., Schlosser, R.: SWIRL: selection of workload-aware indexes using reinforcement learning. In: International Conference on Extending Database Technology (2022)
20. Lan, H., Bao, Z., Peng, Y.: An index advisor using deep reinforcement learning. In: Proceedings of the 29th ACM International Conference on Information & Knowledge Management (2020)
21. Li, G., Zhou, X., Li, S., Gao, B.: Qtune: a query-aware database tuning system with deep reinforcement learning. Proc. VLDB Endow. **12**, 2118–2130 (2019)
22. Licks, G.P., Couto, J.C., de Fátima Miehe, P., de Paris, R., Ruiz, D.D., Meneguzzi, F.: SmartIX: a database indexing agent based on reinforcement learning. Appl. Intell. **50**, 2575–2588 (2020)
23. Ma, L., Aken, D.V., Hefny, A.S., Mezerhane, G., Pavlo, A., Gordon, G.J.: Query-based workload forecasting for self-driving database management systems. In: Proceedings of the 2018 International Conference on Management of Data (2018)
24. Paul, D., Cao, J., Li, F., Srikumar, V.: Database workload characterization with query plan encoders. Proc. VLDB Endow. **15**, 923–935 (2021)
25. Sadri, Z., Gruenwald, L., Leal, E.: DRLindex: deep reinforcement learning index advisor for a cluster database. In: Proceedings of the 24th Symposium on International Database Engineering & Applications (2020)

26. Schlichtkrull, M., Kipf, T., Bloem, P., van den Berg, R., Titov, I., Welling, M.: Modeling relational data with graph convolutional networks. In: Extended Semantic Web Conference (2017)

27. Schlosser, R., Kossmann, J., Boissier, M.: Efficient scalable multi-attribute index selection using recursive strategies. In: 2019 IEEE 35th International Conference on Data Engineering (ICDE), pp. 1238–1249 (2019)

28. Schnaitter, K., Abiteboul, S., Milo, T., Polyzotis, N.: On-line index selection for shifting workloads. In: 2007 IEEE 23rd International Conference on Data Engineering Workshop, pp. 459–468 (2007)

29. Sharma, A.K., Schuhknecht, F.M., Dittrich, J.: The case for automatic database administration using deep reinforcement learning. arXiv abs/1801.05643 (2018)

30. Sun, J., Li, G.: An end-to-end learning-based cost estimator. arXiv abs/1906.02560 (2019)

31. Surajit, C., Narasayya, V.R.: Anytime algorithm of database tuning advisor for Microsoft SQL server (2020)

32. Tang, X., Wu, S., Song, M., Ying, S., Li, F.Y., Chen, G.: PreQR: pre-training representation for SQL understanding. In: Proceedings of the 2022 International Conference on Management of Data (2022)

33. Valavala, M., Alhamdani, W.: Automatic database index tuning using machine learning. In: 2021 6th International Conference on Inventive Computation Technologies (ICICT), pp. 523–530. IEEE (2021)

34. Wu, W., et al.: Budget-aware index tuning with reinforcement learning. In: Proceedings of the 2022 International Conference on Management of Data (2022)

35. Yan, Y., Yao, S., Wang, H., Gao, M.: Index selection for NoSQL database with deep reinforcement learning. Inf. Sci. **561**, 20–30 (2021)

NoGar: A Non-cooperative Game for Thread Pinning in Array Databases

Simone Dominico$^{(\boxtimes)}$, Marco A. Z. Alves, and Eduardo C. de Almeida

Federal University of Paraná, Curitiba, Brazil
{sdominico,mazalves,eduardo}@inf.ufpr.br

Abstract. An array database is a software that uses non-linear data structures to store and process multidimensional data, including images and time series. As multi-dimensional data applications are generally data-intensive, array databases can benefit from multi-processing systems to improve performance. However, when dealing with Non-Uniform Memory Access (NUMA) machines, the movement of massive amounts of data across NUMA nodes may result in significant performance degradation. This paper presents a mechanism for scheduling array database threads based on data movement patterns and performance monitoring information. Our scheduling mechanism uses non-cooperative game theory to determine the optimal thread placement. Threads act as decision-makers selecting the best NUMA node based on each node's remote memory access cost. We implemented and tested our mechanism on two array databases (Savime and SciDB), demonstrating improved NUMA-affinity. With Savime, we observed a maximum speedup of 1.64× and a consistent reduction of up to 2.46× in remote data access during subarray operations. With SciDB, we observed a speedup of up to 1.38× and a reduction of 1.71× in remote data access.

Keywords: Array databases · Thread pinning · Nash equilibrium · Query processing

1 Introduction

An Array Database Management System (Array database) is a software specifically designed for modeling, storing, and processing multidimensional arrays. Array databases have become crucial components for many science and engineering applications due to the significant increase in data volume and use of multidimensional data. For instance, over the last decade, NASA[1] has accumulated almost 32 petabytes of scientific data, while ECMWF[2] has stored 220 petabytes [1]. According to [2], the sheer volume of data generated by scientists has driven the creation and adoption of Array databases. This overwhelming

---

[1] National Aeronautics and Space Administration.

[2] European Center for Medium-Range Weather Forecasts.

This work was supported by Serrapilheira Institute (grant number Serra-1709-16621).

data volume has motivated researchers to develop tools like Array databases, capable of processing vast amounts of data without resource constraints.

Array databases have similar performance requirements to traditional Database Management System (RDBMS), supporting parallel processing and concurrency control. However, they have the added advantage of supporting various multidimensional operations, like geometric and linear algebra operations such as array slicing, transposition, addition, and subtraction. Compared to NoSQL databases, Array databases offer better performance for complex mathematical operations and multidimensional data manipulation, making them a strategic choice for applications requiring advanced and precise analyses.

The execution of multidimensional operations requires high-performance multi-core processor systems like NUMA machines. These machines divide the memory space among the processors into memory modules. Each processor has access to its local memory module and the modules of other processors, referred to as remote memory. Accessing remote memory involves using an cross-chip interconnect with higher latencies than local memory accesses. Ideally, array databases should provide linear speedup when utilizing all the resources from NUMA systems. However, query threads may need to transfer large amounts of data during data processing, resulting in significant performance degradation when accessing remote memory. Hence, it is crucial to distribute query threads and data carefully to avoid performance penalties and leverage the benefits of NUMA.

The execution of query threads in array databases is similar to the ones used by RDBMS relying on the Operating System (OS) to map query data and worker threads to processor cores [3]. The OS uses load balance strategies to spread the threads over the cores and does not take into consideration specific memory access patterns from the interaction between operations running in the database and the multi-core processor architecture. The result is a sub-optimal performance with a sub-linear speedup impact on the overall execution time by 50% in subarray operations and more than 90% in aggregation and array join operations in the Savime database.

In this paper, we detail the implementation of a thread scheduler mechanism in two array databases, Savime and SciDB [4]. Our hypothesis is that mapping query threads to specific CPU cores based on the memory access pattern of each query operator can mitigate data movement across NUMA nodes and thus improve performance. Our thread scheduler uses game theory and NUMA hardware counters to analyze query memory access patterns, guiding thread pinning strategy for optimal performance. Based on Nash Game Theory [5], each thread is an independent, non-cooperative agent following a greedy strategy. Results show increased performance, reduced remote data access, and energy savings in both evaluated array databases. Our study is the first attempt to examine the influence of NUMA architecture on multidimensional array databases, particularly emphasizing thread pinning. While earlier research has concentrated on enhancing task placement, data partitioning, and load balancing in NUMA architectures for query processing in RDBMS, our work takes a novel approach by analyzing the implications specific to multidimensional array databases.

Results show performance speedup, consistent reduction in remote data access, and energy savings in both Array databases evaluated. In summary, our main contributions are the following:

- **Thread allocation mechanism:** we present a mechanism for allocating threads on NUMA machines based on memory access patterns and hardware counter metrics.
- **Analysis of distinct array operations:** we performed an extensive experimental evaluation of the impact of the NUMA architecture on multidimensional query operations.
- **Performance improvements:** our mechanism improves execution time, reduces remote memory accesses, and lowers energy consumption in NUMA architecture for two open-source array databases.

In Sect. 2, we discuss the memory access patterns and overview the NUMA architecture. Section 3 covers the decision-making method for thread pinning using game theory. In Sect. 4, we present our experimental setup and results. Section 5 discusses related work, and we conclude in Sect. 6.

2 Background

In this section, we discuss array database access patterns, their characteristics, and performance implications, followed by an explanation of the NUMA architecture.

2.1 Memory Access Patterns in Array Databases

The n-dimensional array data model is a crucial feature of array databases. This model simplifies data access and analysis through different perspectives, as each cell of the array holds attributes with the same data type. The array is divided into equally-sized chunks for storage, where each is a physical representation of the array. In this paper, we concentrate on two open-source, full-stack array databases, Savime and SciDB, as they implement the array model from scratch without relying on the relational model, unlike other databases like RasdMan [6], ArrayStore [7] and SciQL [8].

The execution of a query in array data models involves extracting data from arrays using nested function calls, and the resulting chunks are pipelined through query operations. Similar to relational databases, the execution of each operation presents one of the following memory access patterns [9]: i) operations with high and low data reuse, ii) operations with non-coalescing memory access, and iii) those with coalescing memory access patterns. We assume that each memory access patterns exercise the memory architecture differently and may point out the effects of our thread placement mechanism.

Subarray/subset operations slice n-dimensional data in a specific range. It is classified as low data reuse with coalescing memory access in 2/3 of dimensions. Unused projected values during subarray processing indicate low data reuse.

Fig. 1. Two examples of the subarray and aggregation operations. (Color figure online)

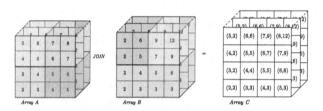

Fig. 2. Example of the join operation in an Array database. The operation joins the values of array A and B. As a result, we have array C.

During the subarray processing, the projected values are no longer used during the operation, characterizing low data reuse. Array databases implement subarray/subset operators in different ways. Savime finds cells between the range using the filter and, as a result, generates many chunks with different sizes. SciDB decodes the compressed binary data into chunks and redistributes the data to produce a new chunk configuration for the results in the interest range.

The subarray operation in the array data model is simple, but query selectivity may require processing all chunks. Figure 1 shows two subarray examples, selecting data from multidimensional array chunks in light colors. Blue represents the subarray, projecting a portion of the entire array.

Another operation considered is an aggregation (Fig. 1). Savime scans the chunks sequentially and saves the results in a buffer. After processing all the chunks, Savime merges the temporary groups [10]. In SciDB, the aggregation goes directly into the cell, which is evaluated independently of the chunk position. The aggregation operation is categorized as high data reuse because of the grouping computation and it is distinct from subarray because of its non-coalescing memory access.

Figure 2 illustrates an example of the join operation based on the array dimensions. Savime performs the join operation by iteratively combining nested pairs of chunks, and uses a sort merge join to combine the chunks. SciDB's join operation combines all chunk dimensions by treating dimension IDs and values as columns, resulting in a new array. By efficiently accessing and processing different cell IDs using a nested loop, the join operation is categorized as a high data reuse operation in the smaller array with coalescing memory access pattern.

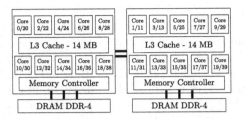

Fig. 3. Example of a NUMA architecture with 2 nodes inspired by Intel Xeon Silver 4114.

2.2 Overview of NUMA Architectures

NUMA is a computing architecture composed of several multi-core processors split into logical nodes. Every node has an Integrated Memory Controller (IMC) and its own designated memory banks. Hence, the main memory is divided among these nodes. Each NUMA node can access all the memory space, yet access time depends on the memory location relative to the processor. A processor accesses its local memory faster than remote memory. The data transfer occurs by interconnection links and may need to transfer through one or more links to reach its destination node. Figure 3 presents a diagram of a NUMA architecture composed of two Intel Xeon Silver 4114 deca-cores processors with a set of DDR-4 memory banks attached to each node. Each processor architecture consists of multiple levels of cache memory, forming a complex memory hierarchy. Cache memories utilize a coherence protocol for a cache-coherent NUMA (ccNUMA) system, ensuring a consistent shared address space view across all nodes.

The most efficient memory access in NUMA architecture occurs when performing local memory node accesses. However, the default data placement policy in the Linux OS is known as the "first-touch". This policy allocates the memory on the NUMA node where the thread is first executed. The OS policy of load balancing between nodes may cause remote access due to thread migration during execution, potentially resulting in increased remote memory access.

3 Nogar: A Non-cooperative Game to Array Database Thread Pinning

We propose a dynamic mechanism based on game theory to optimize thread placement for array databases executing on NUMA systems. Game theory is a mathematical theory for decision-making. We used it to model each thread as a decision-maker or agent in a non-cooperative game. A group of agents occupies each node, and each node's computing advantage varies over time depending on hardware resource consumption.

Each agent has a sequence of independent decisions and analyzes the profit at the start of each decision which depends on the number of other agents at its current location and the cost level in each node.

3.1   Nash Equilibrium

The Nash equilibrium, introduced by John Forbes Nash in [5], is a solution for several decision-makers to find equilibrium in a non-cooperative game. In a Nash equilibrium, players make independent decisions knowing its influence on other players. Each player chooses a strategy to achieve an outcome with the lowest cost possible, and the equilibrium is reached when no player has incentives to change strategy. The formal definition of a Nash equilibrium is:

Let (S, f) be a game with n players, where:

- $S = S_1 \times S_2 \times \ldots \times S_n$ is the strategy set of a profile;
- Player $i \in 1, \ldots, n$;
- $f(x) = f_1(x), \ldots, f_n(x)$ is the set of cost profiles;
- A cost function is evaluated at $x \in S$;
- x_i is a strategy profile of player i and x_{-i} is a strategy profile of all players except player i;
- Each player $i \in 1, \ldots, u$ chooses a strategy x_i, resulting in a strategy profile $x = (x_i, \ldots, x_u)$, then player i with cost $f_i(x)$;
- A strategy profile $x* \in S$ is at Nash equilibrium $\forall i, x_i \in S_i$, namely, $fi(x_i^*, x_{-i}^*) \geq f_i(x_i, x_{-i}^*)$;

When the set of strategies reaches the Nash equilibrium, it is possible that the game has found a solution. The game may have many Nash equilibriums.

3.2   The Thread Pinning Mechanism

To enhance throughput and performance, parallel query processing employs a multi-threading processing approach. At the same time, the operating system applies load-balancing strategies to distribute these threads among cores.

To reduce the negative impact of OS thread migration, our mechanism called **NoGar** performs the thread pinning in NUMA nodes/cores using information regarding the memory access pattern of the operations and the cache miss hardware counter. NUMA cache miss is an important counter that indicates data needs to be moved around nodes, resulting in remote memory access.

The allocation of threads is viewed as a game from a game theory perspective. The threads act as players and use the memory access behavior to define the strategies and determine the thread pinning position. Let $\epsilon = \epsilon_i$ be the allocation profile at each node, where an arbitrary thread i is linked to a specific node γ_{node} through the link $\epsilon_i \in \gamma_{node}$. The all profile set is denoted by δ.

The Nash Equilibrium in the thread allocation game can change depending on the running workload. A Nash equilibrium refers to a possible allocation that satisfies multiple threads and is considered the ideal allocation at a given moment in the game. In this context, the threads seek to minimize the cache miss count of all threads. The objective, which is to reduce cache misses, is expressed as:

$$\min_{\epsilon \in \delta} f(\epsilon) = \sum_{i=1}^{n} m_i(\epsilon)/n \qquad (1)$$

Here, m_i represents the cache misses for thread i under allocation $\epsilon \in \delta$, where m_i depends on the allocation profile ϵ. The details of $m_i(\epsilon)$ are measured using hardware counters, and cache miss measurements are only performed from threads already in the game. Based on these values, the allocation game adjusts to cache measurements and determines the best allocation. The game is represented by the tuple $<T, \delta, m_i i>$. The game aims to achieve a pure Nash equilibrium, denoted by δ^*, where no thread has the incentive to switch to a core of a different NUMA node. Such an allocation is considered a Nash equilibrium if $m_i(\epsilon_i^*, \epsilon_{-i}^*)) \geq m_i(\epsilon_i, \epsilon_{-i}^*)$ for all i and $\epsilon_i \in \epsilon$. This means that when all threads are in the optimal allocation, the Nash equilibrium may change if a new thread starts. The mechanism is designed to identify a set of Nash equilibriums that minimize the total cache miss cost.

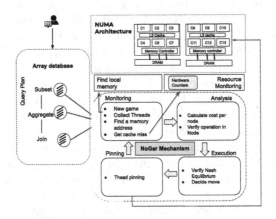

Fig. 4. Overview of the NoGar mechanism.

Figure 4 provides an overview of the NoGar mechanism based on Nash equilibrium theory. When initiating a query, the engine identifies threads by their thread identifier (TID) and instantiates players. The monitoring module then searches the memory addresses associated with each thread used in the player's strategies. During this process, NoGar collects cache misses from each thread. The analysis module then evaluates the costs of each NUMA node and assesses the distribution of threads based on each query operation. The player strategy is based on the cost function, positioning threads at the location with the lowest cost. With different memory access and reuse patterns, it becomes necessary to combine specific strategies, which means thread allocation depends on the operations underway. Operations with high data reuse and coalescent memory access allocate threads on the NUMA node containing the required memory address. Query operations with high data reuse and non-coalescent accesses mirror this behavior. However, operations with low data reuse and non-coalescent access will select the node with the lowest cost, irrespective of data location. This strategy enables operations with conflicting data reuses to choose different nodes. Yet, this might not always be feasible if the Array database executes multiple operations concurrently. Hence, threads primarily select the objective function (Eq. 1) strategy, aiming for minimal cost allocation.

The mechanism is implemented in C language and requires information about the running operation, TID, and accessed addresses. NoGar starts with the array database and collects hardware topology information using the Portable Hardware Locality library (hwloc) [11]. When a query starts, our mechanism starts monitoring hardware counters. The following section presents an experimental analysis of the NoGar mechanism.

4   Experimental Evaluation

The experimental analysis uses two Array databases: Savime version ($v.1.0$) and SciDB version ($v.19.11.5$). The NUMA machine (here called *NUMA-Skylake*) has two nodes, each node with an Intel Xeon Silver 4114 (with Skylake microarchitecture). Each Xeon socket has ten cores with a private L1 (I + D) cache (32 KB each core), a private L2 cache (1 MB each core), and a shared L3 cache (14 MB total per node). The two NUMA nodes are interconnected by a Quick Path Interconnect (QPI) link [12] 4×, with a bandwidth of 21.5 GB/S. The machine includes 128 GB of DDR-4 main memory and 14 TB of disk storage (at 15,000 rpm), running Ubuntu OS version 18.04.01 LTS for Savime and Ubuntu version 14.04.6 LTS for SciDB. We use different OS versions considering the indicated on each Array database documentation. We also evaluated the mechanism on *NUMA-SandyBridge* and *NUMA-NehalemEX* machines, but results are not included due to space limitations. Overall, the results demonstrated that the NoGar engine could perform well on different NUMA architectures. The results showed a speedup of 1.43× and 1.28×, on the NUMA-Sandy-Bridge and NUMA-Nehalem machines respectively.

We used the Performance Counter Monitor (PCM) [13] to measure the hardware performance. The Intel PCM tool estimates the main memory and processor cores' total power consumption. In the experiments, we define the maximum number of available threads for each query execution corresponding to the number of available physical cores. The workload has dense matrices based on data from the seismic benchmark *HPC4e BSC* [14], used in previous work [15]. Finally, we present results considering the average of 10 executions to reduce the variation inherent to OS decisions. The overhead of our mechanism is already included in the presented time results.

4.1   Impact of the Number of Chunks

This section presents the results of the NoGar mechanism for different query operators varying the number of chunks. Operations run simultaneously on a 1 GB database composed of two dense matrices for the join operation and a single dense array for aggregation and subarray. We normalize the results to the values obtained with 20 chunks using the OS scheduler.

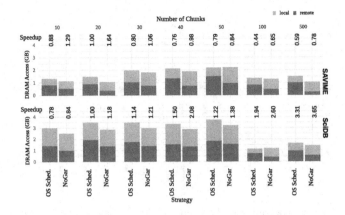

Fig. 5. Subarray: low data reuse and coalescing memory access.

Figure 5 presents the result of the subarray operation. On Savime, NoGar achieved a speedup of 1.64× when the number of chunks equals the number of cores on the machine. A smaller number of chunks increases each chunk size; therefore, threads do not have to change processed data several times. For NoGar, this discovery made game coordination even more accessible. Furthermore, pinning threads to a specific core reduces context switching between CPU cores, which helps maintain cost balance for a longer time, reducing remote access in 2.46× in Savime and 1.71× in SciDB. The results suggest that pinning the thread to a given core improves the use of the NUMA architecture.

Figures 6a and 6b show the results obtained with operations with high data reuse – aggregation and join – normalized by the number of chunks 20. These two operations are positioned by NoGar exactly on the node with the requested data. In Savime, the NoGar mechanism showed a maximum speedup of 1.38× on join and 1.17× on aggregation. Also, the OS performed well with 20 chunks. These two operations in Savime need to go through all the cells of the analyzed dimensions. Considering many chunks, the NoGar had more difficulty maintaining the thread balance among the cores. In SciDB, the join and aggregation obtained the best result, with 500 chunks reaching 4.26× in the aggregation and 2.14× in the join. The thread pinning reduced remote access due to the aggregation implementation of SciDB.

4.2 Evaluating the Behavior of Database Operators

We considered that each operation has different memory access patterns, and the amount of data or chunks processed can change in each operation. Reducing the amount of data processed can further reduce access to remote memory if thread placement is efficient. Therefore, this subsection evaluates each of the database operators separately.

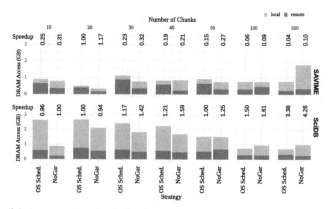

(a) Aggregation: high data reuse and non-coalescing memory access.

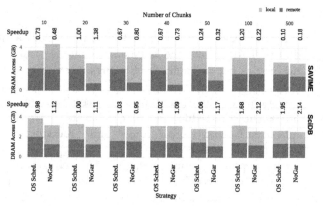

(b) Join: high data reuse and coalescing memory access.

Fig. 6. Speedup and memory accesses comparing the OS scheduler with the NoGar mechanism, varying the number of chunks. The top axis number is the speedup. Values normalized to 20 chunks.

Evaluation of the Subarray Operation: In this experiment, we evaluated the subarray operation with different selectivity. In addition, we varied in this experiment the number of processed chunks, processing all chunks (**M**) and only 20% of chunks (**F**). The selectivity indicates the percentage of data that needs to be filtered to materialize the subarray output. For example, high selectivity (**H** - in this experiment 70%) means that more data is filtered and, as a result, materializes fewer data for output in DRAM. Low selectivity (**L** - in this experiment 20%) indicates the opposite.

Figure 7 shows the results of the subarray operation varying the data selectivity. The graphs are normalized according to the results of the baseline, which is the OS Linux scheduler (*OS Sched.*). The results indicate that the NoGar mechanism, considering high selectivity, achieved a speedup of 2.49× in Savime

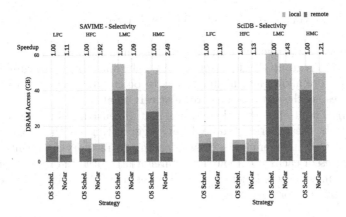

Fig. 7. Speedup and number of memory accesses comparing the OS scheduler with the NoGar engine, varying the selectivity of the **subarray** operator. The number on the top axis is the speedup. Values normalized for 20 chunks.

when pinning the thread to the node with the lowest cost. Remote access was reduced by 7×, indicating efficient utilization of cache memory.

Another interesting point is that thread allocation affects the tested databases differently. While Savime presents the best results with high selectivity, SciDB presents the best results with low selectivity. A query with high selectivity should be faster because it returns a smaller amount of data. However, Savime processes the chunk by selecting precisely the data in the specified range. SciDB, on the other hand, needs to decompress the data and redistribute it among the instances to process them. This behavior may have contributed to NoGar taking a long time to find balance in SciDB and needing to be more efficient than in Savime. The convergence time of our mechanism to find the balance is already included in the achieved times in the results.

As a result, the pinning threads with high selectivity prove to be more efficient in Savime by maximizing data locality and minimizing the extra latency of remote accesses. On the other hand, in Savime, low selectivity selects more data. As a result, the throughput is lower due to NoGar needing to perform more calculations to find the optimal allocation. This happens because the balance can change with the increase in cache misses.

The subarray operation in SciDB allows for selecting specific subsets of data from a multidimensional array, and memory selectivity can affect the performance of this operation. Thread pinning improves performance when dealing with low selectivity by reusing data already loaded into the cache memory. The reusing of data makes it easier for NoGar to find the balance quickly.

Evaluation of Aggregation Operation: The aggregation operation can perform in different dimensions. In this experiment, we compare aggregation performance in the first and last dimensions. NoGar results were similar on both array databases, achieving a speedup of 1.30× and a reduction of 1.50× in remote access.

In Savime, the processing results of each chunk are stored in a buffer that is then utilized to regroup the data and generate the final result. Assigning threads to a specific core allows intermediate data to be stored in the memory of their specific nodes, as the data chunks are relatively small.

In a SciDB multidimensional array, the first dimension has the most elements and represents the outermost data layer. Aggregating on this dimension can cause non-coalescent memory accesses, leading to memory fragmentation and reduced processor cache use. However, aggregating on the last dimension results in better memory access but lower speedup, as it involves high data reuse and non-coalescent memory accesses.

Evaluation of the Join Operation: In Savime, the join operation involves a nested loop to combine pairs of matrices, and the memory access is reduced when the operation is performed in only one dimension. When compared to the OS scheduler, NoGar performs better, achieving a speedup of 1.39×. Since threads work together on a specific chunk in NoGar's placement, unnecessary migrations of threads are prevented and the utilization of cached data in memory is increased. Additionally, in our other work [16], we showed the impact of different thread combinations that demonstrate a reduction in remote accesses, which improves performance.

On the other hand, in the SciDB, all array dimensions are necessarily joined during the join operation. Similarly, in SciDB, NoGar achieved a speedup of 1.38×. SciDB can combine pairs of cells efficiently as the arrays have the same number of chunks and sizes. The thread placement strategy of NoGar reduces remote access by allocating threads precisely on the node where the data is located and taking advantage of contiguous memory accesses.

4.3   Energy Efficiency

The evaluation of energy consumption is based on the *Energy-delay product* (EDP) metric, which measures the relationship between power and performance. This metric is determined by the product of energy and execution time and is an effective indicator of system efficiency, where a smaller value indicates a more efficient system. We consider the processor and main memory power consumption to calculate overall energy consumption.

The results, presented in Fig. 8, confirm the trends observed in the speedup analysis. Specifically, Savime exhibits a better EDP (lower) when the number of chunks is close to the number of cores, while SciDB shows the opposite trend. Notably, the NoGar engine achieved lower EDP in all cases evaluated.

5   Related Work

The need for high-performance parallel computing has motivated several studies to explore the benefits of NUMA architectures in relational query processing. In

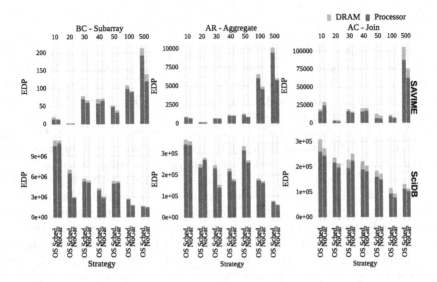

Fig. 8. Power rating including total power (processing + main memory), using the EDP metric.

query processing, Psaroudakis et al. [17] introduced task placement and data partitioning strategies for concurrent scans in NUMA, showing that load balancing between nodes optimizes computational resource use. Similarly, the authors of [18] show that topology awareness of the NUMA architecture improves query processing performance. Gawade and Kersten [19] examine the execution of query plans that are both NUMA architecture-aware and partitioned, taking into account NUMA-awareness. Psaroudakis et al. [20] proposed a prototype adaptive partitioning mechanism to minimize remote accesses and balance NUMA nodes, using processing usage data to detect imbalances between nodes. Morsel leverages data affinity and thread parallelism control in Hyper database [21], dividing data into morsels processed by the same operator pipeline. Several studies explored improving specific relational operators execution in NUMA architecture [3,22–24]. Prior research has presented various approaches to mitigate the impact of the NUMA architecture in a manner that is agnostic to specific applications. Some of these works [25–32] concentrate on thread placement techniques that consider memory communication costs between nodes. These approaches improve the performance of applications that do not share data among processes.

In contrast to other studies, this paper examines the influence of the NUMA architecture on Array databases with distinct storage and memory usage. Our research benefits from the knowledge that the OS conducts first-touch mapping, enabling us to determine the thread and data mapping simultaneously when allocating threads. Unlike other related works that primarily concentrate on data mapping, our study focuses on thread mapping, which has an indirect effect on page mapping.

6   Conclusion

This work considered the NUMA effects on query processing in Array databases. Modern array databases do not take advantage of the potential of the NUMA architecture. Parallel query execution does not control the allocation of threads and leaves this responsibility to the OS, which does not know the exact workload placement of each execution thread.

In this paper, we presented a thread placement mechanism called NoGar that analyzes query workloads to mitigate data movement during the query execution between NUMA nodes. Considering that most query execution scenarios involve multiple threads collaborating or competing to perform a task, we presented a mechanism based on game theory that manages thread pinning. In this work, thread allocation is a multi-agent decision-making game. Our mechanism decisions are made based on information from the memory access pattern of queries and cache misses at each NUMA node. The experimental evaluation showed that the mechanism places threads efficiently most of the time, achieving a performance gain of 48% and energy saving of 37% on average, taking over the scheduling previously performed by the OS.

References

1. Baumann, P., Misev, D., Merticariu, V., Huu, B.P.: Array databases: concepts, standards, implementations. J. Big Data **8**(1), 1–61 (2021)
2. Stonebraker, M., Brown, P., Poliakov, A., Raman, S.: The architecture of SciDB. In: Bayard Cushing, J., French, J., Bowers, S. (eds.) SSDBM 2011. LNCS, vol. 6809, pp. 1–16. Springer, Heidelberg (2011). https://doi.org/10.1007/978-3-642-22351-8_1
3. Dominico, S., de Almeida, E.C., Meira, J.A., Alves, M.A.Z.: An elastic multi-core allocation mechanism for database systems. In: ICDE, pp. 473–484 (2018)
4. Brown, P.G.: Overview of SciDB: large scale array storage, processing and analysis. In: SIGMOD, pp. 963–968 (2010)
5. Nash, J.: Non-cooperative games. Ann. Math. **54**, 286–295 (1951)
6. Baumann, P., Furtado, P., Ritsch, R., Widmann, N.: The RasDaMan approach to multidimensional database management. In: SAC, pp. 166–173 (1997)
7. Soroush, E., Balazinska, M., Wang, D.: ArrayStore: a storage manager for complex parallel array processing. In: SIGMOD, pp. 253–264 (2011)
8. Zhang, Y., Kersten, M., Manegold, S.: SciQL: array data processing inside an RDBMS. In: SIGMOD, pp. 1049–1052 (2013)
9. Kepe, T.R., de Almeida, E.C., Alves, M.A.Z.: Database processing-in-memory: an experimental study. PVLDB **13**(3), 334–347 (2019)
10. Lustosa, H.L.S.: SAVIME: enabling declarative array processing in memory. Ph.D. dissertation, LNCC, Petrópolis - Brasil, Fevereiro, p. 100 (2020)
11. Broquedis, F., et al.: hwloc: a generic framework for managing hardware affinities in HPC applications. In: Euromicro, pp. 180–186 (2010)
12. Intel. Maximizing multicore processor performance (2019). https://www.intel.com/content/www/us/en/io/quickpath-technology/quickpath-technology-general.html
13. Willhalm, T., Dementiev, R., Fay, P.: Intel performance counter monitor (2012). https://software.intel.com/en-us/articles/intel-performance-counter-monitor

14. B. S. Center. HPC4E seismic test suite (2016). https://www.bsc.es/news/bsc-news/new-hpc4e-seismic-test-suite-increase-the-pace-development-new-modelling-and-imaging-technologies
15. Lustosa, H., Porto, F.: SAVIME: a multidimensional system for the analysis and visualization of simulation data. CoRR, vol. abs/1903.02949 (2019)
16. Dominico, S., Alves, M.A.Z., de Almeida, E.C.: On the performance limits of thread placement for array databases in non-uniform memory architectures. Comput. J. **105**, 1059–1075 (2022)
17. Psaroudakis, I., Scheuer, T., May, N., Sellami, A., Ailamaki, A.: Scaling up concurrent main-memory column-store scans: towards adaptive NUMA-aware data and task placement. PVLDB **12** (2015)
18. Kiefer, T., Schlegel, B., Lehner, W.: Experimental evaluation of NUMA effects on database management systems. In: BTW, pp. 185–204 (2013)
19. Gawade, M., Kersten, M.: NUMA obliviousness through memory mapping. In: DAMON, pp. 1–7 (2015)
20. Psaroudakis, I., Scheuer, T., May, N., Sellami, A., Ailamaki, A.: Adaptive NUMA-aware data placement and task scheduling for analytical workloads in main-memory column-stores. PVLDB **2** (2016)
21. Leis, V., Boncz, P., Kemper, A., Neumann, T.: Morsel-driven parallelism: a NUMA-aware query evaluation framework for the many-core age. In: SIGMOD, pp. 743–754 (2014)
22. Albutiu, M.-C., Kemper, A., Neumann, T.: Massively parallel sort-merge joins in main memory multi-core database systems. PVLDB **5** (2012)
23. Li, Y., Pandis, I., Mueller, R., Raman, V., Lohman, G.M.: NUMA-aware algorithms: the case of data shuffling. In: CIDR (2013)
24. Balkesen, C., Alonso, G., Teubner, J., Özsu, M.T.: Multi-core, main-memory joins: sort vs. hash revisited. Proc. VLDB Endow. **7**(1), 85–96 (2013)
25. Diener, M., Cruz, E.H.M., Navaux, P.O.A.: Locality vs. balance: exploring data mapping policies on NUMA systems. In: PDP, pp. 9–16 (2015)
26. Lepers, B., Quéma, V., Fedorova, A.: Thread and memory placement on NUMA systems: asymmetry matters. In: USENIX, pp. 277–289 (2015)
27. Virouleau, P., Broquedis, F., Gautier, T., Rastello, F.: Using data dependencies to improve task-based scheduling strategies on NUMA architectures. In: Dutot, P.-F., Trystram, D. (eds.) Euro-Par 2016. LNCS, vol. 9833, pp. 531–544. Springer, Cham (2016). https://doi.org/10.1007/978-3-319-43659-3_39
28. Di Gennaro, I., Pellegrini, A., Quaglia, F.: OS-based NUMA optimization: tackling the case of truly multi-thread applications with non-partitioned virtual page accesses. In: 16th CCGrid, pp. 291–300 (2016)
29. Wang, W., Davidson, J.W., Soffa, M.L.: Predicting the memory bandwidth and optimal core allocations for multi-threaded applications on large-scale NUMA machines. In: IEEE HPCA, pp. 419–431 (2016)
30. Serpa, M.S., Krause, A.M., Cruz, E.H., Navaux, P.O.A., Pasin, M., Felber, P.: Optimizing machine learning algorithms on multi-core and many-core architectures using thread and data mapping. In: PDP, pp. 329–333. IEEE (2018)
31. Popov, M., Jimborean, A., Black-Schaffer, D.: Efficient thread/page/parallelism autotuning for NUMA systems. In: International Conference on Supercomputing, pp. 342–353 (2019)
32. Cruz, E.H., Diener, M., Pilla, L.L., Navaux, P.O.: Online thread and data mapping using a sharing-aware memory management unit. ACM TOMPECS **5**(4), 1–28 (2021)

LHKV: A Key-Value Data Collection Mechanism Under Local Differential Privacy

Weihao Xue[1], Yingpeng Sang[1(✉)] (ID), and Hui Tian[2]

[1] School of Computer Science and Engineering, Sun Yat-sen University, Guangzhou, China
xuewh5@mail2.sysu.edu.cn, sangyp@mail.sysu.edu.cn
[2] School of Information and Communication Technology, Griffith University, Nathan, Australia
hui.tian@griffith.edu.au

Abstract. Local differential privacy (LDP) is an emerging technology used to protect privacy. Users are required to locally perturb their raw data under the framework of LDP, before they are transmitted to the server. This technology can be applied to various data types, including key-value data. However, in existing LDP mechanisms for key-value data, it is difficult to balance data utility and communication costs, particularly when the domain of keys is large. In this paper we propose a local-hashing-based mechanism called LHKV for collecting key-value data. LHKV can maintain high utility and keep the end-to-end communication costs low. We provide theoretical proof that LHKV satisfies ϵ-LDP and analyze the variances of frequency and mean estimations. Moreover, we employ Fast Local Hashing to accelerate the aggregation and estimation process, which significantly reduces computation costs. We also conduct experiments to demonstrate that, in comparison with the existing mechanisms, LHKV can effectively reduce communication costs without sacrificing utility while ensuring the same LDP guarantees.

Keywords: Local differential privacy · Key-value data · Local hashing

1 Introduction

As the era of information technology continues to flourish, the issue of personal privacy has become increasingly important. The protection of sensitive information from leakage is a primary concern. Differential privacy (DP) is an emerging technology that can effectively safeguard sensitive information while facilitating data intercommunication. Attackers are unable to infer the raw data from published information, thereby ensuring privacy protection. Differential privacy technology consists of central differential privacy (CDP) and local differential privacy (LDP). In CDP, users' raw data is directly uploaded to the server, which then uses the DP mechanism to add noise and publish perturbed data or statistical analysis results. This method requires the server to be fully trusted since it collects all users' data. However, this is not always feasible. To address this issue, the concept of LDP was proposed. Each user uses the LDP mechanism to perturb the raw data before uploading it to the central server, preventing the untrusted server from directly obtaining the real data.

LDP can be applied to many data types and statistical analysis tasks. As technology advances, its applications have expanded from estimating frequencies for categorical data and means for numerical data to more complex data types, such as graph data and key-value data [13]. Key-value data is a common heterogeneous data type, which consists of two parts: the key belonging to categorical data and the value belonging to numerical data. A correlation exists between the key and value, with the value's existence being dependent on the key. Key-value data is prevalent on the Internet, such as movies with corresponding ratings on rating websites, and products with corresponding prices on shopping websites. If key-value data were sent directly to the central server, it could result in sensitive information leakage. A common LDP-based movie rating system is shown in Fig. 1. Each user has one or more key-value pairs, which are first perturbed using LDP mechanisms and then sent to the server. The server aggregates the perturbed data and performs statistical analysis tasks, typically frequency and mean estimations. Importantly, the server cannot deduce the original pairs from the perturbed data, which ensures the privacy of users throughout the process.

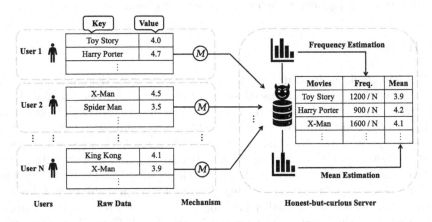

Fig. 1. A Movie Rating System under LDP

There are not many existing LDP mechanisms designed for key-value data. Among them, PrivKVM in [14] and PCKV in [6] are two representatives. The sampling method used in PrivKVM is essentially grouping users by keys, then in each group key-value pairs are perturbed. However, the utility of PrivKVM is relatively low when the domain of keys is large. Furthermore, the mean estimation of PrivKVM requires multiple interactions to be unbiased, leading to an increase in communication costs. PCKV utilizes Generalized Randomized Response and Unary Encoding [10] for the perturbation of key-value pairs. Thus there were two mechanisms, PCKV-GRR and PCKV-UE. However, for collections of key-value data, the size of the key domain is usually very large. In such cases, both mechanisms suffer from certain limitations. PCKV-GRR's utility is significantly reduced when the domain is large, whereas PCKV-UE can improve this issue but leads to a rise in communication costs.

To address the issues mentioned, we present a collection mechanism based on Local Hashing for Key-Value (LHKV) data under LDP. Overall, the main contributions of this paper can be summarized as follows:

1. Considering the shortcomings of existing mechanisms, we design a key-value data collection mechanism called LHKV based on local hashing. Our mechanism can maintain high utility and low communication costs, even when the domain of keys is large. We prove that LHKV guarantees ϵ-LDP, and analyze the variances of frequency and mean estimations.
2. To speed up the procedure of aggregation and estimation, we employ Fast Local Hashing (FLH). Although some theoretical guarantees on accuracy are sacrificed, we achieve computational gains on the server side. Practically, FLH offers significant computational acceleration with minimal loss of utility.
3. We evaluate the proposed mechanism's performance using several datasets. The experimental outcomes indicate that, in comparisons with the existing mechanisms, LHKV can effectively reduce end-to-end communication costs without sacrificing utility under the same LDP guarantees.

2   Related Work

LDP is commonly utilized in various data types and statistical analysis tasks, as it effectively prevents untrusted servers from accessing users' actual data. Currently, most methods in this field mainly focus on categorical and numerical data, with the tasks primarily including the frequency and mean estimations.

For categorical data, Randomized Response is currently the most widely used LDP mechanism due to its high scalability. Warner [12] first proposed Randomized Response for privacy surveys. The main idea behind Randomized Response is to keep sensitive information unchanged with a higher probability and provide a reasonable denial with a smaller probability. As a result, investigators cannot deduce whether the information provided is true or not. Generalized Randomized Response (GRR) is an extension of Randomized Response for multi-value data. In GRR, the raw data remains unchanged with a higher probability and is perturbed to other values with a smaller probability. The basic RAPPOR mechanism, proposed by Erlingsson [5], encodes raw data into a binary vector of the same size as the domain and then flips the 0 and 1 elements in the vector with different probabilities. Wang [10] improved upon RAPPOR by proposing Optimized Unary Encoding (OUE) to minimize approximate variance. In addition, this paper suggested that local hashing can be used to compress the domain of data and thus proposed Binary Local Hashing (BLH) and Optimized Local Hashing (OLH) mechanisms.

For numerical data, the Laplacian mechanism under CDP can be applied in a distributed manner [4]. Mean estimation is a common statistical analysis task for numerical data, and the current randomization methods used for mean estimation can be divided into two categories: extreme values perturbation and distribution perturbation [9]. The former aims to report either of the two extreme values with a certain probability, as demonstrated in [3]. The latter maps the output value to a continuous distribution, where the perturbed value is likely to fall within an interval related to the raw value. An example of this type of method is the Piecewise Mechanism proposed in [8].

Till now, there have not been many methods that apply LDP to the collection of key-value data. Among them, the most representatives are PrivKVM [14] and PCKV [6]. PrivKVM is the pioneering privacy protection mechanism that employs LDP for key-value data. It operates by having each user randomly select a key from the entire key domain, and then perturbing the selected key-value pair based on whether the user holds the key or not. Through multiple rounds of aggregation, it aims to achieve an unbiased estimation of the mean. However, the sampling method of PrivKVM essentially groups users evenly, then each group is used to estimate a specific key. If the domain size is too large, it may lead to low statistical efficiency. Additionally, multiple rounds of interactions in PrivKVM increase communication costs. PCKV introduced the Padding-and-Sampling protocol [7,11] in the sampling phase, where each user adds fake key-value pairs to reach a uniform length of key-value pairs. Then each user samples a key-value pair and perturbs it using GRR and UE, resulting in two mechanisms, PCKV-GRR and PCKV-UE. Notably, if the size of key domain is large, PCKV-GRR may have lower utility, and PCKV-UE may be more effective despite the fact that it incurs higher communication costs.

3 Preliminaries and Problem Definition

3.1 Local Differential Privacy

In contrast to LDP, users directly upload their raw data to the server in CDP. An honest-but-curious server may normally conduct the required analysis tasks over the raw data, but also leak the privacy to others for benefits. This issue can be addressed by LDP, which requires raw data to be perturbed locally by users before being uploaded to the server. This approach guarantees that the raw data is not directly accessible from the untrusted server, thereby preventing data leakage and enhancing the protection of privacy.

Definition 1 (Local Differential Privacy (LDP) [2]). *A randomized mechanism $\mathcal{M}$ is said to satisfy ϵ-LDP for a given $\epsilon \in \mathbb{R}^+$ if and only if, for any two inputs x and x' in the domain and for any output $y \in Range(\mathcal{M})$, the following formula holds:*

$$\frac{\Pr[\mathcal{M}(x) = y]}{\Pr[\mathcal{M}(x') = y]} \le e^\epsilon \qquad (1)$$

Here, the parameter ϵ is known as the privacy budget, which reflects the level of privacy protection provided by the mechanism. A smaller privacy budget corresponds to stronger privacy protection, but it may also result in lower data utility.

3.2 Problem Definition

Consider the model depicted in Fig. 1 and a set of N users $\mathcal{U} = \{u_1, u_2, \cdots, u_N\}$, where u_i holds a set of key-value pairs denoted by S_i. Note that $|S_i| \in [1, d]$. Each key-value pair consists of a key and its corresponding value, denoted as $\langle k, v \rangle$. The key k belongs to the domain of keys $\mathcal{K} = \{1, 2, \cdots, d\}$, and the value v represents the corresponding value. The value v is normalized to the range $[-1, 1]$.

Before uploading data to the server, users apply an LDP mechanism to perturb their key-value pairs. The perturbed data is collected by the server and aggregated to estimate the frequency and mean of each key, which are defined as follows.

Frequency. The frequency of k is the proportion of users that own the key and is denoted as f_k. It is given by:

$$f_k = \frac{\sum_{u_i \in \mathcal{U}} 1_{S_i}(\langle k, \cdot \rangle)}{N} \tag{2}$$

where $1_{S_i}(\langle k, \cdot \rangle)$ is an existence judgement function, which equals 1 if u_i has the key k, and 0 otherwise.

Mean. The mean of key k is the average value among the users who own it and is denoted as m_k. It is given by:

$$m_k = \frac{\sum_{u_i \in \mathcal{U}, \langle k, v \rangle \in S_i} v}{N f_k} \tag{3}$$

The server estimates $\hat{f}_k$ and $\hat{m}_k$ for all keys, which are estimators of f_k and m_k, respectively. Our goal is to minimize the estimation error, so as to maximize the utility of the estimated data.

4   LHKV

LHKV consists of three main steps: data preprocessing, perturbation, and server aggregation and estimation. In Sect. 4.1, we will preprocess the raw data, which involves sampling and discretization of the key-value pairs. In Sect. 4.2, we will present our perturbation method, which ensures ϵ-LDP, and demonstrate how the users perturb the sampled key-value pairs. In Sect. 4.3, we will explain how the server aggregates the perturbed data received from the users and performs frequency and mean estimations of the keys. We will also provide a theoretical analysis of the estimators. Finally, in Sect. 4.4, we will introduce the use of fast local hashing to accelerate the server aggregation and estimation process.

4.1   Data Preprocessing

To perturb the key-value pairs, there are different approaches that can be taken. Dividing the privacy budget equally among each pair is one approach, but this could result in too much noise and low utility. Another is to randomly select one pair and assign the entire privacy budget to it. Following the latter approach, the Padding-and-Sampling protocol is a widely used sampling method that ensures the number of items across all users is uniform before sampling [7,11]. This protocol was first proposed for set-valued data under LDP and later extended to key-value data [6]. In this paper, we use the Padding-and-Sampling protocol.

Under this protocol, if the number of pairs owned by a user is less than a predefined size l, the user adds fake keys with corresponding values set to 0 and then samples one pair from the l pairs. If it is greater than l, the user just randomly samples one from their local pairs. Figure 2 provides an example of $l = 4$.

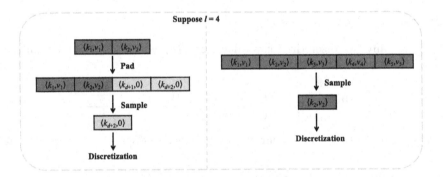

Fig. 2. The Padding-and-Sampling Protocol

After sampling, we need to discretize the sampled pair $\langle k, v \rangle$. We use a probability of $\frac{1+v}{2}$ to map it to $\langle k, 1 \rangle$, and a probability of $\frac{1-v}{2}$ to map it to $\langle k, -1 \rangle$. This ensures that the raw key-value pair is discretized without bias.

4.2   Perturbation

PCKV in [6] employs two basic LDP mechanisms, GRR and UE, for perturbation. As we have discussed, when the domain of keys is large, the utility of PCKV-GRR is low. Though the utility of PCKV-UE is high, its communication costs are high. For categorical data, OLH proposed in [10] uses local hashing and can have high utility and low communication costs, particularly for large domains, thus combine the advantages of GRR and UE. Therefore, in this paper we utilize OLH to perturb pairs.

We define the set of all possible pairs after discretization as $\mathcal{S} = \{\langle k, v \rangle | k \in \{1, \cdots, d + l\}, v \in \{1, -1\}\}$. For any two different key-value pairs $\langle k, v \rangle$ and

$\langle k', v' \rangle$ belonging to the set $\mathcal{S}$, let $\mathbb{H} = \{H | H : \langle k, v \rangle \rightarrow [g], \langle k, v \rangle \in \mathcal{S}\}$ be a universal hash function family, where $[g]$ represents the set of $\{0, 1, ..., g-1\}$ and g is the size of the output space of $\mathbb{H}$. It holds that $\Pr[H(\langle k, v \rangle) = H(\langle k', v' \rangle)] \leq \frac{1}{g}$. Each user randomly picks H from $\mathbb{H}$ to map the sampled pair $\langle k, v \rangle$ to $x = H(\langle k, v \rangle) \in [g]$. In practice, the input of a hash function can be set to $k \times v$ or $3 \times k + v$, both of which can ensure that the input corresponding to each key-value pair is unique. Thus, a tuple (H, x) comprising the hash function and the hash value can represent each key-value pair. To protect privacy, we need to perturb the tuple (H, x). Assuming that the perturbed tuple is (H, y), we perturb (H, x) as follows:

$$\Pr[y = i] = \begin{cases} p = \frac{e^\epsilon}{e^\epsilon + g - 1}, & \text{if } x = i \\ q = \frac{1}{e^\epsilon + g - 1}, & \text{if } x \in [g] \setminus i \end{cases} \tag{4}$$

According to [10], the choice of $g = e^\epsilon + 1$ can minimize the variance of the estimators. In this case, $p = \frac{1}{2}$, $q = \frac{1}{2e^\epsilon}$.

Theorem 1. *Our approach satisfies ϵ-LDP.*

Proof. For any two discretized key-value pairs, $\langle k_1, v_1 \rangle$, $\langle k_2, v_2 \rangle$ and any output (H, y), the following holds true:

$$\frac{\Pr[(H, y) | \langle k_1, v_1 \rangle]}{\Pr[(H, y) | \langle k_2, v_2 \rangle]} = \frac{\Pr[\mathcal{M}(H(\langle k_1, v_1 \rangle)) = y]}{\Pr[\mathcal{M}(H(\langle k_2, v_2 \rangle)) = y]} \leq \frac{\frac{e^\epsilon}{e^\epsilon + g - 1}}{\frac{1}{e^\epsilon + g - 1}} = e^\epsilon \tag{5}$$

According to Definition 1, our approach satisfies ϵ-LDP. ∎

4.3 Aggregation and Estimation

After completing the local perturbation, each user sends the tuple (H, y) obtained from the perturbation to the server for aggregation and analysis. In this paper, we aim to estimate the frequency and value mean of keys, so we need to estimate the frequency of all possible pairs after discretization. To achieve this, we first introduce a helper function $B(\cdot)$ which determines whether a perturbed hash value y and raw key-value pair $\langle k, v \rangle$ are related. The definition of $B(\langle k, v \rangle, (H, y))$ is as follows:

$$B(\langle k, v \rangle, (H, y)) = \begin{cases} 1, & \text{if } H(\langle k, v \rangle) = y \\ 0, & \text{if } H(\langle k, v \rangle) \neq y \end{cases} \tag{6}$$

When the output of the helper function is 1, $\langle k, v \rangle$ is considered supported. In practice, the server uses the hash function of each user to calculate the hash values of all key-value pairs, and then compares them with the hash value the user sent. The server counts the number of times each (k, v) is supported and makes an unbiased estimation on its frequency, i.e., $f_{k,v}$. Finally the server makes an unbiased estimation on the frequency of key k, i.e., f_k. The following theorem gives unbiased estimators for $f_{k,v}$ and f_k.

Theorem 2. *If $l \geq |S_i|$ for all $u_i \in \mathcal{U}$, then the unbiased estimators of $f_{k,v}$ and f_k, are given as $\hat{f}_{k,v}$ in Eq. (7) and $\hat{f}_k$ in Eq. (8), respectively.*

$$\hat{f}_{k,v} = \frac{g \sum_{y \in [g]} B(\langle k, v \rangle, (H, y)) - N}{N(pg - 1)} \cdot l \tag{7}$$

$$\hat{f}_k = \frac{g \sum_{v \in \{1, -1\}} \sum_{y \in [g]} B(\langle k, v \rangle, (H, y)) - 2N}{N(pg - 1)} \cdot l \tag{8}$$

Proof. To prove that $\hat{f}_{k,v}$ is an unbiased estimator of $f_{k,v}$, that is, to prove $\mathbb{E}[\hat{f}_{k,v}] = f_{k,v}$, we first calculate the expectation of $\sum_{y \in [g]} B(\langle k, v \rangle, (H, y))$ as follows:

$$\mathbb{E}\left[\sum_{y \in [g]} B(\langle k, v \rangle, (H, y)) \right] = N \frac{f_{k,v}}{l} p + N(1 - \frac{f_{k,v}}{l})(\frac{1}{g}p + \frac{g - 1}{g}q)$$

$$= N \frac{f_{k,v}}{l}(p - \frac{1}{g}) + N \frac{1}{g} \tag{9}$$

Thus we have

$$\mathbb{E}[\hat{f}_{k,v}] = \frac{g \mathbb{E}\left[\sum_{y \in [g]} B(\langle k, v \rangle, (H, y)) \right] - N}{N(pg - 1)} \cdot l$$

$$= \frac{g \left[N \frac{f_{k,v}}{l}(p - \frac{1}{g}) + N \frac{1}{g} \right] - N}{N(pg - 1)} \cdot l \tag{10}$$

$$= f_{k,v}$$

Therefore, $\hat{f}_{k,v}$ is an unbiased estimator of $f_{k,v}$. For the frequency f_k of key k, an estimator $\hat{f}_k$ can be obtained by adding the estimators of $\langle k, 1 \rangle$ and $\langle k, -1 \rangle$, as shown below:

$$\hat{f}_k = \hat{f}_{k,1} + \hat{f}_{k,-1}$$

$$= \frac{g \sum_{v \in \{1, -1\}} \sum_{y \in [g]} B(\langle k, v \rangle, (H, y)) - 2N}{N(pg - 1)} \cdot l \tag{11}$$

Because $\hat{f}_{k,1}$ and $\hat{f}_{k,-1}$ are unbiased, $\hat{f}_k$ is also an unbiased estimator of f_k. This concludes the proof. ∎

The following theorem gives the variance of the unbiased estimator $\hat{f}_k$.

Theorem 3. *The variance of $\hat{f}_k$ is*

$$Var[\hat{f}_k] = \frac{2(g - 1)l^2}{N(pg - 1)^2} + \frac{f_k(g - pg - 1)l}{N(pg - 1)} \tag{12}$$

Proof. Let's first calculate the variance of $\hat{f}_{k,1}$ and $\hat{f}_{k,-1}$, and the calculation process is as follows:

$$
\begin{aligned}
Var[\hat{f}_{k,1}] &= \frac{g^2 l^2}{N^2(pg-1)^2} Var\left[\sum_{y\in[g]} B(\langle k,1\rangle,(H,y))\right] \\
&= \frac{g^2 l^2}{N^2(pg-1)^2}\left[N\frac{f_{k,1}}{l}p(1-p)+N(1-\frac{f_{k,1}}{l})\frac{1}{g}(1-\frac{1}{g})\right] \quad (13)\\
&= \frac{(g-1)l^2}{N(pg-1)^2}+\frac{f_{k,1}(g-pg-1)l}{N(pg-1)}
\end{aligned}
$$

We can obtain $Var[\hat{f}_{k,-1}] = \frac{(g-1)l^2}{N(pg-1)^2}+\frac{f_{k,-1}(g-pg-1)l}{N(pg-1)}$ in a similar way. Therefore, the variance of $\hat{f}_k$ is given by:

$$
\begin{aligned}
Var[\hat{f}_k] &= Var[\hat{f}_{k,1}]+Var[\hat{f}_{k,-1}]\\
&= \frac{(g-1)l^2}{N(pg-1)^2}+\frac{f_{k,1}(g-pg-1)l}{N(pg-1)}+\frac{(g-1)l^2}{N(pg-1)^2}+\frac{f_{k,-1}(g-pg-1)l}{N(pg-1)}\\
&= \frac{2(g-1)l^2}{N(pg-1)^2}+\frac{f_k(g-pg-1)l}{N(pg-1)}
\end{aligned}
$$

$$(14)$$

This concludes the proof. ∎

As stated in [10], the frequency of most keys is usually very low in practice, implying that f_k is very small. Therefore, the variance can be approximated as:

$$Var^*[\hat{f}_k] \approx \frac{2(g-1)l^2}{N(pg-1)^2} \quad (15)$$

If we set $g = e^\epsilon + 1$, the approximated variance of $\hat{f}_k$ is $\frac{8l^2 e^\epsilon}{N(e^\epsilon-1)^2}$. Notably, this variance is smaller than the corresponding variance of PCKV-UE, which is $\frac{8l^2(e^\epsilon+1)}{N(e^\epsilon-1)^2}$. This theoretical result shows that our proposed mechanism is superior to PCKV-UE in terms of data utility.

The true mean m_k of key k can be calculated as $m_k = \frac{f_{k,1}-f_{k,-1}}{f_{k,1}+f_{k,-1}}$. Therefore, we can use the estimators $\hat{f}_{k,1}$ and $\hat{f}_{k,-1}$ to estimate m_k. The estimator $\hat{m}_k$ is defined as follows:

$$\hat{m}_k = \frac{\hat{f}_{k,1}-\hat{f}_{k,-1}}{\hat{f}_{k,1}+\hat{f}_{k,-1}} = \frac{B(\langle k,1\rangle,(H,y))+B(\langle k,-1\rangle,(H,y))}{B(\langle k,1\rangle,(H,y))+B(\langle k,-1\rangle,(H,y))-2N} \quad (16)$$

Below we analyze the expectation and variance of $\hat{m}_k$. For two random variables X and Y, we can use Taylor expansion to obtain $\mathbb{E}\left[\frac{X}{Y}\right]$ and $Var\left[\frac{X}{Y}\right]$ as following:

$$\mathbb{E}\left[\frac{X}{Y}\right] \approx \frac{\mathbb{E}[X]}{\mathbb{E}[Y]} - \frac{Cov(X,Y)}{\mathbb{E}[Y]^2} + \frac{\mathbb{E}[X]}{\mathbb{E}[Y]^3}Var[Y],$$

$$Var\left[\frac{X}{Y}\right] \approx \frac{Var[X]}{\mathbb{E}[Y]^2} - \frac{2\mathbb{E}[X]\,Cov(X,Y)}{\mathbb{E}[Y]^3} + \frac{\mathbb{E}[X]^2}{\mathbb{E}[Y]^4}Var[Y].$$

For convenience, we define $X = \hat{f}_{k,1} - \hat{f}_{k,-1}$ and $Y = \hat{f}_{k,1} + \hat{f}_{k,-1}$. Then we can calculate the expectation and variance of X and Y, as well as their covariance, as following:

$$\mathbb{E}[X] = f_k m_k, \quad \mathbb{E}[Y] = f_k,$$

$$Var[X] = Var[Y] = \frac{2(g-1)l^2}{N(pg-1)^2} + \frac{f_k(g-pg-1)l}{N(pg-1)},$$

$$Cov[X,Y] = \mathbb{E}[XY] - \mathbb{E}[X]\mathbb{E}[Y]$$
$$= \mathbb{E}[\hat{f}_{k,1}^2 - \hat{f}_{k,-1}^2] - \left(\mathbb{E}[\hat{f}_{k,1}]^2 - \mathbb{E}[\hat{f}_{k,-1}]^2\right)$$
$$= Var[\hat{f}_{k,1}] - Var[\hat{f}_{k,-1}]$$
$$= 0.$$

Therefore, we can obtain the following approximation for $\mathbb{E}[\hat{m}_k]$ and $Var[\hat{m}_k]$:

$$\mathbb{E}[\hat{m}_k] \approx m_k\left[1 + \frac{2(g-1)l^2}{f_k^2 N(pg-1)^2} + \frac{(g-pg-1)l}{f_k N(pg-1)}\right], \tag{17}$$

$$Var[\hat{m}_k] \approx (1+m_k^2)\left[\frac{2(g-1)l^2}{f_k^2 N(pg-1)^2} + \frac{(g-pg-1)l}{f_k N(pg-1)}\right]. \tag{18}$$

It is worth noting that the above analysis is purely theoretical. In practice, we need to make adjustments to the estimated frequency and mean. As pointed out in [6], if $Var[\hat{f}_k]$ is not very small, it is possible that the estimator $\hat{f}_k$ is beyond the range of $[0,1]$ for very high or very low frequency f_k, which is obviously not reasonable. The same is true for $\hat{m}_k$. Therefore, it is necessary to adjust these estimators so that the estimation error can be effectively reduced. Specifically, we first adjust the estimated frequency $\hat{f}_k$ to the range of $[\frac{1}{N}, 1]$. Then, we further adjust the estimated frequencies of $\langle k,1\rangle$ and $\langle k,-1\rangle$ to the range of $[0, N\frac{\hat{f}_k}{l}]$ to calculate $\hat{m}_k$, ensuring the estimated mean value falls within the range of $[-1,1]$.

4.4 Fast Local Hashing

When dealing with a large domain of keys, our mechanism can achieve high utility and low end-to-end communication costs. However, the computational complexity required for aggregation and estimation on the server side can be $O(Nd)$ due to the use of hash functions, leading to significant computation costs. To address this issue, we introduce Fast Local Hashing (FLH) proposed in [1]. Notably, FLH was initially devised for categorical data, and is now being employed on key-value pairs for the first time.

FLH uses a new parameter k', which requires each user to randomly choose one from a set of k' hash functions instead of uniformly sampling a universal hash family [1]. The server pre-computes two $k' \times d$ matrices, M_1 and M_2, where $M_1[i][j]$ denotes the hash value of the key-value pair $\langle j, 1 \rangle$ using the i-th hash function, and $M_2[i][j]$ denotes the hash value of the key-value pair $\langle j, -1 \rangle$ using the i-th hash function. Consequently, the server can traverse the corresponding row based on the users' hash functions, reducing the number of hash function calls from $O(Nd)$ to $O(k'd)$. However, this approach comes at the cost of sacrificing some theoretical accuracy guarantees. In order to gain computational benefits, it is required that $k' \ll N$. In the next section, we will demonstrate that using FLH can significantly accelerate computation while still maintaining high utility.

5   Experiments

5.1   Experimental Settings

We use one synthetic dataset and two real-world datasets to conduct experiments to evaluate LHKV's performance. Table 1 lists the parameters for each dataset. The synthetic dataset, named GAUSS, consists of keys and values that follow a Gaussian distribution. It includes $N = 100000$ users, with a key domain size of $d = 100$, and values ranging from -1 to 1. The Coursera[1] dataset and Clothing[2] dataset were obtained from Kaggle, with their value ranges normalized to $[-1, 1]$.

Table 1. Datasets for experimental evaluation

Datasets	N	d	#Ratings
GAUSS	100000	100	100000
Coursera	140320	1835	140320
Clothing	105508	5850	192544

To evaluate the effectiveness of LHKV, we employ the Mean Square Error (MSE) as a metric to measure the accuracy of frequency and mean estimations. For $k \in \mathcal{K} = \{1, \cdots, d\}$, we use $\hat{f}_k$ and f_k to denote the estimated and true frequency of key k, respectively, and $\hat{m}_k$ and m_k to represent the estimated and true mean of key k, respectively. We calculate the MSE results for both keys and associated values. Each reported result is an average of 10 times of experiments, to ensure the reliability of our results.

---

[1] https://www.kaggle.com/datasets/septa97/100k-courseras-course-reviews-dataset.
[2] https://www.kaggle.com/rmisra/clothing-fit-dataset-for-size-recommendation.

5.2 Experimental Results

We compare our method, LHKV, with several existing LDP mechanisms for key-value data collection, including PrivKVM [14], PCKV-GRR [6] and PCKV-UE [6]. To achieve an unbiased estimation of the mean, PrivKVM requires multiple iterations, and the number of iterations is set to 6, as referred to in [14]. A predefined padding length of l is required for LHKV, PCKV-GRR and PCKV-UE. An appropriate l can be selected for these mechanisms according to the dataset parameters. LHKV based on FLH is also compared. Considering the size of the dataset, the number of hash functions is set to $k' = 1000$ in FLH. The LHKV without FLH can be considered as a special case of $k' = N$. In our experiments, for key-value pair $\langle k, v \rangle$, we take $k \times v$ as the input of the hash functions to ensure that the input corresponding to each pair is unique.

Comparison of Accuracy. Figure 3 and Fig. 4 illustrate the accuracy of the mechanisms for frequency and mean estimations, as the privacy budget ϵ increases from $\ln 2$ to $\ln 128$. For the synthetic dataset, MSE is computed over all keys, while for the real-world datasets, we follow the setting of [6] and select the top 100 frequent keys to calculate MSE. The results reveal a decreasing trend in the MSE of all mechanisms for frequency and mean estimations, with the increase of privacy budget ϵ. Notably, our LHKV outperforms PrivKVM and PCKV-GRR significantly, and shows a slightly better performance than PCKV-UE. Particularly in frequency estimation, the theoretical variance of LHKV and PCKV-UE differs by $\frac{8l^2}{N(e^\epsilon+1)^2}$ as we analyze in Sect. 4.3, which indicates a more significant difference when ϵ is small. These results confirm the accuracy of the theoretical error analysis of LHKV. Additionally, LHKV with FLH acceleration ($k' = 1000$) exhibits an MSE performance similar to that of LHKV without FLH ($k' = N$), suggesting the feasibility of using FLH to expedite the server-side computation.

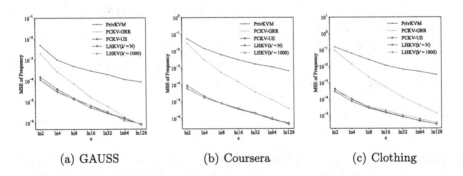

(a) GAUSS   (b) Coursera   (c) Clothing

Fig. 3. MSE results of frequency estimation

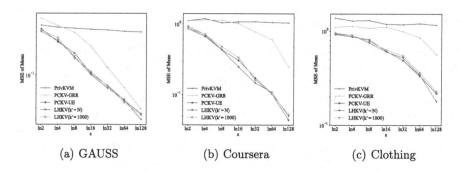

(a) GAUSS   (b) Coursera   (c) Clothing

Fig. 4. MSE results of mean estimation

Comparison of Communication Costs. Table 2 compares the communication costs, measured in terms of the number of bits for end-to-end transmission in the evaluated mechanisms. While PrivKVM and PCKV-GRR exhibit poor accuracy in estimation, their communication costs are relatively low. In contrast, PCKV-UE requires high communication costs, particularly when the domain of keys is large. LHKV demonstrates good accuracy in estimation and can also maintain low communication costs. Furthermore, if FLH is employed to accelerate computations on the server side, the communication costs of LHKV can be further reduced. Notably, the number of bits transmitted by LHKV is independent of the domain size of keys, implying that even for large domains, its communication costs are much lower than PCKV-UE.

Table 2. Comparison on end-to-end communication costs (in bits)

Mechanisms	GAUSS	Coursera	Clothing
PrivKVM	8	12	14
PCKV-GRR	8	12	14
PCKV-UE	202	3672	11704
LHKV($k' = N, \epsilon = \ln 4$)	20	21	20
LHKV($k' = N, \epsilon = \ln 128$)	25	26	25
LHKV($k' = 1000, \epsilon = \ln 4$)	13	13	13
LHKV($k' = 1000, \epsilon = \ln 128$)	18	18	18

Impact of the Number of Hash Functions. Figure 5 illustrates the impact of the number of hash functions on the accuracy of frequency and mean estimations, as well as the server-side computation time. In this experiment, we use only the synthetic dataset GAUSS with $N = 100000$ users and set the privacy budget to $\epsilon = \ln 8$. To facilitate a comprehensive comparison, we also include the corresponding results of PCKV-UE and LHKV without FLH in the figure. According to the experimental results, as the number of hash functions k' increases, the MSEs of frequency and mean estimations gradually approach the corresponding values of PCKV-UE and LHKV. However, an increase in the number of hash functions also results in an increase in the computation time. Nevertheless, even when we use a small k', such as 1000, which is only 1% of the number of users, the accuracy of frequency and mean estimations is still comparable to that of PCKV-UE and LHKV without FLH. Furthermore, the time required for server aggregation and estimation is significantly reduced compared to LHKV without FLH, and is similar to that of PCKV-UE. These observations demonstrate the feasibility of using FLH to accelerate the process of aggregation and estimation.

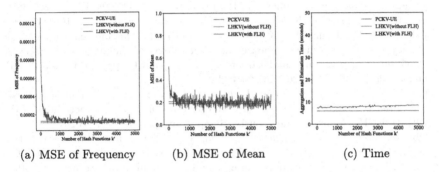

(a) MSE of Frequency   (b) MSE of Mean   (c) Time

Fig. 5. Impact of varying k' in LHKV using FLH ($N = 100000, \epsilon = \ln 8$)

6 Conclusion

In this paper, we propose a new mechanism called LHKV based on local hashing for collecting key-value data under LDP. This mechanism is designed to maintain low communication costs while ensuring high data utility. We theoretically prove that LHKV guarantees ϵ-LDP and analyze the expectation and variance of the estimators. Additionally, we propose to use FLH to accelerate the aggregation and estimation process, significantly reducing the computation time on server. Our experiments demonstrate that LHKV can effectively reduce end-to-end communication costs without sacrificing utility compared with the existing mechanisms. In the future, we plan to investigate how local hashing can be further used to enhance the accuracy of frequency and mean estimations.

Acknowledgement. This work was supported by the Key-Area Research and Development Program of Guangdong Province (No. 2020B010164003), China. The corresponding author is Yingpeng Sang.

References

1. Cormode, G., Maddock, S., Maple, C.: Frequency estimation under local differential privacy. Proc. VLDB Endowment **14**(11), 2046–2058 (2021). https://doi.org/10.14778/3476249.3476261
2. Duchi, J.C., Jordan, M.I., Wainwright, M.J.: Local privacy and statistical minimax rates. In: 2013 IEEE 54th Annual Symposium on Foundations of Computer Science, pp. 429–438. IEEE (2013). https://doi.org/10.1109/FOCS.2013.53
3. Duchi, J.C., Jordan, M.I., Wainwright, M.J.: Minimax optimal procedures for locally private estimation. J. Am. Stat. Assoc. **113**(521), 182–201 (2018). https://doi.org/10.1080/01621459.2017.1389735
4. Dwork, C., McSherry, F., Nissim, K., Smith, A.: Calibrating noise to sensitivity in private data analysis. In: Halevi, S., Rabin, T. (eds.) TCC 2006. LNCS, vol. 3876, pp. 265–284. Springer, Heidelberg (2006). https://doi.org/10.1007/11681878_14
5. Erlingsson, Ú., Pihur, V., Korolova, A.: RAPPOR: randomized aggregatable privacy-preserving ordinal response. In: Proceedings of the 2014 ACM SIGSAC Conference on Computer and Communications Security, pp. 1054–1067 (2014). https://doi.org/10.1145/2660267.2660348
6. Gu, X., Li, M., Cheng, Y., Xiong, L., Cao, Y.: PCKV: locally differentially private correlated key-value data collection with optimized utility. In: Proceedings of the 29th USENIX Conference on Security Symposium, pp. 967–984 (2020)
7. Qin, Z., Yang, Y., Yu, T., Khalil, I., Xiao, X., Ren, K.: Heavy hitter estimation over set-valued data with local differential privacy. In: Proceedings of the 2016 ACM SIGSAC Conference on Computer and Communications Security, pp. 192–203 (2016). https://doi.org/10.1145/2976749.2978409
8. Wang, N., et al.: Collecting and analyzing multidimensional data with local differential privacy. In: 2019 IEEE 35th International Conference on Data Engineering (ICDE), pp. 638–649. IEEE (2019). https://doi.org/10.1109/ICDE.2019.00063
9. Wang, T., Zhang, X., Feng, J., Yang, X.: A comprehensive survey on local differential privacy toward data statistics and analysis. Sensors **20**(24), 7030 (2020). https://doi.org/10.3390/s20247030
10. Wang, T., Blocki, J., Li, N., Jha, S.: Locally differentially private protocols for frequency estimation. In: 26th USENIX Security Symposium (USENIX Security 17), pp. 729–745 (2017)
11. Wang, T., Li, N., Jha, S.: Locally differentially private frequent itemset mining. In: 2018 IEEE Symposium on Security and Privacy (SP), pp. 127–143. IEEE (2018). https://doi.org/10.1109/SP.2018.00035
12. Warner, S.L.: Randomized response: a survey technique for eliminating evasive answer bias. J. Am. Stat. Assoc. **60**(309), 63–69 (1965). https://doi.org/10.1080/01621459.1965.10480775
13. Xiong, X., Liu, S., Li, D., Cai, Z., Niu, X.: A comprehensive survey on local differential privacy. Secur. Commun. Netw. **2020**, 1–29 (2020). https://doi.org/10.1155/2020/8829523
14. Ye, Q., Hu, H., Meng, X., Zheng, H.: PrivKV: key-value data collection with local differential privacy. In: 2019 IEEE Symposium on Security and Privacy (SP), pp. 317–331. IEEE (2019). https://doi.org/10.1109/SP.2019.00018

Investigating Lakehouse-Backbones for Vehicle Sensor Data

Christopher Vox[1]([✉]), David Broneske[2], Jan Piewek[1], Janusz Feigel[3], and Gunter Saake[3]

[1] Volkswagen AG, Wolfsburg, Germany
christopher.vox1@volkswagen.de
[2] German Centre for Higher Education Research and Science Studies, Hannover, Germany
[3] Otto-von-Guericke-Universität Magdeburg, Magdeburg, Germany

Abstract. Through the digitization and automation of vehicles, an increasing amount of data is continuously generated, processed and analyzed. Especially, the storage of this data is of particular importance, since historical vehicle data enables the analysis of driving behavior, the optimization of vehicle functions and the generation of new business models to be able to provide costumers the best experience. However, different communication protocols for inter and intra vehicle communication yield highly complex sensor networks, whose sensor recordings are rarely available synchronously in a central node. This heterogeneous nature of vehicle data requires efficient processing and storage. As a consequence, we benchmark different data structures and metadata concepts in combination with various established databases and file-systems, in order to identify an optimal system for storing vehicle sensor data. Our research shows that the data structure which is embedded in the Lakehouse has to be optimized to achieve the maximum performance for backbones with heterogeneous sensor data. Therefore, we developed the *timestamp partitioned* data structure named as *Schema-2* which shows in combination with TimescaleDB and Druid optimal performance compared to the state-of-the-art time series data structure.

Keywords: Time Series Database · File System · Vehicle Sensor Data Storage · Multivariate Asynchronous Time Series Storage

1 Introduction

The increasing digitization and automation of vehicles is powered by sensors that are integrated into vehicles. Some of these sensors are used for perception but also to control and monitor vehicle components. The majority of sensors generate information, which is transmitted and processed within the vehicle but also between vehicles. Besides the intra and inter vehicle usage of the information, fleet data has a great value creation potential [8]. In order to exploit this potential, the data of vehicles need to be processed, stored and analyzed efficiently.

C. Strauss et al. (Eds.): DEXA 2023, LNCS 14146, pp. 243–258, 2023.
https://doi.org/10.1007/978-3-031-39847-6_17

Especially, the storage of vehicle data is of particular importance, since historical data enables the analysis of driving obligations, the optimization of vehicle functions and the generation of new business models to be able to provide costumers the best experience.

Outgoing from the very first Database Management System (DBMS) for SQL workloads, Data-Lakes and Warehouses have been designed to store and query structured and unstructured data [1]. Armbrust et al. [1] formulated a disadvantage of nowadays data management architectures, which produce transaction overhead due to several systems running in parallel or sequential order. Therefore, they designed a new architecture which they called Lakehouse to support not only standard SQL workloads but also Data Science and Machine Learning (ML) workloads, through a global metadata layer on top of a Data-Lake. Thus, Lakehouse backbones can be used to access the data optimally.

The Data-Lake we investigate consist of vehicle sensor data, which can be characterized as measurements over time, also named as time series. Within literature, various publications have investigated concepts to store time series data optimally. On the one hand, existing databases were extended and also developed to realize optimal access to the stored time series data [15,16]. On the other hand, the suitability of file-systems has been evaluated to store relational data structures [12].

The choice of an optimal backbone is particularly demanding for vehicle sensor data, as the sensor data can be asynchronous in its multivariate space. Due to the fact that the majority of publications focused solely on the search for the optimal system to store synchronous multivariate time series, the minority of publications investigated time series schemata and the impact of the time series schema to the query execution performance.

Hence, to setup the optimal long term storage for vehicle sensor data, we evaluate different systems and schemata to store multivariate asynchronous time series. In addition, we focus on the importance of low selective queries, because Data Science workloads and the training of Deep Neural Networks (DNNs) require large amounts of data.

Accordingly, our contribution is as follows:

- Development of multivariate asynchronous time series schemata with respect to varying metadata over time.
- Investigation of the data ingest and the data retrieval performance for three databases and three file-systems. The retrieval performance is evaluated for unaggregated and aggregated time series materialization.

The remainder is structured as follows. In Sect. 2, we summarize the related work in the fields of non time series and time series databases. In the methodology section, we describe the properties of vehicle sensor data and we introduce relational schemata to store the sensor data (Sect. 3). Finally, we present the results (Sect. 4).

2   Related Work

Time Series Data. Benchmarks for time series databases have been written by [6,9,10]. Liu et al. [9] compared InfluxDB, OpenTSDB, KairosDB and TimescaleDB (TsDB). They showed, that InfluxDB and TsDB are superior to KairosDB and OpenTSDB for exact point queries, time range queries and fetching queries. For aggregation queries, TsDB and OpenTSDB perform similar whereby InfluxDB is significantly faster than the competitors. Hao et al. [6] compared the time series databases InfluxDB, TsDB, Druid and OpenTSDB for workloads such as historical data access and batch based data loading. Their results indicate that InfluxDB and TsDB have the highest ingestion throughput. Based on the results of the six evaluated data analysis queries which fetch data from the bigger dataset, the authors showed that InfluxDB (best system on 3/6 queries) and Druid (best system on 3/6 queries) outperformed TsDB and OpenTSDB. Mostafa et al. [10] evaluated time series databases for the Industrial Internet of Things (IoT). The authors compared InfluxDB, TsDB and Click-House. Based on their highly selective data fetching queries they showed that the column-oriented ClickHouse store with the advanced index mechanism is superior to InfluxDB and TsDB. They showed that a columnar format with advanced data partitioning improves ingest and query performance for time series data. Praschl et al. [11] evaluated relational NoSQL and NewSQL DBMSs on time series data. From the results it can be taken that InfluxDB surpasses MongoDB, TsDB and LeanxcaleSQL in terms of ingest performance whereas MongoDB is slightly faster than InfluxDB in full database materialization.

Non Time Series Data. Qin at al. [12] evaluated SQL engines on Hadoop. They compared the data Warehouse engine Hive, which is based on MapReduce, Impala, which bypasses the MapReduce to enable higher multiprocessing capabilities and SparkSQL, which provides stronger query optimization in comparison to Hive. As file formats the authors chose ORC for Hive and Parquet for Impala and SparkSQL, as these combinations were also found to be optimal by [4,7]. In their experiments they showed that Impala outperformed Hive and SparkSQL within the TPC-H benchmark, due to the in-memory processing of Impala. Advanced Parquet based storage solutions and design paradigms have been developed and investigated by [2,5]. The authors were able to show that improved use-case based partitioning of files in combination with advanced metadata layers can significantly improve query performance.

Ragab et al. [13] extended the research based on file-formats. The authors evaluated the query performance of ORC, Parquet, AVRO and CSV based on three relational schemata and three partitioning techniques in combination with SparkSQL. Their results indicate that vertical partitioned tables are superior to standard relational tables. They showed that the consideration of the relational schemata strongly influences the system performance.

Related Work Problem. Limited research has been conducted in the field of asynchronous time series storage. One of the very few works on asynchronous time series in databases has been conducted by [3]. Colosi et al. [3] evaluated the impact of advanced bucketing in databases where one bucket holds data of a specific sensor. They showed improved query performance compared to the standard time series data schema. In contrast, the most of the introduced time series benchmarks focused on multivariate synchronous time series. Interestingly, when time series data has been evaluated in databases the relational data schema *(timestamp, sensor_ id, value)* was used.

3   Methodology

Different industrial areas search for the optimum database architecture for time series data. Especially, automotive companies need to archive the recorded vehicle information to enable the best user experience and to increase their potential to add value. Hence, in Sect. 3.1 the vehicle sensor data is described in detail. Thereafter, Sect. 3.2 presents the general purpose relational representation of vehicle sensor data. In Sect. 3.3 different relational schemata are introduced and in Sect. 3.4 the simulation conditions for the experiments are presented.

3.1   Vehicle Sensor Data

Vehicle sensors can be classified and separated by the numerical dimensionality of the digitized sensor output. Within this investigation we analyze one-dimensional vehicle sensor data, because the largest share of sensors produce one-dimensional information, such as a single measurement value of the battery temperature. Higher dimensional measurements such as images are considered for future work. Furthermore, the information transmission within vehicles can be event-based and priority-controlled so that sensor information does not necessarily converge at a defined time in a central node [18]. In order to be able to resolve the temporal characteristic of each sensor without losses, a timestamp is assigned to each measurement value and together both values form a tuple. The consecutive tuples can be interpreted as irregular time series [14].

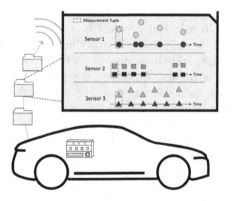

Multiple irregular time series create a multivariate asynchronous time series space. Asynchronicity can be defined as a multivariate time series state, where the measurements of time series are not aligned equally on a

Fig. 1. Buffer based data upload of a digitized vehicle. Each message frame represents a drive recording.

global time domain [17]. This property is illustrated in Fig. 1. Three different sensors are visualized which are recorded irregularly, event-based and with different frequencies. The darker color within a tuple describes the timestamp, whereas the lighter color describes the recorded measurement value. As it can be seen, the measurements of the three sensors are not synchronized. Thereby, the multivariate sensor space is asynchronous, multi-cardinal and irregular but is referred to in simplified terms as asynchronous in the following [19].

Irregularity of vehicle sensor data makes analysis and storage of the data challenging. Especially in literature, asynchronicity of multivariate time series data has been insufficiently considered for databases. Hence, we show possibilities how multivariate asynchronous time series can be prepared for databases in the following.

3.2 Relational Representation of Vehicle Sensor Data

In relational databases data in the form of tuples is organized in relations, which are also named as tables. Through this, relational operations, such as selections, projections and joins, can be applied to the data to enable general-purpose queries such as range queries and aggregation queries.

With the help of primary, natural and foreign keys the relationship between tables is defined and the tables can be linked and merged by DBMS operations appropriately, e.g. when the database is normalized.

For sensor data and thus for time series data a table can be defined by three attributes: The timestamp attribute (T), the value attribute (V) and the sensor identity attribute (SID). Based on these attributes, records from different sensors can be ingested as tuples into a database. The consecutive tuples of multiple modalities are arranged vertically to construct a general purpose table, cf. Schema-1 in Fig. 2. Thus, different optimization strategies such as indexing and search trees for high selective queries can be applied to the general purpose table to increase the DBMS performance.

The raw sensor measurements are mostly integers due to the Analog Digital Conversion (ADC) within vehicles. Metadata is needed to be able to convert the measurements into a unit-based (physical) representation. The conversion rule is defined by the equation: $y_{physical} = y_{raw} * Factor + Offset$. The measurement value y is manipulated by the $Factor$ as well as by the $Offset$, which are sensor specific variables.

A standard procedure of vehicle measurement database ingest can be split into three stages. Firstly, the raw vehicle measurements are parsed. Secondly, the conversion to a unit-based representation takes place. Lastly, the preprocessed measurements are restructured to a table which can be ingested by the majority of general purpose and time series databases, cf. Schema-1 in Fig. 2. The described preprocessing stages are a standard procedure for vehicle sensor data. However, we have to investigate whether the general purpose data structure is suitable for archiving vehicle sensor data.

3.3   Relational Schemata for Vehicle Sensor Data

Based on the idea of storing time series information vertically to enable time queries via a time attribute, as shown in Fig. 2 by Schema-1, we have developed further concepts for suitable data structures. The first competitor is named as Schema-2, cf. Fig. 2. In Schema-2, we create one table for each group of synchronously recorded time series, which share one time axis. For this concept, the linking between tables as well as the access layer of query engines takes place via a global lookup table. In Schema-2 each table consists of a time attribute T and of at least one sensor attribute S_n. If several sensor attributes are combined in a single table, then the sensor attributes are characterized over the same time axis (attribute T) and form a multivariate time series. We assume that Schema-2 will enable a more efficient query of sensor subsets for small and large time selectivities.

If we also take into account use-case specific Data-Lake requirements, such as the support of machine learning-relevant queries, which are rather little selective, then nested data structures could also be suitable for multivariate asynchronous as well as multivariate synchronous time series. For this we have developed Schema-3, cf. Fig. 2. Based on Schema-3, vehicle data (e.g. a one-hour vehicle drive) is structured as table by the sensor identity attribute SID, the measurement attribute V and the time identity attribute TID. A row of Schema-3 describes a measurement period of one sensor entity. Each sensor entity is uniquely described by the attribute value of SID. The measurements of each sensor entity $x_{1,...,n}$ of a defined time range $t_{1,...,n}$ are stored separate from the time information t as a series of measured values nested as an attribute value of V in the respective row. Due to the time identity attribute TID, the time entity, saved as attribute value of TID, clearly refers to the time information of each sensor entity. Moreover, each timestamp series of a measurement series is stored as a sensor entity. The timestamp values are stored in the respective row of the time entity attribute value of SID as an attribute value of V. Also, in Schema-3 several sensor entities can share a time entity. Through this linking, the database memory requirement can be reduced.

In summary, three Schemata have been introduced, which define relational structures for multivariate asynchronous time series. However, each sensor entity possesses supplementary information, which also has to be archived appropriately. Exemplary metadata of vehicle sensors are shown in Table 1.

Schema-1

T	V	SID
t_1	1	S_1
t_3	2	S_1
t_4	5	S_1
t_7	4.6	S_1
t_8	3	S_1
t_1	-0.5	S_2
t_3	-0.5	S_2
t_4	-0.5	S_2
t_7	-1	S_2
t_8	-1	S_2
t_3	103	S_3
t_6	102	S_3
t_8	106	S_3

Schema-2

T	S_1	S_2
t_1	1	-0.5
t_3	2	-0.5
t_4	5	-0.5
t_7	4.6	-1
t_8	3	-1

T	S_3
t_3	103
t_6	102
t_8	106

Schema-3

SID	V	TID
S_1	[1,..., 3]	T_1
S_2	[-0.5,..., -1]	T_1
S_3	[103,..., 106]	T_2
T_1	[t_1,..., t_8]	T_1
T_2	[t_3,..., t_8]	T_2

Fig. 2. Relational schemata for vehicle sensor data.

Table 1. Supplementary information of in-vehicle sensors.

T	SID	Factor	Offset	Description	Anomaly	...
t_1	S_1	0.03	30	Temperature	True	
t_1	S_2	1	0	Light	False	
t_3	S_3	1	0	Control	False	

The table is defined by the attributes T, SID, $Factor$, $Offset$, $Description$ and $Anomaly$ where each row represents a sensor specified by the attribute SID. Attribute values which are not defined or detected are set to $None$ to generate a complete tuple for each sensor. The link between the metadata table and the data table is created by two foreign keys, e.g. the SID and the T attributes. In contrast to the creation of a second table, a flat architecture as single table would cause an increase in database size (disk storage) due to redundant attribute values. However, we do not know in advance whether a flat database architecture or a normalized approach is the optimal vehicle data schema.

Two concepts have been developed for this purpose, which are named as *a)* and *b)*. The concept a) defines a database normalization strategy where data and metadata are separated. The concept b) proposes a denormalization strategy with data and metadata in one flat table.

3.4 Simulation Setup

Ingest. The ingest of vehicle sensor data is carried out based on log files of measurement periods, e.g. test drives. The ingest takes place via two data pipelines, depending on whether a file-system or a database is set up. In the case of a file-system, the log files are converted into a file format suitable for the file-system and are structured accordingly. In the case of databases, the log files are cached in CSV tables structured accordingly and are inserted via a database import routine.

Retrieval. In this paper, two important queries are examined, which are particularly relevant for Data Science workloads. The first query materializes unaggregated sensor data for different T and SID selectivities and can be described schematically using the following SQL query.

```
SELECT 'columns'
FROM 'vehicle_table'
WHERE (t BETWEEN 'start_time' AND 'end_time')
AND sid in 'sid_names'
```

Listing 1.1. Schematic data query.

The value `'columns'` defines the relational attributes according to the three schemata from Fig. 2. The value `'vehicle_table'` defines the implicit table assignment of vehicles. The value t describes the time values of the table and the value sid describes the *SID* attribute values specified with 'sid_names'. The metadata is materialized via the data query, when the metadata concept b) is used, cf. Listing 1.1. In contrast, for concept a), for which the data and the metadata are stored in separate tables, a second query is applied to materialize the metadata.

Another important query for training machine learning algorithms is shown in the use-case of [14]. The authors trained the aging of an engine component based on vehicle sensor statistics calculated over time windows (sliding window). The data aggregation via sliding windows is a relevant usage scenario of vehicle sensor databases, whereby this specific query is also evaluated in this investigation.

For this purpose, 100 time intervals (buckets) are formed for each time selectivity, which are aggregated by the investigated systems by calculating the mean value of the sensor values in each time interval. Thus, the data materialization query shown in Listing 1.1 is extended by a *groupby* operation based on buckets and by an average operation applied to each bucket.

As main performance metric we select the query execution time measured in seconds. Therefore, we measure the total time which includes the database engine operation time as well as the subsequent time until the data is fully materialized as *pandas.DataFrame* in Python.

4   Experiment

All experiments are conducted in the Azure cloud using an Intel Xeon Platinum 8272CL and 32GB RAM and mounted SSDs. The systems and the respective implementation details as well as the data source are presented in the following.

4.1   Data Source

The data source used within this paper is defined by four prototype vehicles whose sensor information are logged over a day. Each vehicle generates 4800 different information, such as real sensor information but also artificial information generated by control units. All information which is recorded within a vehicle is referred to in simplified terms as sensor information even though the information was artificially generated. The sensor measurements are gathered in a central unit, whereby multiple sensors create a multivariate asynchronous time series space as it has been described in Sect. 3.1. Moreover, the four vehicles, which we consider as data source, are described by data collected over the same period of time.

Overall, the data source consists of 1.23 billion rows when we structure the data according to Schema-1. The recorded vehicle data is logged and written to a vehicle-specific, proprietary file format. In addition, the metadata, such as the *Factor*, the *Offset* and the *Unit*, of each sensor is stored as supplementary information within the file. Thus, the logged vehicle data must be parsed and prepared for each system individually.

4.2   Implementation Details

In order to identify which system is suitable to store and query the different schemata optimally, we implement three databases TsDB (v2.8.0), InfluxDB (v2.4.0) and Druid (v24.0.0) and three systems from the Hadoop (v3.3.4) ecosystem: SparkSQL (Spark) (v3.3.0), Hive (v3.1.3) and Impala (v4.1.0). Within all systems, each vehicle receives a unique assignment to a data table or several data tables. Conceptually, this procedure can be understood as an implicit lookup table at vehicle level. We think that such an architecture can be scaled up to larger fleets in the compared systems.

In all systems, tables are created according to the respective schemata shown in Fig. 2. A special feature of Schema-2 is that, in addition to the displayed table, a lookup table is created for each vehicle to enable the database to easily navigate over the multiple tables. The lookup table links the vehicle based time series to the respective tables according to Schema-2. For Schema-3, T_{min} and T_{max} attributes are created for each individual SID attribute value to be able to efficiently filter the table. Furthermore, a *sensor_or_time* attribute is defined which simplifies the distinction between time and sensor value entities for the database. All schemata are sorted according to the SID attribute. Each SID section is again sorted in ascending order, cf. Schema-1.

All investigated systems have been setup in the default configuration. The ingest for TsDB is carried out via *CSV* files, for Hive with *ORC* files and for the other systems with *Parquet* files. Since the Druid version we use only supports nested schemata experimentally, we use *CSV* files to ingest Schema-3 into Druid. All files are structured with regard to the introduced schemata. The default time series data ingest for TsDB, Druid and Influx requires the explicit submission of the time channel attribute. Hence, the systems are able to react optimally on time queries. However, the systems from the Hadoop ecosystem do not support any kind of time indexing based on the default configuration. Lastly, we do not implement Schema-3 for InfluxDB, due to the lacking support of nested data types.

4.3   Data Ingest

In order to evaluate the query performance of the systems, the data source must first be loaded. In the following, we investigate the average ingest performance over all four vehicles.

The average ingest execution times for the six systems which we examine are shown in Fig. 3. It can be seen that Druid, InfluxDB and also TsDB take the longest time to ingest the data, whereby the ingest speed of Druid in combination with Schema-2 is the slowest. Since Schema-3 is not implemented for InfluxDB, the associated measurements are missing. The fastest ingest results through Impala in combination with Schema-3.

In addition to the ingest time, the impact of the ingested data to the database size is shown in the upper graphic of Fig. 3. It can be seen that Druid, Hive, Impala and Spark in combination with Schema-2 cause the lowest increase in

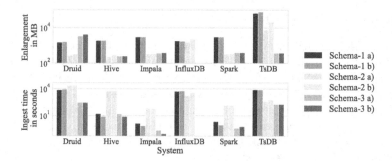

Fig. 3. Representation of the average duration of data ingest (lower graphic) and the enlargement of the database caused by the inserted data (upper graphic) for one vehicle.

size. On the other hand, the size of the TsDB database in combination with Schema-1 and Schema-2 is significantly larger compared to all other systems. We can also see the effect of different schemata to the resulting database size, which varies strongly. Through this, TsDB in combination with Schema-3 can compete with Druid, Hive, Impala and Spark in combination with Schema-2 in terms of efficient data storage.

4.4 Data Retrieval

In this section, we investigate the system behavior based on the data materialization query of Sect. 3.4. We consider different selectivities, which we apply to the T attribute and the SID attribute. We investigate two retrieval types, which are relevant machine learning based use-cases. The first use-case extracts the data without aggregation operations into the main memory. The second use-case aggregates the data based on fixed simulation boundary conditions via mean aggregation. The query performance for the data retrieval over four vehicles is measured in seconds for both use-cases.

Unaggregated Data Retrieval. At first we investigate the system behavior based on the variation of the SID selectivity and the T selectivity in isolation. Consequently, when the T selectivity varies, the SID selectivity is set to 100%. The results of the experiment are shown in Table 2 and in Table 3. From Table 2a it can be seen that TsDB, especially in combination with Schema-1 and Schema-2, is advantageous. Schema-2 enables on average a superior query speed performance, which enables TsDB to respond twice as fast when compared to the combination of TsDB and Schema-1. Moreover, Schema-2 has a positive effect on query speed in combination with Druid. However, the improvement through Schema-2 is not recognizable for every system, which causes a tenfold increase in execution speed for e.g. Hive, due to Schema-2.

The results for the high selective query are reflected in the results of the low selective query shown in Table 2b. The results of the high selectivity value of 50%

Table 2. Visualization of the query execution speed for the time series query for two selectivity values applied to the T attribute. The query execution is measured in seconds. The fastest query execution is highlighted in bold. The second fastest is underlined and the third fastest is dashed.

(a) Examination of a time query (T attribute) which materializes 0.5% of the measurement period of each vehicle.

Schema	Druid	Hive	Impala	InfluxDB	Spark	TsDB
1 a)	252.61	786.36	144.45	207.59	207.64	78.77
1 b)	359.71	977.77	151.95	175.03	233.23	56.58
2 a)	88.70	10217.55	199.81	193.25	262.37	38.86
2 b)	204.90	9591.97	233.38	442.97	446.76	27.02
3 a)	3925.35	2335.38	901.08	-	631.19	539.12
3 b)	3766.25	2331.49	895.32	-	561.32	504.34

(b) Examination of a time query (T attribute) which materializes 50% of the measurement period of each vehicle.

Schema	Druid	Hive	Impala	InfluxDB	Spark	TsDB
1 a)	1514.37	12181.30	8135.43	19417.91	28980.17	3172.70
1 b)	1486.88	21971.95	22338.64	43574.32	46320.58	3617.31
2 a)	4943.06	22567.38	1217.58	4787.50	727.99	832.18
2 b)	15673.23	17434.46	3672.56	23656.99	883.52	2408.92
3 a)	6029.71	3983.30	1700.76	-	3345.58	1089.74
3 b)	6149.03	3739.80	1670.42	-	3085.44	956.61

amplify the performance difference between the metadata structures. Within this benchmark we can derive that Schema-1 is rather unsuitable to extract a large amount of data. In particular, Spark responds sensitively to Schema-2, which results in optimal query execution speed. Interestingly, Schema-2 significantly improves the performance compared to Schema-1 for Impala, InfluxDB, Spark, and TsDB.

The results of two SID selectivities are shown in Table 3. The execution times for the high selective query are visualized in Table 3a. Based on the low selectivity value we show that TsDB is the fastest system in combination with Schema-2. Especially for the low selective query, Schema-3 in combination with TsDB leads to optimal execution speeds, which is demonstrated in Table 3b. Due to the different arrangement of the metadata, we can see that for Schema-1 and Schema-2, the separation of the data and the metadata into different tables tends to result in improved query execution speed for high selectivity values. The improvement is particularly significant for Druid and InfluxDB in combination with Schema-2. Less pronounced for Impala and TsDB. In contrast, the tendency

Table 3. Visualization of the query execution speed for the time series query for two selectivity values applied to the *SID* attribute. The query execution is measured in seconds. The fastest query execution is highlighted in bold. The second fastest is underlined and the third fastest is dashed.

(a) Examination of a *SID* query which materializes 0.5% of the sensors of each vehicle.

Schema	Druid	Hive	Impala	InfluxDB	Spark	TsDB
1 a)	219.98	729.15	232.83	107.80	358.63	2544.86
1 b)	312.19	931.53	256.22	134.54	416.28	3131.79
2 a)	106.26	2181.03	1816.07	108.82	163.56	**16.10**
2 b)	225.47	2071.91	1777.55	242.15	171.94	<u>28.97</u>
3 a)	4414.25	497.44	187.68	-	224.42	90.38
3 b)	4486.41	497.50	185.09	-	209.38	89.20

(b) Examination of a *SID* query which materializes 50% of the sensors of each vehicle.

Schema	Druid	Hive	Impala	InfluxDB	Spark	TsDB
1 a)	1512.76	13657.73	6630.12	20030.87	28018.89	4907.87
1 b)	1484.87	23431.79	18785.34	45250.90	32773.65	13860.23
2 a)	5201.90	13757.18	837.17	5071.64	<u>627.41</u>	835.57
2 b)	16170.47	9092.56	3108.45	22499.68	837.03	2487.85
3 a)	6168.69	2184.70	894.41	-	1129.09	<u>604.23</u>
3 b)	6502.77	1893.64	871.18	-	978.63	**497.21**

found for Schema-2 is not reflected by Schema-3. Schema-3 tends to be rather insensitive to the metadata normalization or denormalization strategy for the majority of systems, which we evaluate. Only for Schema-3 b) in combination with TsDB, Hive and Spark we can see a performance improvement, cf. Table 3b. The query performance for various selectivity values is shown for metadata storage strategy a) in Fig. 4. Interestingly, we see a rather constant behavior of all databases for Schema-3 when the time selectivity is varied. In addition, the illustration shows a lower increase of query execution time for Spark over the increasing selectivity compared to TsDB. Furthermore, we can deduce from the illustration that the state-of-the-art Schema-1 is rather unfavorable for TsDB.

Aggregated Data Retrieval. The aggregation query is of particular importance for training machine learning based algorithms, cf. Section 3.4. Consequently, we investigate the query execution speed for different selectivity values applied to the *SID* Attribute over the whole measurement period of our data source. The results of two selectivity values are shown for Schema-1 and

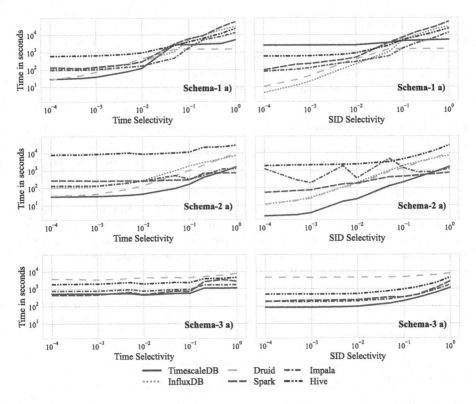

Fig. 4. Visualization of the query execution speed for unaggregated data retrieval. The Time (T) Selectivity and the SID Selectivity are varied in isolation.

Table 4. Visualization of the execution performance of the aggregation query for two selectivity values applied to the SID attribute. The query execution is measured in seconds. The fastest query execution is highlighted in bold. The second fastest is underlined and the third fastest is dashed.

(a) The query materializes the information of 24 sensors (0.5%) for each vehicle.

(b) The query materializes the information of 2400 sensors (50%) for each vehicle.

Sc.	Druid	Hive	Impala	Influx	Spark	TsDB
1 a)	**3.9**	812.8	204.4	5.8	148.7	2504.9
1 b)	5.2	907.0	206.6	4.1	134.5	1968.6
2 a)	4.3	4400.2	35.1	24.4	161.1	5.9
2 b)	5.2	4136.4	38.7	25.8	186.5	8.2

Druid	Hive	Impala	Influx	Spark	TsDB
45.7	1171.7	356.7	140.3	924.5	2243.5
64.3	1415.8	346.3	98.3	885.5	1937.2
38.8	18575.1	790.1	197.8	601.5	183.0
64.9	17254.9	1056.2	220.0	1070.9	554.7

Schema-2 in Table 4. We do no evaluate Schema-3 due to the lacking support of the systems for aggregation operations applied to nested data types.

As it can be seen in Table 4a, Druid in combination with Schema-1 a) shows the fastest database response. InfluxDB in combination with Schema-1 as well as TsDB with Schema-2 are the second and the third fastest architecture respectively. Moreover, the results of the low selective query are shown in Table 4b. Likewise, Druid is the fastest system whereby Schema-2 a) outperforms Schema-1 a) for the selectivity value of 50%. For both selectivity values which we evaluated in Table 4 the normalization strategy tends to be superior.

Data Retrieval Summary. Our research shows that for most of the systems we study, Schema-2 allows a strong performance improvement for time series extraction compared to the standard Schema-1. Especially for high selective unaggregated data queries, the combination of TsDB and Schema-2 is optimal. On the opposite, it can be seen from the results that Spark in combination with Schema-2 and also Schema-3 in combination with TsDB is suitable for low selective queries. Besides the time series extraction, we also investigate aggregation based queries. From the results we can deduce that Druid in combination with Schema-2 seems to be the best system for the aggregation task, due to the faster execution speed of Schema-2 for the low selective query.

If we further analyze the results of the two metadata storage strategies then we can see that the majority of systems perform better with a normalization strategy. However, for some systems a denormalization strategy seems to be rather beneficial.

5   Conclusion

In this paper we elaborated a need for research for storing vehicle sensor data in databases. We identified that the property of asynchronicity of time series data was insufficiently investigated for databases. Therefore, we developed three schemata and two metadata storage strategies, to identify the optimal relational schema to store vehicle sensor data in databases. We evaluated the different data and metadata concepts in combination with six suitable Lakehouse backbones. We show that Schema-3 in combination with Impala achieves Pareto-optimal results regarding ingestion speed and database enlargement. Furthermore, Schema-2 enables optimal query performance for unaggregated and likewise, however, less pronounced for aggregated data retrieval. However, if we consider the rapid extraction of many sensors over very long periods of time, then Schema-3 appears favorable. Consequently, we demonstrate that optimized relational schemata enable a performance improvement for most backbones when compared to the standard time series table performance. Based on these results, the timestamp based partitioning of Schema-2 should further be investigated for other datasets.

Disclaimer. The results, opinions, and conclusions expressed in this publication are not necessarily those of Volkswagen Aktiengesellschaft.

References

1. Armbrust, M., Ghodsi, A., Xin, R., Zaharia, M.: Lakehouse: a new generation of open platforms that unify data warehousing and advanced analytics. In: Conference on Innovative Data Systems Research, CIDAR 11 (2021). http://cidrdb.org/cidr2021/papers/cidr2021_paper17.pdf
2. Chakraborty, J., Jimenez, I., Rodriguez, S.A., Uta, A., LeFevre, J., Maltzahn, C.: Skyhook: towards an arrow-native storage system. In: International Symposium on Cluster, Cloud and Internet Computing, CCGrid 22, pp. 81–88 (2022). https://doi.org/10.1109/CCGrid54584.2022.00017
3. Colosi, M., Martella, F., Parrino, G., Celesti, A., Fazio, M., Villari, M.: Time series data management optimized for smart city policy decision. In: International Symposium on Cluster, Cloud and Internet Computing, CCGrid 22, pp. 585–594 (2022). https://doi.org/10.1109/CCGrid54584.2022.00068
4. Gupta, A., Saxena, M., Gill, R.: Performance analysis of RDBMS and Hadoop components with their file formats for the development of recommender systems. In: International Conference for Convergence in Technology; Annual Conference for Convergence in Technology, IC2T, pp. 1–6 (2018). https://doi.org/10.1109/I2CT.2018.8529480
5. Hansert, P., Michel, S.: Ameliorating data compression and query performance through cracked parquet. Big Data in Emergent Distributed Environments, BiDEDE, pp. 1–7 (2022). https://doi.org/10.1145/3530050.3532923
6. Hao, Y., et al.: TS-benchmark: a benchmark for time series databases. In: IEEE 37th International Conference on Data Engineering (ICDE), pp. 588–599 (2021). https://doi.org/10.1109/ICDE51399.2021.00057
7. Ivanov, T., Pergolesi, M.: The impact of columnar file formats on SQL-on-Hadoop engine performance: a study on ORC and parquet. Concurrency Comput. Pract. Exp. **32**(5), e5523 (2019). https://doi.org/10.1002/cpe.5523
8. Kaiser, C., Festl, A., Pucher, G., Fellmann, M., Stocker, A.: The vehicle data value chain as a lightweight model to describe digital vehicle services. In: Proceedings of the 15th International Conference on Web Information Systems and Technologies, SciTePress (2019). https://doi.org/10.5220/0008113200680079
9. Liu, R., Yuan, J.: Benchmarking time series databases with IoTDB-benchmark for IoT scenarios. http://arxiv.org/pdf/1901.08304v3
10. Mostafa, J., Wehbi, S., Chilingaryan, S., Kopmann, A.: SciTS: a benchmark for time-series databases in scientific experiments and industrial internet of things. In: International Conference on Scientific and Statistical Database Management, SSDBM vol. 34, pp. 1–11 (2022). https://doi.org/10.1145/3538712.3538723
11. Praschl, C., Pritz, S., Krauss, O., Harrer, M.: A comparison of relational, NoSQL and NewSQL database management systems for the persistence of time series data. In: International Conference on Electrical, Computer, Communications and Mechatronics Engineering, ICECCME, pp. 1–6 (2022). https://doi.org/10.1109/ICECCME55909.2022.9988333
12. Qin, X., Chen, Y., Chen, J., Li, S., Liu, J., Zhang, H.: The performance of SQL-on-Hadoop systems - an experimental study. In: International Congress on Big Data, BigData Congress, pp. 464–471 (2017). https://doi.org/10.1109/BigDataCongress.2017.68
13. Ragab, M., Awaysheh, F.M., Tommasini, R.: Bench-ranking: a first step towards prescriptive performance analyses for big data frameworks. In: International Conference on Big Data, Big Data, pp. 241–251 (2021). https://doi.org/10.1109/BigData52589.2021.9671277

14. Sass, A.U., Esatbeyoglu, E., Iwwerks, T.: Data-driven powertrain component aging prediction using in-vehicle signals. In: SOFSEM (2020)

15. Shahid, J.: InfluxDB documentation: release 5.3.1 (2022)

16. timescale: TimescaleDB: SQL made scalable for time-series data (2017)

17. Vox, C., Broneske, D., Shaikat, I., Saake, G.: Data streams: investigating data structures for multivariate asynchronous time series prediction problems. ICPRAM, pp. 686–696 (2023). https://doi.org/10.5220/0011737300003411

18. Woo, S., Moon, D., Youn, T.Y., Lee, Y., Kim, Y.: Can id shuffling technique (CIST): moving target defense strategy for protecting in-vehicle can. IEEE Access **7**, 15521–15536 (2019). https://doi.org/10.1109/ACCESS.2019.2892961

19. Wu, S., et al.: Modeling asynchronous event sequences with RNNs. J. Biomed. Inform. **83**, 167–177 (2018). https://doi.org/10.1016/j.jbi.2018.05.016

Assessing the Effectiveness of Intrinsic Dimension Estimators for Uncovering the Phase Space Dimensionality of Dynamical Systems from State Observations
A Comparative Analysis

Félix Chavelli[1,2]([✉]), Khoo Zi-Yu[1], Jonathan Sze Choong Low[3],
and Stéphane Bressan[1,2]

[1] National University of Singapore, 21 Lower Kent Ridge Rd,
Singapore 119077, Singapore
{chavelli,khoozy}@comp.nus.edu.sg, steph@nus.edu.sg
[2] CNRS@CREATE LTD, 1 Create Way, Singapore 138602, Singapore
[3] Singapore Institute of Manufacturing Technology (SIMTech), Agency for Science,
Technology and Research (A*STAR), Singapore 138634, Singapore
sclow@simtech.a-star.edu.sg

Abstract. Devising a model of a dynamical system from raw observations of its states and evolution requires characterising its phase space, which includes identifying its dimension and state variables. Recently, Boyuan Chen and his colleagues proposed a technique that uses intrinsic dimension estimators to discover the hidden variables in experimental data. The method uses estimators of the intrinsic dimension of the manifold of observations. We present the results of a comparative empirical performance evaluation of various candidate estimators. We expand the repertoire of estimators proposed by Chen et al. and find that several estimators not initially suggested by the authors outperforms the others.

1 Introduction

The study of *dynamical systems* is the modelling and analysis of systems that change over time, as is often of interest in science and engineering. A dynamical system is characterized by its degrees of freedom and corresponding *state variables* and by the rules that govern how its variables change value over time. A dynamical system is typically modelled by a set of equations of its state variables describing its time evolution in phase or *state space* [21]. The study of dynamical systems involves devising and analysing models to understand the system's behaviour under different initial conditions and predict its evolution. One of the first steps in this study, following the collection of raw observations of a system's states and evolution, is the determination of the dimension of the phase space.

In 2021, Boyuan Chen et al. introduced a method for the *automated discovery of fundamental variables hidden in experimental data* [7]. The authors used a two-step approach to identify state variables of a dynamical system from experimental observations. From video recordings of various dynamical systems, Chen et al. estimated the minimum number of independent variables, or *intrinsic dimension* of the state space, needed to describe the system without information loss [6,20]. With such knowledge of the intrinsic dimension, the authors found the state variables of the system that could accurately capture the overall system's dynamics. To discover the intrinsic dimensions of the state space, Chen et al. leveraged intrinsic dimension estimators [4] such as Elizaveta Levina and Peter Bickel's maximum likelihood-based estimator [18], Alessandro Rozza et al.'s Minimum Neighbour Distance-Maximum Likelihood estimator [20] and Matthias Hein's U-statistic-based intrinsic dimension estimator [15]. We expand the repertoire of intrinsic dimension estimators used to discover the intrinsic dimensions of the state space and systematically and empirically compare their performance for the discovery of the intrinsic dimension of a dynamical system from high-dimensional experimental observations.

2   Background and Related Work

Intrinsic dimension estimation is commonly used to represent data in a more compact yet still informative manner and reduce the effects of the 'curse of dimensionality' [3]. The three main classes of intrinsic dimension estimators are projective estimators, graph-based estimators, and topological estimators [6].

Projective intrinsic dimension estimators process high-dimensional data and find a lower-dimensional subspace to project data onto. The dimension of the subspace is the intrinsic dimension estimate [6]. Seminal projective intrinsic dimension estimators include Ian Jolliffe's Principal Component Analysis [16], which searches for the subspace that minimises projection error. The intrinsic dimension estimate is the number of principal components [6]. Such estimators rely on a specific estimated eigenstructure that may not exist in the data.

Graph-based intrinsic dimension estimators build a graph of data points. By employing different distance functions, different graphs can be formed. When the distance function is the Euclidean distance, a k-Nearest Neighbours graph is created [10]. When the weights approximate Geodesic distances, a minimum spanning tree is formed [9]. Various statistics are computed as functions of graph properties, that can be used to approximate the intrinsic dimension of the data [5, 6,8]. However, graph-based intrinsic dimension estimators perform better when manifolds of nonconstant curvature are processed.

Topological intrinsic dimension estimators consider a manifold embedded in higher dimensional space through a proper smooth map and assume that the given data comprises independent and identically distributed points drawn from the manifold through a smooth probability density function. The manifold's topological dimension is the estimated intrinsic dimension. There are two subclasses of topological intrinsic dimension estimators [6].

The first subclass comprises fractal topological intrinsic dimension estimators. They assume a manifold with fractal structure, where the volume of a d dimensional ball of radius r scales with its size $s = r^d$. The intrinsic dimension estimate [6] is the power of the rate of growth, d. Peter Grassberger and Itamar Procaccia's Correlation Integral [13] estimated intrinsic dimensions using a fractal dimension estimator sensitive to local structures, to characterise data from high-dimensional systems. Matthias Hein and Jean-Yves Audibert's U-statistic-based estimator improve the Correlation Integral with a scale-independent kernel function [15]. Amir Massoud Farahmand et al.'s Manifold-Adaptive Dimension estimation Algorithm [12] found a low-dimensional estimate of local high-dimensional data. Combining local estimates produces global estimates.

The second subclass comprises nearest neighbour topological intrinsic dimension estimators. If the sum of angles between each point in the subspace spanned by the k-nearest neighbours and the $k + 1^{th}$-nearest neighbour is less than a threshold parameter, then k is incremented. Otherwise, k is the estimated intrinsic dimension [6]. Elizaveta Levina and Peter Bickel [18] and Gloria Haro et al. [14] treat the neighbours of each point as events in a Poisson process and the distance between the point and its jth nearest neighbour as the event's arrival time. They then estimate the intrinsic dimension by maximising the log-likelihood of the observed process. Haro et al.'s estimator modifies Levina and Bickel's and simultaneously computes the maximum likelihood of a collection of points as a Poisson mixture model instead of an individual point. This accounts for both noise and different Poisson distributions [14].

Some maximum likelihood-based approaches underestimate intrinsic dimensions when data dimensionality is high because nearest-neighbour distances assume that the amount of data becomes unlimited. Therefore, Alessandro Rozza et al. [20] proposed Minimum Neighbour Distance-Maximum Likelihood estimators to estimate intrinsic dimensions with limited data. Furthermore, high-dimensional manifolds are generally twisted and curved, with non-uniformed distributed points. Therefore, Elana Facco et al. [11] propose a topological nearest-neighbour intrinsic dimension estimator that uses only the distance of each point's first and second nearest neighbours in the data. The extreme minimality of this Two Nearest Neighbour intrinsic dimension estimator enabled Facco et al. to reduce the effects of curvature and distribution density variation [11]. Lastly, instead of using maximum likelihood-based approaches, Laurent Amsaleg et al. [1] estimate intrinsic dimensions using the method of moments, which have faster initial convergence and perform well even with limited data.

3 Methodology

The raw observations consist of video recordings of the systems under study. Following Chen et al., consecutive video frames from the video recording (shown in Fig. 1) are encoded into a high-dimensional latent space. To ensure that no information regarding the dynamical system is lost, Chen et al. use a decoder to decode each encoding into a video frame at each time step [7]. They then apply

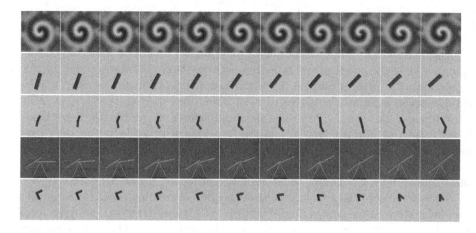

Fig. 1. Sequences of video frames of several dynamical systems: from the top and in each row, the reaction-diffusion system, single pendulum system, double pendulum system, swing stick system and elastic pendulum system.

Levina and Bickel's maximum likelihood-based intrinsic dimension estimator to each encoding to discover the intrinsic dimension of the dynamical system.

We compare the following ten intrinsic dimension estimators: Jolliffe's Principal Component Analysis [16], Costa's k-Nearest Neighbours estimator [10], Grassberger and Procaccia's Correlation Integral [13], Farahmand et al.'s Manifold-Adaptive Dimension estimation Algorithm, Levina and Bickel's [18] and Haro et al.'s [14] maximum likelihood-based estimators, Rozza et al. [20]'s Minimum Neighbour Distance-Maximum Likelihood estimators, Facco et al.'s Two-Nearest Neighbour estimator [11], Amsaleg et al.'s method of moments estimator [1] and Hein and Audibert's U-statistic-based estimator [15].

We compute intrinsic dimensions for the five dynamical systems of Fig. 1. We use the L1-norm between the estimated and the ground truth intrinsic dimension as the estimation loss.[1]

4   Performance Evaluation

Each encoding of a video frame comprises approximately between 1,000 and 10,000 points. We utilise `scikit-dimension`[2] implementations of Jolliffe's Principal Component Analysis, Costa's k-Nearest Neighbours estimator, Grassberger and Procaccia's Correlation Integral, Farahmand et al.'s Manifold-Adaptive Dimension estimation Algorithm, Haro et al.'s maximum likelihood-based estimators, Rozza et al.'s Minimum Neighbour Distance-Maximum Likelihood estimators, Facco et al.'s Two-Nearest Neighbour estimator and Amsaleg et al.'s

---

[1] The code to reproduce the results presented in this paper is available at: https://github.com/fchavelli/id_estimation/tree/main.

[2] The package is available at https://scikit-dimension.readthedocs.io.

Table 1. L1-norm for various intrinsic dimension estimators.

System	Ground Truth	Grassberger et al. [13]	Farahmand et al. [12]	Amsaleg et al. [1]	Rozza et al. [20]	Haro et al. [14]	Facco et al. [11]	Levina et al. [18]	Hein et al. [18]
Reaction-Diffusion	2	0.02 (±0.141)	0.06 (±0.041)	0.20 (±0.093)	0.22 (±0.024)	**0.01** (±0.046)	0.99 (±0.745)	0.16 (±0.113)	1.33 (±1.247)
Single Pendulum	2	0.02 (±0.016)	0.18 (±0.017)	0.10 (±0.017)	0.04 (±0.003)	0.02 (±0.003)	0.11 (±0.013)	0.04 (±0.013)	**0.00** (±0.00)
Double Pendulum	4	0.33 (±0.182)	1.14 (±0.052)	0.80 (±0.018)	0.54 (±0.035)	**0.23** (±0.032)	1.48 (±0.221)	0.66 (±0.028)	1.00 (±0.00)
Swing-stick	4	**0.02** (± 0.121)	1.29 (±0.098)	0.92 (±0.344)	0.20 (±0.056)	0.07 (±0.071)	10.51 (±1.126)	0.91 (±0.354)	1.00 (±0.00)
Elastic Pendulum	6	1.08 (± 0.234)	**0.30** (±0.17)	0.45 (±0.203)	1.07 (±0.094)	1.29 (± 0.124)	1.12 (±0.072)	0.66 (±0.177)	2.00 (±0.00)

method of moments estimator [2]. We use the MATLAB implementation of Hein's U-statistic-based estimator [19]. We utilise Chen et al.'s implementation of Levina and Bickel's maximum likelihood-based estimator.

We report the L1-norm values for various dynamical systems for each estimator in Table 1. All experiments were repeated for 3 unique random seeds. The intrinsic dimension estimator with the lowest L1-norm is presented in bold fonts. We omit the poor results from Jolliffe's Principal Component Analysis and Costa's k-Nearest Neighbours estimator. In general, Haro et al.'s maximum likelihood-based estimator performs best, followed by Grassberger and Procaccia's Correlation Integral, Farahmand et al.'s Manifold-Adaptive Dimension estimation Algorithm and Hein et al.'s U-statistic-based estimator. Levina and Bickel's maximum likelihood-based estimator is never the best-performing estimator. Haro et al.'s estimator outperforms Levina and Bickel's estimator for all systems except the elastic pendulum system.

Rozza et al.'s Minimum Neighbour Distance-Maximum Likelihood estimator and Facco et al.'s Two-Nearest Neighbour estimator perform poorly. They simplify data from high-dimensional manifolds with many twists and curves. However, the selected dynamical systems are Hamiltonian systems. The single pendulum, double pendulum, swingstick and elastic pendulum systems are mechanical systems with conserved total energy while the reaction-diffusion system has a Hamiltonian formalism [17]. This may be because Hamiltonian conservation means that the high-dimensional symplectic manifold is closed, and possibly has fewer twists and curves.

5 Conclusion

We compared various intrinsic dimension estimators for the discovery of the dimension of the phase space of a dynamical system from raw observations of its states and evolution. We find several estimators that outperform Levina and Bickel's maximum likelihood-based estimator initially selected by Chen et al..

We are now studying the interpretation of the latent variable in terms of the known state variables. We first verify that they capture the actual system dynamics and then study the candidate mappings, using symbolic regression, among other tools, between the two sets of variables.

Acknowledgements. This research is supported by Singapore Ministry of Education, grant MOE-T2EP50120-0019, and by the National Research Foundation, Prime Minister's Office, Singapore, under its Campus for Research Excellence and Technological Enterprise (CREATE) programme as part of the programme Descartes.

References

1. Amsaleg, L., et al.: Extreme-value-theoretic estimation of local intrinsic dimensionality. Data Min. Knowl. Disc. **32**(6), 1768–1805 (2018)
2. Bac, J., Mirkes, E.M., Gorban, A.N., Tyukin, I., Zinovyev, A.: Scikit-dimension: a python package for intrinsic dimension estimation. Entropy **23**(10), 1368 (2021)
3. Bellman, R.E.: Adaptive Control Processes. Princeton University Press, Princeton (1961)
4. Bennett, R.S.: Representation and analysis of signals part xxi. the intrinsic dimensionality of signal collections (1965)
5. Brito, M., Quiroz, A., Yukich, J.: Intrinsic dimension identification via graph-theoretic methods. J. Multivar. Anal. **116**, 263–277 (2013)
6. Campadelli, P., Casiraghi, E., Ceruti, C., Rozza, A.: Intrinsic dimension estimation: relevant techniques and a benchmark framework. Math. Probl. Eng. **2015**, 1–21 (2015)
7. Chen, B., Huang, K., Raghupathi, S., Chandratreya, I., Du, Q., Lipson, H.: Automated discovery of fundamental variables hidden in experimental data. Nat. Comput. Sci. **2**, 433–442 (2022)
8. Costa, J., Girotra, A., Hero, A.: Estimating local intrinsic dimension with k-nearest neighbor graphs. In: IEEE/SP 13th Workshop on Statistical Signal Processing, 2005, pp. 417–422 (2005)
9. Costa, J., Hero, A.: Geodesic entropic graphs for dimension and entropy estimation in manifold learning. IEEE Trans. Signal Process. **52**(8), 2210–2221 (2004)
10. Costa, J.A., Hero, A.O.: Determining Intrinsic Dimension and Entropy of High-Dimensional Shape Spaces, pp. 231–252. Birkhäuser Boston, Boston, MA (2006)
11. Facco, E., d'Errico, M., Rodriguez, A., Laio, A.: Estimating the intrinsic dimension of datasets by a minimal neighborhood information. Sci. Rep. **7**(1), 12140 (2017)
12. Farahmand, A.M., Szepesvári, C., Audibert, J.Y.: Manifold-adaptive dimension estimation. In: Proceedings of the 24th International Conference on Machine Learning, pp. 265–272. ICML 2007, Association for Computing Machinery, New York, NY, USA (2007)
13. Grassberger, P., Procaccia, I.: Measuring the strangeness of strange attractors. Physica D **9**(1), 189–208 (1983)
14. Haro, G., Randall, G., Sapiro, G.: Translated poisson mixture model for stratification learning. Int. J. Comput. Vis. **80**, 358–374 (2008)
15. Hein, M., Audibert, J.Y.: Intrinsic dimensionality estimation of submanifolds in R^d, pp. 289–296 (01 2005)
16. Jolliffe, I.T.: Principal component analysis and factor analysis. In: Principal Component Analysis. Springer Series in Statistics, pp. 115–128. Springer, New York (1986). https://doi.org/10.1007/978-1-4757-1904-8_7
17. Kuwamura, M.: The hamiltonian formalism in reaction-diffusion systems. Asymptotic Anal. Singularities-Elliptic Parabolic PDEs Relat. Prob. **47**, 635–646 (2007)
18. Levina, E., Bickel, P.: Maximum likelihood estimation of intrinsic dimension. In: Advances in Neural Information Processing Systems, vol. 17 (2004)

19. Lombardi, G.: Intrinsic dimensionality estimation techniques (2023). https://www.mathworks.com/matlabcentral/fileexchange/40112-intrinsic-dimensionality-estimation-techniques
20. Rozza, A., Lombardi, G., Ceruti, C., Casiraghi, E., Campadelli, P.: Novel high intrinsic dimensionality estimators. Mach. Learn. **89**(1), 37–65 (2012)
21. Stewart, D.E., Dewar, R.L.: Non-linear Dynamics, pp. 167–248. Cambridge University Press, Cambridge (2000)

Towards a Workload Mapping Model for Tuning Backing Services in Cloud Systems

Gaurav Kumar[1], Kshira Sagar Sahoo[2], and Monowar Bhuyan[2(✉)]

[1] Department of Computer Science, GIET University, Gunupur 765022, India
[2] Department of Computing Science, Umeå University, 901 87 Umeå, Sweden
{ksahoo,monowar}@cs.umu.se

Abstract. With the increasing advent of applications and services adopting cloud-based technologies, generic automated tuning techniques of database services are gaining much attraction. This work identifies and proposes to overcome the potential challenges associated with deploying a tuning service as part of Platform-as-a-Service (PaaS) offerings for tuning of backing services. Offering an effective database tuning service requires such tuners whose architecture can support tuning multiple databases and numerous database versions deployed on various types of underlying hardware configurations with varying VM plans. Tuners that offer such capabilities usually attempt to leverage experiences gathered previously. By taking advantage of relevant past experiences, tuners classify the current workload to the most pertinent workload seen recently. In this work, a five-layered, fully connected neural network with ReLU activation function is being employed as the classification model to classify data points into relevant workload classes. The categorical cross-entropy function is employed as the loss function and optimized using Adam optimizer. The work handles the challenges related to the cold-start problem, issues in mapping, and cascading errors. The proposed solution can overcome these issues in a large-scale production environment. The results show that the model has 93.3% accuracy in 93.8% F1-score as compared to the previous model like Ottertune.

Keywords: Knobs · Central Tuner · ANN · Softmax probability · Metric Collector

1 Introduction

With the rising popularity of micro-services in Platform-as-a-Service (PaaS) and Software-as-a-Service (SaaS) architectures in recent years, backing services have become an integral part of a cloud. These architectures become quintessential for consumption by microservices to enable functionalities like data stores and messaging systems [1,2]. Examples of such backing services can be datastores like MySQL, PostgreSQL, Redis, CouchDB, etc., and messaging services like RabbitMQ, Apache Kafka, etc. Platform customers do not have access to tune the configuration knobs for such underlying backing service. The sheer magnitude of the backing service deployments makes it impossible for the service providers to hire a DBA/administrator for real-time tuning of databases. A diverse range of backing service offerings from a platform also adds another layer of complexity [3].

To deal with such situations, it is evident that a highly robust and scalable approach for auto-tuning backing services is required. One such solution is Ottertune [4], which can tune multiple backing services irrespective of different versions and vendors. Ottertune Style Tuners (OST) has been introduced as a generic term used throughout the work to refer to all such tuners employing an architecture similar to Ottertune. OST recommendation engines always have a cost associated with each recommendation request executed for production systems. For service providers, the cost of provisioning a tuner deployment becomes challenging, as a single OST tuner deployment can provide tuning service to a limited number of service-instance. Also, the value of the recommendation becomes important since a customer is charged for each generated recommendation. Thus, an ideal goal is to maximize the number of deployments associated with a single tuning service deployment. These criteria must be maintained while ensuring that the accuracy of the recommendation is not compromised and the performance of the database instances subscribing to the service stays optimal. These tuning-service architectures employ machine learning algorithms for two essential operations: classification and recommendation. As per the model followed by OST recommendation engines and other proprietary tuners, the following issues are observed in the case of real-time production-ready deployment.

- **Cold-start problem:** The accuracy of mapping a workload to the most similar workloads (optimal mapping) increases only with the increasing size of the source workload. This is evident from the fact that the predicted metrics are computed where the length of predicted metrics equals the total knobs attempted in the source workload. The source workload is the workload set being executed on the database instance that is requested for tuning. If the size of the source workload is less (the database has recently been registered to the tuning service), it has observed lesser data points, then the OST recommendation engines suffer higher error percentages in mapping to the target workload. Due to wrong mappings, the obtained recommendations often lead to degradation in transactions per second (tps) on production systems. One such case is when a read-heavy/oriented workload gets mapped to a write heavy/oriented workload, and the obtained recommendations are always inclined to favour write optimal configurations and, in the end, even degrading.
- **Issues of deep mapping:** OST recommendation engine maps an entire source workload to a new target workload which requires a huge number of

data points to be already present in the database instance. The prerequisite is that, the target workloads should have huge observational data points for higher accuracy in mapping. Thus, the workloads seen in the past need to observe all sorts of variations in knobs and metrics for a smoother implementation. This technique becomes not only tedious but also non-deterministic since such training cannot always be guaranteed.

- **Issues of self mapping:** Another major flaw in the OST recommendation engine is that the engine considers the source workload to be one of the potential candidates for the target workload. Since the evaluation of the score depends on the predicted metrics set and its Euclidean distance to the input metric set. There are chances that the score could have come minimum for the target workload, which is the same as the source workload. Such a mapping is completely erroneous and leads to further degradation in the performance of the tuning service for the database instance.
- **Cascading errors:** The impact of the limitations discussed above has a drastic cascading effect. Let us visualize this with an example. Suppose the OST recommendation engine performs an erroneous mapping because of the above limitations and maps the source data point to a wrong target workload. This not only affects the current recommendation request but also corrupts the source workload.

Considering such challenges, in this work, a similar issue has been identified in a real-time deployment employing Ottertune's approach, where the chances of missing SLAs became very high. OST uses previously gained experiences to generate a new recommendation. In [4], authors use Euclidean Distance to map the incoming workload (from any given production workload) to one of the most similar workloads observed in the past. However, as the optimal recommendation generation solely depends upon mapping to such a similar workload, low accuracy of mapping on production systems can even reduce the final throughput drastically. Built upon such challenges, the proposed work identifies such issues with the OST tuning approach deployed for large databases and proposes solutions for the same. The major contributions of this paper are as follows:

- The proposed method maximizes the capability of the tuner by gathering relevant and most recent experiences from the data aggregator.
- A Neural Network is employed as the classifier model to classify data points into relevant workload classes. The classifier used a five-layered, fully connected neural network with ReLU activation. The last layer is a Softmax layer for classification into workload classes.
- The model is trained on the historical set of data for the workloads using categorical cross entropy function as the loss function and optimized using Adam optimizer.
- Further, the work has been compared with Ottertune and measured the F1-score and accuracy.

2   Related Work

There exist multiple works which focus on tuning databases based on physical or logical design [5], index tuning [6], or partitioning schemes [7,8]. Some works focus on tuning a subset of performance-impacting knobs [9]. There are also possible sets of tuners exist that are specific to certain commercial databases, such as DB2's Performance Wizard Tool [10], Oracle's Database Monitoring Tools [11], and Tuner for Microsoft SQL [12]. However, these tuners are limited by specific vendor dependencies and lack a generic approach for tuning certain knob parameters to achieve a specific objective. There are also some tuners that are based on feedback-driven techniques [13,14] which tune certain knobs by executing benchmarks and then observing performance details. Later, the same performance details/statistics are used to recommend changes in the database to make it perform better. Authors used an Ottertune-style knob tuner that leverages large-scale machine-learning techniques to furnish optimal recommendations for a database [15]. OST provide a common platform which can fetch metrics from multiple types of databases based on various VM plans as given by infrastructure providers like AWS, Alibaba Cloud, Azure, Google Cloud Platform, etc. They can use large-scale machine-learning techniques to provide knob recommendations. Keeping in mind the challenges faced by Platform-as-a-Service offerings, this architecture seems to be a scalable solution which can be used for tuning the backing service offerings. The Ottertune Style Tuners (OST) capture the experience gathered from different production databases and then leverages the experiences gained to give an optimal recommendation. Ottertune maps a current workload to one of the closest seen workloads and then uses the experiences (physical features) gained from the mapped workload to provide an optimal recommendation.

There are a couple of other works which use arrival-rate history [16] and query's logical semantics [17] to perform similar clustering of workloads. However, there is a reasonable consensus that OST engines are the best fit for this use case, as these features help in better clustering without understanding the query meaning. Moreover, these features are more sensitive towards database physical design and hardware. However, in [4], authors use Euclidean distance to estimate the closest workload seen previously. Euclidean distance has certain drawbacks to use in the production environment. In this work, we highlight such challenges associated with the mapping of a workload and propose solutions for the same. In [18] author explained how the cold-start could be a large problem in the case of database systems as well as in the field of big data. Also, the authors explained how the cascading error could be detected in the back-propagation while training the model.

Keeping in mind the challenges faced by PaaS offerings, this work proposed an architecture that tunes the backing services. The proposed method captures the experience gathered from different production databases and then leverages the experiences gained to give an optimal recommendation.

3   System Design

3.1   Problem Formulation

At any given time instant t, let's assume D is the set of all database service instances subscribing to the tuning-service deployment provisioned as part of PaaS offerings. N denotes the total number of such instances, such that: $D = \{D_1, D_2, ..., D_n, ...D_N\}$. The set of database service instances denoted by D bind to a single tuning-service deployment, and the tuning recommendations need to be served for all such instances in real-time from the single tuner deployment only. Consequently, the experience gained by the tuning service for tuning database instance, D_1, is stored by the tuning service as workload, W_1. Then, W^P can be defined as the set of all workloads, such that workload W_n^P stores experience obtained from database instance D_n. $W^P = \{W_1^P, W_2^P, ..., W_N^P\}$. Furthermore, each workload stores this experience in the form of multiple points, each point is denoted by a duplet, $\{k_i, m_i\}_{W_n}$, where for each nth workload, k denotes the knobs (physical settings configured), m denotes the metrics captured (statistics representing performance) for database D_n at time instant t. Further, consider that W^O denotes the set of workloads generated offline (non-production system) by executing various SQL workload benchmarks like TPCC, YCSB, Twitter, etc. These offline workloads often use multiple mixtures of a certain benchmark to showcase enough variations, e.g., {Read%, Write%} mixture of TPCC has variations like {50, 50}, {60, 40}, {80, 20}, {20, 80}. Any such workload can be denoted as W_k^O, and the set of all such workloads of total size K can be represented by $W^O = \{W_1^O, W_2^O, ..., W_k^O, .., W_K^O\}$. Let us assume W_n^{source} denotes the workload of a database making a recommendation request. Each time an underlying database service instance sends a request for tuning to the tuning service (at time instance t), the input parameters from the database instance to the tuner service are denoted as $\{k_i, m_i\}_{W_n^{source}}$. The tuning service, then, needs to map the received data-point $k_i, m_{i W_n^{source}}$ to any given previously observed workload $W_{k \neq n}$. This mapping is essential because the data points present in the mapped workload, $W_{k \neq n}$, are used to train the Gaussian Process Regression (GPR) engine in order to provide a recommendation to the requesting database instance.

3.2   Mapping Workload

Here, the problem of mapping a source workload to an existing workload is formulated. The section helps in understanding how the mapping of a workload is tackled by an OST recommendation engine. Continuing from the previous section, if for a time instant t, a tuning request is triggered by a database service instance to the tuning service, and the tuning service receives a new data point, $\{k_i, m_i\}_{W_n^{source}}$. Now, the OST engine employs a ranking/purging methodology to determine top-ranked knobs (characterized by knob ids) and metrics (characterized by metric ids). The engine further filters the current data point, i.e. $\{k_i, m_i\}_{W_n^{source}}$ and keeps the dimensions of knobs and metrics as per the newly obtained knob-ids and metric-ids. The OST engine then proceeds with the pre-processing tasks like binning, normalization, computing deciles, binning, etc. Now, leveraging the workloads experienced by the tuning agent denoted by W^P and the workloads that are generated offline denoted by W^O, the tuning engine has a total of $W^P \cup W^O$ workloads. For each workload in $(W^P \cup W^O)$, the algorithm trains a (Gaussian Process Regression) GPR and obtains a new predicted metrics set denoted by Mpred from the trained GPR for all knob set $(\forall k \in W_n^{source})$ attempted in workload W_n^{source}. Finally, as per the algorithm, the Euclidean distance is calculated and averaged for corresponding metrics, i.e., M^{pred} and $M^{original}$ where $M^{original}$ is the set of all actual metrics observed in workload W_n^{source}. The Euclidean distance is calculated for all predicted metrics and actual metrics. Similarly, for each workload, the Euclidean distance is calculated, and the workload with the least score wins.

3.3   Proposed Classification Approach

As discussed in the previous section, the Euclidean distance-based approach for workload mapping in the OST engine has drawbacks and can cause Service Level Agreement (SLA) violations for production scenarios. It requires a huge number of data points in W_n^{source} before the model converges to an optimum solution. The time of convergence increases with an increase in the number of data points as well. So, on the one hand, we require a large amount of data points before the model can accurately predict the workload class for incoming data points. Having a large number of data points leads to an increase in the time required to predict relevant classes for new test data points. To solve all these shortcomings, we have used a Neural Network as the classifier model to classify data points into relevant workload classes. This significantly reduces the test time required to optimally predict the workload class.

The used classifier is a five-layered, fully connected neural network with ReLU activation. The last layer is a Softmax layer for classification into workload classes (for example, in one of our test cases, into five workload classes). The structure of the ANN model has shown in Fig. 1. The model is trained on the historical set of data for the workloads using categorical cross entropy function as the loss function and optimized using Adam optimizer. We have used accuracy as the optimizing metric and time for one test run as the satisfying metric. The classifier is trained with a total of n workloads (at any time instance t), which create a total of n classes. Now with the proposed classification approach, any new point $\{k_i, m_i\}_{W_n^{source}}$, which comes from a W_n^{source}, gets classified to one of the classes using the trained model. As the classification approach classifies a point instead of classifying the whole source workload, this seems to minimize the highlighted mapping drawback. The dimensions of training data depend on the globally ranked knobs and metrics. This ranking is done by the background tasks such as factor analysis, clustering, and Lasso's feature extraction algorithm.

Theoretically, there is also a high chance that with each trigger of a background task, the globally ranked knobs get changed, which changes the dimensions of the data and ultimately leads to the retraining of the neural network.

The retraining is triggered only when a change is encountered in the set of globally ranked metrics and knobs. However, in production environments, it is observed that this change is seen only when a new database service instance gets added up.

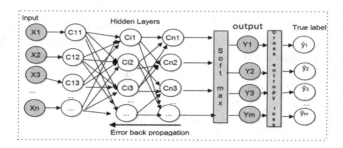

Fig. 1. Structure of ANN having softmax as activation function and CE as loss function

4   System Architecture

This section explains the architecture of the proposed Autonomous Database-as-a-Service model (ADBaaS) when deployed by a PaaS service provider. Figure 2 represents ADBaaS in which the tuning service is modularized into various components. It can be noted that the tuning service is completely de-coupled from the platform architecture and functions like plugins to the database instances subscribing to the service while ensuring minimum interference for any other component.

4.1   Database Service Instances

In a Platform-as-a-service (PaaS) architecture deployed in the cloud, database service instances indicate the physical resources (like VM, containers, etc.) in which a database process, e.g., Postgresql, MySQL, MongoDB etc. runs [19,20]. These database instances are provisioned and managed by the platform service provider, and the applications of the end user bind to these database processes for leveraging the data-persistence features. As showcased in Fig. 2, there could be multiple database instances, each having its own database process running within. The primary roles of a database service agent are to ensure the database process does not crash, ensure the high availability of a cluster, monitor the health of the cluster, etc. Additionally, if the database service instances choose to subscribe to the tuning service, the agent will read the metrics of the database process and send it over to the tuning service as per the configured frequency. The agent also communicates the current knob configuration of the database process to the tuning service. The agent places a request to the tuning service and then applies the recommendation (recommended knobs) received from the tuning service to the database process.

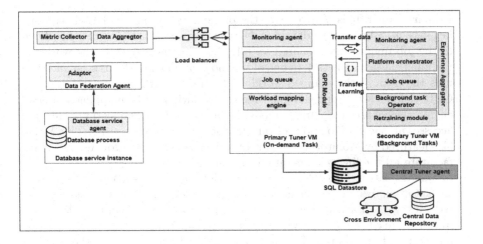

Fig. 2. Proposed ADBaaS architecture

4.2 Data Management Module

The data-management module is the component solely responsible for any interaction between the tuning service and the database service instance. This module contains Data Federation Agent, which further consists of various types of adapters for each database type. Adapters are the components that facilitate communication of any generic module with various database types, agnostic of the specific nuances of the various database types. Since the tuning service is a generic component that can cater to various types of databases, hence this adapter becomes an essential component.

4.3 Tuning Service

The tuning service deployment is divided into two modules, primary tuner and secondary tuner. There has been a conscious rationale behind segregating the service deployment into two separate modules. One obvious reason is the segregation of concerns, separating the execution of on-demand tasks on one system and the execution of background tasks on another system. On-demand tasks will include operations like obtaining recommendations from the GPR module, whereas background tasks will include operations like retraining the models and aggregating experiences gathered from other global installations of the tuning service. Thus, both categories perform very cost-intensive operations. Isolating the two operations in separate resources having identically computing capacity seems justified and prevents unnecessary spikes in physical resource requirement. Lastly, since operations like retraining can occur at any point and could reduce the performance of the tuning service, separating the process into another system ensures that the primary tuning service always performs at optimum levels, irrespective of the retraining tasks. The standard platform auto-scaler responsible for scaling the (primary and secondary) VM's based on load conditions.

- Primary Tuner This VM receives the on-demand recommendation requests, i.e., $<k_i, m_i> W_n$, from a database. For each request, a task is created buffered in the job queue, and executed serially. Now in the first step, the workload mapping classifies the task to one of the most similar workloads seen and then uses the data of the mapped workload to train a GPR. Upon training, the GPR yields a set of optimal knob recommendations. The mapping workload engine always uses the current model in the cache for classification. This VM accesses all the workload data from the common SQL data store. This VM is also featured by Monitoring Agents and Platform Orchestrator Agents, which facilitate all the lifecycle operations and health monitoring.
- Secondary Tuner This VM is responsible for running background tasks such as factor analysis, clustering, and feature extraction. These tasks help in reducing data dimensions for increasing the overall efficacy of tasks triggered from Primary tuners. This VM also has a re-tuner module that performs the retraining of the proposed Neural Network and then transfers the learning to the primary tuner VM in JSON formats. This VM has a job scheduler, which periodically schedules the background tasks. The retraining module reads the flags from common data stores and initiates retraining. This VM is also featured by monitoring agents and platform Orchestrators.

4.4 Central Tuning Agent

Since the tuning service can be deployed in multiple landscapes and environments, it is expected that the service must be exposed to diverse learning experiences in multiple environments. The idea here is to leverage the learning gained by the tuning service in one landscape by all other deployments of the service. This is the primary role of the Central Tuning agent. The agent is globally connected to all the deployments of the tuning service in different environments. The global deployment could be restricted to a certain platform provider, and the discretion of learning can be decided by the platform service provider. The central tuning agent collects the learning models and data points of the tuning service across environments and stores this accumulated data in a central data repository. This new data is fed to the secondary tuner for updating the model with the new data points via a periodically triggered retraining module execution.

5 Overcoming Training Challenges

It is important to figure out the optimal timestamps when re-training the neural network. From production-based scenarios, the following reasons could be identified and attributed to why retraining is required.

- *Add of new database instances*: When a new database service instance, D_{n+1}^{new} subscribes to the tuning service, a new workload W_n^{new} is created in the tuner repository to store the experiences from D_{n+1}^{new}. Now as the total workloads stored have changed i.e., n to $n+1$, the total number of classes for the ANN also changes.

- *Change of global ranked metrics and knobs*: When the background task in the secondary tuner-VM runs, it produces a new set of global ranked metrics and knobs. This causes a change in data dimensions. Empirically, the change in data dimensions also impacts the accuracy and error rate of the model.
- *Classification error rate increases*: Known data points are sent to the classifier periodically, upon which the accuracy rate and error rate are compared against thresholds. In this case, it is possible that retraining by considering the points in a buffer may increase the accuracy rate or reduce the error rate.
- *Evenly distribution of Softmax probabilities*: The Softmax activation function at the classification layer of the model returns probabilities for a data point that belong to a class. In many cases, it is observed that the probabilities are evenly distributed. One such example is, let's consider, there are 3 classes, and the softmax probabilities are {0.323, 0.30, 0.314}. Here the class with a probability of 0.323 clearly wins the margin. However, the other classes also have a very near chance of winning. In such a situation again, it is assumed that retraining with all new possible points will make the probability distribution uneven and avoid certain scenarios.

The background of pruning/ranking tasks responsible for changing global ranked knobs and metrics run periodically. If the globally ranked knobs and metrics change with the trigger of background pruning/ranking, then at the same instance, retraining is also triggered. Apart from the dimension change issues, a common unified solution to other identified problems can be measured the probability distributions using entropy. Shannon entropy [21] of a discrete random variable X can be defined as:

$$H_n(X) = -\sum_{i=1}^{n} p(x_i) log(p(x_i)) \tag{1}$$

where, $p(x_i)$ is the probability of ith outcome of X. We define the similarity index as a measure of the probability distribution, which returns the number of elements (in % scale) having nearly similar probability. Figure 3 depicts the variation of entropy based on a similarity index for a different set of classes. In the production system, the threshold value for entropy is determined statically by fixing the number of classes and similarity index.

During classification, each time entropy value for the probabilities is calculated and then compared against the threshold. If the entropy value is greater than the threshold, the system retrains the model with the sole condition that at least one of the workloads (W_n) has one new point apart from the source workload. The incoming data point $<k_i, m_i> W_n^{source}$, for which the entropy value has been violated, is statically mapped to the workload W_n^{target}, that has maximum mapping frequency for workload W_n^{source}. The maximum value of Shannon entropy for X, with a total n possible outcomes, is $log(n)$. This happens when the probability of a data point in each class becomes equal. The value of $\eta(X)$ is ranging from 0 to 1, i.e., $\eta(X) \in [0, 1]$. This helps in determining the threshold value of entropy.

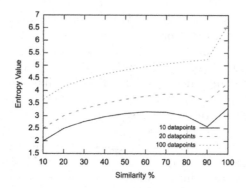

Fig. 3. Entropy variations

6 Performance Evaluation

The experiments were conducted on the AWS landscape, where the infrastructure resources were provisioned by Cloud Foundry [22] managed by Bosh [23]. The tuner deployment consists of 12 tuner instances - m4.xlarge with 4vCPU and 16 GB memory, 5 config-director instances - m4.xlarge. A total of 80 live-database deployments (spawned through t2.small, t2.medium, m4.large, t2.large, and m4.xlarge VM types) to the tuning have been considered. For evaluating the experiments, PostgreSQL (v9.6) was used. All the tuner instances collected data from one common data repository, which is shared by all tuner instances. This instance also has an m4.xlarge configuration. The bare service replicas were created, one for each plan and were used to test the recommendations. We considered 10 min observation time for YCSB and Wikipedia and 5 min observation time for TPCC. Table 1 shows F1 scores and accuracy obtained under different scenarios. Here W represents the number of workloads, and m is the data points per workload. For instance, in 100 workloads and 500 data points/ workload scenarios, the F1-score is 93.8% and accuracy is 93.3% which is $\approx$8.0% improvement over Ottertune (State-of-the-art model).

Table 1. Performance summary

Production Data	Ottertune		Proposed Model	
	F1-Score	Accuracy	F1-Score	Accuracy
W = 100 m = 30	0.435	0.441	0.867	0.877
W = 100 m = 500	0.871	0.866	0.938	0.933
W = 200 m = 30	0.323	0.321	0.865	0.852
W = 200 m = 500	0.914	0.922	0.895	0.903
W = 300 m = 30	0.241	0.242	0.822	0.822
W = 300 m = 500	0.821	0.841	0.856	0.887

Figure 4 shows the frequency of re-training requests triggered when the tuner deployment was hooked with a different set of production systems subjected to varying entropy values. Here, it is observed that with lower values of entropy, more re-training requests are getting triggered. This data can become extremely helpful for an admin to make decisions for optimal values of entropy (based on the similarity index) following the cost of each retraining w.r.t the provisioning cost perspective. Thus, depending on the performance trade-off, the admin can decide on the threshold. Figure 5 showcases another scenario. Here, the tuner had already trained offline with a maximum of write-oriented workloads and significantly fewer read-oriented workloads (110 write workloads and 8 read-oriented workloads). Thus, the setup remains the same as the previous experiment. However, the source workload is highly randomized with a non-uniform mixture of

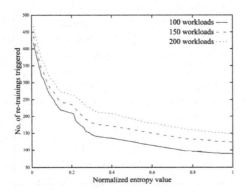

Fig. 4. Retraining variations on production environment.

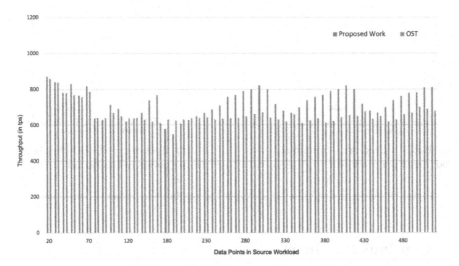

Fig. 5. Average throughput for different data point sets in source workload

read-write queries. OST engine, owing to its limitation of mapping the entire source workload to the target, tries to match all the data points in the source workload and, quite obviously, doesn't find any suitable target workload. Thus, owing to the other limitation of including the source workload as the target workload, it maps to the source workload. Thus, the throughput always remains the same, which is average for all scenarios. The proposed work also suffers from the same limitation and performs poorly initially due to incorrect mapping. However, the retraining module is triggered to ensure proper training of the model. Thus, gradually the performance of the proposed work improves and can adjust to different scenarios.

Figure 6 illustrates the general cold-start problem. Here, the tuner had already seen a maximum of write-oriented workloads and very few read-oriented workloads (110 write workloads and 8 read-oriented workloads). Here the workload is the execution of some preset load like read, write, or a combination of both, which helps the tuner learn and give appropriate knob configuration in real scenarios. In one case the source workload starts from no data points, where we observed reduced throughput initially for the OST approach, as shown in Fig. 6a. Due to the paucity of data, there are many wrong mappings. Hence non-optimal recommendations are generated. With the OST approach, due to incorrect mappings, the throughput of the database decreases, and later all the experiences in the source workload go with the cascading effect of sub-optimal experiences ending at self-mapping. However, with the proposed approach, we see better throughput as we minimize the cold start problem and restrict self-mapping. However, when the data points increase in the source workload, the accuracy of mapping increases, as shown in Fig. 6b. Hence, the GPR engine produces optimal recommendations causing an expected increase in throughput.

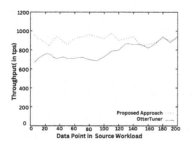 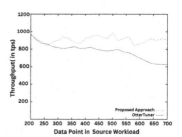

(a) Bootstrapped with minimum 200 points per workload.

(b) Bootstrapped with minimum 700 points per workload.

Fig. 6. Average throughput for different data-point sets in source workload.

7   Concluding Remarks

This work presents a Neural Network based classification approach for optimally mapping the production workloads to the closest resembling target workloads.

The challenges associated with the current OST's workload mapping approach are identified, and then a robust solution is proposed which can handle real-time production scenarios. While introducing a classification approach, an additional cost of retraining is encountered. The proposed approach is modified such that the cost of retraining is minimized from the perspective of provisioning and scalability challenges. Further, the solution is deployed on various cloud systems, and the performance is evaluated and compared with the prior solutions. A significant improvement in the accuracy of classification is achieved in the current approach. The solution is considered to be feasible for large-scale deployment of databases on a real-time system for keeping the database performance optimal at all times. Although the proposed work optimizes the performance of the classification engine, improving the scalability of the tuner to accommodate more database instances could be an interesting problem to tackle.

Acknowledgement. This work was supported by the Kempe fellowship via project no. SMK21-0061, Sweden. Additional support was provided by the Wallenberg AI, Autonomous Systems and Software Program (WASP) funded by Knut and Alice Wallenberg Foundation and the European Commission through the Horizon Europe project SovereignEdge.COGNIT (grant no. 101092711).

References

1. Backing services. https://12factor.net/backing-services. Accessed 01 Feb 2023
2. Babou, C.S.M., et al.: Hierarchical load balancing and clustering technique for home edge computing. IEEE Access **8**, 127593–127607 (2020)
3. Tiwary, M., Mishra, P., Jain, S., Sahoo, K.S.: AutoDBaaS: autonomous database as a service for managing relational database services. In: EDBT, pp. 600–610 (2021)
4. Van Aken, D., Pavlo, A., Gordon, G.J., Zhang, B.: Automatic database management system tuning through large-scale machine learning. In: Proceedings of the 2017 ACM International Conference on Management of Data, pp. 1009–1024 (2017)
5. Chaudhuri, S., Narasayya, V.: Self-tuning database systems: a decade of progress. In: Proceedings of the 33rd International conference on Very Large Data Bases, pp. 3–14 (2007)
6. Gupta, H., Harinarayan, V., Rajaraman, A., Ullman, J.D.: Index selection for OLAP. In: Proceedings 13th International Conference on Data Engineering, pp. 208–219. IEEE (1997)
7. Agrawal, S., Narasayya, V., Yang, B.: Integrating vertical and horizontal partitioning into automated physical database design. In: Proceedings of the 2004 ACM SIGMOD International Conference on Management of Data, pp. 359–370 (2004)
8. Pavlo, A., Jones, E.P.C., Zdonik, S.: On predictive modeling for optimizing transaction execution in parallel OLTP systems. arXiv preprint arXiv:1110.6647 (2011)
9. Sullivan, D.G., Seltzer, M.I., Pfeffer, A.: Using probabilistic reasoning to automate software tuning. ACM SIGMETRICS Perform. Eval. Rev. **32**(1), 404–405 (2004)
10. Kwan, E., Lightstone, S., Storm, A., Wu, L.: Automatic configuration for IBM DB2 universal database. In: Proceedings of IBM Perf Technical report (2002)
11. Yagoub, K., Belknap, P., Dageville, B., Dias, K., Joshi, S., Yu, H.: Oracle's SQL performance analyzer. IEEE Data Eng. Bull. **31**(1), 51–58 (2008)

12. Narayanan, D., Thereska, E., Ailamaki, A.: Continuous resource monitoring for self-predicting DBMS. In: Proceedings of the 13th IEEE MASCOTS 2005, pp. 239–248. IEEE (2005)
13. Brown, K.P., Carey, M.J., Livny, M.: Goal-oriented buffer management revisited. ACM SIGMOD Rec. **25**(2), 353–364 (1996)
14. Trummer, I.: Demonstrating DB-BERT: a database tuning tool that "reads" the manual. ACM SIGMOD Rec. 2437–2440 (2022)
15. Wang, X., Nedjah, N., Zhang, P., Shi, H., Ye, F., Li, Y.: Parameters tuning of multi-model database based on deep reinforcement learning. J. Intell. Inf. Syst. (2022)
16. Ma, L., Van Aken, D., Hefny, A., Mezerhane, G., Pavlo, A., Gordon, G.J.: Query-based workload forecasting for self-driving database management systems. In: Proceedings of the 2018 International Conference on Management of Data, pp. 631–645 (2018)
17. Pavlo, A., et al.: Self-driving database management systems. In: CIDR, vol. 4, p. 1 (2017)
18. Tey, F.J., Wu, T.-Y., Lin, C.-L., Chen, J.-L.: Accuracy improvements for cold-start recommendation problem using indirect relations in social networks. J. Big Data **8**(1), 1–18 (2021). https://doi.org/10.1186/s40537-021-00484-0
19. Amsterdamer, Y., Callen, Y.: Provenance-based SPARQL query formulation. In: Strauss, C., Cuzzocrea, A., Kotsis, G., Tjoa, A.M., Khalil, I. (eds.) DEXA 2022. LNCS, vol. 13426, pp. 116–129. Springer, Cham (2022). https://doi.org/10.1007/978-3-031-12423-5_9
20. Lisa, et al.: Data integration, management, and quality: from basic research to industrial application. In: Kotsis, G., et al. (eds.) DEXA 2022. CCIS, vol. 1633, pp. 167–178. Springer, Cham (2022). https://doi.org/10.1007/978-3-031-14343-4_16
21. Shannon, C.E.: A mathematical theory of communication. Bell Syst. Tech. J. **27**(3), 379–423 (1948)
22. Cloud-foundry. https://www.cloudfoundry.org/. Accessed 11 Jan 2023
23. Bosh. https://bosh.io/docs/. Accessed 18 Jan 2023

Compliance and Data Lifecycle Management in Databases and Backups

Nick Scope[1], Alexander Rasin[1]([✉]), Ben Lenard[1], and James Wagner[2]

[1] DePaul University, Chicago, IL 60604, USA
nscope52884@gmail.com, arasin@cdm.depaul.edu, blenard@anl.gov
[2] The University of New Orleans, New Orleans, LA 70148, USA
jwagner4@uno.edu

Abstract. From the United States' Health Insurance Portability and Accountability Act (HIPAA) to the European Union's General Data Protection Regulation (GDPR), there has been an increased focus on individual data privacy protection. Because multiple enforcement agencies (such as legal entities and external governing bodies) have jurisdiction over data governance, it is possible for the same data value to be subject to multiple (and potentially conflicting) policies. As a result, managing and enforcing all applicable legal requirements has become a complex task. In this paper, we present a comprehensive overview of the steps to integrating data retention and purging into a database management system (DBMS). We describe the changes necessary at each step of the data lifecycle management, the minimum functionality that any DBMS (relational or NoSQL) must support, and the guarantees provided by this system. Our proposed solution is 1) completely transparent from the perspective of the DBMS user; 2) requires only a minimal amount of tuning by the database administrator; 3) imposes a negligible performance overhead and a modest storage overhead; and 4) automates the enforcement of both retention and purging policies in the database.

Keywords: Databases · Privacy Compliance · Retention · Purging

1 Introduction

Organizations are subject to a variety of data management rules for how data must be archived, preserved, or destroyed. As new legislation is passed, these requirements are becoming more expansive and more strictly enforced. For example, Europe's General Data Protection Regulation (GDPR) privacy rules extend to organizations with customers in Europe (even when the organization is based outside of Europe). Organizations that fail to adhere to these policies risk their customers' privacy and are subject to potentially large fines. Thus, databases must incorporate the features and functionality necessary to remain compliant.

For purposes of this paper, we define *policy* as the set of rules an organization must follow with respect to data preservation and destruction. These policies can be the result of internal requirements, other business partners, or government agency mandates. Failure to comply with these policies could result in large fines, a loss of customers, and an irrecoverable breach of customer data privacy.

Although current industry tools (see Sect. 2) offer some important data governance capabilities, database management systems (DBMS) must be updated to support compliance in data storage. DBMSes do not currently include native retention or purging functionality that can be applied at record level. Google, Amazon, Oracle, and IBM all offer various object-storage compliance functionality for their remote storage. These all use date-criteria for defining policies timelines, but none of these offer tuple, cell, or value based policy enforcement in a database. Instead, objects are placed into "buckets" and policies are applied at the bucket-level. Because databases contain intermixed records which are subject to different policies, applying the policy at the bucket level risks non-compliance with one policy at the cost of another.

However, DBMS storage (relational or NoSQL) is much more complex and fine-grain, representing the data at individual record and value level. Currently, with respect to databases, organizations are forced to create ad-hoc solutions to meet policy compliance requirements; these solutions are typically developed by either re-purposing other existing tool functionality or manually performing the steps to enforce compliance.

Governance policies depend on multiple factors and can be surprisingly complex. The Office of the National Coordinator for Health Information Technology provides a summary overview with examples for how many states in the United States have their own requirements for retaining and destroying healthcare data [30]: Oregon requires hospitals to retain all records for 10 years after the date of the last discharge; Hawaii requires the full medical record history to be retained for 7 years after the last data entry. Adding to the complexity, the data of minors and adults can be governed by different policies. For example, in North Carolina, hospitals are required to retain adult patients data for 11 years following discharge, while the data of patients who are minors (at the time of record creation) must be retained until the patient's 30th birthday. Thus, the policy expiration must reference patient's date of birth, with different rows or columns of a database table governed by different requirements.

Adding to the complexity, database administrators must consider the possible conflict between multiple requirements (e.g., retention versus destruction of the same data item). For example, GDPR's Article 17 requires that an organization purge personal data "the personal data are no longer necessary in relation to the purposes for which they were collected [5]", but if the same data item was pertinent to an impending or an ongoing lawsuit, an organization must retain the data until it is no longer required to be retained (i.e., the lawsuit has been resolved). Therefore, any organization relying on manual solutions for their compliance must consider the high labor cost of enforcing compliance.

1.1   System Overview

Ataulla et al. [9] first proposed the idea of defining data governance policies through a SQL query (see Sect. 4.1) as a first step towards native DBMS pol-

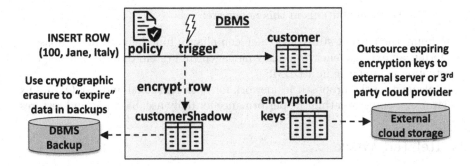

Fig. 1. Data lifecycle workflow changes in a DBMS to support data purging policies. Retention policies will use triggers and defined policies to retain data in an additional `customerArchive` table

icy support. Scope et al. [27] proposed leveraging DBMS triggers (natively supported by all major database vendors) and revising the backup process to support policy-based data purging. Scope et al. [26] also prototyped the same strategy in the context of NoSQL (MongoDB) databases. In this paper, we present and evaluate an end-to-end approach to offer a native support for data governance (retention and purging) in relational and NoSQL DBMSes.

Figure 1 summarizes the integration of our data purging mechanism steps into a DBMS (retention mechanism details are not pictured). Policies are defined with database queries, SQL or NoSQL, such as "rows inserted into a customer table must be retained for a duration of 5 years". Each inserted (or updated) row is checked by a trigger against applicable purging policies, if any. The values covered by purging requirement are encrypted with a corresponding policy-based key, and inserted into the `customerShadow` table (an encrypted counterpart copy of the original `customer` table).

The `customerShadow` table is backed up instead of backing up the `customer` table, to enable "remote" erasure by destroying the corresponding key upon policy expiration. In order to fully satisfy purging requirements, the database must also securely delete encryption keys from backups [22]. Towards that end, the encryption key table is backed up separately with an independent storage service, to ensure that keys are expunged upon expiration. The encryption key table is itself encrypted to minimize the impact of a potential data breach. However, we note that the mechanisms described here are not designed to be a security solution but are a governance compliance mechanism. Thus, if the encryption keys were somehow compromised (or inadvertently copied), this framework can re-create a new encryption key table (and underlying keys) and re-encrypt the backups and shadow tables. This would not address the data theft, but once all of the data is encrypted with new keys, it would restore data storage compliance.

Our corresponding retention mechanism (not pictured in Fig. 1) checks deleted rows for values that are currently protected by retention policy. Such values are stored in an archive table and purged through the same means (see Sect. 4.1).

In sum, the contributions in this paper are:

- Defining the current state of privacy compliance functionality in databases
- Outline an external encryption key management system that guarantees data retention compliance in a DBMS
- Implementing the proposed framework for both relational and NoSQL JSON databases and evaluating the performance for daily use, backups, and restores

2   Related Work

Kamara and Lauter [16] concluded that using cryptography can improve privacy protections when using remote storage. Furthermore, their research has shown that erasing an encryption key can be a means to rendering remote data irrecoverable. We leverage cryptographic erasure mechanism to remotely purge database values to ensure data privacy purging compliance. Kamara and Lauter's work does not discuss how to manage encryption keys or to apply them at a fine-grained level necessary for compliance.

Reardon et al. provided a extensive overview of secure deletion [22]. The authors defined three user-level approaches to secure deletion: 1) execute a secure delete feature on the physical medium 2) overwrite the data before unlinking or 3) unlink the data to the OS and fill the empty capacity of the physical device's storage. Their methods require the ability to directly interact with the physical storage device, which may not be possible for all database backups. Offline backups (e.g., backup tapes in a warehouse) are still subject to purging and retention policies. Thus, destroying (either physically or with a complete deletion wipe) an entire backup to guarantee purging compliance, the destruction would come with sacrificing retention compliance.

Scope et al. [27] presented a generalized data purging workflow which supports "remote" destruction of expired data (e.g., inaccessible records stored in a backup) in a relational database via cryptographic erasure. Encryption keys are chosen based on the purging duration and policy; values not subject to purging are stored without encryption. When the purge criteria has been met, the corresponding encryption key is deleted, rendering all encrypted data permanently irrecoverable (i.e., purged). Additionally, research was conducted on only purging compliance in NoSQL JSON databases [26]. Neither paper addressed how to guarantee retention compliance while implementing purging functionality.

Scope et al. [25] later expanded the previous work to incorporate functionality that simultaneously considered both retention and purging policies. Although these papers did leverage encryption, they did not provide a framework to manage the encryption keys (i.e., how to store the encryption key backups). Additionally, this paper focused exclusively on relational databases. This paper aims to incorporate a compliant approach for managing the encryption keys (regardless of the database logical layout, including both relational and NoSQL).

On the industry side, Amazon S3 offers an object life-cycle management tool [8]. S3 is file-based and lacks the granularity to fully support retention and purging at the individual tuple level. Furthermore, NoSQL stores (e.g., MongoDB evaluated in this paper) also require a value-level granularity to implement data governance policies.

Google Cloud Platform (GCP) offers a similar tool to Amazon S3 by supporting file-level compliance [3]. GCP's Bucket Lock offers a retention solution which guarantees all files are protected until the *retention lock* has expired. Conversely, GCP's Object Lifecycle Management tool uses rules which trigger an automated deletion of files. Overall, real-world retention and purging policies require fine-grain destruction and retention of data which is currently not supported by current industry tools.

3   Data Governance and Compliance

Business Records are the units for organizational rules and requirements for data management. United States federal law refers to a business record broadly as any "memorandum, writing, entry, print, representation or combination thereof, of any act, transaction, occurrence, or event [that is] kept or recorded [by any] business institution, member of a profession or calling, or any department or agency of government [...] in the regular course of business or activity" [31]. In other words, business records describe any interaction or transaction resulting in new data.

Policy is any formally established rule for organizations. Policies can originate from a variety of sources such as legislation or as a byproduct of a court ruling. Companies may also establish their own internal data retention policies to protect confidential data. In practice, database administrators work with domain experts and sometimes with legal counsel to define business records and retention requirements based on the written policy. Policies can use a combination of time and external events as the criteria for data retention and destruction.

Retention is the preservation of all data subject to a policy. Retention requirements supersede the requirement to destroy data.

Purging is the permanent and irreversible destruction of data in a business record [15]. A business record purge can be accomplished by physically destroying the device, fully erasing all data on the device, or encrypting and erasing the decryption key (although the ciphertext still exists, destroying the decryption key makes it irrecoverable). If any part of a business record's data remains recoverable or accessible, then the data purge is not considered successful. If a purging policy overlaps with a retention policy, the data must not be purged until after all retention policies have expired.

Problem Statement: All encryption used by this framework is deployed with the intention of facilitating compliance and not for security purposes. Thus, all security considerations are beyond the scope of this paper. Additionally, data processing compliance (i.e., only using customer data where consent has been given for processing) is beyond the scope of this paper. Our goal is to implement automated retention and purging policy enforcement procedures during database transactions, backups, and restores, agnostic of DBMS logical layout.

4 System Overview

In this section, we describe our system that offers a comprehensive support for data governance policy compliance in DBMSes. We first describe the components that were previously proposed and then discuss changes and new components introduced as part of this paper. In this paper, we use the term *table* to refer to both a relational database table and a collection in JSON NoSQL database.

4.1 Background

The policies are defined using SQL or NoSQL queries (the idea originally pioneered by Ataullah et al. [9]). Therefore, the database could return the rows and columns that were subject to any particular policy by executing the corresponding query. For example, the following SQL query expresses a policy to retain all data from the tables `customerPayment` and `orderShipping` minimally 90 days after the payment date.

```
SELECT * FROM customerPayment NATURAL JOIN orderShipping
WHERE DATEDIFF(day, orderShipping.paymentDate,
       date_part('day', CURRENT_DATE)) < 90;
```

Each table containing data subject to retention rule has a corresponding *shadow archive table*. The shadow archive table stores data which was deleted (i.e., no longer needed by users) but that is protected by retention policy (for some duration or indefinitely). Similarly, each table with data subject to a purging rule has a corresponding *shadow table*. For records subject to purging, the record's values are encrypted, before a copy of the record is placed into the shadow tables; data not subject to purging is copied into shadow table in its original form. The shadow tables replace the original tables in backup; they also contain columns that provide a mapping to the corresponding encryption key.

For all defined policies, we store encryption keys and corresponding policies in the `policyOverview` table; the DDL (using Postgres) for the `policyOverview` table can be found below. The `policyOverview` table contains the date on which each key will be purged. Purging the key would purge all corresponding encrypted values across all of the shadow tables and shadow archive tables.

```
CREATE TABLE public.policyOverview (
    policyid integer NOT NULL,
    policy character varying(50),
    expirationDate date,
    encryptionkey character varying(50));
```

Whenever a user executes an INSERT, DELETE, or UPDATE, the framework determines if any of the data is subject to a retention or purging policy. Because retention takes priority over purging, data which is subject to both must be retained until the retention policy requirements have been met. During a restore, the shadow tables are restored and then loaded into the user-facing tables (e.g., `customer` is loaded from `customerShadow`). The data for which has not been purged (i.e., encryption key is still available) is decrypted. If the encryption key has been deleted due to a purging requirement, the values are restored as a `NULL`. For relational databases, if the primary key of a tuple cannot be restored, the entire tuple is deleted. With JSON NoSQL databases, when a key has been purged, all associated values are not restored.

One of the major challenges to guaranteeing compliance is the problem of handling encryption keys. Backing up the keys would interfere with being able to purge data (because database backups cannot be edited to selectively remove data). Scope et al. [27] proposed storing the encryption keys in a separate linked database to reliably support purging. However, the question of how to manage encryption keys was not considered in prior work.

4.2   Leveraging External or Third Party Servers

In order to successfully apply cryptographic erasure, we must guarantee that the deleted encryption keys have been irrecoverably erased. Otherwise, deleted encryption keys may be restored from a backup and decrypt purged data. Many industry tools (e.g., AWS S3) provide the ability to automatically "expire" objects at the file granularity, but any external storage which provides automated file-level time-based erasure would satisfy the requirements of this framework. We propose using such a system for automatic deletion of files to purge encryption keys (based on expiration date).

Because this framework depends on using external servers to backup the encryption keys, there is a risk of a server outage. The ability to access encryption keys is only needed during a restore of the `policyOverview` table (which would only occur during a database restore). Therefore, if restores are not common, the risk would be minimal and acceptable.

In instances where restores are common and must not be delayed or where high availability of backups is required, leveraging multiple external servers can be used for storing the encryption key backups in parallel. AWS, Google, and IBM all offer file-level automatic deletion at a set time [3,4,8].

4.3   Encryption Keys During the Backup and Restore Process

During the creation of the standard database backups, our framework leverages backup scripts to create the backup of the encryption keys and automatically uploads them to the external server(s) designated (using whichever scheduler an organization deploys, e.g., CRON job). These backup scripts must be revised to first upload the encryption keys to the third-party servers before executing the database backup procedure. Keys which have already been uploaded and have their automatic deletion criteria set do not need to be re-uploaded again. In our `policyOverview` table, we store the date at which an encryption key was uploaded to prevent it from being redundantly re-uploaded during future backups. Thus, only newly created keys would require being backed up in addition to the standard database backup procedure.

Because the `policyOverview` table is not backed up with the other tables, this framework requires some capability to backup individual table/collection spaces. In instances where an encryption key's purged date has passed, we can assume that the corresponding key has been automatically deleted (due to the system automatically removing the file based to the expiration date). All remaining encryption keys which have not been purged are downloaded from the remote server and used to decrypt their associated business records (using the original framework outlined by Scope et al. in [25]) during the restore process. Once restored, our framework moves the records which are either not encrypted or for which a decryption key is still available from the shadow tables into the active tables.

4.4 NoSQL Process Considerations

Although the functionality in this framework remains consistent between a relational database and a NoSQL JSON database, there are some additional factors to consider. For purposes of this discussion, we use terminology and commands from Postgres and MongoDB. When generating backups for a relational database, using a `pg_dump` command targeting the shadow tables guarantees that neither the unencrypted data nor the encryption keys is placed into standard backups (which would prohibit the keys from being purged). With MongoDB, we leverage `collections` during backups to limit the backup to only the shadow collections (using the command `mongodump`).

Any NoSQL JSON database which uses this framework must support triggers. Only MongoDB Atlas (the cloud version) offers trigger functionality, while the local (free) version of MongoDB does not currently support triggers.

With relational databases, during the restore, any tuple with a purged primary key is removed from the table. In NoSQL JSON databases this translates into removing all values when a corresponding key has been purged. If a subset of the values of a key have been purged, instead of `NULL`ing out the keys, we simply remove the value from the corresponding key-value.

Although in Postgres we leverage `PGP_SYM_ENCRYPT` (a default supported module) to apply encryption, in MongoDB we utilize the `ClientEncryption` functionality found within the `Explicit Encryption` framework. Our proposed framework requires any database to support some level of encryption functionality which can be incorporated into a trigger.

5 Experiments

We implemented and evaluated a prototype of our framework with the active tables in Fig. 2. This schema reflects only the tables needed for the policies that we define and does not show all tables in the database schema. In practice, we expect most policies to cover data in one or two tables/collections. Other tables which do not directly apply to a policy will not impact policy-enforcing performance. We demonstrate that our approach can be implemented without

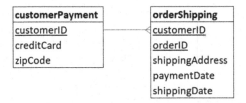

Fig. 2. Tables used in our policy definitions and experimental evaluation

changing the original (user-facing) database schema and by extending backup procedures using only natively available DBMS backup functionality.

5.1 Experimental Setup

Hardware: We used a server with dual Intel Xeon E5645, each with 6 physical cores and Hyper Threading enabled, 64 GB of ram, and an SSD for storage. The server was running CentOS 8 Stream x86_64 with Kernel Virtual Machine [17] (KVM) as the hypervisor software. We used two Virtual Machines (VMs) to carry out the experiments; since a majority of database interactions operate in a client-server model, we deployed two independent VMs to represent client and server. Both VMs were built with CentOS 8 Stream x86_64, Postgres 14.5, MongoDB 4.1, 1 x vNIC and a 25 GB QEMU copy-on-write [6] (QCOW2) file on an SSD. The client VM has 4 GB of RAM and 4 vCPUs and the server VM was allocated 8 GB of RAM and 4 vCPUs. QCOW2 file was partitioned into: 350 MB/boot, 2 GB swap space, with the remaining storage used for the / partition, using standard partitioning and ext4 file system. Only these two VMs were running on the hypervisor to minimize runtime fluctuations.

Policies and Keys: We created one retention and one purging policy; both policies covered all columns in tables from Fig. 2. We then generated 30,000 business records which approximately equally fell under 1) neither policy, 2) only the retention policy, 3) only the purging policy, or 4) both policies.

In Sect. 5.2, we evaluate framework performance overhead using a real-world simulated query workload on a local database. We use synthetic data and the additional generated encryption keys (total of 21 encryption keys) in our experiments. In Sect. 5.3 we analyze the performance overhead of our framework during the backup and restore of a relational and NoSQL database. Our experiments confirmed the framework enforces retention and purging compliance.

5.2 Query Overhead Imposed by the Framework

Relational Databases: SELECTs do not incur any retention or purging overhead in our framework. Because real-world data warehouse workloads are typically 90% SELECTs [14], in practice the compliance overhead would apply a relatively small fraction of queries. To mirror data warehouse workloads reported by Hsu

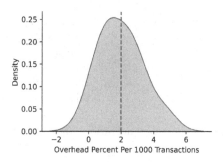

Fig. 3. Overhead of our framework during a simulated query workload

et al. [14], our query workload consisted of 9,000 (90%) SELECT queries, 700 (7%) UPDATE queries, and 300 (3%) DELETE queries. Because our framework runs the same process for UPDATEs and INSERTs, we use UPDATEs for performance evaluation.

In order to measure the runtime overhead, we ran an identical query workload on two identical databases, one with our framework enabled and one without any additions. Different types of queries were mixed in randomly in our workload; we measured the elapsed time after each 1,000 query transactions. We manually verified that the the encryption keys and archival processes were correctly applied to the data after running the simulated workload. The overall overhead distribution can be seen in Fig. 3. On average, our proposed framework had a 2% overhead compared to the database without the compliance framework.

Our workload replicates the average expected query distribution observed by Hsu et al. [14]. In practice, the overhead will depend on the policy sizes, the frequency which queries trigger a policy action, and the types of queries run. The evaluation of each of the such factors is beyond the scope of this paper.

The closest related research conducted by Ataullah et al. [9] uses triggers to determine whether or not an UPDATE or DELETE would violate a retention policy. In instances where a query would result in non-compliance, the query is blocked. Thus, both our framework and the research by Ataullah et al. require a trigger initiating and the code required to determine whether or not a query would result in a compliance violation. We have a small overhead of archiving data compared to their solution of blocking a query; this results in the trade-off of not having to adjust queries at the cost of automatic archiving overheads.

NoSQL Databases: In this paper, we focus on evaluating local (non-cloud) databases to minimize the number of factors outside of our control that may affect performance. MongoDB only supports triggers in a cloud-based version; thus we do not evaluate the query overhead performance in this paper. Scope et al. [26] verified the functionality of using triggers and cryptographic erasure to support purging in MongoDB Atlas (cloud-based version of MongoDB).

5.3   Backup and Restore Overhead Imposed by the Framework

In this experiment, we evaluate the cost of backing up and restoring the encryption keys for our framework. We discuss the overhead cost of backing up a single

Table 1. Backup and Restore File Sizes (in bytes)

File	Relational	NoSQL JSON
Full Database	64,011,916	149,151,434
Single Encryption Key File	2,810	3,437
Separate Encryption Key (21 Files)	72-73	165

key file (i.e., the `policyOverview` table) versus backing up each key independently, in addition to executing a full backup/restore from a local file. We use the encryption keys and data described in Sect. 5.2 (where 21 encryption keys were generated). Table 1 summarizes the file sizes in this experiment.

This experiment evaluates the additional overhead cost of backing up the encryption keys to an external server. To measure the upper-bound overhead of our framework, we disabled key caching during the restore.

Relational Backup and Restore: We ran 10 backups with and without our framework enabled to evaluate the overhead cost of backing up a single key file (i.e., backing up the entire encryption key table as a single file) and all 21 encryption keys independently. The observed overheads of these backups are shown in Figs. 4 and 5 (with the storage costs outlined in Table 1). For the single key file backup, the performance overhead was 72%, and for backing up the individual 21 keys the overhead was 100%.

We then evaluated the cost of restoring a relational database, introducing the additional step of restoring of the encryption keys from an external server. Figure 6 presents the overhead of restoring a single file, while Fig. 7 shows the overhead of restoring all 21 keys. For a single key file, the average restore overhead was approximately 105%. For the full restore of individual 21 keys, the overhead was 92%.

NoSQL Backup and Restore: To verify our framework backup and restore works with a NoSQL JSON logical layout, we performed the same evaluation

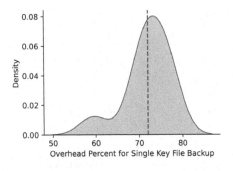

Fig. 4. Relational single-file backup overhead

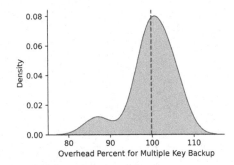

Fig. 5. Relational individual 21-file backup overhead

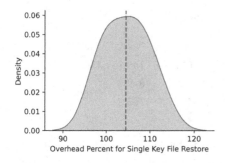

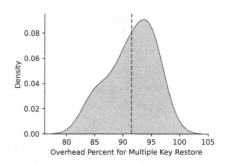

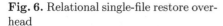

Fig. 6. Relational single-file restore over-head

Fig. 7. Relational 21-file restore over-head

(using the same data in NoSQL JSON collections) with a local MongoDB. Thus, the *shadow collections* are backed up using standard MongoDB backup procedures (`mongodump`) and our `encryptionKeys` collection is backed up to external servers. As with the relational database, we evaluate the overhead for backing up and restoring a single encryption key file as well as for separately backing up and restoring all of the keys generated during our analysis of the overhead.

Figures 8 and 9 provide an overview of the overhead incurred by backing up the encryption keys in MongoDB. The average overhead of a single key file was 105%, and for backing up all of the 21 keys separately was 115%.

Figures 10 and 11 summarize the overhead impact of the framework on the restore process. Our framework adds a 107% overhead to restoring a single key file from a remote server; when restoring each key independently, the overhead is increased to 114%.

5.4   External Backup and Restore Performance Considerations

As the industry moves to the cloud environment, the backup and restore process can be influenced by a number of factors, including the internet connection speed, hypervisor load, or disk type. In our experiments, we used an external

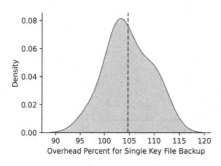

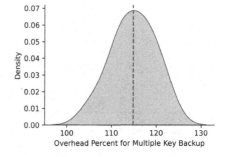

Fig. 8. NoSQL 1-file backup overhead

Fig. 9. NoSQL 21-file backup overhead

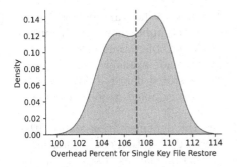

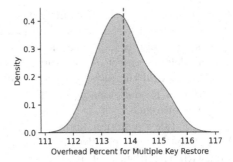

Fig. 10. NoSQL 1-file restore overhead   **Fig. 11.** NoSQL 21-file restore overhead

server offsite from our university as described in Sect. 5.1. We also used a NAS appliance that was located within the University, so the server and NAS are roughly 60 miles apart. The network connection from the server to the Internet is 1Gbps symmetrical and the NAS appliance also has a 1Gbps symmetrical connection. However, due to the nature of the Internet the speed for which these files were uploaded or downloaded can fluctuate based on Internet congestion. Moreover, the number of files that are uploaded or downloaded can influence the duration of the transfer as well as the load of the NAS device.

We use a simple server running KVM, but a cloud service would introduce additional complexity. For example, AWS offers different tiers, some tiers share the hypervisors resources while other tiers are dedicated. Larger organizations with dedicated on-premise hardware running this framework may be able to leverage their existing architecture resulting in a lower overhead percent. Furthermore, a higher ratio of business records to encryption keys will result in a smaller overhead during backup and restore.

Almost all modern file-transfer services require authentication (which is part of AAA [7]), and our SFTP connection authentication incurred a time cost. In other words, uploading the keys to a separate external server requires additional time even if the volume of the data remains the same. Moreover, the authentication time for a connection could vary based on a multitude of factors ranging from the load of the target server to the load and response time of the authentication service. Thus, the performance of backup and restore processes can fluctuate based on a multitude of factors outside of the control of our framework and systems.

6 Discussion

6.1 Third Party Server Vendor Considerations

For cryptographic erasure to fully satisfy purging requirements, all pertinent encryption keys must be rendered irreversibly irrecoverable. Many remote storage options do not offer the ability to implement a Secure Deletion process on specific files. Amazon's documentation [1,33] states, "When an object is deleted from Amazon S3, removal of the mapping from the public name to the object starts immediately [...] Once the mapping is removed, there is no external access to the deleted object. That storage area is then made available only for write operations and the data is overwritten by newly stored data [...] AWS Backup randomizes its deletions within 8 h following recovery point expiration to maintain performance."

GCP takes a similar approach [2,3], where the documentation states "[the] Google backup cycle is designed to expire deleted data within data center backups within six months of the deletion request. Deletion may occur sooner depending on the level of data replication and the timing of Google's ongoing backup cycles [...] After the data is marked for deletion, an internal recovery period of up to 30 days may apply depending on the service or deletion request."

Thus, the secure deletion guarantees provided would depend on the vendor. If one were to use an "in-house" external server for encryption key storage, a guaranteed secure deletion could be implemented. Organizations must balance these considerations to decide between a vendor or an in-house service.

6.2   Managing the Size of the Archive Tables

Although this framework supports the restoration of all archived data, in practice, many organizations may choose to limit the amount of data in the archive outside of backups. Because data in the archive is not expected to be used regularly (otherwise it would not be in an archive), many organizations may not want to use their high performance storage on data which is not frequently accessed. Thus, this framework supports a parameter which limits the restore of data from the archive backups during the restore process to a specified time range.

With larger databases, we would consider partitioning the archive tables to enable more efficient deletions. For example, we could partition tables on a policy date field and make a new partition every month. Before we drop the month that has expired, we would export it to a non-proprietary format for easy retrieval at a later date. When keeping records for 10+ years, it is simpler to recover the data in a non-proprietary format since versions of software, hardware, and operating systems change over the years. Many organizations use CSV, JSON, XML, or even HDF5 file formats for long term software-independent storage.

6.3   Reclaiming Unused Storage Space

After marking the data for deletion, databases will flag the row for deletion without actually purging the data from disk; in the case of an UPDATE, the database operation will often mark the old row for deletion and insert a new (updated) row. In that case, the row's pre-update data will still exist in the underlying database pages. This old data remains on disk until the RDBMS reuses the tablespace space or a reorganization of the tablespace purges the old data from database storage [18]. The challenge with management or purging such deleted data are due to DBMSes not providing tools or mechanisms to monitor or modify their internal storage.

6.4   Concerns with Forensically Recoverable Data

Deleted data that still remains on a storage medium but is no longer referenced by a file system or a DBMS can still be reconstructed using a variety of forensic methods (e.g., [10,24,34,35]). Although this paper addresses how to manage encryption keys used for purging data across backups, we consider forensically recoverable data to be beyond the scope of this paper. Lenard et al. [18] analyzed how various types of databases and their defrag options are able to remove the surviving deleted data from backups. We therefore recommend regularly running a defrag on a database to expedite the process of clearing out deleted data from database pages, particularly before these pages are placed in a backup.

There are data sanitization techniques that seek to destroy deleted data so that it can no longer be forensically reconstructed. Although Lenard et al. [18,19] investigated the data left forensically recoverable in different parts of database system and Wagner et al. [36] developed the API to interact with low-level database storage, there are currently no solutions available to sanitize database storage. Most research and tools for data sanitization involve overwriting blocks at the disk level (e.g., [11–13,23]) and cannot overwrite individual database records. SQLite is the only DBMS that supports data sanitization with the secure_delete setting [28], which is disabled by default due to a negative impact on performance. If enabled, secure_delete explicitly overwrites deleted data with zeros. Stahlberg et al. presented a similar method for MySQL [29].

Although the laws do not explicitly detail the technical steps of comprehensive data destruction, this level of data destruction is typically described by individual organizations or government agencies (e.g., NSA [21], NIST [20], or IRS [32]). Data sanitization is a problem that we consider to be outside the scope for this paper; our encryption protects data covered by purging policies, even from forensic recovery, but a more general data sanitization approach may compliment the work in this paper.

7   Conclusion

Data management research must continue to address and refine the support for database compliance functionality with respect to customer privacy. Although some research has begun to address current shortcomings, an increasing proliferation of new rules, requirements, and complexity will result in increased compliance pressures. Current purging and retention compliance support is limited to either coarse-grained (i.e., file-level) applications or does not consider both retention and purging simultaneously when enforcing compliance policies. Fine-grained compliance functionality must be researched and implemented in database systems to automatically enforce compliance. This paper outlines a comprehensive compliance support framework that implements retention and purging support throughout databases and their backups; our experiments demonstrate that our framework can guarantee compliance requirements with an acceptable performance overhead and with minimal additional infrastructure requirements.

References

1. Amazon web services: Overview of security processes. https://d1.awsstatic.com/whitepapers/aws-security-whitepaper.pdf
2. Data deletion on google cloud documentation. https://cloud.google.com/docs/security
3. GCP object lifecycle management. https://cloud.google.com/storage/docs/lifecycle
4. IBM cloud object storage - overview, https://www.ibm.com/cloud/object-storage
5. Regulation (eu) 2016/679 of the European parliament and of the council (2020). Accessed June 2021. https://gdpr.eu/tag/gdpr/
6. Qcow (2022). https://en.wikipedia.org/wiki/Qcow
7. AAA (computer security) (2023). https://en.wikipedia.org/wiki/AAA_(computer_security)
8. Amazon: Aws s3 (2020). Accessed Aug 2020. https://aws.amazon.com/s3/
9. Ataullah, A.A., Aboulnaga, A., Tompa, F.W.: Records retention in relational database systems. In: Proceedings of the 17th ACM Conference on Information and Knowledge Management, pp. 873–882 (2008)
10. Carrier, B.: The sleuth kit (2011). http://www.sleuthkit.org/sleuthkit/
11. Chow, J., Pfaff, B., Garfinkel, T., Rosenblum, M.: Shredding your garbage: Reducing data lifetime through secure deallocation. In: USENIX Security Symposium, pp. 22–22 (2005)
12. Garfinkel, S.L., Shelat, A.: Remembrance of data passed: a study of disk sanitization practices. IEEE Secur. Priv. **99**(1), 17–27 (2003)
13. Gutmann, P.: Secure deletion of data from magnetic and solid-state memory. In: Proceedings of the Sixth USENIX Security Symposium, vol. 14, pp. 77–89. San Jose, CA (1996)
14. Hsu, W.W., Smith, A.J., Young, H.C.: Characteristics of production database workloads and the TPC benchmarks. IBM Syst. J. **40**(3), 781–802 (2001)
15. International Data Sanitization Consortium: Data sanitization terminology and definitions (2017). Accessed Feb 2021. https://www.datasanitization.org/data-sanitization-terminology/
16. Kamara, S., Lauter, K.: Cryptographic cloud storage. In: Sion, R., et al. (eds.) FC 2010. LNCS, vol. 6054, pp. 136–149. Springer, Heidelberg (2010). https://doi.org/10.1007/978-3-642-14992-4_13
17. KVM. https://www.linux-kvm.org/page/Main_Page
18. Lenard, B., Rasin, A., Scope, N., Wagner, J.: What is lurking in your backups? In: Jøsang, A., Futcher, L., Hagen, J. (eds.) SEC 2021. IAICT, vol. 625, pp. 401–415. Springer, Cham (2021). https://doi.org/10.1007/978-3-030-78120-0_26
19. Lenard, B., Wagner, J., Rasin, A., Grier, J.: SysGen: system state corpus generator. In: Proceedings of the 15th International Conference on Availability, Reliability and Security, pp. 1–6 (2020)
20. National Institute of Standards and Technology: Guidelines for media sanitization (2006)
21. National Security Agency Central Security Service: NSA/CSS storage sanitization manual (2014)
22. Reardon, J., Basin, D., Capkun, S.: Sok: secure data deletion. In: 2013 IEEE Symposium on Security And Privacy, pp. 301–315. IEEE (2013)
23. Reardon, J., Capkun, S., Basin, D.: Data node encrypted file system: efficient secure deletion for flash memory. In: Proceedings of the 21st USENIX Conference on Security symposium, pp. 17–17. USENIX Association (2012)

24. Richard III, G.G., Roussev, V.: Scalpel: a frugal, high performance file carver. In: DFRWS. Citeseer (2005)
25. Scope, N., Rasin, A., Lenard, B., Heart, K., Wagner, J.: Harmonizing privacy regarding data retention and purging. In: Proceedings of the 34th International Conference on Scientific and Statistical Database Management, pp. 1–12 (2022)
26. Scope, N., Rasin, A., Lenard, B., Wagner, J., Heart, K.: Purging compliance from database backups by encryption. J. Data Intell. **3**(1), 149–168 (2022)
27. Scope, N., Rasin, A., Wagner, J., Lenard, B., Heart, K.: Purging data from backups by encryption. In: Strauss, C., Kotsis, G., Tjoa, A.M., Khalil, I. (eds.) DEXA 2021. LNCS, vol. 12923, pp. 245–258. Springer, Cham (2021). https://doi.org/10.1007/978-3-030-86472-9_23
28. SQLite: PRAGMA statements (2018). https://www.sqlite.org/pragma.html#pragma_secure_delete
29. Stahlberg, P., Miklau, G., Levine, B.N.: Threats to privacy in the forensic analysis of database systems. In: Proceedings of the 2007 ACM SIGMOD International Conference on Management of Data, pp. 91–102. ACM, Citeseer (2007)
30. The Office of the National Coordinator for Health Information Technology: State medical record laws: Minimum medical record retention periods for records held by medical doctors and hospitals (2022)
31. United States Congress: 28 U.S. code §1732 (1948). https://www.law.cornell.edu/uscode/text/28/1732
32. U.S. Internal Revenue Service: Media sanitization methods (2017). https://www.irs.gov/privacy-disclosure/media-sanitization-methods
33. Vliet, J.V., Paganelli, F., Geurtsen, J.: (2012). https://docs.aws.amazon.com/aws-backup/latest/devguide/deleting-backups.html
34. Wagner, J., Rasin, A., Grier, J.: Database forensic analysis through internal structure carving. Digit. Investig. **14**, S106–S115 (2015)
35. Wagner, J., Rasin, A., Grier, J.: Database image content explorer: carving data that does not officially exist. Digit. Investig. **18**, S97–S107 (2016)
36. Wagner, J., Rasin, A., Heart, K., Malik, T., Grier, J.: Df-toolkit: interacting with low-level database storage. Proc. VLDB Endowment **13**(12) (2020)

A Real-Time Parallel Information Processing Method for Signal Sorting

Xiaofang Liu, Chaoyang Wang[✉], and Xing Fan

Wuhan Digital Engineering Institute, Wuhan, China

Abstract. With the increasing complexity of electronic countermeasures, the sorting and identification of radar signals have become an important part of the information processing system. To meet the requirements of shipborne information infrastructure, combined with the research of maritime information systems, we propose a multi-computer information processing architecture. Further, based on a timestamp-based parallel processing scheduling strategy and task priority processing method, we design a real-time parallel signal scheduling algorithm for this architecture. When the single node of the shipboard information processing platform is insufficient in computing resources, this algorithm coordinates the idle resources from adjacent nodes, which can solve the large-scale signal sorting synchronization problem. Experiment results show the effectiveness of our method.

Keywords: electronic countermeasures · signal sorting · parallel processing · task coordination · multi-computer architecture

1 Introduction

Radar pulse sorting refers to the process of separating various radar pulse trains from randomly staggered pulse streams [6]. Nowadays, in the domain of information processing system (IPS for short), the signal sorting ability has become one of the fundamental signs of whether a reconnaissance system can adapt to the modern electronic warfare signal environment. For the wide use of new system radars, the radar, jamming signal styles, jamming, and anti-jamming measures are becoming increasingly diverse.

Because of the complex electromagnetic environment, signal sorting faces the following key problems: Firstly, the number of radiation sources and the intensity of signals in the time, space and frequency range have increased sharply [3,5]. Secondly, the radar signal modulation is complex and the modulation parameter conversion is flexible [8]. Thirdly, the signal recognition is required to be accurate and real-time [1]. Combined with the research of maritime information systems, we propose a multi-computer information processing architecture (McIP for short). Moreover, based on a timestamp-based parallel processing scheduling strategy and task priority processing method, we design a real-time parallel scheduling signal sorting algorithm (PS^3 for short) for this architecture.

C. Strauss et al. (Eds.): DEXA 2023, LNCS 14146, pp. 298–303, 2023.
https://doi.org/10.1007/978-3-031-39847-6_21

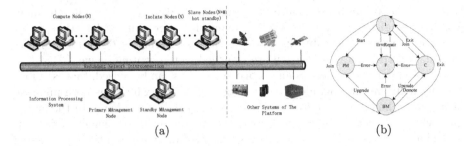

Fig. 1. (a) Architecture of McIPS; (b) Node State Transition of McIPS.

2 Multi-computer Information Processing Architecture

2.1 McIP Architecture Operation Mechanism

By utilizing the information infrastructure in the platform to support multiprocessors to complete information processing, the first problem is to determine the architecture and working mechanism of the multi-computer system, that is, to determine the functional allocation and relationship between modules on different nodes, as well as the node interconnection mode and topology. The architecture of the dynamically configurable and highly available multi-computer information processing system designed in this paper is shown in Fig. 1(a).

On each node, the working state is divided into four types: primary management node state (PM), standby management node state (BM), computing node state (C) and isolated node state (I). While the non-working state is the fault node state (F). The working state of all the nodes is not fixed, and can be dynamically migrated during operation. The node state transition in the above high availability information processing system is shown in Fig. 1 (b). The transition of node state is a dynamic reconstruction process for the whole information processing system.

2.2 McIP Architecture Design

Based on the above analysis, we design a dynamically configurable and highly available multi-computer intelligence processing architecture that meets the requirements of shipborne information infrastructure. Its architecture is shown in Fig. 1(a) that mainly includes:

Management Nodes. They are used to divide the tasks to be processed into multiple subtasks. Based on the processing data volume and priority of each subtask, as well as the task stock, processing efficiency and subtask execution time of each processing node, management nodes allocate each subtask to the processing node that can meet the delay requirements for parallel processing.

Processing Nodes. They are used to detect the conflicts of the subtasks after receiving the subtasks. The conflicting subtasks enter the blocking queue and wait for the preset time before entering the waiting queue, while the non-conflicting subtasks enter the waiting queue for synchronous detection in the order of priority. The subtasks that pass the synchronous detection are processed, and the subtasks that fail are returned to the waiting queue.

Above management nodes and processing nodes can be reused, that is, a processing node can be upgraded to a management node, which has both management and scheduling functions while completing data processing. In the actual project case, we use four servers, including one main management node (this node has both management and processing node functions) and three processing nodes. Moreover, McIP uses COTS (Commercial Off The Shelf [2]) components to construct the highly available infrastructure.

3   Parallel Scheduling Signal Sorting Algorithm

3.1   PS³ Sorting and Scheduling Process

Some technical definitions and conventions to be followed for this algorithm include: 1) The computing nodes of IPS in the server are connected through a high-speed bus. 2) The computing nodes of the ship's information processing infrastructure can only run one task at the same time. After the task is submitted to the computing node for operation, the demand of the task on the computing node will not change within a computing cycle. 3) A task can only be started when all its predecessor tasks have been completed and the required data has arrived. Once the task starts, it cannot be interrupted.

Following the definitions and conventions, we propose a real-time parallel scheduling signal sorting algorithm. This algorithm divides the task to be processed into multiple subtasks. The processing node will detect the conflict of the subtask, and the conflicting subtask will enter the blocking queue and wait for the preset time before entering the waiting queue, while the non-conflicting subtasks will enter the waiting queue for synchronous detection in the order of priority. The subtasks that pass the synchronous detection are processed, and the subtasks that fail are returned to the waiting queue. As shown in Fig. 2(a), the system environment for this algorithm is composed of a management node and several processing nodes, on which the task scheduler runs, the specific steps of the PS³ algorithm are described as follows:

When the management node receives task A be processed, the first step is to allocate the task. Firstly, the scheduler on the management node determines the priority of task A. Secondly, divide task A into subtasks Ai, where $A = \sum Ai$. Finally, calculate the execution time TAi_m of the subtask Ai,

$$TAi_m = SAi_m \times \frac{\omega}{n_m} \sum_{q_m=1}^{n_m} \frac{TAq_m}{SAq_m} \tag{1}$$

where TAi_m is the processing time of the subtask Ai on the processing node m, SAi_m is the processing data amount of the subtask Ai on the processing node m, n_m is the number of subtasks successfully submitted on the processing node

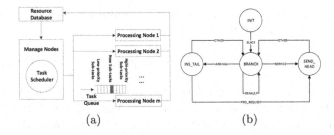

Fig. 2. (a) Timestamp-based Real-time PS^3; (b) Simulation Model.

m, TAq_m is the execution time of the q_m-th subtask on the processing node m, SAq_m is the processing data amount of the q_m-th subtask on the processing node m, and ω is the processing efficiency of the parallel processing of the processing node.

The second step is to verify the task scheduling ability according to the priority and processing time of the subtasks. Firstly, determine the priority of the subtasks that have been successfully submitted to the processing node m based on the priority of the task A. Set Uq_m as the priority of subtask ($Uq_m = 1, 2, \ldots, \rho$), where $Uq_m = 1$ and $Uq_m = \rho$ indicate the highest and the lowest priority respectively. Then, calculate the processing waiting time Wai_m of the subtask Ai,

$$WAi_m = \frac{\sum_{Uq_m=1}^{\rho}\left(\gamma_{Uq_m}\sigma_m^2 + \gamma_{Uq_m}\sum_{q_m=1}^{n_m}P_{q_m}TAq_m\right)}{2\left(1 - \sum_{Uq_m=1}^{\rho-1}\gamma_{Uq_m}\sum_{q_m=1}^{n_m}P_{q_m}TAq_m\right)\left(1 - \sum_{Uq_m=1}^{\rho}\gamma_{Uq_m}\sum_{q_m=1}^{n_m}P_{q_m}TAq_m\right)} \tag{2}$$

where Wai_m is the processing waiting time of the subtask Ai on the processing node m, γ_{Uq_m} is the arrival interval of the subtask with priority Uq_m on the processing node m, δ is the variance of the average processing time of the subtask on the processing node m, and Pq_m is the execution probability corresponding to the q_m-th subtask on the processing node m ($\sum_{q_m=1}^{n_m}P_{q_m} = 1$). At the same time, calculate the total processing delay FAi_m of the subtask Ai on the processing node m,

$$FAi_m = \text{Max}\left(TAi_m + WAi_m\right) + T_C \tag{3}$$

where T_C is the scheduling time of the task. In the actual system, according to the statistical results, the scheduling time of the same scheduler is basically the same when scheduling a large number of information tasks. Here we set T_C as a constant value. Finally, judge whether $FAi_m \leq TDAi - TBAi$ is true ($TDAi$ is the processing deadline of the subtask Ai, and $TBAi$ is the arrival time of the subtask Ai). If it is not, return to the previous step to redivide the task A. Otherwise, assign the subtask Ai to the processing node m.

The last step is to send the subtask Ai to the processing node m to execute the task. Firstly, the data conflict detection is performed on the subtask. The new subtask Ai that arrives at the processing node m is denoted as Ai_m, while

Am_0 denotes the subtask that has entered the waiting queue. If the priority of Am_0 is lower than Ai_m and meets,

$$TRAi_m \bigcap WA_{m0} \neq \emptyset \ or \ WAi_m \bigcap TRA_{m0} \neq \emptyset \ or \ WAi_m \bigcap WA_{m0} \neq \emptyset \quad (4)$$

Then the subtask Ai_m enters the blocking queue and assigns the value of Pam_0 to Pai_m. If the priority of Am_0 is higher than Ai_m and meets Eq. 4, then the subtask Ai_m enters the blocking queue. Otherwise, the subtasks Ai_m will enter the waiting queue and be sorted from the highest to the lowest priority. Where $TRAm_0$ is the estimated execution time of the subtask Am_0 when there is no conflict, WAm_0 indicates the processing waiting time of the subtask Am_0 on the processing node m, Wai_m is the processing waiting time of the subtask Ai on the processing node m, $TRAi_m$ is the estimated execution time of the subtask Ai_m when there is no conflict, PAm_0 and PAi_m are priorities for Am_0 and Ai_m respectively. Then, perform synchronous detection on the subtasks Ai_m that is wait for the queue head. If Ai_m meets,

$$TDAi_m - TBAi_m \geq \mathrm{Max}\,(TAi_m + WAi_m) + T_C \quad (5)$$

Then the subtask Ai_m enters the execution queue and completes data processing. Otherwise, the subtask Ai_m returns to the waiting queue. Where Wai_m is the processing waiting time of the subtask Ai on the processing node m, Tai_m is the execution time of the subtask Ai on the processing node m, T_C is the scheduling time of the task, $TDAi_m$ is the processing deadline of the subtask Ai on the processing node m, $TBAi_m$ is the arrival time of the subtask Ai on the processing node m.

4   Simulation Experiments

In this section, we compare our algorithm with the classical scheduling algorithms: the synchronous scheduling algorithm (Synchronous for short) [4] and the adaptive highest priority first policy (AHPFP for short) algorithm [7]. For simplified calculation, we assume that each node has the same processing capacity in the test. Figure 2(b) shows the process model for processing the task queue in the node model.

Based on this simulation model, we conducted sufficient experiments. As shown in Fig. 3(a), when the system is at the normal load, i.e., task arrival ratio is no more than 24 trans/sec, our algorithm and AHPFP algorithm perform better. With the increase of system load, the miss ratio of AHPFP algorithm begins to be higher than our algorithm. The performance degradation of AHPFP algorithm may be due to the high system load resulting in the high ratio of task conflicts and restarts. While our algorithm abandons the tasks that have a low complication possibility in advance, which can avoid the waste of resources.

Figure 3(b) shows that these three algorithms all face performance degradation when the system exits a service data skew problem. This is because the

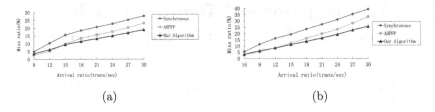

Fig. 3. (a) Processing Task "Arrival ratio-Miss ratio" (no data skew); (b) Processing Task "Arrival ratio-Miss ratio" (20% data skew).

bottleneck of data skew restricts the improvement of the overall system. However, our algorithm adopts the coordination method of timestamp, which can not only ensure the correct control of the concurrent execution of subtasks, but also avoid increasing the communication cost, so its performance is still better than the other two algorithms.

5   Conclusion

Based on a timestamp-based parallel processing scheduling strategy, and introduces the processing method of task priority, we proposes a method for information processing using a multi-computer system, which well solves the synchronization and scalability problems of signal processing. In the future, machine learning methods such as reinforcement learning can be used to judge the expected execution results more accurately, and thus improve our method.

References

1. Ali, A., Yangyu, F.: Unsupervised feature learning and automatic modulation classification using deep learning model. Phys. Commun. **25**, 75–84 (2017)
2. Brownsword, L., Carney, D., Oberndorf, T.: The opportunities and complexities of applying commercial-off-the-shelf components. Crosstalk **11**(4), 4–6 (1998)
3. Kishore, T.R., Rao, K.D.: Automatic intrapulse modulation classification of advanced LPI radar waveforms. IEEE Trans. Aerosp. Electron. Syst. **53**(2), 901–914 (2017)
4. Li, K., Ganesan, V., Sivakumar, A.: Synchronized scheduling of assembly and multi-destination air-transportation in a consumer electronics supply chain. Int. J. Prod. Res. **43**(13), 2671–2685 (2005)
5. Li, X.: Research on radar emitter sorting and recognition technology based on machine learning (in chinese) (2020)
6. Shakor, M.: Scheduling and synchronization algorithms in operating system: a survey. J. Stud. Sci. Eng. **1**, 1–16 (2021)
7. Sonia, T., Dandamudi, S.: An adaptive scheduling policy for real-time parallel database systems. In: International Conference on Massively Parallel Computing Systems, pp. 8–19 (1998)
8. Wang, X.: Electronic radar signal recognition based on wavelet transform and convolution neural network. Alex. Eng. J. **61**(5), 3559–3569 (2022)

Learning Optimal Tree-Based Index Placement for Autonomous Database

Xiaoyue Feng, Tianzhe Jiao, Chaopeng Guo, and Song Jie[(✉)]

Northeastern University, Shenyang, China
{neufengxiaoyue,jiaotianzhe}@stumail.neu.edu.cn,
guochaopeng@swc.neu.edu.cn, songjie@mail.neu.edu.cn

Abstract. The distributed tree-based index has been widely adopted to process queries on large-scale data. For the same index, the placement of the index determines the query performance. Current research on index placement is based on predefined rules, such as user-supplied functions, load balance, and data location. However, they cannot retain high performance since they fail to adapt their policies to the user's query habit. In this paper, we propose a method to learn the optimal index placement (LiP) for different workloads to solve the problem. Our method improves the query efficiency by evaluating the locality but quantifying the locality is challenging. Experiments prove that our approach reduces the average query time by 6% compared to previous methods.

Keywords: Distributed index · Index placement · DRL · Locality · Autonomous database

1 Introduction

Nowadays, with the wide application of distributed database systems, the distributed tree-based index has been widely adopted to process queries for large-scale data due to its good scalability, stable performance, and strong cache locality [1,2]. For the same index, the placement of the index determines the query performance. For instance, if the block of index and query data are on the same server, it is local access; otherwise, it is remote access that needs more access time. Index placement today is a hot research topic in distributed index management.

Current researches on the distributed tree-based index are to distribute the tree nodes to different servers [3,4]. Most of them place the nodes according to a predefined rule, such as user-supplied function [5], load balance [6,7], and data location [8–10]. The first means that the customized function places the tree nodes on servers. The second distributes tree nodes to data servers evenly. The third ensures index and index attributes are on the same server. However, they cannot retain high performance when the workload changes since they fail to adapt their policies to the user's query habit. The essence is that it cannot consider the locality comprehensively. Locality refers to the distance between

C. Strauss et al. (Eds.): DEXA 2023, LNCS 14146, pp. 304–309, 2023.
https://doi.org/10.1007/978-3-031-39847-6_22

the position of an index and corresponding data. Finding the optimal locality is challenging since it relies on various queries-related factors, such as the target attributes of queries, join tables of queries, condition attributes of queries, query frequency, and so on.

To solve the problem, we propose a method to learn the optimal index placement (LiP). We leverage deep reinforcement learning (DRL) [11] to address the complexity and dynamics of queries. Designing the reward function that fits the objective of the DRL problem is a challenge. In our method, we adopt the query cost to reflect the effect of index placement. We use the What-if tool [12] to estimate the query processing cost, which was used to design the reward function. Our method has three contributions:

(1) We define the distributed tree-based index placement problem and solve it based on DRL technology that improves query efficiency.
(2) We propose a machine learning-based index tuning method to improve the query performance for autonomous databases.
(3) In this paper, we discuss the tree-based index placement and use the machine learning-based method to solve the problem. It is a reference function for future research on other index structures with machine learning methods.

2   Methodology

In this section, we will introduce our LiP method, which is a two-step method, including the learning step and the placing step. In the learning step, We adopt the DRL to select a response server to process every query request. In the placing step, we construct the tree nodes in corresponding servers based on the result of the learning step. The learning step is a core part of our LiP method, and we will explain it in detail.

2.1   DRL Framework Overview

This section discusses the DRL framework in the learning step of LiP. As shown in Fig. 1, our framework consists of several modules, including input, agent, environment, and cost estimation. In our case, the query, attributes in a query, data location, and server layout are inputs to the DRL framework. The data location is that the data is placed on which servers. The server layout refers to the number and location of servers.

The DRL agent takes the responsibility of selecting the response server for processing every query request. It learns to find the optimal response server for each query by interacting with the environment. Before training the DRL model, the environment interacts with the cost estimation module and obtains the query processing cost to compute the reward. This module receives the response server for each query as input. Then, it estimates the query cost of each query using the What-if optimizer tool, and we design the reward function according to the result. Then, the environment calculates the reward by the reward function as the average query cost of the workload in the training process. The output of our DRL framework is the optimal response server for each query.

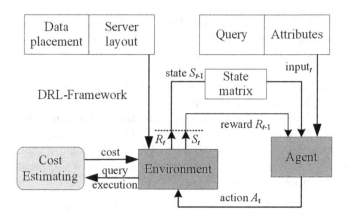

Fig. 1. The DRL framework of LiP

2.2   DRL Components

Environment. The components of the environment include states, actions, and a reward function.

- State Representation. Queries and their response servers represent the state. We represent the state by a matrix of size $r \times m$, where r is the number of queries and m is the number of data servers. In the state matrix, an entry is set to 1 if the optimal response server for query $q_i (i \in [1, r])$ is $s_j (j \in [1, m])$; otherwise, it is set to 0. For instance, we have two queries q_1, q_2 and two servers s_1, s_2, where s_1 is the response server of q_1 and s_2 is the response server of q_2, then the state matrix will be the following:

$$StateMatrix = \begin{pmatrix} 1 & 0 \\ 0 & 1 \end{pmatrix} \qquad (1)$$

- Set of actions. An action changes a response server for a query. The action is represented by the query number and the server number. For example, if we want to change the response server of q_1 to s_2, the action is represented by $[1, 2]$.
- Reward function. We design the reward function with two objectives: average access cost and load balance. The reward function is described as follows: first, we compute the average access cost of the workload using Eq. (2), where Q is the set of queries, q is a query, n denotes the number of queries, AC is the access cost of query q, and AAC is the average access cost of the workload Q; second, the reward value is computed by Eq. (3), where r is the reward value, var is the variance of the set of numbers which represents the number of queries assigned to every data server.

$$AAC(Q, n) = \frac{\sum_{q \in Q} AC(q)}{n} \qquad (2)$$

$$r = 100 \times \tanh\left(10 \times AAC(Q,n)\right) \times \tanh\left(\frac{10}{a^{var(q,s)}}\right) \qquad (3)$$

Agent. The agent is responsible for learning the optimal response server for each query in the query workload by maximizing the reward function. It follows the off-policy deep Q-learning algorithm [13,14] to predict the next action a and decides how to adjust the response data servers for each query according to the reward. It chooses the action with the highest reward value in its current state at 80% probability and chooses a random action at 20% probability. The agent acts in a predefined number of episodes. An episode starts with an initial state S_0 where each query corresponds to a random data server, and the agent selects an action at each time step. In our work, an episode ends when the provided condition is achieved.

3 Experiment

3.1 Setup

We use the database schema and query workload from the standard TPC-H benchmark [15]. The training workload consists of 500 queries. Also, we generated 100 queries to test the trained agent and 5000 queries to validate LiP. Our experiments set four data servers and a master server. The database system is PostgreSQL on each data server. LiP and PostgreSQL run on Ubuntu Linux 18.04 LTS with 128 GB of DDR4 main memory, and two Intel Xeon Silver 4114 10-core CPUs. We compare the average query time of LiP with Baseline and CG-index methods in our experiment. The Baseline is to distribute the tree nodes to data servers randomly and evenly. CG-index ensures index and index attributes are on the same server.

3.2 Results

Our experiment evaluates the performance of the recommended nodes placement by LiP in comparison with baseline and CG-index. Our LiP method reduces the average query time by 6% with a learning rate of 0.0005 compared to the baseline. We calculate the average growth percentage across all the number of queries at 3G data volumes as the improvement percentage. In our experiment, we changed the learning rate in DRL for a more detailed comparison. We set the learning rate at 0.0001, 0.0005, and 0.001, respectively. We compare the average query time for different data volumes and queries. As Fig. 2 shows, LiP with different learning rates performs better than the baseline and CG-index for different data volumes. And the performance is the best when the learning rate is 0.0005. For the baseline method, the average query time increases slightly decreases with the number of queries, because the load on the data server increases as the number of queries increases. For the small number of queries, all approaches show similar behavior. As the number of queries increases, LiP shows a significant improvement over baseline and CG-index because the influence of locality becomes more pronounced as the number of queries increases.

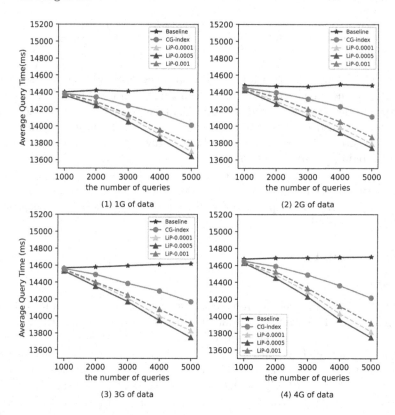

Fig. 2. Average query time of the workload for three methods.

4   Conclusion

In this paper, we introduce a novel method to learn distributed index placement through DRL, which improves query performance. The main idea is that the DRL agent learns its decisions based on the experiences by monitoring the rewards via trying different actions. We show that our approach can select the well-performed distributed index placement via experiments. In the future, we plan to extend the LiP method to solve other index-tuning issues.

References

1. Mol, R.D., Barranco, C.D., Tré, G.D.: Indexing possibilistic numerical data using interval B+-trees. Fuzzy Sets Syst. **413**, 138–154 (2020)
2. Zhang, W., Yan, Z., Lin, Y., et al.: A high throughput B+tree for SIMD architectures. IEEE Trans. Parallel and Distrib. Syst. **31**(3), 707–720 (2019)
3. Ziegler, T., Vani, S.T., Binnig, C., et al.: Designing distributed tree-based index structures for fast RDMA-capable networks. In: Proceedings of the 2019 International Conference on Management of Data, pp. 741–758 (2019)

4. Huang, B., Yuxing, P.: An efficient distributed B-tree index method in cloud computing. Open Cybern. Syst. J. **8**(1), 302–308 (2014)
5. Aguilera, M.K., Golab, W., Shah, M.A.: A practical scalable distributed B-tree. Proc. VLDB Endowment **1**(1), 598–609 (2008)
6. Sowell, B., Golab, W., Shah, M.A.: Minuet: a scalable distributed multiversion B-tree. Proc. VLDB Endowment. **5**(9), 884–895 (2012)
7. Bochmann, G.V., Asaduzzaman, S.: Distributed B-tree with weak consistency. In: Gramoli, V., Guerraoui, R. (eds.) NETYS 2013. LNCS, vol. 7853, pp. 159–174. Springer, Heidelberg (2013). https://doi.org/10.1007/978-3-642-40148-0_12
8. Wu, S., et al.: Efficient B-tree based indexing for cloud data processing. Proc. VLDB Endowment **3**(1–2), 1207–1218 (2010)
9. Zhou, W., et al.: SNB-index: a SkipNet and B plus tree based auxiliary Cloud index. Cluster Comput. **17**, 453–462 (2014)
10. Singh, H., Bawa, S., et al.: A MapReduce-based scalable discovery and indexing of structured big data. Future Gener. Comput. Syst. **73**, 32–43 (2017)
11. Du, W., Ding, S.: A survey on multi-agent deep reinforcement learning: from the perspective of challenges and applications. Artif. Intell. Rev. **54**, 1–24 (2020)
12. Chaudhuri, S., Narasayya, V.: AutoAdmin "what-if" index analysis utility. ACM SIGMOD Rec. **27**(2), 367–378 (1998)
13. Lin, J., Li, Y.Y., Song, H.B.: Semiconductor final testing scheduling using Q-learning based hyper-heuristic. Expert Syst. Appl. **187**, 115978 (2022)
14. Ahmed, F., Cho, H.S.: A time-slotted data gathering medium access control protocol using Q-learning for underwater acoustic sensor networks. IEEE Access. **9**, 48742–48752 (2021)
15. TPC: TPC-H benchmark. http://www.tpc.org/tpch/

Social Links Enhanced Microblog Sentiment Analysis: Integrating Link Prediction and Sentiment Connection Weights

Xiaomei Zou[1], Taihao Li[1(✉)], and Jing Yang[2]

[1] Zhejiang Lab, Hangzhou, Zhejiang, China
{zouxiaomei,lith}@zhejianglab.edu.cn
[2] School of Computer Science and Technology, Harbin Engineering University, Harbin, Heilongjiang, China
yangjing@hrbeu.edu.cn

Abstract. The emerging microblogging service provides a new channel for people to share opinions and sentiment. As a result, microblog sentiment analysis has become a cutting-edge and popular research field, which has many important applications. Existing methods mostly extract sophisticated features from microblog texts without considering that microblogs are networked data, which suffer from poor performance. To address this issue, we propose a new model that assumes microblogs are interconnected and that connected microblogs are more likely to share the same sentiment. We leverage two types of information to model the connections between microblogs: user information and friend information. Our assumption is supported by two sociological theories: sentiment consistency and emotional contagion. The connections between microblogs based on user and friend information are often sparse and noisy, which can limit the effectiveness of sentiment analysis. To mitigate this issue, we use link prediction to identify potential connections between microblogs and introduce a sentiment connection weights matrix to quantify the degree of sentiment difference between connected microblogs. We then integrate potential social links and sentiment connection weights into our content-based sentiment model using a Laplacian regularization term. To demonstrate the effectiveness, sufficient experiments are conducted on two real datasets to show that exploring potential links and introducing sentiment connection weights can improve the performance of microblog sentiment analysis significantly.

Keywords: Microblog sentiment analysis · Social information · Link prediction

1 Introduction

Microblogging platforms such as Sina Weibo and Twitter are increasingly popular as user-centered websites in the era of Web 2.0. In contrast to traditional blogs

C. Strauss et al. (Eds.): DEXA 2023, LNCS 14146, pp. 310–325, 2023.
https://doi.org/10.1007/978-3-031-39847-6_23

or news platforms where users can only browse pre-edited content, microblogging platforms encourage users to actively participate in content creation. This has led to the posting of numerous texts related to users' personal lives in a fast and convenient way. Additionally, these platforms offer various interaction mechanisms, including following, commenting, liking, and reposting, which further enhances their popularity. Take the Chinese leading online social network Sina Weibo as an example, Sina Weibo had an average of 252 million daily active users in June 2022. As a result, microblogging websites have become a huge resource bank. Microblog sentiment analysis has become a hotspot from academia to industry in recent years as its great significance and wide range of applications in recommendation systems [1], stock prediction [2], customer relation management [3], crisis management [4] and so on.

Existing methods of microblog sentiment analysis can be divided into lexicon-based methods and machine learning methods. In lexicon-based methods, sentiment lexicons such as SentiWordNet or SenticNet [5,6] are utilized to assign sentiment scores to words. In machine learning methods, microblogs are represented as feature vectors, then machine learning methods such as Recurrent Neural Networks are applied to classify microblogs. However, these two methods usually have low effectiveness as they exploit microblog content information only. Microblogs are very noisy and short, they only contain one or two sentences. Besides, people have various expression styles, abbreviations and repeated words such as 'lol' and 'cooooool' usually occur in microblog content. These inherent characteristics of microblogs hinder the extraction of effective features of content for sentiment analysis, resulting in suboptimal performance.

Microblogs lie in the environment of online social networks. In this environment, users are influenced by others (Social Influence [7]), and they tend to follow others who share similar interests or opinions (Homophily [8]). This phenomenon means that microblogs posted by connected users tend to share similar sentiment labels, and it is called emotional contagion [9]. The sentiment of a person about a certain topic is usually consistent within a period. Reflecting this information on users' behavior, it means that microblogs posted by the same user also have consistency in their sentiment labels. This phenomenon is called sentiment consistency [10]. Motivated by these observations, some researchers study how to model and incorporate these latent connections between microblogs with content information for microblog sentiment analysis [11,15,16]. However, the social connections between microblogs in these methods are usually very sparse. Besides, they fail to extract more precise social contexts. Therefore, they have limited improvement in sentiment analysis results. To alleviate these problems, a novel approach is proposed to expand and refine connections between microblogs in this paper. Specifically, an original microblog graph is constructed on our two theoretical bases mentioned above: sentiment consistency and emotional contagion. Then the connections of the original microblog graph are extended by link prediction methods which can predict the probability of a new link between unconnected nodes. To better model the sentiment coherence in the microblog graph, this paper also introduces the sentiment connection weights of microblogs. These approaches guarantee that we can get more accurate representations of social contexts and extract the potential sentiment connections between microblogs fully. At last, social contexts and microblog content information are integrated into a unified sentiment analysis framework by a Laplacian regularization term. The main contributions of this paper include:

1. A microblog graph is constructed based on the theory of sentiment consistency and emotional contagion. Both statistical methods and mathematical methods are used to verify these two theories in online social networks.
2. Link prediction methods are applied to alleviate the sparse problem of microblog connections and sentiment connection weights between microblogs are proposed to remove noises from social contexts.
3. Experiments are conducted on two public datasets and show that exploiting link prediction and sentiment weights can lead to statistically significant improvements in microblog sentiment analysis.

2   Related Work

Lexicon-based methods and machine learning methods are two commonly used approaches in microblog sentiment analysis. Lexicon-based methods used lexicons as knowledge to classify texts. Usually, they use a sentiment lexicon for polarity prediction through matching words in the text and their associated sentiment. Many scientists are working to build better lexicons to improve sentiment analysis performance. [17] proposed an unsupervised microblog sentiment analysis method by constructing a lexicon that was represented as a co-occurrence graph. [18] exploited word emoticons relationships, word sentiment relationships, and existing sentiment lexicons to get the new lexicon for Chinese microblog sentiment analysis. Besides, an enhanced mutual information based data-driven method was used to catch new words in microblogs. [19] built a novel cognitive-inspired approach that explored the wrongly predicted texts for new sentiment words detection and polarity score assignment. There are also many machine learning methods. Text context information was exploited in [20]. They used microblogs in the same conversation as external information. A hierarchical structure of the Long Short-Term Memory network and two attention layers were used to model word-level vectors and microblog-level vectors. Some methods combined both. [21] represented the lexicon SenticNet into continuously distributed vectors and used them as features for microblog sentiment analysis. In [22], the authors proposed a convolutional neural network framework with lexicon embeddings to analyze sentiment. However, all these methods focus on how to extract delicate text content features from microblogs and ignore the widely existing social contexts in microblogging platforms.

Social contexts are explored in many social network analysis tasks such as recommendation systems [12], information retrieval [13], and fake news detection [14]. Recently, some researchers study the correlation between social contexts and microblog sentiment and how to integrate social contexts with content-based sentiment analysis methods. [15] exploited user-user relationships and user-microblog relationships to model the internal sentiment connection between microblogs. [16] explored social contexts at the label prediction stage to regularize and refine sentiment labels. [23] built a repost network between users and microblogs, they studied the influence of sentiment diffusion on the predicted microblog sentiment labels. [24] explored user information and proposed a personalized approach, which also considered friend information to make the personalized model weights of connected users as close as possible. [25] build an approval network between users based on homophily and constructuralism of social networks, then they took both the text content and network information into account for aspect-level sentiment analysis by an unsupervised probabilistic model. [26] argued that both positive and negative user interactions could contain useful signals for sentiment analysis and proposed an unsupervised approach to incorporate textual information and network information. However, these methods only consider direct relations between microblogs or users, which are usually very sparse and noisy. They fail to extract internal sentiment connections between microblogs fully and precisely.

3   Methodology

In this section, we propose our method of extending and refining social connections between microblogs on the basis of sentiment consistency and emotional contagion. In particular, we construct a microblog graph by these two theories, then link prediction methods are implemented on this graph to acquire potential connections between microblogs and alleviate the sparse problem of connections. Besides, we introduce a sentiment connection weights matrix to represent the sentiment difference degree between connected microblogs. A bigger element value indicates a lower sentiment difference and vice versa. The new microblog network and sentiment connection weights matrix are integrated into microblog sentiment analysis by Laplacian regularization.

3.1   Problem Statement

Since the sentiment labels of microblog texts in datasets are usually discrete, the sentiment analysis problem can be formalized as a classification problem. Each microblog and its corresponding label form a tuple $(\mathbf{x}, \mathbf{y})$. $\mathbf{x} \in R^{1 \times d}$ is a row vector and denotes the microblog content feature vector, where d represents the dimension of the feature space. Vector $\mathbf{y} \in R^{1 \times c}$ represents the sentiment label vector, where c denotes the number of sentiment classes. As the classification problem in this paper is polarity, $c = 2$. Then the entire training set can be denoted by $(\mathbf{X}, \mathbf{Y})$, where the feature vectors of all microblogs constitute the matrix $\mathbf{X} \in R^{n \times d}$, n is the number of microblogs. Similarly, $\mathbf{Y} \in R^{n \times c}$ represents the sentiment label matrix. The i-th row $\mathbf{x}^i$ of the matrix $\mathbf{X}$ represents the feature vector of the i-th microblog, and $\mathbf{y}^i$ is its corresponding label vector.

3.2   Modeling Text Information

In this subsection, we introduce how to model text information of microblogs. In a sentiment prediction problem, pre-processing is the first step. We use the unigram model with term presence and do not remove punctuation and stop words as they carry sentiment information. Different from formal texts, emoticons are a popular way for people to convey their nuanced meaning in online social networks. So emoticons are extracted as features by regular expressions. Usually, the problem of microblog sentiment analysis is represented as a loss minimization problem. A loss function $f(\mathbf{W}; \mathbf{x}^i, \mathbf{y}^i)$ is defined to quantify the difference between the prediction label $\hat{\mathbf{y}}^i = h_{\mathbf{W}}(\mathbf{x}^i)$ of a sentiment analysis model and its true sentiment label $\mathbf{y}^i$, where $\mathbf{W} \in R^{d \times c}$ represents the feature weight parameters, function $h_{\mathbf{W}}(\mathbf{x}^i)$ represents the sentiment analysis model. The loss function is also called the empirical risk. The optimal set of parameters $\mathbf{W}^*$ is desired to find such that:

$$\mathbf{W}^* = \arg \min_{\mathbf{W}} \sum_{i=1}^{n} f(\mathbf{W}; \mathbf{x}^i, \mathbf{y}^i)$$
$$= \arg \min_{\mathbf{W}} f(\mathbf{W}; \mathbf{X}, \mathbf{Y}) \tag{1}$$

The loss function can be squared losses, cross entropy, or other losses. In this paper, we choose the squared loss as an example, where $f(\mathbf{W}; \mathbf{X}, \mathbf{Y}) = ||h_{\mathbf{W}}(\mathbf{X}) - \mathbf{Y}||_F^2$. Besides, the classification model $h_{\mathbf{W}}(\mathbf{X})$ can be other classifiers such as convolutional neural networks or recurrent neural networks.

It usually leads to an overfitting problem by only minimizing the empirical risk. Besides, not all features are helpful for microblog sentiment analysis, so it is vital to select features that are beneficial to the task. To handle these two problems, Lasso is introduced which has been successfully applied in many text mining fields [27, 28]. If two models produce similar empirical risks, the "simpler" model should be selected. This characteristic can make for a more robust model and improve model performance. As a sparse model, Lasso can automatically select sentiment features and filter neutral ones. We also utilize the Lasso model in this paper, it can be represented by:

$$\mathbf{W}^* = \arg \min_{\mathbf{W}} f(\mathbf{W}; \mathbf{X}, \mathbf{Y}) + \beta ||\mathbf{W}||_1, \tag{2}$$

where β is a positive hyper-parameter which controls the contribution of the L_1 norm. To analyze the sentiment of an unseen microblog, we can compute:

$$\arg \max h_{\mathbf{W}}(\mathbf{X}) = \arg \max_{i \in P, N} \mathbf{x} \mathbf{w}_i \tag{3}$$

where $\mathbf{w}_i$ represents the i-th column of $\mathbf{W}$. As microblog sentiment analysis is regarded as a binary classification problem, the ground truth label of a positive microblog can be represented by $y = [+1, -1]$. For a negative microblog, its corresponding label is $y = [-1, +1]$. Using Eq. 3 to classify new microblogs, we provide a detailed description in this case:

$$\mathbf{y} = \begin{cases} +1 & \mathbf{xw}_1 > \mathbf{xw}_2 \\ -1 & \mathbf{xw}_1 < \mathbf{xw}_2 \\ +1 \quad or \quad -1 \quad randomly & \mathbf{xw}_1 = \mathbf{xw}_2 \end{cases} \quad (4)$$

3.3 Modeling Social Relations

Social relations between microblogs are explored in this section. In the environment of social networks, the sentiment of microblogs is highly related to their corresponding users. Besides, users are influenced by others, the influence brings changes in users' attitudes and sentiment. So in this paper, we take these two pieces of information into consideration: user information and friend information. To use user information for microblog sentiment analysis, we assume that the sentiment of a certain user holds similar in a period. As a result, the sentiment of their published texts is also similar in this period. This is supported by the sentiment consistency theory [10]. We construct a microblog-microblog matrix $\mathbf{A}_{sc} \in R^{n \times n}$ to extract user information from social networks. The $\mathbf{A}_{sc_{ij}}$ entry of the matrix represents whether there is a connection between the i-th microblog and the j-th microblog. In other words, it represents whether two microblogs are published by the same user. Given a set of users $\mathbf{u} = \{u_1, u_2, \ldots, u_m\}$, we can get a user-microblog matrix $\mathbf{U} \in R^{m \times n}$ in which rows represent different users, columns represent different microblogs, and the entry $\mathbf{U}_{ij} = 1$ if and only if user i posted microblog j. m denotes the number of users of the dataset. Then we use $\mathbf{U}$ to compute $\mathbf{A}_{sc}$ by $\mathbf{A}_{sc} = \mathbf{U}^T \times \mathbf{U}$.

Similarly, to integrate friend information with microblog sentiment analysis, we assume that users tend to hold the same opinion as their friends through the social influence process. This assumption can be supported by the theory of emotional contagion [9]. This phenomenon is called "birds of a feather flock together". A microblog-microblog matrix $\mathbf{A}_{ec} \in R^{n \times n}$ is defined to extract this information. Given the friendships between users represented by $\mathbf{F} \in R^{m \times m}$, $\mathbf{A}_{ec}$ can be constructed by $\mathbf{A}_{ec} = \mathbf{U}^T \times \mathbf{F} \times \mathbf{U}$. The entry $\mathbf{F}_{ij} = 1$ represents user i and user j are friends. Therefore, if an element of matrix $\mathbf{A}_{ec_{ij}} \neq 0$, the microblog i and microblog j are posted by friends.

After modeling user information and friend information, we integrate them by $\mathbf{A}_{dr} = \mathbf{A}_{sc} + \mathbf{A}_{ec}$ and get a microblog graph $\mathcal{G} = (\mathcal{V}, \mathcal{E})$ whose adjacency matrix is $\mathbf{A}_{dr}$, where $\mathcal{V}$ is the node set, each node in $\mathcal{V}$ represents a microblog. $\mathcal{E}$ represents the edge set. However, the adjacency matrix of this graph is usually very sparse. In addition, the sentiment correlation between microblogs is not considered. To alleviate the sparse problem, we utilize link prediction to discover the internal connections between microblogs. The structural or node properties are usually exploited to compute the similarity score between a node pair. To show the effectiveness of link prediction for microblog sentiment analysis, we use three popular link prediction methods: Adamic Adar algorithm (AA), Preferential Attachment algorithm (PA), Resource Allocation algorithm (RA). The three methods predict the potential relations between nodes by their structural information, which is very efficient.

Adamic Adar algorithm [29] is a method for measuring intimacy based on common neighbors between nodes. It considers the influence of each neighbor by its corresponding degree. The influence of a node with a lower degree should be greater than that of a node with a higher degree. Therefore, the weight values of different nodes can be represented by:

$$\mathbf{A}_{lp_{ij}} = \sum_{u \in N(i) \cap N(j)} \frac{1}{\log |N(u)|}, \tag{5}$$

where $|N(u)|$ represents the number of the neighbour set of node u. A higher value of $\mathbf{A}_{lp_{ij}}$ implies larger intimacy between two microblogs. When $\mathbf{A}_{lp_{ij}} = 0$, it means that nodes x and y are not close.

The Preferential Attachment algorithm has been presented by [30] on a scale-free network. Preferential Attachment is a metric used to calculate the tightness of nodes based on their shared neighbors. This method assumes that the possibility of a new edge is proportional to the degrees of its two corresponding nodes. So the problem can be formulated as:

$$\mathbf{A}_{lp_{ij}} = |N(i)| * |N(j)|, \tag{6}$$

where $|N(u)|$ represents the size of the neighbour set of node u. The same as in the Adamic Adar algorithm, a higher $\mathbf{A}_{lp_{ij}}$ value indicates that two nodes are closer.

The Resource Allocation algorithm is proposed by [31]. This method regards the common neighbors between two nodes as the medium of transmission. In the beginning, each node has a unit of resources. Given a node pair a and b that are unconnected, there are some resources needing to transmit from a to b and the resource of each node is divided equally to its neighbors. Then the whole resources of node b received from node a can be calculated by:

$$\mathbf{A}_{lp_{ij}} = \sum_{u \in N(i) \cap N(j)} \frac{1}{|N(u)|}, \tag{7}$$

After link prediction, the final relationship matrix of microblogs can be calculated by:

$$\mathbf{A} = a^1 * \mathbf{A}_{sc} + a^2 * \mathbf{A}_{ec} + a^2 * \mathbf{A}_{lp}. \tag{8}$$

where a^1, a^2, and a^3 are the weights which controls the computation $\mathbf{A}$. In this paper, al the weights are set to 1.

Sentiment correlation between microblogs is not considered in the above matrix A. Relations in matrix A may have some noises. To represent the sentiment relationship between microblogs and remove noises, a weight matrix C is established in this paper. The element of matrix $\mathbf{C}$ can be computed by:

$$\mathbf{C}_{ij} = \begin{cases} 1 - \dfrac{|\mathbf{y}_0^i - \mathbf{y}_0^j|}{2}, & \text{if } i \text{ and } j \text{ are connected,} \\ 0, & \text{if } i \text{ and } j \text{ are unconnected.} \end{cases} \tag{9}$$

where $\mathbf{y}_0^i$ represents the sentiment value of the i-th mciroblog. As what we are concerned about is a polarity classification problem in this paper, if the sentiment label of a microblog i is positive, $\mathbf{y}^i = [1, -1]$ and vice versa. Therefore, the range of $|\mathbf{y}_0^i - \mathbf{y}_0^j|$ is between $[0, 2]$, and the value of $\mathbf{C}_{ij}$ lies in $[0, 1]$. A high value of $\mathbf{C}_{ij}$ indicates a small difference between the sentiment of two microblogs. We integrate the sentiment correlation matrix $\mathbf{C}$ and the enhanced microblog matrix $\mathbf{A}$ by $\mathbf{M} = \mathbf{C} \circ \mathbf{A}$, then get the final microblog-microblog graph whose adjacency matrix is $\mathbf{M}$, where $\circ$ means the Hadamard product.

We model the relationships between social contexts and the sentiment of microblogs by utilizing a Laplacian regularization term. The main idea of the Laplacian regularization is to assign similar sentiment scores to two connected microblogs as they may convey the same sentiment. In other words, we assume that the constructed microblog graph is smooth in the view of graph signal processing. So the following model is established:

$$
\begin{aligned}
&\min \frac{1}{2} \sum_{i=1}^{n} \sum_{j=1}^{n} \mathbf{M}_{ij} \|\hat{\mathbf{y}}^i - \hat{\mathbf{y}}^j\|_F^2 \\
&= \min \frac{1}{2} \sum_{i=1}^{n} \sum_{j=1}^{n} \sum_{k=1}^{c} \mathbf{M}_{ij} (\hat{y}_{ik} - \hat{y}_{jk})^2 \\
&= \min \sum_{i=1}^{n} \sum_{j=1}^{n} \sum_{k=1}^{c} \mathbf{M}_{ij} (\hat{y}_{ik})^2 - \sum_{i=1}^{n} \sum_{j=1}^{n} \sum_{k=1}^{c} \mathbf{M}_{ij} \hat{y}_{ik} \hat{y}_{jk} \\
&= \min \sum_{k=1}^{c} \hat{\mathbf{Y}}_{*k}^T (\mathbf{D} - \mathbf{M}) \hat{\mathbf{Y}}_{*k} \\
&= \min \mathbf{Tr}(\hat{\mathbf{Y}}^T \mathbf{L} \hat{\mathbf{Y}})
\end{aligned}
\tag{10}
$$

where $\hat{y}$ represents the predicted sentiment label, $\mathbf{L} = \mathbf{D} - \mathbf{M}$ denotes the Laplacian matrix of the constructed microblog graph, $\mathbf{D} \in R^{n \times n}$ denotes a diagonal matrix whose entries represent the degree of corresponding microblogs. If the graph is smooth, the value of this equation is small. In the extreme situation, when the value is zero, the sentiment value of all microblogs is the same. Therefore, the final microblog sentiment analysis model can be formulated as:

$$
W^* = \underset{\mathbf{W}}{\arg\min} \, f(\mathbf{W}; \mathbf{X}, \mathbf{Y}) + \beta \|\mathbf{W}\|_1 + \frac{\alpha}{2} \mathbf{Tr}(\hat{\mathbf{Y}}^T \mathbf{L} \hat{\mathbf{Y}}),
\tag{11}
$$

where α, β are non-negative coefficients which control the weights of social contexts and model parameters respectively.

4   Experiment

The microblog sentiment analysis performance in this paper is measured by accuracy and the macro average F_1-score as the datasets are unbalanced. The ratios of positive to negative microblogs are 1:2.705 and 1:1.697 in HCR and OMD dataset respectively.

Two real public datasets (OMD and HCR) are used to verify the effectiveness of our proposed method integrating link prediction methods and sentiment connection weights into sentiment analysis. The sentiment labels of both datasets are labeled manually. OMD dataset consists of tweets about the 2008 Obama and McCain presidential debate, and the HCR dataset consists of tweets about the health care reform of America. These two datasets also contain user information about who posted the tweets. To get the user friend information, we use the complete follower graph built by [33] in 2009.

4.1 Observations

Since the major motivation of this work is the intuition that exploiting social information can promote the effectiveness of microblog sentiment analysis, we first study the degree of the correlation between network structure and microblog tags. Both statistical methods and mathematical methods are investigated to verify our motivation.

The probabilities of whether two connected microblogs have the same sentiment in the final graph are computed on both datasets. Table 1 shows the results. "AA", "PA" and "RA" represent the graph built by the Adamic Adar algorithm, the Preferential Attachment algorithm, and the Resource Allocation algorithm correspondingly. From Table 1, we can see that the possibility of two connected microblogs predicted by link prediction methods is much larger than two randomly selected microblogs.

Table 1. The shared sentiment possibility (%) of different link prediction methods and randomly selected microblogs

Method	Dataset	
	HCR	OMD
random	60.564	53.300
AA	77.536	54.752
RA	77.397	54.456
PA	77.593	54.368

The possibilities of connectedness between microblogs with the same sentiment and microblogs with the different sentiment in the final graph are also computed to verify the correlation between network links and microblog sentiment. Figure 1 show the results. "AA", "PA" and "RA" represent different graphs built by the Adamic Adar algorithm, the Preferential Attachment algorithm, and the Resource Allocation algorithm correspondingly. "same" denotes the possibility of connectedness between microblogs with same sentiment. "diff" denotes the possibility of connectedness between microblogs with different sentiment. Figure 1 shows that microblogs with the same sentiment are more likely to connect each other.

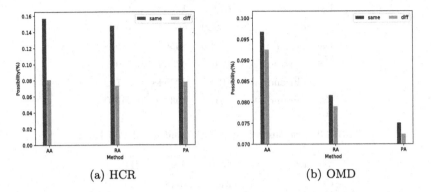

(a) HCR   (b) OMD

Fig. 1. Connectedness ratio conditioned on labels of different methods

Besides two statistical methods, we also use a mathematical method to verify our motivation. Follow [34], we compute the smoothness of different graphs by $smoothness = \mathbf{Y}^T \mathbf{L} \mathbf{Y} = \frac{1}{2} \sum_{i=1}^{n} \sum_{j=1}^{n} \mathbf{M}_{ij} ||\mathbf{Y}_i - \mathbf{Y}_j||_F^2$, where $\mathbf{L}$ is the Laplacian matrix of $\mathbf{M}$. If the smoothness values are small, it indicates that the sentiment of two connected microblogs is more likely the same. From the view of signal processing, a smaller value means there is a strong correlation between the microblog graph and its label signal. We compare the smoothness between the graphs we built and their corresponding random graphs. To be fair, the two graphs have the same number of nodes and edges. A normalized difference between the random graph and its corresponding graph is defined as follows:

$$ND = \frac{RS - GS}{RS} \tag{12}$$

where RS and GS represent the smoothness of random graphs and microblog graphs built by link prediction methods respectively. A larger ND denotes that there is a stronger sentiment correlation between two connected microblogs. Table 2 and 3 show the results. "User" represents the microblog graph built by user information. "Friend" represents the microblog graph built by friend information. "User+Friend" represents the graph built by user and friend information without link prediction enhancement. "User+Friend+AA", "User+Friend+PA" and "User+Friend+RA" represent the microblog graphs enhanced by the Adamic Adar algorithm, the Preferential Attachment algorithm, and the Resource Allocation algorithm respectively. From Table 2 and 3, we can see that the sentiment signals of all microblog graphs are smoother than their corresponding random graphs. This phenomenon validates our hypothesis of sentiment consistency and emotional contagion in online social networks to some extent. Besides, exploiting link prediction methods can make the graph smoother than other graphs on both datasets as demonstrated in the fourth column, which is consistent with our assumption.

Table 2. The smoothness of different graphs on OMD

Method	GS	RS	ND
User	6440	12448	0.48265
Friend	43220	86264	0.49898
User+Friend	43220	86564	0.50072
User+Friend+AA	121627	433028	0.71912
User+Friend+RA	103887	427548	0.75702
User+Friend+PA	95193	493480	0.80710

Table 3. The smoothness of different graphs on HCR

Method	GS	RS	ND
User	2488	8648	0.71230
Friend	62148	225056	0.72386
User+Friend	62148	225096	0.72390
User+Friend+AA	127385	634656	0.79928
User+Friend+RA	119497	591092	0.79784
User+Friend+PA	131240	599928	0.78124

4.2 Model Effectiveness

To show the effectiveness of our intuition, we first conduct experiments without text content information and only use relations between microblogs to analyze microblog sentiment. The results are shown in Table 4. From this table, we can see that using social relations between microblogs does help sentiment analysis. The performance of all methods using social information is better than randomly guessing. Besides, the accuracy of methods exploiting link prediction is higher than the "original" microblog graph. This result indicates the effectiveness of alleviating sparsity of microblog graphs.

Table 4. Microblog sentiment analysis accuracy (%) without text content information

Method	Dataset	
	OMD	HCR
User+Friend	61.486	73.291
User+Friend+AA	62.753	73.639
User+Friend+RA	62.585	74.120
User+Friend+PA	63.007	73.570

4.3   Performance Evaluation

To further show the effectiveness of our proposed model, 5-fold cross-validation is used to compare our proposed method with other state-of-art microblog sentiment analysis methods.

Five widely used classifiers: Least Squares (LS), Lasso, Logistic regression (LR), Support Vector Machine (SVM), and Naive Bayes (NB) are used as baseline methods [28,35]. Several deep learning methods CNN [37], LSTM [38] and GNN [39] are also used for comparison. We also compare the performance of different methods which explore social contexts, such as SANT [15] and PMSA [24]. VADER, a social media sentiment analysis tool is also compared [36]. AASA, RASA and PASA are our proposed method using graphs built by user, friend, link prediction and sentiment correlation weights in Sect. 3. They exploit different link prediction methods the Adamic Adar algorithm, the Resource Allocation algorithm and the Preferential Attachment algorithm respectively.

Results are shown in Table 5. The following observations can be drawn from this table:

1. Methods that explored social contexts have better results than traditional methods. This result indicates that it is effective to integrate microblog connections into microblog sentiment analysis. There is a correlation between social contexts and microblog sentiment analysis.
2. Lasso outperforms the least squares method. These results show that not all text features are helpful for microblog sentiment analysis. Lasso can filter out many words without emotional tendency as word distribution is very sparse on the microblogging platform.
3. The personalized method PMSA outperforms SANT on OMD, but it has a lower result than SANT on HCR. This is mainly due to that the average number of microblogs posted by users on OMD is larger than that on HCR. It indicates that data sparsity limits the performance of PMSA.
4. Among all methods, our method has achieved the best performance. There are three reasons: 1) our method utilizes l_1 norm regularization which can select emotional words automatically. 2) link prediction is applied in our method, which can solve the sparse problem of microblog connections. 3) sentiment weights are proposed to make connections between microblogs more precise. Three link prediction methods have achieved comparable results on two datasets, which need deep analysis.

Table 5. The sentiment analysis results of different methods

Method	OMD		HCR	
	Accu(%)	F1(%)	Accu(%)	F1(%)
LS	69.504	67.455	65.205	59.318
LASSO	75.336	71.343	73.711	56.431
NB	77.363	73.614	74.474	67.231
SVM	75.760	74.067	74.132	66.476
LR	78.208	75.780	75.875	65.950
VADER	64.021	63.557	55.090	53.320
LSTM	78.390	75.962	74.476	66.582
CNN	75.424	72.901	76.923	61.786
GNN	78.294	76.355	76.780	68.454
SANT	78.545	75.893	76.156	69.749
PMSA	78.375	75.780	76.008	65.364
AASA	80.066	77.422	77.825	**71.388**
RASA	**80.574**	**77.958**	**77.895**	70.955
PASA	80.150	77.653	77.547	70.681

5   Conclusion

Based on two sociological theories, sentiment consistency and emotional contagion, this paper assumes that microblogs published by the same user or connected users have a high possibility to have similar sentiment in a period and build a microblog graph by user information and friend information. However, the built graph is usually very sparse. To alleviate this problem, we explore the potential links between microblogs by link prediction methods. Both statistical methods and mathematical methods are used to study the correlation between the links extended by link prediction methods and microblog sentiment labels. Besides, we also introduce sentiment connection weights to refine microblog connections. After modeling the connections between microblogs, we integrate them with microblog text information. Experiments on two real-world microblog sentiment datasets show the effectiveness of exploring potential links between microblogs other than user information and friend information. This work can suggest some interesting directions for future work. For example, it would be promising to predict potential links by utilizing sentiment information of users about different topics.

Acknowledgement. This paper is supported by 1) Zhejiang Provincial Natural Science Foundation of China under Grant No. LQ23F020039, 2) National Science and Technology Major Project of China under Grant No. 2021ZD0114303, 3) National Natural Science Foundation of China under Grant No. 62176087.

References

1. Yang, D., Zhang, D., Yu, Z., Wang, Z.: A sentiment-enhanced personalized location recommendation system. In: Proceedings of the 24th ACM Conference on Hypertext and Social Media, HT 2013, Paris, France, pp. 119–128 (2013)
2. Bollen, J., Mao, H., Zeng, X.: Twitter mood predicts the stock market. J. Comput. Sci. **2**(1), 1–8 (2011)
3. Cambria, E., Schuller, B., Xia, Y., White, B.: New avenues in knowledge bases for natural language processing. Knowl.-Based Syst. **108**, 1–4 (2016)
4. Wu, Y., Liu, S., Yan, K., Liu, M., Wu, F.: OpinionFlow: visual analysis of opinion diffusion on social media. IEEE Trans. Vis. Comput. Graph. **20**(12), 1763–1772 (2014)
5. Cambria, E., Liu, Q., Decherchi, S., Xing, F., Kwok, K.: SenticNet 7: a commonsense-based neurosymbolic AI framework for explainable sentiment analysis. In: Proceedings of the Thirteenth Language Resources and Evaluation Conference, pp. 3829–3839 (2022)
6. Baccianella, S., Esuli, A., Sebastiani, F.: SentiWordNet 3.0: an enhanced lexical resource for sentiment analysis and opinion mining. In: Proceedings of the Seventh International Conference on Language Resources and Evaluation (LREC 2010), vol. 10, pp. 2200–2204 (2010)
7. McPherson, M., Smith-Lovin, L., Cook, J.M.: Birds of a feather: homophily in social networks. Ann. Rev. Sociol. **27**(1), 415–444 (2001)
8. Marsden, P.V., Friedkin, N.E.: Network studies of social influence. Sociol. Methods Res. **22**(1), 127–151 (1993)
9. Hatfield, E., Cacioppo, J.T., Rapson, R.L.: Emotional contagion. Curr. Dir. Psychol. Sci. **2**(3), 96–100 (1993)
10. Abelson, R.P.: Whatever became of consistency theory? Pers. Soc. Psychol. Bull. **9**(1), 37–64 (1983)
11. Speriosu, M., Sudan, N., Upadhyay, S., Baldridge, J.: Twitter polarity classification with label propagation over lexical links and the follower graph. In: Proceedings of the First Workshop on Unsupervised Learning in NLP, EMNLP 2011, pp. 53–63 (2011)
12. Song, C., Wang, B., Jiang, Q., Zhang, Y., He, R., Hou, Y.: Social recommendation with implicit social influence. In: Proceedings of the 44th International ACM SIGIR Conference on Research and Development in Information Retrieval, pp. 1788–1792 (2021)
13. Khalifi, H., Dahir, S., El Qadi, A., Ghanou, Y.: Enhancing information retrieval performance by using social analysis. Soc. Netw. Anal. Min. **10**, 1–7 (2020)
14. Mehta, N., Pacheco, M.L., Goldwasser, D.: Tackling fake news detection by continually improving social context representations using graph neural networks. In: Proceedings of the 60th Annual Meeting of the Association for Computational Linguistics (Volume 1: Long Papers), pp. 1363–1380 (2022)
15. Hu, X., Tang, L., Tang, J., Liu, H.: Exploiting social relations for sentiment analysis in microblogging. In: Proceedings of the 6th ACM International Conference on Web Search and Data Mining, pp. 537–546 (2013)
16. Wu, F., Huang, Y., Song, Y.: Structured microblog sentiment classification via social context regularization. Neurocomputing **175**(PartA), 599–609 (2016)
17. Cui, A., Zhang, M., Liu, Y., Ma, S.: Emotion tokens: bridging the gap among multilingual Twitter sentiment analysis. In: Salem, M.V.M., Shaalan, K., Oroumchian, F., Shakery, A., Khelalfa, H. (eds.) AIRS 2011. LNCS, vol. 7097, pp. 238–249. Springer, Heidelberg (2011). https://doi.org/10.1007/978-3-642-25631-8_22

18. Wu, F., Huang, Y., Song, Y., Liu, S.: Towards building a high-quality microblog-specific Chinese sentiment lexicon. Decis. Support Syst. **87**, 39–49 (2016)
19. Xing, F.Z., Pallucchini, F., Cambria, E.: Cognitive-inspired domain adaptation of sentiment lexicons. Inf. Process. Manag. **56**(3), 554–564 (2019)
20. Feng, S., Wang, Y., Liu, L., Wang, D., Yu, G.: Attention based hierarchical LSTM network for context-aware microblog sentiment classification. World Wide Web **22**(1), 59–81 (2019)
21. Ma, Y., Peng, H., Khan, T., Cambria, E., Hussain, A.: Sentic LSTM: a hybrid network for targeted aspect-based sentiment analysis. Cogn. Comput. **10**(4), 639–650 (2018)
22. Shin, B., Lee, T., Choi, J.D.: Lexicon integrated CNN models with attention for sentiment analysis, arXiv preprint arXiv:1610.06272 (2016)
23. Wang, L., Niu, J., Yu, S.: SentiDiff: combining textual information and sentiment diffusion patterns for Twitter sentiment analysis. IEEE Trans. Knowl. Data Eng. **32**(10), 2026–2039 (2019)
24. Wu, F., Huang, Y.: Personalized microblog sentiment classification via multi-task learning. In: Proceedings of the Thirtieth AAAI Conference on Artificial Intelligence, pp. 3059–3065 (2016)
25. Fersini, E., Pozzi, F., Messina, E.: Approval network: a novel approach for sentiment analysis in social networks. World Wide Web **20**(4), 831–854 (2017)
26. Cheng, K., Li, J., Tang, J., Liu, H.: Unsupervised sentiment analysis with signed social networks. In: 31st AAAI Conference on Artificial Intelligence, pp. 3429–3435 (2017)
27. Skianis, K., Rousseau, F., Vazirgiannis, M.: Regularizing text categorization with clusters of words. In: Proceedings of the 2016 Conference on Empirical Methods in Natural Language Processing, pp. 1827–1837 (2016)
28. Hastie, T., Tibshirani, R., Friedman, J.: The elements of statistical learning. J. R. Stat. Soc. **167**(1), 192 (2001)
29. Adamic, L.A., Adar, E.: Friends and neighbors on the web. Soc. Netw. **25**(3), 211–230 (2003)
30. Barabási, A.-L., Albert, R.: Emergence of scaling in random networks. Science **286**(5439), 509–512 (1999)
31. Zhou, T., Lü, L., Zhang, Y.-C.: Predicting missing links via local information. Eur. Phys. J. B **71**(4), 623–630 (2009)
32. Beck, A., Teboulle, M.: A fast iterative shrinkage-thresholding algorithm for linear inverse problems. SIAM J. Imaging. Sci. **2**(1), 183–202 (2009)
33. Kwak, H., Lee, C., Park, H., Moon, S.: What is Twitter, a social network or a news media? In: Proceedings of the 19th International Conference on World Wide Web, pp. 591–600 (2010)
34. Keramatfar, A., Amirkhani, H., Bidgoly, A.J.: Modeling tweet dependencies with graph convolutional networks for sentiment analysis. Cogn. Comput. **14**, 2234–2245 (2022)
35. Go, A., Bhayani, R., Huang, L.: Twitter sentiment classification using distant supervision. Cs224n Project Report (2009)
36. Hutto, C., Gilbert, E.: VADER: a parsimonious rule-based model for sentiment analysis of social media text. In: Proceedings of the International AAAI Conference on Web and Social Media, vol. 8, pp. 216–225 (2014)
37. Kim, Y.: Convolutional neural networks for sentence classification. In: 2014 Conference on Empirical Methods in Natural Language Processing, EMNLP 2014, Doha, Qatar, pp. 1746–1751 (2014)

38. Hochreiter, S., Schmidhuber, J.: Long short-term memory. Neural Comput. **9**(8), 1735–1780 (1997)
39. Kipf, T.N., Welling, M.: Semi-supervised classification with graph convolutional networks. In: International Conference on Learning Representations, pp. 1–14 (2017)

Discovering Diverse Information Considering User Acceptability

Yuki Ito[1]([✉]) and Qiang Ma[2]([✉]) [iD]

[1] Kyoto University, Kyoto, Japan
ito.yuki.46m@st.kyoto-u.ac.jp
[2] Kyoto Institute of Technology, Kyoto, Japan
qiang@kit.ac.jp

Abstract. Filter bubbles occur when search algorithms selectively show users information that aligns with their preferences, leading to a limited view of the world. Many studies aim to address this problem by seeking diverse information, but they often overlook the aspect of acceptability. We propose a method to discover documents (tweets, reviews, or so) that provide diverse viewpoints while considering their acceptability. Our method captures each opinion's features such as its polarity, aspects, etc., and identifies documents encompassing both empathy and diversity by analyzing the features. Evaluation results confirm the effectiveness of our method in discovering acceptable and diverse opinions.

Keywords: Diversity · Filter Bubbles · Acceptability

1 Introduction

The filter bubble is a challenge, where the information provided to users is too biased [12]. To solve this, current filter bubble reduction methods attempt to offer various information and opinions to users [2,3,6,11]. However, some of such opinions may conflict with users' preferences [16]. Feeling uncomfortable with these opinions may result in behaviors such as ignoring them, assuming they lack credibility, or seeking opinions that match one's thinking [1]. This may hinder efforts to reduce the filter bubble. To address this issue, we propose a method that focuses on acceptability (whether a text is easily accepted by the user) in addition to diversity. From documents such as tweets and reviews, our proposed method search for texts that express different opinions from users while also showing empathy for their opinions. Our method first transforms documents into a document representation model (DRM), which formally represents the features of the opinion. Subsequently, it evaluates the acceptability (Is it easy for users to accept?) and diversity (Is its opinion different from the user's?) of each text by the degree of empathy, objectivity, opposition, and diversity, which

This work was partly supported by JSPS KAKENHI (19H04116, 23H03404).

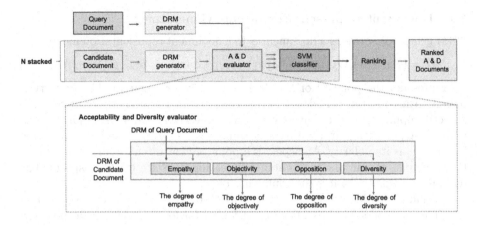

Fig. 1. Overview of proposed method

are defined in Sect. 3. From these results, it classifies texts based on whether or not they provide various viewpoints in acceptable ways and rank them.

Our contributions are as follows. (1) We propose a document representation model (Sect. 3.1). (2) We propose novel concepts of empathy, objectivity, opposition, and diversity as criteria for acceptable opinions that offer different viewpoints (Sect. 3.2). (3) The proposed method's effectiveness is validated by performance metrics on the dataset obtained from the user experiment (Sect. 4).

2   Related Work

Many algorithms aim to reduce filter bubbles by recommending diverse information. For instance, [16] proposed a method that suggests contextually natural opposing opinions in a dialogue generation system. Approaches encouraging behavior change have also been explored. [10] graphed relationships and evaluations of stakeholders in news articles to reveal biases. Another effective approach is making different opinions easily accessible, as demonstrated by the NewsSalad app [6]. Metacognitive abilities, such as recognizing cognitive biases and seeking diverse viewpoints, have also been studied [8]. While existing studies focus on reducing filter bubbles, none of them consider acceptability. To the best of our knowledge, such a method has not yet been developed.

3   Proposed Method

The overview of the proposed method is shown in Fig. 1. Our method uses a query document (user's preferred opinions) and a candidate set (N documents on the same topic as the query document) as inputs, and outputs the ranking of these candidates based on acceptability and diversity.

3.1   Document Representation Model Generator

DRM generator converts a document S to a document representation model M_S.

$$M_S = (P_S, wps), O_S, ((a_1, p_1, aps_1), (a_2, p_2, aps_2), \ldots (a_m, p_m, aps_m)) \quad (1)$$

P_S represents the polarity of S, which is positive or negative. The polarity is paired with wps, the confidence of P_S. O_S denotes the objectivity of S. Since objective opinions tend to have fewer verbs, adjectives, and adverbs [9], we estimate their percentage in the target document as O_S. a_i is an aspect appearing in S and p_i is its polarity. An aspect is a point of view often discussed in a topic. In this paper, for simplicity, all nouns appearing in S are used as aspects. The real number aps_i indicates the confidence of p_i.

P_S and p_i are determined by using a BERT-based sentiment analysis method [15]. This method generates a two dimension vector for representing the polarity; one for negative and the other for positive. The max value in the vector is denoted as wps or aps_i. P_S is identified by inputting S to the method. When identifying p_i, we first separate S at *the boundary of polarity* into $(S_1, S_2, \ldots, S_m)$. The boundary of polarity means the position where the polarity is likely to switch (such as punctuation marks and conjunctions/conjunctive particles in inverse clauses found by morphological analysis tool MeCab [7]). We consider the polarity of S_j ($\ni a_i$) as the polarity of a_i.

3.2   Acceptability and Diversity Evaluator

The acceptability and Diversity evaluator judges "how acceptable a candidate S' is to users whose opinion is same as S (Acceptability)" and "whether it offers new perspectives to them (Diversity)".

Acceptability is decided by the degree of empathy and objectivity. The degree of empathy evaluates if S' shows the attitude of understanding users' opinions, which makes it easier for users to accept the document even if its opinion differs from their values [1, 4, 14]. The degree of empathy of S' is scored based on whether S' satisfies the following three conditions. (1) S' has the boundary of polarity. (2) Before the boundary, S' represents the same opinion as the user (Judged by P_S and $P_{S'}$). (3) The part of S' before the boundary shows similar polarities for similar aspects as the query document (Judged by aspect-polarity sequences in M_S and $M_{S'}$). The degree of objectivity $O_{S'}$ is considered because objective opinions tend to be more easily accepted than subjective ones.

Diversity is estimated by the degree of opposition and diversity. The degree of opposition refers to the level of disagreement that S' has with the user's opinion, which is calculated from (P_S, wps) and $(P_{S'}, wps')$ This is considered because showing opposing opinions are important to mitigate filter bubbles. The degree of diversity refers to the various viewpoints in the document. By estimating this, we aim to discover documents that provide users with diverse views in a time-saving way. We defined the degree of diversity by an average of semantic similarity of arbitrary combinations of aspects in S'. Semantic similarity is determined using the cosine similarity of the embedding vectors obtained from the final layer output of BERT in the sentiment analysis model.

3.3 SVM Classifier and Ranking

The candidates are classified by an SVM model (LinearSVC of scikit-learn [13]), which takes the four scores obtained up to this point as inputs. Its output results from a binary classification of whether or not the candidate S' is an acceptable and diverse opinion.

We rank candidate documents classified as acceptable and diverse documents by the following procedure. (1) Choose S' whose $P_{S'}$ is different from P_S. (2) Sort the candidates in descending order based on the number of boundaries of polarity. In case of a tie, prioritize the longer documents. (3) Choose S' whose $P_{S'}$ is the same as P_S and sort them as same as step 2. (4) Concatenate the ranking results of step 3 to the end of the results of step 2.

4 Evaluation

4.1 Training

We implement the system proposed in Sect. 3 and conduct experiments to evaluate the proposed method's classification and ranking performance. To train the sentiment analysis method [15] in the DRM generator, we use the Twitter Japanese reputation analysis dataset [17]. This dataset collects tweets by topics and each tweet's opinion as classified as follows: (1) Positive (2) Negative (3) Both positive and negative (4) Neither positive nor negative. We use tweets with labels of (1), (2), and (3) for training. Note that (3) were relabeled as positive or negative according to what polarity their writers most wanted to assert in the tweet. To reduce the training time, we selected about 1000 documents from this dataset, ensuring that the ratio of negative and positive documents was 1:1.

Next, to train the SVM classifier, we selected query documents and candidates from [17]. Using the proposed method, we calculated candidates' degrees of empathy, objectivity, opposition, and diversity. They are labeled *correct* (which means that the document is acceptable and diverse to the query document) if they are classified into both (1) and (2), or classified into (3), otherwise labeled *incorrect*. We set the ratio of incorrect cases to correct ones to 1:1.

4.2 User Experiments

To evaluate the effectiveness of our method, we collected documents from Twitter and the Amazon review corpus [5]. We collected documents on four topics from Twitter and one topic from the review corpus, with 50 documents for each topic. From each topic, We selected two query documents. One has a negative polarity, and the other is positive. Then we obtain ground truths for classification and the gain for ranking by user experiments. In the experiments, we asked subjects to rate each candidate document on a five-point scale from the viewpoints of "whether it is acceptable to users who share the same ideas as the query document" and "whether it offers diverse perspectives to such users" respectively. The documents with an average score of acceptability and diversity greater than three were labeled as *correct*, following Sect. 4.1.

Table 1. Evaluation results of the classification and ranking performances

evaluation target	Accuracy	Precision	Recall	F1	nDCG
prop-emp	0.466	0.78	0.166	0.267	0.956
prop-obj	0.57	0.619	0.824	0.695	0.923
prop-opp	0.46	0.571	0.451	0.494	0.917
prop-div	0.618	0.784	0.534	0.618	0.939
proposed	0.644	0.715	0.698	**0.695**	0.927
neural net	0.532	0.737	0.361	0.478	**0.930**

4.3   Result

We conducted an ablation study of the proposed method by evaluating the following baselines. prop-emp (A method using only the degree of empathy), prop-obj (using only the degree of objectivity), prop-opp (using only the degree of opposition), prop-div (using only the degree of diversity). In addition, we prepared a neural network-based method as a comparative method, reproducing the proposed method's processing flow.

In the evaluation process, we first input query and candidate documents made in Sect. 4.2 to each target model and calculate the SVM classification results' precision, recall, and F1 score. Let S_c be the set of candidates classified as acceptable and diverse opinions. We also calculate the rank and gain of the documents in S_c and compute nDCG. Each ranked document's gain is estimated based on its polarity label and the number of the boundary of polarity. A gain of 5 is assigned for documents with opposite polarity to the query and boundary number > 3, and a gain of 4 is for documents with opposite polarity and boundary number ≤ 3. A gain of 3 is assigned for documents with the same polarity as the query document. The gain is defined between 3 and 5 because it aligns with our definition in the user experiment (Sect. 4.2), where we define documents as *correct* if their average scores exceed 3.

Table 1 shows the evaluation results.

Our proposed method has the highest F1 score and nDCG over 0.9, which shows that the proposed method achieves high classification performance without sacrificing ranking performance. It can also be said that using all four indicators results in higher validity for classification and ranking than using only some of them from their low evaluation result. The high nDCG value of the proposed method shows that documents obtained by the method are accessible for users to accept and can provide new perspectives. While the neural net method achieves the highest nDCG value, it does not necessarily imply superiority over the proposed method. Its low recall and F1 score suggest that it includes fewer documents in its ranking compared to other methods. This limitation hinders its effectiveness in countering the filter bubble, which requires encountering a greater variety of opinions.

5 Conclusion

In this paper, we propose a method for discovering opinions that are easy for users to accept and bring diverse perspectives to them. The experimental results confirmed that the proposed method enables the discovery of acceptable and diverse opinions. Moreover, the proposed method outperforms the comparative method primarily in classification. Suppose the proposed method is applied to display opinions that differ from the user's preferences to reduce filter bubbles. In that case, user-friendly documents are expected to be prioritized, and the phenomenon of unconsciously excluding different opinions can be reduced.

References

1. Ecker, U.K., et al.: The psychological drivers of misinformation belief and its resistance to correction. Nat. Rev. Psychol. **1**(1), 13–29 (2022)
2. Gharahighehi, et al.: Making session-based news recommenders diversity-aware. In: Proceedings of OHARS, pp. 60–66 (2020)
3. Grossetti, Q., et al.: Reducing the filter bubble effect on twitter by considering communities for recommendations. IJWIS **17**(6), 728–752 (2021)
4. Hyland-Wood, B., et al.: Toward effective government communication strategies in the era of COVID-19. Humanit. Soc. Sci. Commun. **8**(1), 30 (2021)
5. Keung, P., et al.: The multilingual amazon reviews corpus. arXiv:2010.02573 (2020)
6. Kiritoshi, K., et al.: Named entity oriented difference analysis of news articles and its application. IEICE Trans. Inf. Syst. **99-D**(4), 906–917 (2016)
7. Kudo, T., et al.: Applying conditional random fields to Japanese morphological analysis. In: EMNLP 2004, pp. 230–237 (2004)
8. Liu, G.: Moving up the ladder of source assessment: expanding the CRAAP test with critical thinking and metacognition. C RL News **82**(2), 75 (2021)
9. Mingyong, Y., et al.: A subjective expressions extracting method for social opinion mining. Discret. Dyn. Nat. Soc. **2020**, 2784826 (2020)
10. Ogawa, T., et al.: News bias analysis based on stakeholder mining. IEICE Trans. Inf. Syst. **94-D**(3), 578–586 (2011)
11. Pardos, et al.: Combating the filter bubble: Designing for serendipity in a university course recommendation system. arXiv:1907.01591 [cs.IR] (2019)
12. Pariser, E.: The Filter Bubble: How the New Personalized Web is Changing What We Read and How We Think. Penguin, London (2011)
13. Pedregosa, F., et al.: Scikit-learn: machine learning in python. J. Mach. Learn. Res. **12**, 2825–2830 (2011)
14. Santos, L.A., et al.: Belief in the utility of cross-partisan empathy reduces partisan animosity and facilitates political persuasion. Psychol. Sci. **33**(9), 1557–1573 (2022)
15. Yang, H., Li, K.: A modularized framework for reproducible aspect-based sentiment analysis. CoRR abs/2208.01368 (2022)
16. Yoshida, S., Ma, Q.: Generating dialogue sentences to promote critical thinking. In: Hartmann, S., Küng, J., Kotsis, G., Tjoa, A.M., Khalil, I. (eds.) DEXA 2020. LNCS, vol. 12391, pp. 354–368. Springer, Cham (2020). https://doi.org/10.1007/978-3-030-59003-1_23
17. Yu, S.: Filtering method for twitter streaming data using human-in-the-loop machine learning. J. Inf. Process. **27**, 404–410 (2019)

Confidential Truth Finding
with Multi-Party Computation

Angelo Saadeh[1,6], Pierre Senellart[2,4,5,6,7]($\boxtimes$), and Stéphane Bressan[3,6,7]

[1] LTCI, Télécom Paris, IP Paris, Palaiseau, France
angelo.saadeh@telecom-paris.fr
[2] DI ENS, ENS, PSL University, CNRS, Paris, France
pierre@senellart.com
[3] National University of Singapore, Singapore, Singapore
steph@nus.edu.sg
[4] Inria, Paris, France
[5] Institut Universitaire de France, Paris, France
[6] CNRS@CREATE LTD, Singapore , Singapore
[7] IPAL, CNRS, Singapore , Singapore

Abstract. Federated knowledge discovery and data mining are challenged to assess the trustworthiness of data originating from autonomous sources while protecting confidentiality and privacy. Truth-finding algorithms help corroborate data from disagreeing sources. For each query it receives, a truth-finding algorithm predicts a truth value of the answer, possibly updating the trustworthiness factor of each source. Few works, however, address the issues of confidentiality and privacy. We devise and present a secure secret-sharing-based multi-party computation protocol for pseudo-equality tests that are used in truth-finding algorithms to compute additions depending on a condition. The protocol guarantees confidentiality of the data and privacy of the sources. We also present a variants of a truth-finding algorithm that would make the computation faster when executed using secure multi-party computation. We empirically evaluate the performance of the proposed protocol on a state-of-the-art truth-finding algorithm, 3-Estimates, and compare it with that of the baseline plain algorithm. The results confirm that the secret-sharing-based secure multi-party algorithms are as accurate as the corresponding baselines but for proposed numerical approximations that significantly reduce the efficiency loss incurred.

Keywords: truth finding · secure multi-party computation · secret-sharing · uncertain data · privacy

1 Introduction

Truth-finding algorithms [6] help corroborate data from disagreeing sources. For each query it receives, a truth-finding algorithm predicts a truth value of the answer, possibly updating the trustworthiness factor of each source. Few works, however, address the issues of confidentiality and privacy. We consider the design

© The Author(s), under exclusive license to Springer Nature Switzerland AG 2023
C. Strauss et al. (Eds.): DEXA 2023, LNCS 14146, pp. 332–337, 2023.
https://doi.org/10.1007/978-3-031-39847-6_25

and implementation of truth-finding algorithms that protect the confidentiality of sources' data, using secret-sharing-based secure multi-party computation [3], or simply secure multi-party computation (MPC).

We devise and present a secure multi-party pseudo-equality protocol that securely computes additions depending on a condition – we call them conditioned additions – for truth-finding algorithms. In particular, we present a secure equality test alternative that uses a polynomial evaluation to reduce the number of communication; this is used for conditioned additions, an operation that is an essential building block of many truth-finding algorithms. The protocol guarantees the confidentiality of the data. We also devise several variants of privacy-preserving truth-finding algorithms; ones that implement the truth-finding algorithms without changes, and others with modifications that aim to make the computation more efficient.

The secure multi-party protocols are then implemented with two servers. We empirically evaluate the performance of the proposed protocol on a state-of-the-art truth-finding algorithm, 3-Estimates [4, Algorithm 4] (see also [2,5] for further experiments on this algorithm), and compare it with that of the non-secure baseline algorithms. The results confirm that the secure multi-party algorithm is as accurate as the corresponding baseline except for proposed modifications to reduce the efficiency loss incurred.

Set $n \in \mathbb{N}^*$, and let $\mathcal{V}$ be a set of n sources. The client would like to label k queries (or facts) $\{f^1, ..., f^k\}$. A truth-finding algorithm outputs a truth value for a query when different data sources (or sources) provide disagreeing information on it. Concretely, the truth-finding algorithm takes $v^1, ..., v^n$ as input with $v^i \in \{-1, 0, 1\}^k$, and outputs estimated truth values in $[-1, 1]^k \subset \mathbb{R}^k$ or $[0, 1]^k \subset \mathbb{R}^k$ depending on the truth-finding algorithm.

Truth-finding (or truth discovery) algorithms [6] are usually run by the client in order to know the truth value of a given query when the sources give disagreeing answers. That is, for each of the client's queries, each source in $\mathcal{V}$ delivers an answer v^i such that an output of 1 corresponds to a positive answer, -1 to a negative one, and 0 if the source does not wish to classify the data point. 3-Estimates [4] is a truth-finding algorithms that given a number of queries k, output a truth value in the range $[-1, 1]^k \subset \mathbb{R}$ and a trust coefficient in each of the sources, or sources. In addition, 3-Estimates computes an estimate of the difficulty of each query.

The goal of this work is to execute truth-finding algorithms that protect sources' data using secure multi-party computation (MPC) [1,3]. More generally, given a function F and a set of private inputs $x^1, ..., x^m$ respectively owned by $P_1, ..., P_m$, MPC is a cryptographic approach that makes it possible to compute the output of the function $F(x^1, .., x^m)$ without resorting to a third party that would compute the function F and would send the result back. MPC will be used to implement the 3-Estimates algorithm without having any source disclose their answer.

Because of lack of space, details are ommitted. An extended version is available as [8], which also covers another truth finding algorithm, Cosine, from [4].

2   Proposed Approach

The first task we wish to achieve is private voting, i.e., the client sends queries to each source, and the source classifies the query. In the case where the query is a vector of features and the models are logistic regressions, existing MPC works [7] can keep the query private. We suppose that the answers are already computed and secret-shared on two servers P_1 and P_2 using a two-party additive secret sharing. In other words, P_1 holds v_1^{ij} and P_2 holds v_2^{ij} such that $v^{ij} = v_1^{ij} + v_2^{ij}$ is the ith source's answer for the query f^j and is equal to $-1, 0$, or 1.

The second step which is the aggregation of the data (the answers) is computed on the two servers P_1 and P_2. The problem is now constructing a secure two-party computation algorithm with additively shared data that implements the truth-finding algorithms using their arithmetic circuits. Once the circuits are evaluated, the two servers (P_1 and P_2) send their share of the output to the client who reconstructs it by adding the received shares together.

Other than additions and multiplications, the truth-finding algorithm we implement – 3-Estimates – uses existing real-number operations like division, and square root, which are dealt with in standard ways [8]. We focus on computing conditioned sums by replacing equality tests with degree-two polynomial evaluations.

The truth-finding algorithms we use require conditioned additions. Given two vectors of same size $t = (t^1, ..., t^k) \in \mathbb{R}^k$, $z = (z^1, ..., z^k) \in \{-1, 0, 1\}^k$, and an element $\kappa \in \{-1, 0, 1\}$, we define the following operation: $S := \sum_{i:z^i=\kappa} t^i$. In other words, the ith element of t, t^i, is added to the sum only if the ith element of z, z^i, is equal to κ. The difficulty is that even though κ is public, z^i is private. To achieve this in MPC we start by defining the following function, for $i \in \{1, ..., r\}$:

$$\mathcal{E}(z^i, \kappa) = \begin{cases} 1 \text{ if } z^i = \kappa \\ 0 \text{ if not.} \end{cases}$$

A naive way to compute the sum S is as follows: $S = \sum_i \mathcal{E}(z^i, \kappa) \cdot t^i$. This way to compute S requires an equality test which is costly in MPC. To this end, we propose an alternative that makes good use of the fact that $z^i, \kappa \in \{-1, 0, 1\}$. The goal is to express the function $\mathcal{E}$ as a polynomial so that it can be computed using the smallest number of additions and multiplications possible. We define and use the following expressions of $\mathcal{E}(z^i, \kappa)$.

If $\kappa = -1$, we compute S as follows: $S = \sum_i \frac{1}{2}((z^i)^2 - z^i) \cdot t^i$. We have:

$$\frac{1}{2}((z^i)^2 - z^i) = \begin{cases} 1 \text{ if } z^i = -1 \\ 0 \quad \text{if } z^i = 0 \\ 0 \quad \text{if } z^i = 1 \end{cases}$$

Hence, by multiplying $\frac{1}{2}((z^i)^2 - z^i)$ by t^i, the only elements considered in the sum are the ones such that $z^i = -1$. The function $\frac{1}{2}((z^i)^2 - z^i)$ is equal to $\mathcal{E}(z^i, -1)$. If $\kappa = 0$ we similarly compute S as: $S = \sum_i (1 - (z^i)^2) \cdot t^i$. It is also straightforward that the function $1 - (z^i)^2$ is equal to $\mathcal{E}(z^i, 0)$ because it outputs 1 if $z^i = 0$ and 0 otherwise. If $z = 1$, in the same way, S is computed as: $S = \sum_i \frac{1}{2}((z^i)^2 + z^i) \cdot t^i$.

Lemma 1 (Conditioned additions). *Denote by $\Pi_{\mathcal{E}}$ the MPC protocol implementing the function $\mathcal{E}$ using the three previously defined degree-2 polynomials. $\Pi_{\mathcal{E}}$ does not reveal information about the other's player's share.*

Proof. The three conditioned sums defined in this section do not need comparisons and they are expressed using only additions and multiplications, so their security level is the same as Π_{add} and Π_{mul}. □

See [8] for how this allows us to reformulate the truth finding algorithm for 3-Estimates with MPC.

Normalization in 3-Estimates. In the 3-Estimates algorithm, the truth value, trust factor, and difficulty score need to be normalized at each step. This could be done using a secure comparison protocol to securely compute the minimum and the maximum of each value, and then normalize them as it is done in [4]. Secure comparisons however are very costly in MPC. To reduce the amount of communication we replace the normalization based on finding the maximum and minimum by a pre-computed linear transformation which forces the values to stay between 0 and 1. Concretely we apply the function $h(x) = 0.5x + 0.25$ to all the values after each update. We evaluate the impact of this change in the experiments. The chosen function, h, is not perfect. Indeed, if we have information about the distribution of the parameters, we can pre-compute a linear normalization for every iteration. Using any public pre-computed or pre-defined normalizing function improves the efficiency of the algorithm because it would translate to using multiplication and addition by public constants, which is communication-free.

3   Experimental Results

We evaluate our protocol on two computing servers. We suppose that the sources have already answered and secret-shared their answers. We use the ring $\mathbb{Z}_{2^{60}}$ with 20 bits of fixed precision. The two servers communicate via a local socket network implemented in Python on an Intel Core i5-9400H CPU (2.50 GHz × 8) and a RAM of 15.4 GiB. For the sake of the experiment, these communications are not encrypted or authenticated.

We implement our solution using the dataset Hubdub from [4].[1] This dataset is constructed from 457 questions from a Web site where users had to bet on future events. As the questions had multiple answers, they have been increased to 830 questions to obtain binary questions with answers $-1, 0$ or 1. The client sends the 830 queries to be classified by each source, and after the classification, the sources secret-share them on two servers to evaluate using MPC the 3-Estimates truth-finding algorithm. At the end of the evaluation, the results are reconstructed by the client. The results include the truth value for each query

---

[1] Datasets used, as well as the source code of our implementation, are available at https://github.com/angelos25/tf-mpc/.

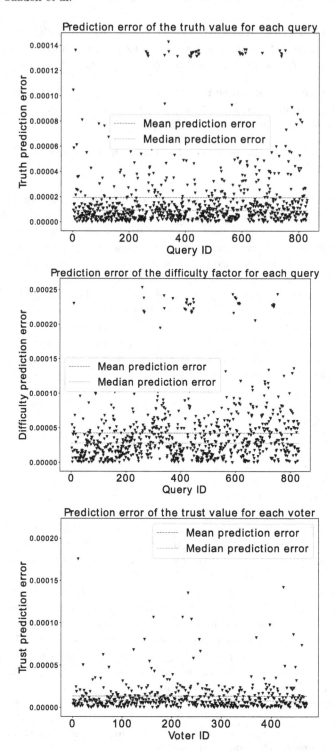

Fig. 1. Prediction errors between secure multi-party computation and the base model results with 3-Estimates on Hubdub dataset.

(the label), a difficulty score for each query, and a trustworthiness factor for each of the 471 sources. In Fig. 1 we show the difference between the predictions from the base model and the predictions from the MPC evaluation. The base model corresponds to the 3-Estimates algorithm implemented without MPC on the plain data. The MPC evaluation contains errors compared to the base model, and these errors are mostly below 10^{-4}. To evaluate the impact of the errors induced by MPC, we look at label prediction. The MPC method labels all the questions exactly the same way as the baseline method, so both methods made the same number of errors, i.e., 269 (as shown in [4], this is less than majority voting and some other methods). On average, the execution of each iteration took $52.85\,s$ wall-clock time, or $39.58\,s$ CPU time. The MPC model is $2\,000$ times slower than the base model, this is due to the high number of comparisons that should be made to normalize the three factors.

If we use the pre-computed linear function h presented at the end of Sect. 2 the outputs will be very different of course because of the aforementioned reasons, but wall-clock time of each iteration is reduced to $0.58\,s$ and the CPU time to $0.48\,s$ making it almost 100 times faster. This normalization alternative increases the number of queries labeled differently by the MPC to 5, however, it yields 266 errors in total. For this specific dataset, the pre-computed normalization used happens to gives better results than the original baseline.

Acknowledgments. This research is part of the program DesCartes and is supported by the National Research Foundation, Prime Minister's Office, Singapore under its Campus for Research Excellence and Technological Enterprise (CREATE) program.

References

1. Ben-Or, M., Goldwasser, S., Wigderson, A.: Completeness theorems for non-cryptographic fault-tolerant distributed computation. In: STOC (1988)
2. Berti-Équille, L.: Data veracity estimation with ensembling truth discovery methods. In: BigData (2015)
3. Cramer, R., Damgård, I., Nielsen, J.B.: Secure Multiparty Computation and Secret Sharing. Cambridge University Press (2015)
4. Galland, A., Abiteboul, S., Marian, A., Senellart, P.: Corroborating information from disagreeing views. In: WSDM (2010)
5. Li, X., Dong, X.L., Lyons, K., Meng, W., Srivastava, D.: Truth finding on the deep Web: Is the problem solved? PVLDB 6(2) (2013)
6. Li, Y., Gao, J., Meng, C., Li, Q., Su, L., Zhao, B., Fan, W., Han, J.: A survey on truth discovery. SIGKDD Explorations (2016)
7. Mohassel, P., Rindal, P.: Aby^3: A mixed protocol framework for machine learning. In: CCS (2018)
8. Saadeh, A., Senellart, P., Bressan, S.: Confidential truth finding with multi-party computation (extended version). CoRR abs/2305.14727 (2023)

A Key-Value Based Approach to Scalable Graph Database

Zihao Zhao[1,2], Chuan Hu[1,2], Zhihong Shen[1(✉)], Along Mao[1,2], and Hao Ren[1]

[1] Computer Network Information Center, Chinese Academy of Sciences,
Beijing, China
{zhaozihao,huchuan,bluejoe,almao,rh}@cnic.cn
[2] University of Chinese Academy of Sciences,
Beijing, China

Abstract. An increasing number of applications are modeling data as property graphs. In various scenarios, the scale of data can differ significantly, ranging from thousands of nodes/relationships to tens of billions of nodes/relationships. While distributed native graph databases can cater to the management and query requirements of large-scale graph data sets, they tend to be relatively cumbersome for small-scale data sets. This motivates us to develop a lightweight, scalable graph database capable of handling data across different scales. In this paper, we propose a method for constructing a graph database based on key-value storage, outlining the process of mapping graph data to key-value storage and executing graph queries on the key-value storage. We implemented and open-sourced a graph database based on RocksDB, namely KVGDB, which can manage data in an embedded fashion and be easily scaled to distributed environments. Experimental results demonstrate that KVGDB can effectively meet the management and query requirements of graph data sets, even at the scale of billions of nodes/relationships.

Keywords: Graph Database · Graph data · Key-Value Database

1 Introduction

Graph databases have emerged as a powerful tool for modeling and analyzing complex relationships between data entities in various applications, such as social networks [6] and knowledge graphs [2]. The data scale in these applications varies greatly, ranging from thousands of nodes/relationships to tens of billions of nodes/relationships. Traditional distributed native graph databases, such as TigerGraphDB [3] and ByteGraph, are based on distributed environments and can be cumbersome when dealing with small-scale datasets. Cloud databases like Amazon NeptuneDB [1] cannot be deployed locally and are difficult to meet the requirements of embedded applications. We are motivated to

This work was supported by the National Key R&D Program of China(Grant No.2021YFF0704200) and Informatization Plan of Chinese Academy of Sciences(Grant No.CAS-WX2022GC-02).

develop a lightweight, scalable graph database capable of handling data across different scales. Key-value databases store data in the form of key-value pairs, boasting excellent scalability. They can be used in embedded environments as well as expanded to distributed environments to address storage and query requirements of large-scale datasets. Additionally, key-value databases exhibit high performance, with the cost time of prefix search unaffected by the scale of data.

This paper proposes a key-value based approach for building scalable graph databases that can efficiently manage and query graph data of various scales.

2 Methodology

In key-value databases, both the key and the value are byte arrays. Taking RocksDB as an example [4], it manages data based on Log-Structured Merge Trees (LSM). Each write operation generates a memtable in memory, which, upon reaching a certain size, is written to an SST file on disk. By default, the key-value pairs in the SST file are sorted by their keys. This sorting scheme enables key-value databases to achieve good performance in precise search and prefix search operations. Typically, a prefix iterator is used for prefix search with a time complexity of O(m+k), where m is the number of keys satisfying the prefix condition, and k is the length of the longest key. Therefore, when designing storage formats and retrieval methods for graph data on key-value databases, it is crucial to fully exploit the inherent data order and the fast prefix search capabilities of key-value databases.

2.1 Storage

Suppose a property graph could be simply represented as $G =< N, R >$, where N is the set of nodes and R is the set of relationships (a.k.a. edges). The key-value storage model (as illustrated in Fig. 1) of G could be represented as $KVG =< NS, NLS, RS, RTS, ORI, IRI, PID, PI >$, where:

- NS: NodeStore, where the key is the combination of LabelID and NodeID, and the value is a byte array containing all the property information of the node. If a node has m labels (m >1), then in the NodeStore, the node is stored as m key-value pairs, each corresponding to a label. Specifically, if a node has no label, the storage engine will set its LabelID to an ID representing an empty label.
- NLS: NodeLabelStore, it stores the label information of nodes, where the key is the combination of NodeID and LabelID, and the value is blank.
- RS: It is the RelationshipStore, where the key is the RelationshipID (i.e. RelID in Fig. 1), and the value is a byte array containing all the property information of that relationship.
- RTS: RelationTypeStore, where the key is the combination of TypeID and RelationshipID of a relationship, and the value is an empty byte array.

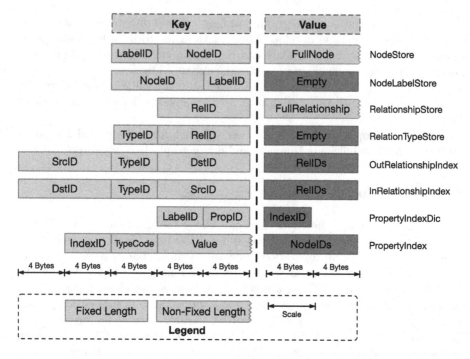

Fig. 1. Mapping Graph Data to Key-Value Storage

- *ORI*: OutRelationIndex, it is the outgoing edge index, built for relationships to accelerate graph query processing for a specific relationship direction. The key consists of source node ID (SrcID), relationship type ID (TypeID) and destination node ID (DstID) in order, and the value is the relationship IDs for all the relationships correspond to the key.
- *IRI*: InRelationIndex, it is the incoming edge index, which modifies the order of the key in the OutRelationIndex to destination node ID, relationship type ID and source node ID, with the rest remaining unchanged.
- *PID*: PropertyIndexDic, it is the embedded property index dictionary, storing IndexIDs of embedded property indexes. An index is uniquely identified by a LabelID and a PropID (i.e., the ID of the property name). The query engine can determine whether an index exists based on the LabelID and PropID through the PropertyIndexDic; if it exists, further property filtering can be performed in the PropertyIndex (i.e. *PI*).
- *PI*: PropertyIndex, it is the embedded property index, used for storing property indexes. In PropertyIndex, the key is a combination of IndexID, Type-Code, and Value, where IndexID refers to the property index ID described in PropertyIndexDic, TypeCode refers to the type encoding of the property value (such as integers, floating-point numbers, etc.) and Value refers to the actual value of the property. The value contains the IDs of all nodes with property values equal to Value under the constraints of the given LabelID and PropID.

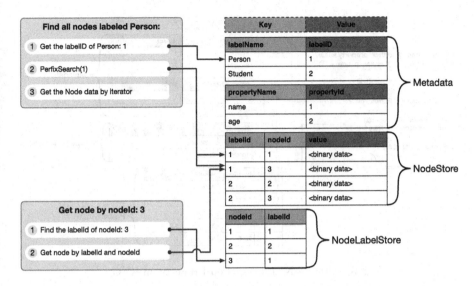

Fig. 2. Find Nodes based on label and ID on KVG

2.2 Query

Figures 2, 3 present three query operations on *KVG*, namely finding nodes by label, finding nodes by ID, and finding nodes by property. The retrieval operations of relationships share the similar process. More generally, other graph operations can be derived from the steps:

- Find all nodes labeled as *Person*. The process is shown in Fig. 2. First, obtain the LabelID of *Person* (assumed to be 1) from the metadata. Then, in the NodeStore, perform a prefix search to find the starting position with LabelID = 1. Next, traverse the data downwards until the first node with a different LabelID is encountered.
- Find a node with a specific ID (assumed to be 3). The process is shown in Fig. 2. First, in the NodeLabelStore, perform a prefix search to find the first key-value pair with NodeID = 3, and get the node's LabelID(1). Then, based on the LabelID(1) and NodeID(3), perform a precise search in the NodeStore to find the corresponding data item and get the full node.
- Filter nodes by the property value, as shown in Fig. 3. Suppose the query condition is to find all nodes labeled as *Person* with the *age* value of 31. First, obtain the LabelID(1) and PropID(2) from the metadata. Then, in the PropertyIndexDic, get the indexID(2). Finally, perform a prefix search on the PropertyIndex based on the IndexID(2), TypeCode(2), and PValue(31) to find the corresponding nodeIDs.

3 Implementation and Experiments

We implemented KVG based on RocksDB [4] and named it KVGDB. KVGDB has been open-sourced and adopted as the storage engine of PandaDB[1] [7].

---

[1] https://github.com/grapheco/pandadb-v0.3.

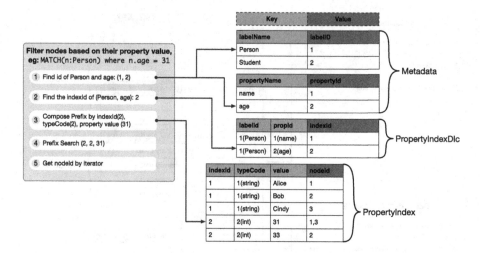

Fig. 3. Filtering Nodes Based on Property on KVG

KVGDB adopts Cypher [5] as the query language. We evaluate the performance of KVGDB on the LDBC-SNB dataset. LDBC-SNB [8] is currently the most popular property graph benchmark, which includes a scalable social-network dataset. The datasets used in this study are detailed in Table 1. The experiment is carried out on a server with 38 4GB memory, 28 CPU-cores and 10 TB hard disks.

Table 1. Details of Dataset

Dataset	Num of nodes	Num of edges	Size on the disk
D1	83,298,515	507,720,806	38 GB
D2	2,523,446,454	17,016,067,035	1.24 TB

Table 2 lists the basic graph query operations tested in this experiment and their execution times on different datasets. In the table, *KVGDB on D1* and *KVGDB on D2* represent the cost times for KVGDB to execute operations on datasets D1 and D2 (see Table 1), respectively. The *Baseline* represents the cost time for Neo4j-community-3.5.6 (one of the most successful graph databases) to execute the queries on dataset D1. We did not evaluate Neo4j on D2, because Neo4j failed to load D2 within 12 h.

The experimental results show that KVGDB performs better than the baseline, the execution time of most operations is within 10ms, and the execution time of each operation on the two datasets are quite similar. This indicates that the operation time does not increase significantly with the growth of data size, which is consistent with the characteristics and design expectations of KV databases.

Notably, according to the data in the first row of the table, the execution time for obtaining all nodes (getAllNodes) and all relationships (getAllRelationships) is much higher than that for other operations. This is because the operation to retrieve all nodes requires deserialization of all node data (and similarly for relationships), making it a traversal operation. The execution time is already close to the limit under existing hardware conditions.

Table 2. Cost Time of Graph Operation on the KVGDB

Operation	Baseline on D1	KVGDB on D1	KVGDB on D2	Operation	Baseline on D1	KVGDB on D1	KVGDB on D2
getAllNodes	396 s	36.8 s	1225 s	getAllRelationships	1218 s	123 s	6972 s
allLabels	3 ms	12 ms	15 ms	getRelationType	9 ms	<1 ms	<1 ms
addLabel	9 ms	4 ms	4 ms	addRelationType	15 ms	4 ms	21 ms
allPropertyKeys	1 ms	<1 ms	<1 ms	allPropertyKeys	1 ms	<1 ms	<1 ms
getPropertyKey	8 ms	<1 ms	<1 ms	getPropertyKey	10 ms	<1 ms	<1 ms
addPropertyKey	11 ms	<1 ms	2 ms	addPropertyKey	7 ms	<1 ms	<1 ms
getNodeById	8 ms	6 ms	7 ms	getRelationById	7 ms	<1 ms	12 ms
hasLabels	15 ms	<1 ms	<1 m	relSetProperty	7 ms	16 ms	88 ms
nodeAddLabel	12 ms	2 ms	6 ms	relRemoveProperty	8 ms	3 ms	13 ms
nodeRemoveLabel	9 ms	6 ms	31 ms	findToNodeId	7 ms	<1 ms	<1 ms
nodeSetProperty	11 ms	4 ms	7 ms	findFromNodeId	7 ms	<1 ms	8 ms
nodeRemoveProperty	9 ms	5 ms	8 ms	addRelation	10 ms	8 ms	9 ms
addNode	34 ms	<1 ms	2 ms	deleteRelation	12 ms	3 ms	2 ms
deleteNode	16 ms	<1 ms	2 ms	findOutRelations	145 ms	3 ms	2 ms
				findInRelations	901 ms	3 ms	5 ms

4 Conclusion

In this paper, we proposed a method for mapping graph data to key-value storage and implemented a scalable graph database based on RocksDB, namely KVGDB. It can manage data in an embedded fashion and be easily expanded to distributed environments for large-scale datasets. In the future, we will study graph pattern matching algorithms suitable for the features of KVGDB.

References

1. Bebee, B.R., et al.: Amazon neptune: graph data management in the cloud. In: ISWC (P&D/Industry/BlueSky) (2018)
2. Bollacker, K., Evans, C., Paritosh, P., Sturge, T., Taylor, J.: Freebase: a collaboratively created graph database for structuring human knowledge. In: Proceedings of the 2008 ACM SIGMOD International Conference on Management of Data, pp. 1247–1250 (2008)

3. Deutsch, A., Xu, Y., Wu, M., Lee, V.: Tigergraph: a native MPP graph database. arXiv preprint arXiv:1901.08248 (2019)
4. Dong, S., Kryczka, A., Jin, Y., Stumm, M.: Rocksdb: evolution of development priorities in a key-value store serving large-scale applications. ACM Trans. Storage (TOS) **17**(4), 1–32 (2021)
5. Francis, N., et al.: Cypher: an evolving query language for property graphs. In: Proceedings of the 2018 International Conference on Management of Data, pp. 1433–1445 (2018)
6. Myers, S.A., Sharma, A., Gupta, P., Lin, J.: Information network or social network? the structure of the twitter follow graph. In: Proceedings of the 23rd International Conference on World Wide Web, pp. 493–498 (2014)
7. Shen, Z., Zhao, Z., Wang, H., Liu, Z., Hu, C., Zhou, C.: `PandaDB`: intelligent management system for heterogeneous data. Int. J. Softw. Inform. **11**(1), 69–90 (2021)
8. Szárnyas, G., et al.: The LDBC social network benchmark: Business intelligence workload. Proc. VLDB Endowment **16**(4), 877–890 (2022)

Bitwise Algorithms to Compute the Transitive Closure of Graphs in Python

Xiantian Zhou[1](✉), Abir Farouzi[2], Ladjel Bellatreche[2], and Carlos Ordonez[1]

[1] University of Houston, Houston, USA
xiantianzhou@gmail.com
[2] LIAS/ISAE-ENSMA, Chasseneuil-du-Poitou, France

Abstract. The transitive closure (TC) of a graph is a core problem in graph analytics. There exist many High Performance Computing (HPC) and database solutions to solve the TC problem for "big graphs". However, they generally require the graph to fit in main memory and they require converting into specific binary file formats. To solve such limitations, this paper presents a novel solution to solve TC within the Python library ecosystem, combining HPC techniques and database system algorithms. We introduce two complementary algorithms removing HPC memory limitations: (1) an algorithm that efficiently converts edges into bit vectors and (2) a database-oriented, bit-vector, highly parallel matrix algorithm, which processes the graph in blocks. An experimental evaluation shows our solution provides better performance than state-of-the-art Python libraries.

1 Introduction

Transitive closure (TC) is one of the most computationally intensive tasks in data science research, primarily due to the large size and complex structure of graphs. It plays a crucial role in various graph problems. For instance, triangle enumeration represents the initial two steps in TC [4]. Consequently, several solutions leveraging HPC technologies have been proposed, including graph engine solutions, SQL-based solutions, and Python libraries [5]. SQL-based solutions offer elegance and memory limitations freedom, while Python is a popular language for data analysis, offering numerous libraries and packages for graph analysis such as GraphBLAS, Scikit-network, and NetworkX. These libraries provide state-of-the-art graph algorithms [2,3]. However, Python may be slow when analyzing large graphs, particularly those that cannot fit in RAM. Moreover, significant research progress has been made on efficient analytic algorithms for TC, some of which are based on relational algebra operations. For example, some algorithms employ hash-based fragmentation or fragmentation based on the semantic content of data. [7] explored a double-hash data fragmentation scheme. Another class of parallel TC algorithms is based on matrix manipulation. [1] presented parallel algorithms for computing the transitive closure of a database relation, applicable on both shared-memory and message-passing architectures. Generally, these parallel transitive closure algorithms operate directly on the adjacency list. However, parallel TC algorithms that work on a matrix representation can be more efficient. In fact, our experiments with different input graphs have shown that the TC graph density can exceed 70%, even if the input graph density is below 10%.

In this paper, we introduce disk-based distributed TC solutions that operate on the bit-matrix. We study how to develop TC algorithms within the Python ecosystem while adhering to principles of database systems. Our ultimate aim is to enable efficient processing of large graphs without memory limitations and with acceptable response times. Our implementations are suitable for reachability and path problems, and our experimental study shows the superiority of our solutions over existing popular analysis systems, suggesting potential advancements in bridging high-performance computing and Python.

2   Background

2.1   Graph

Let $G = (V, E)$ be a directed graph with $n = |V|$ vertices (V is the set of vertices) and $m = |E|$ edges (E is the set of edges). An edge in E links two vertices in V, and has a direction. The adjacency matrix of G is a $n \times n$ matrix where a 1 is stored in the entry (i, j) if there exists an edge from vertex i to vertex j. Storing the adjacency matrix in this sparse form helps conserve space and CPU resources. In our work, we do not use weight since we are solving TC problem. Thus, the input graph is represented as an edge list $E(i, j)$ and can be sorted either by i (E_i) or j (E_j). Since only existing edges are stored, the space complexity is $O(m)$. In sparse matrices, we assume $m = O(n)$.

The TC graph G^* compute all vertices reachable from each vertex in G. It is defined as: $G^* = (V, E^*)$, where $E^* = \{(i, j)$ s.t. there is a path between i and $j\}$. TC graph is stored as an adjacency matrix, because TC graph is much denser than the sparse input graph. The entries of TC matrix can be stored in one bit. Our paper uses a 64-bits integer in C to store 64 edges.

2.2   Classical Algorithms in Main Memory

Warshall's algorithm is recognized as the best algorithm for computing TC. It performs perfectly with HPC in main memory. However, when dealing with large graphs that cannot fit in main memory, it fetches random edges from disk and reads the entire matrix into memory at least N times. To address this issue, Warren proposed an improvement of Warshall's algorithm that reduces the number of I/O for large graphs by processing the matrix elements in a row order in two passes [6]. So we think Warren is better for large graphs because it has lower I/O: it requires loading block fewer times from disk into RAM. Thus, it is chosen as the base algorithm for developing our solution.

Input: E
Output: E^*

```
1 for i ← 2 to n do
2     for j ← 1 to i − 1 do
3         if E[i, j] = 1 then
4             | set E[i, *] = E[i, *] ∨ E[j, *]
5         end
6     end
7 end
8 for i ← 1 to n − 1 do
9     for j ← i + 1 to n do
10        if E[i, j] = 1 then
11            | set E[i, *] = E[i, *] ∨ E[j, *]
12        end
13    end
14 end
```

Algorithm 1: Warren's Algorithm

3   TC Solved in Python with Database Algorithms

Inspired by database systems, our solutions process the input graph by blocks instead of reading the entire graph into main memory.

3.1   Transforming the Edge Data Set into a Bit Matrix

Storing the TC graph in a matrix has many advantages such as using one bit to store each edge, and doing a world-parallel "OR" instruction by storing the matrix in packed row-major order. Therefore, our solution will pre-process the input graph by transforming it into a bit-matrix.

For large graphs, we read and process the input graph by blocks in the main memory. We summarize the conversion of a block into a bit-matrix block below.

1. Initialize a bit-matrix block to zero.
2. For each neighbor j of a source vertex i, find the position of j bit in row i.
3. Set the bit entry (i, j) to 1.

We call the vector of bits performing OR together as a *bit-vector*. A bitmask will be used to extract or set a bit in the bit-vector. In our solution, each block is read from disk, converted to a bit-matrix, and written back to disk. Figure 1 shows the workflow of converting the input data into a bit-matrix, whose size is $n/8$ bytes instead of $n * 4$ bytes (edges are stored with integer values).

3.2   Our Scalable Warren's Algorithm

We choose Warren's algorithm as the base algorithm to develop our solution for parallel systems, and we use the Numpy library in Python to implement

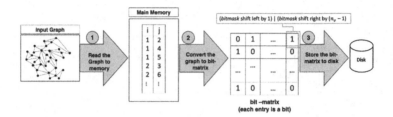

Fig. 1. Converting the input graph into a bit-matrix.

Input: $E_b, n_v, j_{start}, j_{end}$
Output: E_b^*

```
1  while not end of bit-matrix do
2  |    read a block, block_end ← min(block_start + block_size, n)
3  |    for i ← block_start to block_end do
4  |    |    bitmask ← 1
5  |    |    for j ← j_start to j_end do
6  |    |    |    bitmask ← (bitmask shift left by 1 ) | ( shift right by (n_v − 1))
7  |    |    |    if E_b(i, j/bsize) & bitmask = 1 then
8  |    |    |    |    for k ← 1 to n/n_v do
9  |    |    |    |    |    E_b[i, k] = E_b[i, k] ∨ E_b[j, k]
10 |    |    |    |    end
11 |    |    |    end
12 |    |    end
13 |    end
14 |    write the block to disk; block_start ← block_end + 1
15 end
```

Algorithm 2: *block_tc* algorithm.

this algorithm. The bit-matrix is stored on disk in a row-wise manner, allowing for continuous access during bitwise OR operations, which promotes bit-level parallelism. In our solution, we employ parallel processing to read a bit-vector, thereby enhancing computational speed. The size of the vector is determined by the processor's word size, which is dictated by the CPU. The parallel processes reduce the number of instructions that the system must execute.

Furthermore, we read and process a block once, since the entire input graph can be too large to fit in main memory. A block that contains several continuous rows is a plain array of bits. When one block finishes processing, a new one will replace the old one. At anytime during execution, there are two blocks in main memory, so the memory space needed is much smaller than $O(n^2)$. Algorithm 3 shows our improved TC algorithm. A variable *bitmask* which has the same bit size as *bitvector* will be used and initialized as 1 (the leftmost bit is 1). Note that TC will be dense for any connected graph, since each iteration makes the partial result denser. That is why a pure sparse solution becomes slow. If we use a dense matrix format from the beginning, the competing time for each loop is related to the number of vertices which is a constant.

Input: $E_b, bitvector, block_{size}$
Output: E_b^*

1   $n \leftarrow |V|$, $bitmask \leftarrow 1$, $n_v \leftarrow |bitvector|$, $block_{start} \leftarrow 2$
2   $\text{bloc\_tc}(E_b, n_v, 1, i-1)$
3   $block_{start} \leftarrow 1,$
4   $\text{bloc\_tc}(E_b, n_v, i+1, n)$

Algorithm 3: Our block-based TC system.

3.3 Time, Space and I/O Cost Analysis

Let us consider the limiting cases of a complete graph and a totally disconnected graph. That means the bit-matrix is an all-one matrix and an all-zero matrix. The time complexity of our solution is between $O(n^2/(n_v))$ and $O(n^3/(n_v^2))$, where n_v is the size of bit-vector. We can perform an 'or' operation on $vector\_size$ (n_v) bits at a time, which is the size of the $bitvector$. Then limiting cases can give n^2/n_v to n^3/n_v^2 number of 'Or' operations. Our solution is easy to be paralleled for the Numpy library in Python, the time complexity can be $O(n^2/(n_v * P))$ to $O(n^3/(n_v^2 * P))$ for the limiting cases where p is the number of processors. For large graphs, it will be processed by blocks. Note that the entire graph will be read to main memory once, even our solution processes by blocks. Suppose a block contains n_b rows, the number of I/O is n/n_b.

For space complexity, since TC computes whether a vertex i can reach another vertex j $(i, j \in V)$, each entry of the TC bit matrix is represented by one bit. The size of the TC bit matrix is $n^2/8$ byte since each bytes contains 8 bits. When processing by blocks, there are at most two blocks in main memory at the same time. Thus, the space complexity of our solution is $O(2 * n_b * n/8)$.

4   Performance Evaluation

4.1   Experimental Setup

Software and Hardware. For the competing system, we choose Python NetworkX; a popular graph analytics in Python. It is used for the creation, manipulation, and study of the complex graphs. We execute each experiment five times on a virtual machine which has 8 cores, 20 GB RAM, 1 TB disk. The size of the bit-vector is 64 to match the number of bits of CPU registers (a 64 bit int).. The computing time of our solutions includes I/O, transforming the input table to a binary matrix, and computing TC.

Data Sets. The used data sets are summarized in Table 1, obtained from the konect network data collection[1]. We chose graphs with different numbers of vertices and edges. Moreover, we use the density to indicate the graph structure, where graph density is $\frac{m}{n*n}$. We choose graphs with different density to evaluate our algorithm.

---

[1] http://konect.cc/.

4.2 Comparing with Python NetworkX Graph Library

The NetworkX library in Python has a function *transitive_closure*() that returns TC of a directed graph. Table 1 shows the average time measures. Our solution demonstrates significant performance advantages over NetworkX across various types of graphs. Particularly for large graphs like Marvel and Gnutella, our solution successfully completes the execution while NetworkX fails. Notably, our solution exhibits efficiency not only for dense graphs but also for sparse graphs. It is important to note that the running time is influenced by the density of the graph, as observed during the analysis of WekiLinks and Marvel's execution.

Table 1. Comparing with NetworkX, time in seconds.

Data set				Competitor	Our solution
	n	m	*density*	NetworkX	*block_tc*
Hamster	1859	12K	< 0.36%	191	4
WekiLinks	6K	439K	< 0.67%	Stop	1200
Gnutella	10K	39K	< 0.034%	Stop	112
Marvel	19K	96K	< 0.025%	Fail	439

"Stop" when computation is more than 30 min.

5 Conclusions and Future Work

In this paper, we explore bitwise algorithms to study TC with Python libraries. We believe Python language is more feasible, even other tools are efficient. Inspired by the database processing and HPC, We presented a disk-based solution to compute the TC of graphs. Our experiments show that it consistently outperformed popular analytic platforms, NetworkX library in Python. For future work, we will explore the logarithmic algorithm, and exploit more HPC techniques, such as multicore CPUs and GPUs to solve graph path problems.

References

1. Agrawal, R., Dar, S., Jagadish, H.V.: Direct transitive closure algorithms: design and performance evaluation. ACM Trans. Database Syst. **15**, 427–458 (1990)
2. Bonald, T., de Lara, N., Lutz, Q., Charpentier, B.: Scikit-network: graph analysis in Python. J. Mach. Learn. Res. **21**, 185:1–185:6 (2020)
3. Chamberlin, J., Zalewski, M., McMillan, S., Lumsdaine, A.: PyGB: GraphBLAS DSL in Python with dynamic compilation into efficient C++. In: IPDPS (2018)
4. Farouzi, A., Bellatreche, L., Ordonez, C., Pandurangan, G., Malki, M.: A scalable randomized algorithm for triangle enumeration on graphs based on SQL queries. In: Song, M., Song, I.-Y., Kotsis, G., Tjoa, A.M., Khalil, I. (eds.) DaWaK 2020. LNCS, vol. 12393, pp. 141–156. Springer, Cham (2020). https://doi.org/10.1007/978-3-030-59065-9_12

5. Ordonez, C.: Optimization of linear recursive queries in SQL. IEEE Trans. Knowl. Data Eng. **22**, 264–277 (2010)
6. Warren, H.S.: A modification of Warshall's algorithm for the transitive closure of binary relations. ACM Commun. **18**, 218–220 (1975)
7. Zhou, X., Zhang, Y., Orlowska, M.E.: Parallel transitive closure computation in relational databases. Inf. Sci. **92**, 109–135 (1996)

Discovering Top-K Partial Periodic Patterns in Big Temporal Databases

Palla Likhitha$^{(\boxtimes)}$ and Rage Uday Kiran

The University of Aizu, Fukushima, Japan
likhithapalla7@gmail.com, udayrage@u-aizu.ac.jp

Abstract. Partial periodic pattern mining involves discovering all the patterns in a temporal database that satisfy the specified *minimum periodic support* (*minPS*) and *period* (*per*) constraints. The *minPS* controls the minimum times a pattern must occur periodically in a database. The *per* controls the maximum inter-arrival time within which a pattern must reappear to consider its reoccurrence to be periodic in a database. Setting appropriate *minPS* and *per* values for any database is an open research problem. This paper addresses this open problem by proposing a solution to discover top-k partial periodic patterns in temporal databases. Top-k partial periodic patterns represent a total of *k* number of partial periodic patterns having the highest *minPS* value in a database. An efficient depth-first search algorithm, called top-k Partial Periodic Pattern Miner (*k*-3PMiner), which takes *k*, and *per* thresholds as an input was presented to find all desired patterns in a database. Experimental results on synthetic and real-world databases demonstrate that our algorithm is memory and runtime efficient and highly scalable.

Keywords: Data mining · Pattern mining · Periodic · Temporal Database

1 Introduction

Partial periodic pattern mining is an important knowledge discovery technique to find all patterns exhibiting partial periodic behavior in a temporal database. The basic model of partial periodic pattern mining is as follows [4]: Let $I = \{i_1, i_2, ..., i_n\}$ be the set of n items appearing in a database. A set of items $X \subseteq I$ is called an itemset. An itemset containing m items is called a m-itemset. The length of this itemset is m. A transaction t consists of timestamp, and an itemset. That is $t = (ts, Y)$, where ts represents the transaction time and Y is an itemset. A temporal database TDB is an ordered collection of transactions, i.e. $TDB = \{t_1, t_2, \cdots, t_k\}$, where $k = |TDB|$ represents the total number of transactions. Let ts_{min} and ts_{max} be the minimum and maximum timestamps of all the transactions in TDB, respectively. For a transaction $t = (ts, Y)$, such that $X \subseteq Y$, it is said that X occurs in t and such a timestamp is denoted as ts^X. The total number of transactions containing X in TDB is defined as the frequency of X and denoted as $freq(X)$. That is, $freq(X) = |TS^X|$. Let ts_j^X, $ts_k^X \in TS^X$,

C. Strauss et al. (Eds.): DEXA 2023, LNCS 14146, pp. 352–357, 2023.
https://doi.org/10.1007/978-3-031-39847-6_28

$1 \leq j < k \leq m$, denote any two consecutive timestamps in TS^X. An **inter-arrival time** of X denoted as $iat^X = (ts_k^X - ts_j^X)$. Let $IAT^X = \{iat_1^X, iat_2^X,$ $\cdots, iat_k^X\}$, $k = sup(X) - 1$, be the list of all inter-arrival times of X in TDB. An inter-arrival time of X is said to be **periodic** (or interesting) if it is no more than the user-specified *period* (*per*). A $iat_i^X \in IAT^X$ is said to be **periodic** if $iat_i^X \leq per$. Let $\widehat{IAT^X}$ be the set of all inter-arrival times in IAT^X with $iat^X \leq per$. That is, $\widehat{IAT^X} \subseteq IAT^X$ such that if $\exists iat_k^X \in IAT^X : iat_k^X \leq per$, then $iat_k^X \in \widehat{IAT^X}$. The period-support of X, denoted as $PS(X) = |\widehat{IAT^X}|$. Given a temporal database (TDB), *period* (*per*), and *minimum period-support* (*minPS*), the problem of partial periodic pattern mining is to find all patterns in TDB that have periodic-support no less than *minPS*.

Uday et al. [4] described a pattern-growth algorithm to find desired patterns in a temporal database. Ravi et al. [6] extend this model to discover the partial periodic patterns in columnar temporal databases. However, this model's widespread adoption and successful industrial application were hindered by this obstacle: "*minPS and per are two key constraints that make partial periodic pattern mining practicable in real-world applications. They are used to prune the search space and limit the number of patterns generated. Unfortunately, setting these two constraints for an application is an open research problem and may require a profound knowledge of the application's background.*" With this motivation, this paper proposes a solution of finding top-k partial periodically occurring patterns in a temporal database.

The contribution of this paper is as follows. First, we propose an extended model of finding top-k partial periodic patterns in a temporal database. Two constraints, namely k and *per*, were employed to find the interesting top-k partial periodic patterns having the highest *minPS* value in the database. A novel concept known as *dynamic minimum periodic support* was introduced to reduce the search space and computational cost-effectively. We also introduce an efficient algorithm, called top-k Partial Periodic Pattern Miner (k-3PMiner), to find all the desired patterns. Experimental results on synthetic and real-world databases demonstrate that our algorithm is memory and runtime efficient.

The rest of the paper is organized as follows. Section 2 presents the extended model of top-k partial periodic patterns and the proposed algorithm. Section 3 reports the experimental results. Finally, Sect. 4 concludes the paper with research directions.

2   Proposed Algorithm

2.1   Basic Idea: Dynamic Minimum Periodic-Support

Reducing the enormous search space is challenging as our model does not employ any constraint to reduce the search space. Finding candidate items (or 1-items) play a crucial role in discovering complete set of top-k partial periodic patterns. Algorithm 1 describes finding all candidate items that exist in a database to construct *c3PList*. Algorithm 2 descibes the procedure of finding all the top-k partial periodic patterns in a database.

Algorithm 1. PartialPeriodicItems(Temporal Database (TDB), K (k), period (per):

1: Let's say that the c3PList=($Y, TS\text{-}list(Y)$) is a dictionary that keeps track of temporal information about a pattern that occurs in a TDB. First, let's create a temporary list called TS_l and use it to keep track of the *timestamp* of the last time an item appeared in the database. Let PS be a temporary list to record the *periodic support* of an item in the database. Let *topkPatterns* be a list to record the top items with highest periodic support value. Let *dMinPS* be a variable to store the dynamic minimum periodic support *dMinPS* among *topkPatterns*.
2: **for** each transaction $t \in TDB$ **do**
3:   **if** ts_{cur} is i's first occurrence **then**
4:     Insert i and its timestamp into the c3P-list.
5:     Set $TS_l[i] = ts_{cur}$ and $PS^i = 0$.
6:   **else**
7:     Add i's timestamp in the c3P-list.
8:     **if** $(ts_{cur} - TS_l[i]) \leq per$ **then**
9:       Set $PS^i + +$.
10:    Set $TS_l[i] = ts_{cur}$.
11: Sort the items in the c3P-list in ascending order of their periodic support.
12: **for** each item i in c3P-list **do**
13:   **if** $length(topkPatterns) < K$: **then**
14:     Store the item into *topkPatterns*
15: $dMinPS = min$(periodic support of all items in *topkPatterns*)
16: Call k-3PMiner(c3P-List).

Algorithm 2. k-3PMiner(c3P-List)

1: **for** each item i in c3P-List **do**
2:   Set $tp = \emptyset$ and $X = i$;
3:   **for** each item j that comes after i in the c3P-list **do**
4:     Set $Y = X \cup j$ and $TS^Y = TS^X \cap TS^j$;
5:     Calculate $minPS$ of Y;
6:     **if** $PS(TS^Y) \geq dMinPS$ **then**
7:       Add Y to tp and Y is considered as candidate top-k partial periodic itemset;
8:     Check(Y, TS^Y)
       (to check if pattern can make in to top-k partial periodic pattern)
9:   k-$PFPMiner(tp)$

Algorithm 3. Check(X, TS-List)

if $minPS(TS - List) > dMinPS$ **then**
  Pop the Last pattern and insert X in *topkPatterns*.
  $dMinPS = min$(periodic support of all items in *topkPatterns*)

3   Experimental Results

Since there exists no algorithm to find Top-k partial periodic patterns in temporal databases using k constraint, we evaluated our algorithm k-3PMiner with naïve algorithm, The naïve algorithm involves the following two steps: (i) finding all partial periodic patterns in a temporal database using 3P-Growth algorithm [5] and (ii) generating top-k partial periodic patterns from all partial periodic patterns by performing another sorting.

3.1   Experimental Setup

Our k-3PMiner algorithm was developed in Python 3.7 and executed on a Gigabyte R282-z94 rack server machine containing two AMD EPIC 7542 CPUs and 600 GB RAM. The operating system of this machine is Ubuntu Server OS 20.04. The experiments have been conducted on both synthetic (**T10I4D100K**) and **BMS-WebView-1** and real-world **Pollution** databases.

The **T10I4D100K** is a sparse synthetic database generated using the procedure described in [2]. This database contains 870 items and 100,000 transactions. The *minimum, average*, and *maximum* transaction lengths of this database are 1, 10, and 29 respectively. The **BMS-WebView-1** is a sparse database containing 59,602 transactions and 497 items. The *minimum, average*, and *maximum* transaction lengths of this database are 1, 10, and 76 respectively.

The Pollution database is a high dimensional real-world database provided by Japanese Ministry of the Environment developed the Atmospheric Environmental Regional Observation System (AEROS) [3] to tackle air pollution problems. Each transaction contained the following information: *timestamp in hours, station identifiers that have recorded $PM_{2.5}$ values no less than 16 $\mu g/m^3$*. The resulting database, **Pollution**, contained 1600 items and 720 transactions. The minimum, average, and maximum transaction lengths are 11, 460, and 971, respectively. The k3P-miner code and the databases were provided at [1] of our experiments.

3.2   Evaluation of both the Algorithms by Varying only k

Figures 1a, 1b, 1c shows the top-k partial periodic patterns discovered on different T10I10D100K, BMS-WebView-1 and Pollution databases by varying k value, respectively. The *per* values are set at 2000, 1000 and 250 (in count) respectively. For Naïve algorithm the *minPS* values are set at 100, 30, 250 (in count) respectively. As k increases, the number of top-k patterns also increases.

Figures 2a, 2b, and 2c show the time consumed at a different number of k values in T10I10D100K, BMS-WebView-1 and Pollution databases, respectively. It can be observed that an increase in k increases the runtime to find all top-k partial periodic patterns being generated at different k values. As k increases, the number of patterns to be mined increases, resulting in time consumption.

Figures 3a, 3b, and 3c show the memory consumed at a different number of k values in T10I10D200K, BMS-WebView-1 and Pollution databases, respectively.

It can be observed that an increase in k increases the memory to find all top-k partial periodic patterns being generated at different k values.

3.3   Scalability Test

In this experiment, we have used the Kosarak database, which is a huge database having 9,90,000 transactions (in count). We have divided this database into five segments, each consisting of 200,000 transactions. We have evaluated the performance of k-3PMiner by adding each successive segment to the ones that came before it. The runtime requirements and memory consumption k-3PMiner for each segment of the Kosarak database are shown in Fig. 4a and 4b, when $k = 200$. The following are some noteworthy findings that can be derived from these figures: (i) runtime requirements of k-3PMiner increases almost proportionally as database size grows. (ii) memory requirements of k-3PMiner where we can observe same as 4a.

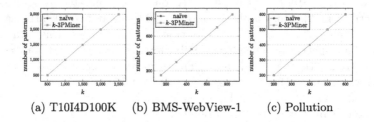

(a) T10I4D100K   (b) BMS-WebView-1   (c) Pollution

Fig. 1. top-k patterns on various databases by varying k

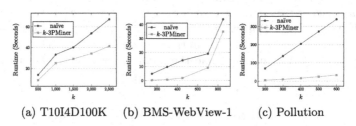

(a) T10I4D100K   (b) BMS-WebView-1   (c) Pollution

Fig. 2. Runtime evaluation on various databases by varying k

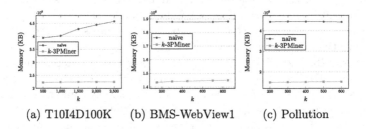

(a) T10I4D100K   (b) BMS-WebView1   (c) Pollution

Fig. 3. Memory evaluation on various databases by varying k

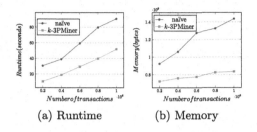

(a) Runtime   (b) Memory

Fig. 4. Scalability of k-3PMiner

4 Conclusions and Future Work

In this paper, we have proposed an efficient depth-first search algorithm, called top-k Partial Periodic Pattern Miner (k-3PMiner), to find all desired patterns in big temporal databases. We have solved the open research problem of setting $minPS$ and *per* constraints by introducing a novel upper-bound measure named *dynamic minimum periodic support*. An in-depth examination of the proposed k-3PMiner approach on four synthetic and real-world databases revealed that its memory consumption and runtime are efficient and highly scalable. As for future work, we will work on discovering top-k partial periodic patterns in uncertain databases.

References

1. k3pminer and datasets to verify repetability. https://github.com/udayRage/codeData/DEXA_2023
2. Agrawal, R., Imieliński, T., Swami, A.: Mining association rules between sets of items in large databases. In: SIGMOD, pp. 207–216 (1993)
3. Ministry of Environment, J.: Atmospheric environmental regional observation system (2021). http://soramame.taiki.go.jp/ Accessed 1 June 2021
4. Kiran, R.U., Shang, H., Toyoda, M., Kitsuregawa, M.: Discovering partial periodic itemsets in temporal databases. In: International Conference on Scientific and Statistical Database Management, pp. 30:1–30:6 (2017)
5. Kiran, R.U., Venkatesh, J., Toyoda, M., Kitsuregawa, M., Reddy, P.K.: Discovering partial periodic-frequent patterns in a transactional database. J. Syst. Softw. **125**, 170–182 (2017)
6. Kiran, R.U., et al.: Efficient discovery of partial periodic patterns in large temporal databases. Electronics **11**(10), 1523 (2022). https://doi.org/10.3390/electronics11101523,https://www.mdpi.com/2079-9292/11/10/1523

Query Optimization

Dexteris: Data Exploration and Transformation with a Guided Query Builder Approach

Sébastien Ferré[(✉)]

Univ Rennes, CNRS, Inria, IRISA, 35000 Rennes, France
ferre@irisa.fr

Abstract. Data exploration and transformation remain a challenging prerequisite to the application of data analysis methods. The desired transformations are often ad-hoc so that existing end-user tools may not suffice, and plain programming may be necessary. We propose a guided query builder approach to reconcile expressivity and usability, i.e. to support the exploration of data, and the design of ad-hoc transformations, through data-user interaction only. This approach is available online as a client-side web application, named Dexteris. Its strengths and weaknesses are evaluated on a representative use case, and compared to plain programming and ChatGPT-assisted programming.

1 Introduction

In recent years, data has become ubiquitous and it is increasingly important for stakeholders to derive value from them. To achieve this objective, data analysis methods have been developed to help stakeholders gain insight into their data. However, it is almost always necessary to explore and transform data before applying data analysis methods. *Data exploration* is crucial to help data analysts understanding the data, and choosing the transformations to apply. *Data transformation* refers to the process of converting data from one format, structure, or type to another to meet the requirements of a particular use case or analysis. Examples of data transformation operations include filtering, aggregating, sorting, merging, pivoting, and applying mathematical or statistical functions to the data. In this paper, we focus on fine-grained data transformations, like converting between ad-hoc CSV and JSON files, and extracting or aggregating information from such files.

When choosing a tool for data exploration and transformation, data stakeholders are left with a trade-off between their expressive power and their usability. Expressive power is the range of questions and transformations that can be applied to the data. Usability is the degree of technical skills required to use it, as well as the level of guidance provided by the tool. At one end of the spectrum there is full-fledged programming, e.g. programming in the rich environment

---

This research is supported by the CominLabs project MiKroloG.

of Python. This obviously offers the highest expressive power but this requires advanced programming skills and offers little guidance. At the other end of the spectrum there are end-user intuitive applications with a GUI (Graphical User Interface). For instance, spreadsheets are obviously more usable, as testified by their widespread usage. However, they are limited to tabular data, and computations are mostly cell-wise. In between there are ETL tools (Extract, Transform, Load), e.g. Talend, Pentaho. They offer high-level features for common data formats and common data transformations, e.g. merging data, removing duplicate data. However, in many cases, they require to write SQL queries to extract data from relational data tables, or JSON paths to extract data from JSON files, or other kinds of code. They thus require advanced technical skills, although to a lesser degree than full-fledged programming skills.

The N<A>F design pattern [4] has been shown to help reconciling expresivity and usability. It does so by relying on a formal language, and by bridging the gap between the end-user and the formal language with a *guided query builder* approach. Complex queries are incrementally and interactively built with data feedback and guidance at every building step. The design pattern has already been applied to the querying of SPARQL endpoints [5], the authoring of RDF descriptions and OWL ontologies [4], and data analytics on RDF graphs [6].

In this paper, we present the application of the N<A>F design pattern to the task of data exploration and transformation. We choose JSONiq [7] – the JSON query language – as the target formal language because it combines several advantages. It is a high-level declarative query language and yet a Turing-complete programming language; it lies on W3C standards; and although it uses JSON as a pivot data format, it is interoperable with other data formats such as text, CSV or XML. JSONiq is to JSON data what XQuery is to XML, and what SQL is to relational data. As a declarative and expressive language, it is a good fit for data exploration, data extraction, data transformation, and even data generation. The contributions of this work are:

1. a guided query builder approach to data exploration and transformation, based on the N<A>F design pattern, where arbitrary computations can be achieved, and only primitive inputs are required from end-users;
2. an online prototype called Dexteris[1], running as a client-side web application.

The paper is organized as follows. Section 2 motivates our approach with a concrete example. Section 3 discusses related work, and describes the N<A>F design pattern. Sections 4, 5, and 6 defines the three parts of N<A>F for data exploration and transformation: the intermediate language, the machine side, and the user interface. Section 7 evaluates the strengths and weaknesses of our approach, comparing it with plain programming and ChatGPT-assisted programming. Supplementary materials are available online[2] for the motivating example and evaluation use case (input and output files, Python programs and JSONiq queries, chat logs, and a Dexteris screencast).

---

[1] Freely available at http://www.irisa.fr/LIS/ferre/dexteris/.
[2] http://www.irisa.fr/LIS/ferre/pub/dexa2023/.

2   Motivating Example

As a motivating example, let us consider a scenario where the available input file is a CSV file describing projects by their ID, name, and members.

```
project id,project name,members
P1,Alpha,"Alice, Charlie"
P2,"Beta 2","Bob, Charlie"
```

The objective is to obtain a JSON file organized as a list of unique project members, describing each member by the list of projects she takes part in, and the number of such projects.

```
[ {"member": "Alice",
   "projects": [{"id": "P1", "name": "Alpha"}],
   "project number": 1},
  {"member": "Charlie",
   "projects": [{"id": "P1", "name": "Alpha"},
                {"id": "P2", "name": "Beta 2"}],
   "project number": 2},
  {"member": "Bob",
   "projects": [{"id": "P2", "name": "Beta 2"}],
   "project number": 1} ]
```

Any system that works on tabular data only will have a hard time generating the expected output data because the latter has a nested structure. Indeed, it is a list of objects that have a field ("projects") whose values are again lists of objects. The transformation from the available input to the expected output requires at least the following processing steps, in informal terms:

- reading the tabular data structure (CSV) in the input file;
- iterating over projects, i.e. over rows;
- splitting the lists of members, in the third column;
- grouping all (project, member) pairs by member;
- collecting all projects of each member;
- counting the number of projects per member;
- generating a JSON object for each member;
- collecting them and writing the whole in JSON format.

A concise Python program that performs the transformation is about 30 lines. Using JSONiq, we can make the transformation shorter (12 lines) and higher-level, in particular without assignments to mutable variables, hence without having to reason about computation states [10]. It relies on a JSON view of a CSV file, where each row is represented as a JSON object with CSV columns as fields.

```
for $row in collection("input.csv")
  let $project := {
      "id" : $row."project id"
```

```
    "name" : $row."project name" }
for $member in split($row."members", ", ")
  group by $member // $project is now a sequence of objects
  return {
      "member" : $member,
      "projects" : [ $project ],
      "project number" : count($project) }
```

Using the Dexteris tool that implements our approach, it is possible to build the above JSONiq program by starting from the input file, and then by applying suggested elementary transformations one after another. The only required inputs are elementary values: field names, new variable names, and the splitting separator. The number of required steps is 25, which includes 25 selections and 8 short inputs. Any part of a built transformation can be edited a posteriori. For instance, if one wants the JSON output to be in alphabetical order of members, only 2 additional steps are needed.

3   Related Work and Background

To cope with the difficulty to write data transformations in general-purpose programming languages, high-level data-oriented languages have been defined and even standardized. Notable examples are XQuery for XML data [14], JSONiq for JSON data [7] – which was strongly inspired by XQuery –, and formula languages that back graphical tools, such as the M language in Power BI [2]. Graphical tools like Power BI ambition to make all transformations doable in a graphical way but they recognize that *"there are some transformations that can't be done in the best way by using the graphical editor."* [9] The latter is also limited to tabular data, although it can cope with varied formats. The closest work to ours is probably the educative platform Scratch by MIT [12], which enables users to build arbitrarily complex programs in a purely graphical way, by assembling blocks with syntax-based shapes. However, its programming language is not appropriate for data transformations. Another related domain is *program synthesis* [8], which ambitions to generate programs solely by providing examples of input-output pairs. For instance, given strings like *"Dr. Helen Smith (1999)"*, it can learn to output strings like *"Smith H."* from a few examples. It is however yet too limited in the size of the examples and in the complexity of generated programs to be largely applicable to data transformations. Moreover, some transformations are one-off and therefore producing an example implies producing the expected output. It is also sometimes simpler to specify the transformation in an intentional way rather than by providing examples.

The purpose of the N<A>F design pattern [4] is to bridge the gap between an end user speaking a natural language (NL) and a machine understanding a formal language (FL), as summarized in Fig. 1. The design pattern has for instance been instantiated to the task of semantic search with SPARQL as the formal language [5]. The central element of the bridge is made of the Abstract Syntax Trees (AST) of an Intermediate Language (IL), which is designed to

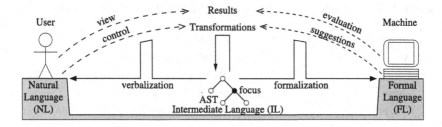

Fig. 1. Principle of the N<A>F design pattern

make translations from ASTs to both NL (*verbalization*) and FL (*formalization*) as simple as possible. IL may not have any proper concrete syntax, NL and FL playing this role respectively for the user and for the machine.

N<A>F follows the query builder approach, where the structure that is incrementally built is precisely an AST. Unlike other query builders, the generated query (FL) and the displayed query (NL) may strongly differ in their structure thanks to the mediation of IL. The AST is initially the simplest query, and is incrementally built by applying *transformations* (not to be confused with the data transformations this paper is about). A transformation may insert or delete a query element at the *focus*. The *focus* is a distinguished node of the AST that the user can freely move to control which parts of the query should be modified. Results come from the evaluation of the formalized query, and are viewed by the user. Transformations are suggested by the machine based on the formalized query and actual data, and controlled by the user. Both results and transformations are verbalized in NL for display to the user. At each step, the user interface shows: (a) the verbalization of the current query with the focus highlighted, (b) the query results, and (c) the suggested transformations.

4   Intermediate Language and Transformations

Given that the target formal language (FL) JSONiq is already high-level and declarative, we also use it as the Intermediate Language (IL). As shown in the next section, this does not make the formalization step from IL to FL void, and it therefore still makes sense to distinguish between IL and FL. Indeed, a key feature of IL is the notion of *focus* that impacts both the results and the suggested query transformations. In particular, the position of the focus modifies the semantics of the query in order to show the internals of the computations performed by the query at the focus point.

In this section, we first define the FL/IL as a large subset of JSONiq plus a few extensions of our own. We then introduce the list of elementary query transformations such that all queries can be built through finite sequences of such transformations.

Table 1. JSONiq constructs: expressions, FLWOR clauses, and syntactic sugar.

Expression constructs (*expr*)	Semantics
VAR	variable (prefixed by $)
JSON	JSON value
{ *expr* : *expr* , ...}	object construction
{| *expr* |}	object sequence to object (merge)
[*expr*]	sequence to array
expr []	array to sequence
FUNC (*expr* , ...)	function calls and operators
()	the empty sequence
expr , *expr*	sequence concatenation
expr . *expr*	object lookup by field
expr [[*expr*]]	array lookup by index
if *expr* then *expr* else *expr*	conditional expression
flwor ↵ ... return *expr*	expression nested in FLWOR clauses

FLWOR clauses (*flwor*)	Semantics
let VAR := *expr*	binding a new variable
def FUNC (VAR , ...) = *expr*	defining a new function
for VAR in *expr*	iterating on a sequence
where *expr*	filtering an iteration
order by *expr* [asc | desc] , ...	ordering an iteration
group by VAR , ...	grouping and concatenating
count VAR	variable for the position in iteration

Syntactic sugar

$expr_1$! $expr_2$ = for $$ in $expr_1$ ↵ return $expr_2$
$expr_1$ [$expr_2$] = for $$ in $expr_1$ ↵ where $expr_2$ ↵ return $$

4.1   A Language Based on JSONiq

Table 1 lists the constructs of JSONiq that are used in our IL (symbol ↵ represents a carriage return). They are presented in concrete syntax for readability and for consistency with the standard JSONiq syntax. However, they are used under abstract syntax only for IL, and verbalized in a slightly different way to make them more intuitive to end users (see Sect. 6).

The table gives the semantics of each construct in an informal way. An original aspect of JSONiq is that values are *sequences* of *items*, where items are JSON values. For recall, JSON values are one of: strings delimited by double quotes, numbers, Boolean values (**true** and **false**), arrays delimited by square brackets, objects with named fields and delimited by curly braces, and the **null** value.

Array members and object field values can be arbitrarily nested JSON values. Another original aspect of JSONiq expressions is the FLWOR clauses (pronounce "flower") that help working with sequences in a declarative way. They are inherited from XQuery [14], and they are analogous to clauses found in SQL and SPARQL. FLWOR is an acronym for the constructs: for, let, where, order by, and return. The combination of for and where clauses can express joins like in relational databases. The combination of for, group by, and aggregation functions – i.e. functions from sequences to items – can express analytical queries, similar to OLAP cubes [3].

An important ingredient of the semantics and evaluation of JSONiq expressions is the *environment*. It defines for each sub-expression the set of variables that are in scope. Variables are added to the environment by the FLWOR clauses let, for, and count, and are in the scope of the FLWOR clauses coming after, until the expression after return. An environment maps each variable in scope to its value, a sequence of JSON values.

Our language has a few differences with JSONiq. First, it currently misses a few constructs, left for future work, namely: anonymous functions and partial function application, switch and try-catch expressions, and type-related expressions (e.g., instance-of). Second, it adds two convenient FLWOR clauses to explode JSON objects in as many variable bindings are there are object fields:

- let * := *expr* assumes that the expression returns an object, and is then equivalent to have a let-binding for each object field, making them directly available as local variables;
- for * in *expr* is equivalent to for $$ in *expr* ↪ let * := $$, thus combining iteration on objects, and exploding them into let-bindings.

Here is the query that defines the expected data transformation in the motivating example (Sect. 2), in our JSONiq-based language.

```
printJSON(
    for * in parseCSV(file <example_input.csv>)
      let $project := {
          "id" : $(project id),
          "name" : $(project name) }
      for $member in split($members, ", ")
        group by $member
        return {
            "member" : $member,
            "projects" : [ $project ],
            "project number" : count($project) })
```

The expression file <example_input.csv> evaluates to the raw contents of the input file, a string. A query can use any number of files. Function parseCSV turns this raw string into a JSON representation of the tabular data, i.e. a sequence of JSON objects, each object representing a CSV row with column headers as object fields. The for * clause iterates over the CSV rows, and bind each column as a variable (e.g., $(project id)). After grouping by member, variable `$project` maps to a sequence of projects, all projects related to the current member. This can be seen as a default aggregation, from which any other aggregation can be computed. For instance, [`$project`] aggregates the sequence of projects as a JSON array, and `count($project)` counts the number of projects. Finally, function `printJSON` turns the JSON result, a sequence of JSON objects, into a raw string in JSON format, ready for writing into a file.

4.2   Query Focus and Query Transformations

A *query focus* splits a query into the sub-expression at focus, and the *context* of that sub-expression. For instance, in the expression `split($row.members,",")`, if the focus is on the sub-expression `$row.members` then the context is [`split(•, ", ")`], where • (called *hole*) locates the focus position. If the focus is on `$row` then the context is [`split(•.members, ", ")`], which can be seen as the nesting of two elementary contexts: [`•.members`] and [`split(•, ", ")`].

A *query transformation* modifies the expression around the focus or moves the focus. The empty sequence () serves as the initial query, and also to fill in new sub-expressions introduced along with constructs. A transformation belongs to one of the following kinds, beside focus moves.

- An *elementary expression* that replaces the sub-expression at focus: a scalar value (e.g. `"id"`), a variable (e.g. `$member`), etc.
- An *elementary context* that is inserted between the sub-expression at focus and its context: e.g., [`•, ()`], [`(), •`], [`for * in •`]. The hole is located at one sub-expression, and other sub-expressions are initialized to ().
- The *addition* of an element other than a sub-expression, for which there is no focus: e.g., adding a field to an object, adding a variable to a **group by** clause, adding a parameter to a defined function.
- The *deletion* of the sub-expression at focus, i.e. its replacement by (), or the *deletion* of the elementary context at focus.

Some transformations have editable parts, which have to be filled in by the user. This is the case for scalar values, for field names, and for the name of new variables introduced by binding constructs. The query in the above section can be built through the following sequence of transformations (25 steps separated by /, up is for moving the focus up in the AST).

```
file <example_input.csv>  /   parseCSV(•)   /    for * in •   /
{"id": ()} / $(project id) / "name": () / $(project name) /
up / let $project := • / $members / split(•, ()) / ", " / up /
for $member in • / group by $member / {"member": ()} / $member /
"projects": () / $project / [ • ] / "project number": () /
$project / count(•) / up⁶ / printJSON(•)
```

It can be proved by induction that all expressions can be built in a finite number of steps, proportional to the syntactic size of the expression. The next section explains how the user receives data feedback and guidance at every step, so that what here seems like a purely syntactic process is actually a data-centered and guided incremental process.

5   Formalization and Suggestions (Machine Side)

The formalization process determines the evaluation to be actually performed, and hence the results to be displayed to the user. An important point is that it depends on the focus position. To motivate and illustrate this dependency, suppose we have the following expression

```
for $i in 1 to 3
  return 1 to $i
```

where the operator a to b valuates to the sequence a, $a + 1$, ..., b. The result of the whole expression is the sequence of integers 1, 1, 2, 1, 2, 3. Now, suppose that the focus is on the sub-expression after **return**, then the expected result is the value of that sub-expression 1 to $i. However, this value depends on variable $i, which is introduced in the focus context. Therefore, a useful result is a mapping from each value of $i to the value at focus, bound to an implicit variable $focus.

$i	$focus
1	1
2	1, 2
3	1, 2, 3

Each row of such a table is actually an environment, i.e. a mapping from variables in the focus scope to their values.

5.1   Formalization by Expression Rewriting

Formalization is performed by rewriting the query expression and focus position into a new expression whose evaluation results in a sequence of environments. Let the query be the expression $e = C(f) = c_k(\ldots c_1(f)\ldots)$, where f is the sub-expression at focus, C is the context of the focus. The context C can be decomposed into a series of elementary contexts c_i that need to be applied to the sub-expression f bottom-up in order to get the whole expression e. In the above example, $f = [1$ to $i]$, $k = 2$, $c_1 = [$return $\bullet]$, and $c_2 = [$for $i in 1 to 3 $\hookleftarrow \bullet]$.

The rewriting process starts by initializing the rewritten expression e' as

$$e' := \text{let } \$focus := f \hookleftarrow \text{return } \$env$$

where variable $focus is bound to the sub-expression at focus, and the environment is returned through a special variable $env. Then each elementary context is processed bottom up, from c_1 to c_k.

- FLWOR clauses are applied unmodified, so that iterations, bindings, filtering, grouping and ordering are kept in the focus-dependent evaluation. They determine the rows of the table of results.

- Conditional expressions with the hole in one of the branches are simplified by replacing the other branch by the empty sequence. For instance, context [if e_1 then • else e_3] applied to the rewritten expression leads to $e' :=$ if e_1 then e' else ().
- All other contexts are ignored, and hence excluded from the rewritten expression. For instance, ignoring context $func(e_1, •)$ enables to focus on the second argument of the function, temporarily ignoring the first argument and the function application.

The rewritten expression is therefore a chain of FLWOR clauses (and conditionals) ending with the return of an environment. The evaluation result is therefore a sequence of environments. Given that the environments bind the same set of variables, the result can be presented as tabular data, with one row for each environment – each iteration step –, and a column for each variable. The last column is the $focus variable, which plays a central role for computing the suggested query transformations. Note that although the view on results is tabular, the data can be arbitrarily nested JSON data. The table of results contains JSON values, and its shape automatically adapts to the current query and focus.

5.2   Computation of Suggestions

In the N<A>F design pattern, the set of suggestions is the subset of query transformations that are well-defined and relevant given the current query, the current focus, and the results of the query formalization. First, a static analysis of the current query and focus is performed in order to identify variables and functions in scope and to derive type constraints. The considered types are the JSON types: numbers, strings, booleans, arrays, and objects. For instance, the focus context [• . e_2] calls for JSON objects, while the focus context [e_1 . •] calls for strings (field name). Also, if the focus sub-expression is a comparison, then its type is boolean, which suggests to insert contexts such as [if • then () else ()] (conditional expression) or [where •] (filtering).

Second, a dynamic analysis of the results is performed in order to identify which data types are available at focus. For instance, the presence of numbers suggests to apply arithmetic operators; and the presence of arrays suggests to apply a lookup-by-index operator, for instance. We also collect the fields defined in objects, in order to suggest the lookup-by-field operator with pre-defined field names. The dynamic analysis also looks at the number of rows, and at the sequence lengths of focus values. The former conditions the insertion of FLWOR clauses, which are only relevant when there are multiple rows, i.e. in the scope of a for clause. The latter conditions the insertion of iterations and aggregations, which are only relevant with non-singleton sequences.

For the sake of efficiency, dynamic analysis is only performed on a sample evaluation of the results, bounding the number of rows, and bounding the number of computed items per sequence. This relies on a lazy evaluation of expressions.

Fig. 2. Screenshot of Dexteris in the course of building the example query.

6 Verbalization and Control (User Interface)

Verbalization of the query and results, and control of the suggested transformations characterize the user interface, and hence the user experience. Figure 2 shows a screenshot of the Dexteris tool. The query and focus can be seen at the top left. The results can be seen in the table at the bottom. The suggested transformations can be seen in the three lists at the top right: functions and operators (left), variables and JSON value constructors and accessors (middle), FLWOR clauses (right).

The verbalization remains close to the original concrete syntax of JSONiq, as given in Table 1, and hence it is less natural than in previous N<A>F applications. To justify this, note that verbalizing the expression e_1 [[e_2]] as *"the e_2-th member of e_1"*, or the expression {"id" : $(project id), "name" : $(project name) } as *"an object whose id is the project id, and whose name is the project name"*, is closer to natural language but it does not make it easier to read. By the way, JSONiq already uses explicit keywords like where or if, and most special characters correspond to JSON notations (e.g., square brackets and curly braces). We apply the following changes to the concrete syntax of JSONiq in order to lift its more unnatural aspects.

- The dollar sign in front of variables and the double quotes surrounding strings are removed. Colors are used to distinguish numbers (dark blue), strings (dark green), booleans (green/red), variables (dark red), and function names (purple). In particular, this avoids the need for escaping characters in strings.
- The implicit $$ variable in syntactic sugar is renamed as this.
- A mixfix syntax is used for some functions, which includes the usual infix notation for arithmetic and logical functions: e.g., [e_1 + e_2], [not e_1], [split

e_1 by e_2]. This makes the role of the different function arguments more explicit and readable.

– The semi-colon ; is used as the sequence separator to avoid confusion with other usages of the comma (arrays, objects, function arguments).

Moreover, as queries are built by applying query transformations rather than edited as text, there is no issue with operator priorities and other ambiguities so that grouping brackets become useless. Those groupings are made visible by indentation and through focus moves because the sub-expression at focus is highlighted (in light green, see Fig. 2).

Suggestions are controlled simply by clicking them. When a suggestion has input widgets, they may be filled beforehand. Those inputs are for primitive values in the middle list, and for variable/function names in the right column. The focus can be moved up with one of the suggestions, and in all directions with key strokes (Ctrl+arrows). Any focus can also be selected directly by clicking on it in the query area. For advanced users, the text input below the query provides a command line interface for quickly inserting data values, and applying query transformations. When a suggestion is clicked, its command is displayed in the text input as hint in order to help the user learning them.

7   Evaluation

Because data transformation tasks are often ad-hoc and incompletely specified, it is difficult to conduct a systematic evaluation similar to what is done for fully automated approaches, e.g. supervised classification tasks. Moreover, we do not know of tools comparable to Dexteris, tools that would support the design of almost arbitrarily complex transformations without requiring the user to write some code at some point. We therefore choose to report on a representative use case that was encountered in a real setting. It involves two file formats (JSON and CSV), and many features of the JSONiq language. It is relatively simple to informally describe while being non-trivial to implement. We compare the user experience and result with plain programming and using ChatGPT [11], which is known for its capability to generate code from textual prompts in a versatile and multi-turn way.

Task. The objective is to transform the JSON files of the Mintaka dataset [13] by extracting some information and formatting it into CSV files. The end goal was to prepare a training set for question answering [1]. A JSON file is a list of questions, where each question is described by an ID, the question in English and other languages, Wikidata entities, answers, and the complexity type of the question (e.g., ordinal, count). Figure 3 shows an excerpt of a JSON file. It has up to 5 levels of nested lists and objects. It also features some heterogeneity in the representation of entities and answers, depending on their type (e.g., Wikidata entities, numbers). The output CSV file should have 5 columns: ID, Type, Question, Entities, and Answer. The three first columns directly correspond to fields of the question objects. However, the last two columns are string aggregations

```
[  {    "id": "9ace9041",
        "question": "What is the fourth book in the Twilight series?",
        "translations": {"fr": "Quel est le quatrième livre de la série Twilight ?", ...}
        "questionEntity": [
            {    "name": "Q44523",
                 "entityType": "entity",
                 "label": "Twilight", ...}, ...],
        "answer": {
            "answerType": "entity",
            "answer": [ { "name": "Q53945", "label": "Breaking Dawn" } ],
            "mention": "Breaking Dawn" },
        "category": "books",
        "complexityType": "ordinal"
   }, ... ]
```

```
ID,Type,Question,Entities,Answer
9ace9041,ordinal,What is the fourth book in the Twilight series?,Q44523: Twilight,Q53945: Breaking Dawn
...
```

Fig. 3. Excerpt of the input file (top), a JSON list of Mintaka question descriptions and of the expected output file (bottom), a CSV file with one row per question.

of respectively the list of entities, and the list of answers. Moreover, Wikidata entities should be formatted so as to combine their name and their label, while only the name should be used for literal values. Finally, the rows should be sorted by question type. The task therefore involves nested iterations - on questions, on question's entities, and on question's answers, navigation in JSON objects, string building and aggregation, conditional expressions, and ordering.

Plain Programming. We described the task to four experienced Python programmers (students in our lab), provided two example questions, and asked them to write a program for the desired transformation. Two of them overlooked missing fields in some questions so that their program failed on the whole input file. However, they could quickly fix their program for those exceptional cases. The total time they needed to complete the task was consistent, between 30 and 45 min. Their programs were between 36 and 68 lines of code. They declared that they would need about 15min to rewrite their program from scratch.

Programming with ChatGPT. On a first attempt[3], we gave it a representative excerpt of the JSON file (2 questions of different types), and the expected output, then we asked it to generate a Python program to do the translation. After 15min and 3 turns, we stopped because it had a very shallow understanding of the task, and when prompted to look better at the data, it started to hallucinate code irrelevant to the task. On a second attempt, we precisely described in text the structure of the input file, and the structure and contents of the output file. On each turn, we tested the generated Python program, sometimes correcting it for obvious small errors (e.g., field names). After 40min and 8 turns of rather constructive chat, we obtained an almost correct program but then it started to diverge on the last remaining bug related to the missing fields. It should be noted that our successive prompts were strongly based on reading the generated

---

[3] The chat logs can be found at http://www.irisa.fr/LIS/ferre/pub/dexa2023/.

code because that code could not be run, and so errors could not be reported in terms of errors in the generated data.

Using Dexteris. After opening the input file and parsing it as JSON data, the results in Dexteris shows a sequence of JSON objects. An obvious step is therefore to iterate over them, the suggestion `for *` is selected in order to expose each object field as a result column. From there, a JSON object is created to define a CSV row, with one field for each expected column. The three first fields (columns) are simply defined by picking the right variable among the exposed object fields (e.g. `$id`). For entities, the user starts from the value of field `questionEntity`, iterates over them as this is a list, then builds a string by concatenating two fields of each entity, with a colon in between. The aggregation `concat_with_separator` can then be applied to the sequence of formatted entities in order to have one string value for the CSV column"Entities". A similar process is applied to the answers, this time using the mapping construct `!`, and a conditional depending on the answer type. Finally, the ordering construct is inserted, and the function `printCSV` is applied to the whole in order to convert the generated JSON objects into CSV rows. A screencast of the whole building sequence is available on the supplementary material page.

The whole query can be built in 49 steps, including focus changes. When the need for the transformation arose, it took us less than 30 min to build the query and output the transformed data, including the exploration of data, and thinking about what to generate exactly in the output file. Indeed, in real situation, the task is often incompletely specified, and gets refined when exposed to the actual data and unexpected cases. Building again the query in a straight way, knowing exactly what to do, takes 5 min, hence a building speed of about 10 steps/min. For recall, the Python programmers declared that they would need 15min to recode their program from scratch.

Strengths and Weaknesses. Comparing the user experience between Dexteris, plain programming, and ChatGPT reveals the strengths and weaknesses of our approach. The main strengths of our approach are:

- *safeness*: the program (the JSONiq query) is valid at all time, there are no issues with syntax errors or runtime errors;
- *program introspection*: every part of the program can be introspected (and modified) by moving the focus around, like consulting values extracted from the inputs, verifying some sub-computation, or unfolding an iteration;
- *data-centric view*: no need to go forth and back between the data and the program, no need to switch between the programming language and textual prompts, the data are right there and determine what program constructions can be inserted, e.g. a JSON list suggests to insert an iteration, only the variables that are in scope and of the right type are suggested;
- *robustness*: peculiar cases in the input data are smoothly handled whereas they trigger runtime errors in Python programs, human- or machine-generated, e.g. an empty sequence is generated in case of a missing field.

On the weakness side, our approach is not immediately usable, unlike ChatGPT, although it provides more control and does not expose the user to a general programming language. It is also less versatile and scalable than plain programming.

Efficiency. Dexteris is a prototype, and it runs entirely in the browser as a client-side application. Its efficiency and scalability are therefore limited. However, the Mintaka use case demonstrates that it is efficient enough to cope with many practical use cases that data stakeholders encounter. The smallest Mintaka file is 3.8 M, and the 280 kB output is generated in about 1.5 s in Firefox 88.0.1 on Fedora 32 with an Intel Core i7 × 12 and 16 GB RAM. The largest file is 26.7 MB and the 1.9 MB output (a CSV with 14k rows) is generated in about 15 s.

8   Conclusion and Perspectives

We have defined and implemented Dexteris, a tool for data exploration and transformation based on the N<A>F design pattern. It allows the end-user to define complex data transformations in a data-centric way, without having to write any piece of code. Compared to plain programming, or ChatGPT-assisted programming, it features a safer and more robust process. In the future, Dexteris will be improved by covering the missing JSONiq constructs, and extending the set of functions and supported data formats.

References

1. Affolter, K., Stockinger, K., Bernstein, A.: A comparative survey of recent natural language interfaces for databases. VLDB J. **28**(5), 793–819 (2019). https://doi.org/10.1007/s00778-019-00567-8
2. Becker, L.T., Gould, E.M.: Microsoft power bi: extending excel to manipulate, analyze, and visualize diverse data. Ser. Rev. **45**(3), 184–188 (2019)
3. Codd, E., Codd, S., Salley, C.: Providing OLAP (On-line Analytical Processing) to User-Analysts: An IT Mandate. Codd & Date Inc, San Jose (1993)
4. Ferré, S.: Bridging the gap between formal languages and natural languages with zippers. In: Sack, H., Blomqvist, E., d'Aquin, M., Ghidini, C., Ponzetto, S.P., Lange, C. (eds.) ESWC 2016. LNCS, vol. 9678, pp. 269–284. Springer, Cham (2016). https://doi.org/10.1007/978-3-319-34129-3_17
5. Ferré, S.: Sparklis: an expressive query builder for SPARQL endpoints with guidance in natural language. Semant. Web: Interoperability, Usability, Applicability **8**(3), 405–418 (2017). http://www.irisa.fr/LIS/ferre/sparklis/
6. Ferré, S.: Analytical queries on vanilla RDF graphs with a guided query builder approach. In: Andreasen, T., De Tré, G., Kacprzyk, J., Legind Larsen, H., Bordogna, G., Zadrożny, S. (eds.) FQAS 2021. LNCS (LNAI), vol. 12871, pp. 41–53. Springer, Cham (2021). https://doi.org/10.1007/978-3-030-86967-0_4
7. Florescu, D., Fourny, G.: JSONiq: the history of a query language. IEEE Internet Comput. **17**(5), 86–90 (2013)
8. Gulwani, S.: Automating string processing in spreadsheets using input-output examples. In: Symposium on Principles of Programming Languages, pp. 317–330. ACM (2011)

9. Microsoft: PowerQuery. https://learn.microsoft.com/en-us/power-query/
10. Moseley, B., Marks, P.: Out of the tar pit. Software Practice Advancement (2006)
11. OpenAI: ChatGPT. http://chat.openai.com
12. Resnick, M., et al.: Scratch: programming for all. Commun. ACM **52**(11), 60–67 (2009)
13. Sen, P., Aji, A.F., Saffari, A.: Mintaka: A complex, natural, and multilingual dataset for end-to-end question answering. In: International Conference on Computational Linguistics, pp. 1604–1619. International Committee Computational Linguistics (2022)
14. XQuery 3.0: An XML query language (2013). http://www.w3.org/TR/xquery-30/, http://www.w3.org/TR/xquery-30/. W3C Proposed Recommendation

A Neighborhood Encoding for Subgraph Queries in Graph Databases

Chems Eddine Nabti[1], Thamer Mecharnia[2], Salah Eddine Boukhetta[2], Karima Amrouche[2], and Hamida Seba[1]([✉])(iD)

[1] Univ Lyon, UCBL, CNRS, INSA Lyon, LIRIS, UMR5205, 69622 Villeurbanne, France
hamida.seba@univ-lyon1.fr
[2] Ecole nationale Supérieure d'Informatique (ESI), Alger, Algeria

Abstract. Subgraph isomorphism search is a fundamental problem in querying graph-like structured data. It consists to enumerate the subgraphs of a data graph that match a query graph. It is an NP-complete problem that knows several investigations. Most of them extend Ullmann's backtracking algorithm and rely on filtering and pruning mechanisms to reduce the search space. Most of these solutions focus on how to alleviate the searching step of the algorithm, identified as the most costly part, with various techniques and data structures. However, little effort is devoted to reduce the cost of the filtering step. In this paper, we take a completely different approach that relies on a constant time pruning mechanism while keeping Ullman's backtracking subgraph search subroutine. The main idea is to aggregate the semantic and topological information that surround a vertex into a simple integer. This simple neighbourhood encoding reduces the time complexity of vertex filtering from cubic to quadratic. We evaluate our approach on several real-word datasets and compare it with the state of the art algorithms.

Keywords: subgraph queries · subgraph isomorphism search · graph databases

1 Background and Motivation

Subgraph isomorphism search, also known as exact subgraph matching, is a fundamental task on which are based search and querying algorithms on graph data. It is the problem of enumerating all the occurrences of a query graph within a larger graph called the data graph (see Fig. 1). Subgraph isomorphism search is an NP-complete problem that knows extensive investigations. We can cite without being exhaustive Ullmann's algorithm [15], VF2 [5] and its extension VF3 [4], QuickSI [13], GraphQL [8], GADDI [17], SPath [18], Turbo$_{ISO}$ [7] and its extension CECI [1], and CFL-Match [2] and its extension DAF [6]. One can find in [9–11,16] useful studies that survey and compare most of these methods on several aspects of query processing.

C. Strauss et al. (Eds.): DEXA 2023, LNCS 14146, pp. 377–391, 2023.
https://doi.org/10.1007/978-3-031-39847-6_30

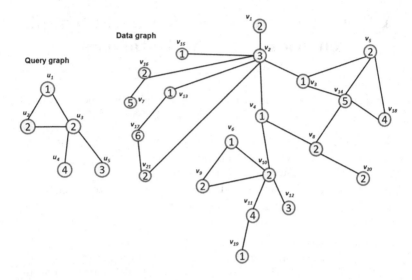

Fig. 1. Running Example.

Existing algorithms for subgraph isomorphism search are built onto two basic tasks: Filtering and Searching. Filtering is an important step and determines the efficiency of the algorithm. The searching step is generally based on the Ullmann's backtracking subroutine [15] that searches in a depth-first manner for matching between the query graph and the filtered data graph resulting from the filtering step. So, the aim of the filtering step is to reduce the search space on which the searching step operates. By construction of these algorithms, they all base their correctness on Ullman's.

Our analysis of the filtering and searching steps within existing algorithms highlighted three weaknesses as follows:

Weakness 1: *High filtering cost.* The main pruning mechanism used by existing methods is the features of the $k-$neighbourhood of query vertices. This is the amount of information used when matching a query vertex with data vertices. The more information is used, i.e., k is big, the more the pruning of the search space can be important. However, representing compactly the $k-$neighbourhood for practical comparisons is a challenging issue. In fact, the representation of this information has a direct impact on its cost which increases with the value of k. Besides filtering with the vertex label and the vertex degree, the lightest $k-$neighbourhood filter is to consider the features of the one-hop neighbourhood, i.e., $k = 1$. For this, approaches such as Turbo$_{ISO}$ [7] and CFL-Match [2] use the Neighbourhood Label Frequency (NLF) filter [19]. NLF ensures that a data vertex v is a candidate for a query vertex u only if the neighbourhood of v, denoted $N(v)$, includes the neighbourhood of u (see lines 5–9 of Algorithm 1). This test concerns the set of labels of the neighbours of vertex v, denoted $\ell(N(v))$.

Algorithm 1: NLF and MND filters.

Data: A potential candidate vertex v for a query vertex u
Result: TRUE if v is candidate for u and FALSE otherwise
1 **begin**
2   **if** $mnd_G(v) < mnd_Q(u)$ **then**
3 | **return** (FALSE);
4   **end**
5   **foreach** *label* $l \in \ell(N(u))$ **do**
6   **if** $|\{w \in N(v)|\ell(w) = l\}| < |\{w \in N(u)|\ell(w) = l\}|$ **then**
7 | **return** (FALSE);
8   **end**
9   **end**
10   **return** (TRUE);
11 **end**

However, NLF is expensive: it is $\mathcal{O}(|V(Q)||V(G)||\mathcal{L}(Q)|)$ where $|V(Q)|$ is the number of vertices in the query, $|V(G)|$ is the number of vertices in the data graph and $\mathcal{L}(Q)$ is the set of unique labels of the query graph. This means that NLF is $\mathcal{O}(|V(Q)|^3)$ in the worst case. So, to avoid applying NLF systematically on each vertex, CFL-Match [2] proposes the Maximum Neighbours-Degree (MND) filter, which can be verified in constant time for each candidate data vertex. This means that the MND filter for all vertices is $\mathcal{O}(|V(Q)||V(G)|)$ and hence $\mathcal{O}(|V(G)|^2)$ in the worst case. The maximum neighbour-degree of a vertex u in a graph G, denoted $mnd_G(u)$, is the maximum degree of all its neighbours [2]. A data vertex v is not a candidate for a query vertex u if $mnd_G(v) < mnd_Q(u)$. As MND is not as powerful as NLF, the idea is to apply it before applying NLF as detailed in Algorithm 1 (see lines 2–3). However, MND is not always effective as we can see in the example depicted in Fig. 2 where only 3 vertices (among the 13 prunable ones) are pruned with the MND filter and consequently NLF must be applied for each of the remaining vertices. Furthermore, MND needs to compute the degrees of all the vertices prior to its use. So, it is less scalable than NLF that only requires the knowledge of the labels of the neighbours of each vertex.

Weakness 2: *Global filtering vs local filtering.* Our study of existing algorithms shows that filtering mechanisms can be classified into two categories depending on their scope: local or global. A local filtering mechanism prunes the set of vertices that are candidates for a single vertex. A global pruning operates on the whole search space. However, local pruning is predominant in existing solutions. Some mechanisms allow global pruning but they require extra passes of the data graph to be effective. The matching order is such a mechanism. However, it is a very difficult problem to choose a robust matching order mainly because the number of all possible matching orders is exponential in the number of vertices. So, it is expensive to enumerate all of them. For example, Tuorbo$_{ISO}$ relies on vertex ordering for pruning. However, to compute this order, it needs to compute for each query vertex a selectivity criteria based on the frequency of its label in the data graph. However, this means extra passes on the data graph for computing these frequencies.

Weakness 3: *Late filtering.* Our analysis of how filtering and searching are undertaken with respect to each other in the state of the art algorithms revealed

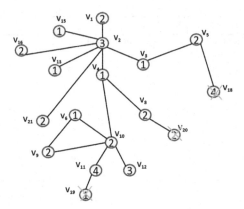

Fig. 2. MND Filter on the running example. The vertices that do not match query labels have been pruned beforehand).

that most algorithms apply their filtering mechanisms during subgraph search. In fact, little filtering, reduced mainly to label or degree filtering, is undertaken prior to subgraph search. This means that, the first cartesian products involved in the subgraph search task are costly. To tackle this, CFL-Match [2] applies the MND-NLF filter, locally, prior to subgraph search. However, as we can see in Fig. 3, the amount of achieved pruning depends on the order within which vertices are parsed. In our example, if v_2 is processed before v_{16} the amount of pruning is less than the one obtained with the reverse order. To get caught up, existing solutions rely on additional mechanisms and data structures during subgraph search such as NEC tree in Turbo$_{ISO}$ [7] and *CPI* in CFL-Match [2] that both use path-based ordering during subgraph search. Modifications and

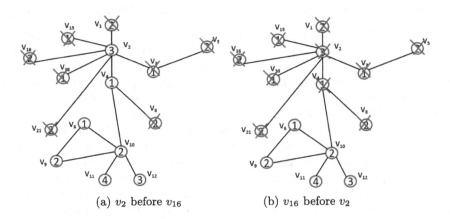

(a) v_2 before v_{16} (b) v_{16} before v_2

Fig. 3. NLF filtering with two different vertex parsing orders

extensions of these data structures are also proposed in CECI [1] and its DAF [6].
However, the underlying data structures are time and space exponential [2,16].

In this paper, we propose to address these limitations as follows:

- simplify the encoding of the k-neighborhood so as: (1) to reduce its cost for
  filtering and (2) to be able to simply update it after each local pruning to
  ensure a global pruning of the search space as early as possible without the
  need of complex data structures. To do so, we rely on a new vertex encoding
  method.
- achieve filtering prior to subgraph search. For this, we introduce the Iterative
  Local Global Filtering mechanism (ILGF), a simple way to achieve global
  punning relying on local pruning filters.

Our main contribution is a novel encoding of vertices, called Compact Neighbourhood Encoding (CNE) that distills all the information around a vertex in
a single integer leading to a simple but extremely efficient filtering scheme for
processing subgraph isomorphism search. The whole filtering process is based
on integer comparisons. CNEs are also easily updatable during filtering. Moreover, this encoding can be twined with any subgraph matching algorithm and
aims to reduce as early as possible the search space by achieving global pruning.
We conduct extensive experiments using real datasets in different application
domains to attest the effectiveness and efficiency of the proposed scheme used
in combination with Ullmann's algorithm.

The remainder of this paper is organised as follows: Sect. 2 first formalises the
problem of subgraph isomorphism search and defines the notation used throughout the paper, then, it introduces our main contribution: the compact neighbourhood encoding and how it is used to solve subgraph isomorphism search. Section 3
presents a comprehensive experimental study on several datasets. Section 4 concludes the paper.

2 Proposed Approach

2.1 Definitions and Notation

A data graph G is a 4-tuple $G = (V(G), E(G), \ell, \Sigma)$, where $V(G)$ is a set of
vertices (also called nodes), $E(G) \subseteq V(G) \times V(G)$ is a set of edges connecting
the vertices, $\ell : V(G) \cup E(G) \rightarrow \Sigma$ is a labelling function on the vertices and
the edges where Σ is the set of labels that can appear on the vertices and/or
the edges. We use $|V(G)|$ and $|E(G)|$ to represent respectively the number of
vertices and the number of edges in G.

An undirected edge between vertices u and v is denoted indifferently by (u, v)
or (v, u). A neighbour of a vertex v is a vertex adjacent to v. The degree of a
vertex v, denoted $deg(v)$, is the number of its neighbours. We also use $deg_S(v)$
to denote the number of neighbours of v that have a label in the set S. We use
$N(v)$ to represent the neighbours of vertex v. $\ell_G(u)$ (or simply $\ell(u)$ when there
is no ambiguity) represents the label of vertex u in G and $\ell((u, v))$ is the label
of the edge (u, v) in G.

A graph that is contained in another graph is called a subgraph and can be defined as follows:

Definition 1. *A graph $G_1 = (V(G_1), E(G_1), \ell_1, \Sigma)$ is a subgraph of a graph $G_2 = (V(G_2), E(G_2), \ell_2, \Sigma)$ if $V(G_1) \subseteq V(G_2)$, $E(G_1) \subseteq E(G_2)$, $\ell_1(x) = \ell_2(x)$ $\forall x \in V(G_1)$, and $\ell_1(e) = \ell_2(e)$ $\forall e \in E(G_1)$.*

Definition 2. *A graph $Q = (V(Q), E(Q), \ell, \Sigma)$ is subgraph isomorphic to a graph $G = (V(G), E(G), \ell, \Sigma)$ if and only if there exists an injective mapping h from $V(Q)$ to $V(G)$ such that:*

1. $\forall x \in V(Q) : \ell(x) = \ell(h(x))$
2. $\forall (x, y) \in E(Q) : (h(x), h(y)) \in E(G)$

For presentation convenience, we do not show edge labels in our examples but these labels are considered in our algorithms and datasets.

2.2 Compact Neighbourhood Encoding (CNE)

In our method, the high-level idea is to put into a simple integer the neighbourhood information that characterise a vertex. Matching two vertices is then a simple comparison between integers. Given a vertex u, the compact neighbourhood encoding of u, denoted $cne(u)$, distils the whole structure that surrounds the vertex into a single integer. It is the result of a bijective function that is applied on the vertex's neighbourhood information. This function ensures that two given vertices u and v will never have the same compact neighbourhood encoding if they have the same label and the same number of neighbours unless they are isomorphic at one-hop. To compute CNEs, we use pairing functions. A pairing function on a set A associates each pair of members from A with a single member of A, so that any two distinct pairs are associated with two distinct members [14]. It is a bijection and according to Fueter-Pólya theorem [14], the only quadratic pairing function is the Cantor polynomial $f : \mathbb{N} \times \mathbb{N} \to \mathbb{N}$ defined by $f(k_1, k_2) = \frac{1}{2}(k_1 + k_2)(k_1 + k_2 + 1) + k_2$. It assigns consecutive numbers to points along diagonals in the plane.

To pair more than two numbers, pairings of pairings can be used. For example $f(i, j, k)$ can be defined as $f(i, f(j, k))$ or $f(f(i, j), k)$, but $f(i, j, k, l)$ is defined as $f(f(i, j), f(k, l))$ to minimise the size of the produced number. So, by composing $k - 1$ times the bijection of $\mathbb{N}^2$ on $\mathbb{N}$, we obtains a bijection of $\mathbb{N}^k$ on $\mathbb{N}$ which is a polynomial of degree k [14].

To use this bijection on vertices' labels, we first assign a unique integer to each vertex label. This assignment can be simply achieved by numbering labels parting from 1 or by using an associative array to store the query labels. Let $ord(\ell(u))$ be the subroutine used to retrieve the integer associated to the label of vertex u. $ord(\ell(u))$ will return 0 if vertex u has a label that does not belong to $\mathcal{L}(Q)$. This will systematically prune the neighbours that do not verify the label filter and avoid to consider them in the computation of the CNE of a vertex. In our case, the parameter $k = |\mathcal{L}(Q)|$, i.e., the number of distinct labels in the query.

So, the compact neighbourhood encoding of vertex u in G is given by:

$$cne(u) = \hbar(1, x_1) + \hbar(2, x_1 + x_2) + \cdots + \hbar(k, x_1 + x_2 + x_3 + \cdots + x_k).$$

$$cne(u) = \sum_{j=1}^{k} \hbar(j, x_1 + \ldots + x_j) \text{ where } \hbar(p, s) = \binom{s + p - 1}{p} = \frac{(s + p - 1)!}{p!(s - 1)!}$$

To compute $\hbar(j, x_1 + \ldots + x_j)$, j corresponds to a query label index, and x_j is the number of occurrences of label j in the direct neighbourhood of vertex u. This provides CNEs with the same filtering capacity as the NLF filter.

Example: Figure 4(a) illustrates the CNEs of the query graph of our running example. These CNEs are computed as follows:
For this query graph, the integers used to represent the labels are: 1, 2, 3, and 4, i.e., $k = 4$.
$cne(u_1) = \hbar(1, 0) + \hbar(2, 0 + 2) + \hbar(3, 0 + 2 + 0) + \hbar(4, 0 + 2 + 0 + 0) = 0 + 3 + 4 + 5 = 12$.
In fact, we can see that labels 1, 3 and 4 do not appear in the neighbourhood of of u_1 and label 2 appears 2 times. The remaining CNEs are computed similarity:

$cne(u_2) = g_4(1, 1, 0, 0) = \hbar(1, 1) + \hbar(2, 1 + 1) + \hbar(3, 1 + 1 + 0) + \hbar(4, 1 + 1 + 0 + 0) = 13$.
$cne(u_3) = g_4(1, 1, 1, 1) = \hbar(1, 1) + \hbar(2, 1 + 1) + \hbar(3, 1 + 1 + 1) + \hbar(4, 1 + 1 + 1 + 1) = 49$.
$cne(u_4) = g_4(0, 1, 0, 0) = \hbar(1, 0) + \hbar(2, 0 + 1) + \hbar(3, 0 + 1 + 0) + \hbar(4, 0 + 1 + 0 + 0) = 3$.
$cne(u_5) = g_4(0, 1, 0, 0) = \hbar(1, 0) + \hbar(2, 0 + 1) + \hbar(3, 0 + 1 + 0) + \hbar(4, 0 + 1 + 0 + 0) = 3$.

A data vertex v is a candidate for a query vertex u if it verifies the label filter, i.e., $\ell(v) = \ell(u)$, the degree filter, i.e., $deg_{\mathcal{L}(Q)}(v) < deg_{\mathcal{L}(Q)}(u)$ and the CNE filter given by :

Lemma 1 (CNE filter). *Given a query Q and a data graph G, a data vertex $v \in V(G)$ that verifies the label and degree filters is not a candidate of $u \in V(Q)$ if $cne(v) < cne(u)$.*

Proof. We prove the lemma by contradiction. Assume v is a candidate of u with $cne(v) < cne(u)$. That is, there is an embedding M that maps u to v. This means that $\ell(v) = \ell(u)$ and $deg(v) \geq deg(u)$ and $\ell(N(u)) \subseteq \ell(N(v))$. Let $deg(u) = k$ and $deg(v) = k + t$, $t \geq 1$. Let $(l_1, l_2, \cdots, l_k)$ be the labels of the neighbours of u according to the order given by function $ord()$. Similarly, let $(l_1, l_2, \cdots, l_k, l_{k+1}, \cdots, l_{k+t})$ be the labels of the neighbours of v. By construction, we have $cne(v) = g_{k+t}(l_1, l_2, \cdots, l_{k+t}) = g_k(l_1, l_2, \cdots, l_k) + \hbar(k + 1, l_1 + \ldots + l_{k+1}) + \cdots + \hbar(k + t, l_1 + \ldots + l_{k+t})$. So, $cne(v) = cne(u) + \hbar(k + 1, l_1 + \ldots + l_{k+1}) + \cdots + \hbar(k + t, l_1 + \ldots + l_{k+t})$. As $t > 0$, we reach a contradiction. Thus, the lemma holds.

Note that, verifying one candidate vertex v for a query vertex u takes $\mathcal{O}(1)$ time versus $\mathcal{O}(|\mathcal{L}(Q)|)$ for NLF.

Theorem 1. *The CNE filter is $\mathcal{O}(n^2)$ in the worst case where $n = |V(G)|$.*

Proof. As verifying one candidate vertex v for a query vertex u takes $\mathcal{O}(1)$, the CNE filter on all the query vertices is $\mathcal{O}(|V(G)||V(Q)|)$. So, in the worst case, it is $\mathcal{O}(|V(G)|^2)$.

2.3   Iterative Local Global Filtering Algorithm (ILGF)

The aim of the Iterative Local Global Filtering Algorithm (ILGF) is to reduce globally the search space using CNEs. It relies on the fact that $cne(v)$ can be easily updated after a local filtering giving rise to new filtering opportunities. Algorithm 2 details this iterative filtering process. To verify the CNE filter on a candidate data vertex, the algorithm uses the *cneVerify()* subroutine that implements Lemma 1 and consequently allows to verify that a data vertex is a candidate for a given query vertex according to the label, degree and CNE filters defined above. The ILGF algorithm removes iteratively from G the vertices that do not match a query vertex using the label, the degree and the CNE filters (see lines 5–15 of the algorithm). Each time a vertex is removed by the filtering process the degree and CNE of its neighbours are updated (lines 10–13) giving rise to new filtering opportunities. However, it is important to note that filtering iterations do not parse all the remaining vertices in the data graph. In fact, the set *nextFilter* keep track of the neighbours of the vertices pruned during the current iteration. The next iteration parses only the vertices contained in *nextFilter*. Filtering stops when no further vertices are removed, i.e., *nextFilter* is empty.

Example: Figure 4 illustrates the ILGF algorithm on our running example. Figure 4 (b) shows the CNEs computed for the data vertices. During degree and CNE computation, the label filter is applied and the vertices that do not verify this filter are pruned and are not considered in the degree and CNE of their neighbours. This is the case for vertices v_7, v_{14} and v_{17}. So, only the data vertices that verify the label filter are considered when computing degrees and CNEs. The first iteration, of the ILGF algorithm (see Fig. 4 (c)), finds out that vertices v_1, v_3, v_5, v_{13}, v_{15}, v_{16}, v_{19}, v_{20} and v_{21} cannot be mapped to any query vertex because:

- v_1, v_{13}, v_{15}, v_{16}, v_{19}, v_{20} and v_{21} do not pass the degree filter,
- v_3 and v_5 do not pass the CNE filter. We can clearly see that according to the label and degree filters, v_3 can be mapped to u_1 and v_5 can be mapped to u_2 however their CNEs do not mach these vertices: $cne(v_3) < cne(u_1)$ and $cne(v_5) < cne(u_2)$

After removing these vertices and updating the degree and CNE of their neighbours a new filtering iteration is triggered (see Fig. 4(d)). We note here that this second filtering iteration concerns only vertices v_2, v_8, v_{11}, and v_{18} whose degrees and CNEs have been modified consequently to the previous filtering iteration. The second filtering iteration reveals that vertices v_8 and v_{18} can also be pruned. They do not pass the degree filter. The resulting filtered data graph is depicted in Fig. 4(e) that also shows the new filtering opportunities triggered by the second filtering iteration. A third filtering iteration is launched on the vertices whose degrees and CNEs have been modified. Consequently, the third filtering iteration verifies only vertex v_4. v_4 does not verify the CNE filter and can be pruned leading to the graph depicted on Fig. 4(f). This figure shows also that the degree and CNE of vertices v_2 and v_{10} are updated. Consequently these two vertices will be the target of the final filtering iteration that prunes vertex v_2 which no longer verifies the degree filter.

Algorithm 2: $ILGF$.

Data: A set of vertices $S \subseteq V(G)$ candidate to be pruned in a graph G.
Result: A filtered version of G

```
1  begin
2  |   toFilter ← S;
3  |   nextFilter ← ∅;
4  |   repeat
5  |   |   foreach vertex v ∈ toFilter do
6  |   |   |   if (∀u ∈ V(Q), !cneVerify(v, u)) then
7  |   |   |   |   affected ← N(v);
8  |   |   |   |   nextFilter ← nextFilter ∪N(v);
9  |   |   |   |   remove v from V(G) and the corresponding edges from E(G);
10 |   |   |   |   foreach x ∈ affected do
11 |   |   |   |   |   update deg(x);
12 |   |   |   |   |   update cne(x);
13 |   |   |   |   end
14 |   |   |   end
15 |   |   end
16 |   |   toFilter ← nextFilter;
17 |   |   nextFilter ← ∅;
18 |   until (toFilter=∅);
19 end
```

Algorithm 3: Function $cneVerify(v,u)$.

Data: A data vertex v and a query vertex u.
Result: returns true if v is a candidate for u according to the label, degree and CNE filters.

```
1  begin
2  |   return (ℓ(u) = ℓ(v) ⋀ deg_{ℒ(Q)}(u) < deg_{ℒ(Q)}(v) ⋀ cne(u) < cne(v)) or
   |          (ℓ(u) = ℓ(v) ⋀ deg_{ℒ(Q)}(u) = deg_{ℒ(Q)}(v) ⋀ cne(u) = cne(v)))
3  end
```

2.4   Subgraph Search

After filtering, the data graph contains only the vertices that are candidates for query vertices, i.e., the vertices map at one-hop according to the CNE filter. Subgraph search allows to verify the mapping at k-hops. Algorithm 4 implements this step. It is a depth first search subroutine that parses the filtered data graph and lists the subgraphs of the filtered data graph that are isomorphic to the query by verifying the adjacency relationships. This step allows also to handle edge labels by discarding those that do not match the query labels. The subroutine *neighborCheck()* verifies that a mapping (v, u) is added to the current partial embedding M only if v and u have neighbors that also map.

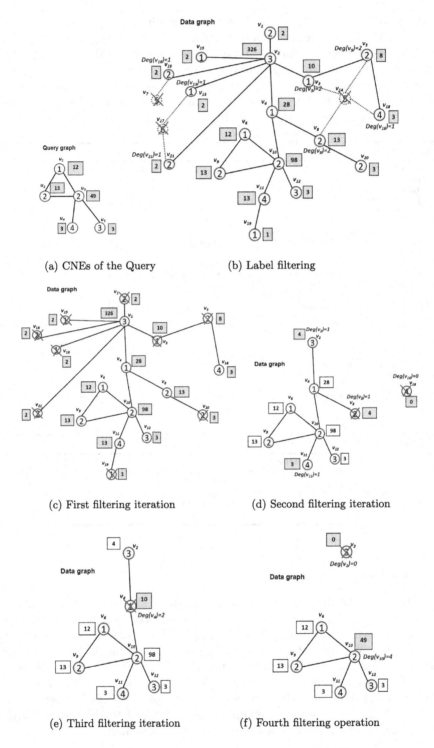

Fig. 4. Filtering iterations of our running example.

Algorithm 4: SubgraphSearch.

Data: a partial embedding M.
Result: All embeddings of Q in G.

```
 1  begin
 2      if |M| = |V(Q)| then
 3          Report M;
 4      end
 5      Choose a non matched vertex u from V(Q);
 6      C(u) ← { non matched v ∈ V(G) ,such that cneVerify(v, u));
 7      foreach v ∈ C(u) do
 8          if neighborCheck(u,v, M) then
 9              M ← M ∪ {(u, v)};
10              SubgraphSearch(M);
11              Remove (u, v) from M ;
12          end
13      end
14  end
```

Algorithm 5: Function $neighborCheck(u, v, M)$.

Data: a partial embedding M, a query vertex u and a data vertex v.
Result: returns true if u and v have neighbours that match.

```
 1  begin
 2      return (∀(u', v') ∈ M, ((u, u') ∈ E(Q) → (v, v') ∈ E(G) ⋀ ℓ((u, u')) = ℓ((v, v')))
 3  end
```

Our subgraph isomorphism search algorithm, denoted by CNI-Match (for Compact Neighbourhood encoding Isomorphism Search), is given by Algorithm 6. It first filter the data graph by invoking the ILGF algorithm on the set of vertices of the data graph. Then it call the *SubgraphSearch* subroutine to find the embeddings of the query graph.

Algorithm 6: CNI-Match.

Data: A data graph G and a query graph Q
Result: All the occurrences of Q in G

```
 1  begin
 2      ILGF(V(G), G);
 3      foreach  vertex u ∈ V(Q) do
 4          C(u) ← {v ∈ V(G) such that cneVerify(v, u)};
 5          if C(u) = ∅ then
 6              return (∅);
 7          end
 8      end
 9      M ← ∅;
10      SubgraphSearch(M);
11  end
```

---

3   Experiments

We evaluate the performance of our algorithm, *CNI-Match* (for Compact Neighborhood encoding idex based Matching), over various types of graphs, sizes of queries and number of labels. We also compare it with one of the most efficient state of the art algorithm, CFL-Match [2]. Note that CFL-Match is compared to the other existing solutions, such as Turbo$_{ISO}$, QuickSI and SPath, and showed to be more efficient in [2,7,12]. Our aim is to show that our encoding scheme is effective and can be coupled with existing algorithms to enhance performance. Using it with the basic approach of Ullmann's algorithm allowed us to outperform a more sophisticated scheme, i.e., CFL-Match.

Table 1. Graph Dataset Characteristics.

Dataset	$|V|$	$|E|$	$|\Sigma|$	average degree
HUMAN	4,675	86,282	44	36.9
HPRD	9,460	37,081	307	7.8
YEAST	3,112	12,519	71	8.1

$|\Sigma|$ is the number of distinct labels.

For a fair comparison, we implemented the two algorithms[1] in the same environment and framework using C++ and the SNAP library[2] We also used compiling option $-O3$. For CNI-Match, we used the GMP[3] specialised library to compute factorials and store them. All experiments are performed on an *Intel i*5 3.50 GHz, 64 bits computer with 8 GB of RAM running windows 7. The source code of our approach is available in https://gitlab.liris.cnrs.fr/hseba/cne.

We first describe the datasets used in the experiments, then we present our results.

3.1 Datasets

We use three main datasets which are known datasets used by almost all existing methods in their evaluation process (cf. Table 1). So, we mainly use them as comparative datasets. The underlying graphs represent protein interaction networks coming from three main organisms: human (HUMAN and HPRD datasets) and yeast (YEAST dataset). The HUMAN dataset is available in the RI database of biochemical data [3]. HPRD and YEAST come from the work of [11] and [2].

To query the HUMAN, HPRD and YEAST datasets, we constructed a set of sparse and dense queries for each dataset. Each query is a connected subgraph of the data graph obtained using a random walk on the data graph where the next vertex is selected according to the sparsity of the query. For a sparse query, the next vertex is selected among the neighbors that have the least number of neighbors. For a dense query, the next vertex is selected among the neighbors that have the greatest number of neighbors. For sparse queries, we provide 20

---

[1] The source code of the two algorithms is available on Git and will be provided.
[2] Stanford Network Analysis Platform. http://snap.stanford.edu/.
[3] https://gmplib.org/.

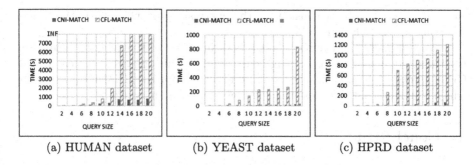

Fig. 5. Time performance on sparse queries: varying $|V(Q)|$.

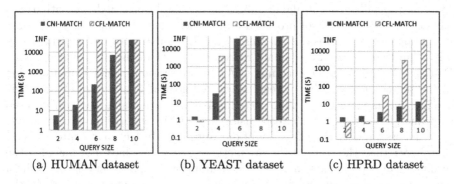

Fig. 6. Time performance on dense queries: varying $|V(Q)|$ (Results are in log scale).

query sets for each dataset, each containing 100 query graphs of the same size. For dense queries, we provide 10 query sets for each dataset, each containing 100 query graphs of the same size.

3.2   Results

In this subsection, we report and comment the results obtained by comparing our algorithm with the state of the art algorithm CFL-Match [2]. Our main metric is the time performance by varying $|V(Q)|$, i.e., the number of vertices in the query, the density of the queries, and the amount of memory used by the algorithms. We present the obtained results according to these metrics. We note also that all the algorithms output the same sets of isomorphic subgraphs for each query graph.

Figure 5 shows the average total processing time of the two algorithms on the three datasets when processing sparse queries. On the y-axis, INF means that the processing of the set of queries exceeded 12 h execution time and has been aborted. According to this figure, CNI-Match is, compared to CFL-Match, on average 12 times faster on the YEAST dataset and 17 times faster on the HPRD dataset. For the dense and difficult dataset HUMAN, CNI-Match is 4 times faster than CFL-Match only by considering the query sizes for which CFL-Match has not reached the INF threshold.

Figure 6 shows the average total processing time of the two algorithms on the three datasets when processing dense queries. This figure shows clearly that CFL-Match is too slow for dense queries especially for the HUMAN dataset for which it has not obtained less than 12 h even for the query size 2.

4   Conclusion

Subgraph isomorphism search is an NP-complete problem. This means a processing time that grows with the size of the involved graphs. Pruning the search space is the pillar of a scalable subgraph isomorphism search algorithm and has been the main focus of proposed approaches since Ullmann's first solution. In this paper, we proposed *CNI-Match*, a simple subgraph isomorphism search algorithm that relies on a compact representation of the neighbourhood, called Compact Neighbourhood Encoding (*CNE*), to perform an early global pruning of the search space. *CNE* aggregates the topological information of each vertex into an integer. This vertex encoding is easily updatable and can be used to prune globally the search space using an iterative algorithm. Our extensive experiments validate the efficiency of our approach. In fact, our approach is 12 time faster than the state of the art method. Even if we have not included the results of our experiments on memory consumption, for paper length constraints, our approach also registers far better results than the state of the art. In the proposed approach, the neighbourhood encoding scheme, CNE, is coupled with Ullmann's basic algorithm, but it is a general encoding method that can be used with other algorithms for better performance. So, an interesting future extension is to couple it with other sophisticated schemes and also compare it with other frameworks.

Acknowledgements. This work is funded by the French National Research Agency under grant ANR-20-CE23-0002 and INFO-Bourg department, IUT Lyon 1.

References

1. Bhattarai, B., Liu, H., Huang, H.H.: CECI: compact embedding cluster index for scalable subgraph matching. In: Proceedings of the 2019 International Conference on Management of Data, SIGMOD 2019, pp. 1447–1462. Association for Computing Machinery, New York (2019). https://doi.org/10.1145/3299869.3300086
2. Bi, F., Chang, L., Lin, X., Qin, L., Zhang, W.: Efficient subgraph matching by postponing cartesian products. In: Proceedings of the 2016 International Conference on Management of Data, SIGMOD 2016, pp. 1199–1214 (2016)
3. Bonnici, V., Giugno, R., Pulvirenti, A., Shasha, D., Ferro, A.: A subgraph isomorphism algorithm and its application to biochemical data. BMC Bioinform. 14(Suppl 7), S13 (2013)

4. Carletti, V., Foggia, P., Saggese, A., Vento, M.: Challenging the time complexity of exact subgraph isomorphism for huge and dense graphs with VF3. IEEE Trans. Pattern Anal. Mach. Intell. **40**(4), 804–818 (2018). https://doi.org/10.1109/TPAMI.2017.2696940
5. Cordella, L.P., Foggia, P., Sansone, C., Vento, M.: A (sub)graph isomorphism algorithm for matching large graphs. IEEE Trans. Pattern Anal. Mach. Intell. **26**, 1367–1372 (2004)
6. Han, M., Kim, H., Gu, G., Park, K., Han, W.S.: Efficient subgraph matching: harmonizing dynamic programming, adaptive matching order, and failing set together. In: Proceedings of the 2019 International Conference on Management of Data, SIGMOD 2019, pp. 1429–1446. Association for Computing Machinery, New York (2019). https://doi.org/10.1145/3299869.3319880
7. Han, W.S., Lee, J., Lee, J.H.: Turboiso: towards ultrafast and robust subgraph isomorphism search in large graph databases. In: ACM SIGMOD International Conference on Management of Data, pp. 337–348. SIGMOD 2013 (2013)
8. He, H., Singh, A.K.: Graphs-at-a-time: query language and access methods for graph databases. In: ACM SIGMOD International Conference on Management of Data, SIGMOD 2008, pp. 405–418 (2008)
9. Katsarou, F., Ntarmoset, N., Triantafillou, P.: Subgraph querying with parallel use of query rewritings and alternative algorithms. In: EDBT (2017)
10. Kim, H., Choi, Y., Park, K., Lin, X., Hong, S.H., Han, W.S.: Versatile equivalences: speeding up subgraph query processing and subgraph matching. In: Proceedings of the 2021 International Conference on Management of Data, SIGMOD 2021, pp. 925–937 (2021). https://doi.org/10.1145/3448016.3457265
11. Lee, J., Han, W.S., Kasperovics, R., Lee, J.H.: An in-depth comparison of subgraph isomorphism algorithms in graph databases. In: 39th International Conference on Very Large Data Bases, pp. 133–144 (2013)
12. Ren, X., Wang, J.: Exploiting vertex relationships in speeding up subgraph isomorphism over large graphs. Proc. VLDB Endow. **8**(5), 617–628 (2015)
13. Shang, H., Zhang, Y., Lin, X., Yu, J.X.: Taming verification hardness: an efficient algorithm for testing subgraph isomorphism. Proc. VLDB Endow. **1**(1), 364–375 (2008)
14. Stein, S.K.: Mathematics: The Man-Made Universe. McGraw-Hill, New York (1999). Dover Publications; 3rd Revised edn. (21 March 2013)
15. Ullmann, J.R.: An algorithm for subgraph isomorphism. J. ACM **23**(1), 31–42 (1976)
16. Zeng, L., Jiang, Y., Lu, W., Zou, L.: Deep analysis on subgraph isomorphism (2021)
17. Zhang, S., Li, S., Yang, J.: GADDI: distance index based subgraph matching in biological networks. In: EDBT 2009, pp. 192–203 (2009)
18. Zhao, P., Han, J.: On graph query optimization in large networks. PVLDB **3**(1), 340–351 (2010)
19. Zhu, G., Lin, X., Zhu, K., Zhang, W., Yu, J.X.: TreeSpan efficiently computing similarity all-matching. In: Proceedings of the 2012 ACM SIGMOD International Conference on Management of Data, SIGMOD 2012, pp. 529–540 (2012)

MIRS: [MASK] Insertion Based Retrieval Stabilizer for Query Variations

Junping Liu[1], Mingkang Gong[1], Xinrong Hu[1], Jie Yang[2(✉)], and Yi Guo[3]

[1] School of Computer Science and Artificial Intelligence,
Wuhan Textile University, Wuhan, China
{jpliu,mkg,hxr}@wtu.edu.cn
[2] School of Computing and Information Technology, University of Wollongong,
Wollongong, Australia
jiey@uow.edu.au
[3] School of Computer, Data and Mathematical Sciences,
Western Sydney University, Penrith, Australia
y.guo@westernsydney.edu.au

Abstract. Pre-trained Language Models (PLMs) have greatly pushed the frontier of document retrieval tasks. Recent studies, however, show that PLMs are vulnerable to query variations, *i.e.*, queries containing misspellings or word re-ordering of original queries, and *etc.*. Despite the increasing interest to robustify the retriever performance, the impact of the query variations is not fully exploited. To effectively address this problem, this paper revisits the Masked-Language Modeling (MLM) and proposes a robust fine-tuning algorithm, termed [MASK] Insertion based Retrieval Stabilizer (MIRS). The proposed algorithm differs from existing methods via the **injection** of [MASK] tokens into query variations and further encouraging the representation similarity between the pair of original queries and their variations. In comparison to MLM, the traditional [MASK] substitution-then-prediction is less emphasized in MIRS. Additionally, an in-depth analysis of our algorithm is also provided to reveal: (1) the latent representation (or semantic) of the original query forms a convex hull, while the impact of the query variation is then quantified as a "distortion" to this hull via deviating the hull vertices; and (2) inserted [MASK] tokens play a significant role in enlarging the intersection between the newly-formed hull (after variations) and the original one, thereby preserving more semantic from original queries. With the proposed [MASK] injection, MIRS exhibits a relative 1.8 MRR@10 absolute point enhancement on average in the retrieval accuracy, verified using 5 baselines across 3 public datasets with 4 types of query variations. We also provide intensive ablation studies to investigate the hyperparameter sensitiveness, to breakdown the model into individual components to manifest their efficacy, and further, to evaluate the out-of-domain model generalizability.

Keywords: Document Retrieval · Masked-Language Modeling · Model Robustness · Query Variations · Query Representation

This work is partially supported by the Australian Research Council Discovery Project (DP210101426) and the Australian Research Council Linkage Project (LP200201035).

C. Strauss et al. (Eds.): DEXA 2023, LNCS 14146, pp. 392–407, 2023.
https://doi.org/10.1007/978-3-031-39847-6_31

1 Introduction

The task of document retrieval has gained significant popularity over the past few years. Dense retriever systems [5,11,17], inherited from the Pre-trained Language Models (PLMs), have achieved remarkable success compared to traditional (bag-of-words) sparse retriever, such as BM25 [14]. However, PLM based retriever systems are also fragile to query variations [1,19]. That is, state-of-the-art models could exhibit a surprisingly 35% performance reduction with the presence of even a single-character misspelling within queries [12], showing their non-robust retrieval matching between queries and documents/passages. A few work have successfully applied data-augmentation training to the retrieval task [15,19,20], from which the key component is to involve query variations (as augmentation) in the training process. In other words, query variations are leveraged as the normal queries to fine-tune the retriever model, learning informative representation to tolerate perturbed queries. More details are provided in Sect. 2.

Despite the increasing research effort in robustifying the retriever systems, the impact of the query variation is still unclear, and more investigations are required to improve the model generalizability and robustness. To take one step towards this goal, we introduce **MIRS**, a [MASK] Insertion based Retrieval Stabilizer for enhancing the sustainability and flexibility of retrieval systems (shown in Fig. 1). Similar to existing work, MIRS also includes query variations as augmented training samples for the model fine-tuning. The proposed method, however, is characterized by the **injection** of [MASK] tokens directly in the query variations. Additionally, a discrepancy loss is further incorporated in the proposed method to ensure that masked query variations remain semantically close to original ones. Theoretic analysis is provided to justify the benefit of inserting [MASK] tokens as a performance stabilizer. We first quantify the latent representation of the original query, from the geometry perspective, and observe that it creates its own convex hull in the Euclidean space. We further argue that the query variation leads to a distortion to this original convex hull due to deviating or even removing existing vertices. The proposed [MASK] injection, accordingly, enlarges the intersection between the original hull and that of the related variation, thereby preserving more original representations and further robustifying the retrieval performance.

Notably, the [MASK] token is traditionally applied in the Masked-Language
Modeling (MLM) for pre-training, *i.e.*,
predicting tokens that are **replaced**
by [MASK]. In the retrieval context, **ColBERT** is reported in [6],
where [MASK] tokens are inserted
into query for input padding purposes.
In [18], query tokens are substituted
by [MASK] before recovering them to
create another "crafted" query. In the
proposed MIRS, by contrast, neither
padding nor substitution-then-prediction
task is emphasized. Instead, [MASK]
tokens are **inserted** directly into query
variations as the "wild cards" to create the masked variation, for which
are trained simultaneously with original
ones.

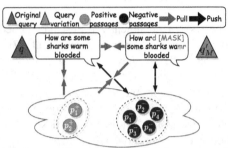

Fig. 1. An illustration case of the proposed
[MASK] Insertion based Retrieval Stabilizer
(MIRS), where words ("*are*" and "*warm*")
from the original query are perturbed as "*ard*"
and "*wamr*", respectively. MRIS, accordingly,
injects [MASK] tokens to this query variation,
and pulls the pair of the original query and
masked variation together in the latent space.

The main contributions of the proposed work are summarized as follows: (i) A novel
[MASK] Insertion based Retrieval Stabilizer (termed MIRS) is introduced in this paper
to address the problem associated with the query variation. (ii) The proposed MIRS is
characterized by inserting additional [MASK] tokens to variations and maximizing the
representation similarity between the pair of original queries and their variations. (iii)
Theoretical analysis reveals that injected [MASK] creates a large convex hull overlapping with that of the original query to increase the probability of persevering original
semantic. (iv) Empirically, our model outperforms existing approaches on three highly-
competitive retrieval datasets with four types of query variations, advancing the best
state-of-the-arts by 1.8 absolute MRR@10 points on average in accuracy.

2   Related Work

Document Retrieval. Recent decades have witnessed a significant research interest in
information retrieval systems to search for evidences (relevant documents/passages)
from a large-scale corpus matching a given query. The two-stage retriever-reader
paradigm has been extensively studied [8, 11], where relevant evidences are first
recalled by the retriever from the whole corpus in a coarse manner, before applying
the reader to re-rank them more carefully.

As the upper bound of the subsequent-reader performance is determined by the
first-stage retriever [8], there has been a rich literature for investigating either sparse or
dense retriever models. The sparse retrievers typically leverage Term Frequency-Inverse
Document Frequency (TF-IDF) or BM25 [14] to encode candidate documents/passages
with sparse representations. Despite the lightweight, sparse methods rely on lexical
overlapping or exacting matching to identify candidates, thereby failing to perform

the retrieval in the semantic level. With the recent development of Pre-trained Language Models (PLMs), dense retrieval models, compared to sparse ones, have significantly advanced the frontier by retrieving semantically relevant but lexically different evidences [5, 17]. Typically, the dual-encoder architecture is adopted to encode input queries and documents separately into low-dimensional (dense) vectors. The similarity of produced vectors (or the pair of query-and-document embeddings) are then utilized as the matching score. The flexibility of the dual-encoder architecture has spurred further research to improve its efficiency, such as augmenting document representations with interpolation and perturbation [4], exploiting document relations [13], and generating pseudo query embeddings for the low-resource training [16], *etc.*.

Vulnerability and Robustness. On the other hand, abundant evidences [7, 10] also indicate that dense retrievers are vulnerable to noisy (perturbed) samples (*i.e.*, the model performance can be dramatically impacted by (even) small perturbations to the input query). Note that those perturbations explicitly refer to those of **syntax-changing but not-semantic** modifications. The perturbed query is also known as the **query variation** and is the focus of this paper.

Towards this end, Penha *et al.* [12] provide a systematical review on four variations (including Misspelling, Naturality, Order and Paraphrasing), and comprehensively investigate their impacts on the downstream retrieval tasks. The finding reveals a varying extent of the model vulnerability towards to different variations. In particular, Misspelling (or queries contain typos) empirically has the largest effect via degrading an averaged 35% retrieval performance. Accordingly, multiple works have attempted to robustify retrievers with the presence of query variations and have achieved demonstrative results. Zhuang *et al.* particularly investigate the Misspelling variation [19]. They further introduce **DRTA** to employ variations (with a coin-flipping strategy) as the data augmentation for the model training. Later, in [20], they provide a further analysis on queries with typos, and argue that the performance degrade is caused by the token-distribution change during the process of the typo-word tokenization. They further propose **CBST** to maximize the model score distribution obtained from the query with/without typos. Concurrently, an improvement work of DRTA, termed **DACL**, is found in [15]. In addition to leveraging variation samples for training, DACL also introduces the concept of the contrastive learning, to enforce the latent representation of query variations stays close to those of original queries and far apart from other distinct queries. Another similar work can also be found in [7] based on the contrastive learning. The difference is that the correct pairs of query-passage is grouped in the latent space, instead of queries (only) in DACL. Additionally, Chen *et al.* propose **RoDR** [1] from which a standard retrieval model (say DR_O) is trained first to initialize another model (say DR_N). Then, query variations are further employed to train DR_N and simultaneously maximize the score similarity of query-passage pairs obtained by the DR_O and DR_N.

The proposed method is different from existing approaches in the sense that it manipulates query variations via injecting [MASK] tokens as a placeholder, while others remain variations unchanged. Importantly, those additional [MASK] tokens play a critical role in increasing the chance of semantic-overlapping between the original queries and their variations.

3 Proposed Method

This section presents a simple yet effective algorithm to stabilize the retriever performance w.r.t. query variations, termed [MASK] Insertion based Retrieval Stabilizer (MIRS). Our method consists of three components, including the original retrieval, masked augmentation, and semantic alignment.

3.1 MIRS

<u>O</u>riginal <u>R</u>etrieval. Let q and p^+ represent the given **clean** query and its positive (relevant) passage, respectively. Given a set of n negative (irrelevant) passages $\{p_1^-, \cdots, p_n^-\}$ and an encoder $\mathbf{F}$ (usually implemented using PLMs such as BERT), the original retrieval component aims to rank p^+ higher than p_j^- ($\forall j \in [1, n]$) via optimizing the following Negative Log-Likelihood loss:

$$\mathcal{L}_{OR} \triangleq -\log \frac{e^{\text{sim}(\mathbf{F}(q), \mathbf{F}(p^+))}}{e^{\text{sim}(\mathbf{F}(q), \mathbf{F}(p^+))} + \sum_{j=1}^{n} e^{\text{sim}(\mathbf{F}(q), \mathbf{F}(p_j^-))}}, \tag{1}$$

where $\mathbf{F}(\cdot)$ represents the extracted latent representation, and $\text{sim}()$ is a similarity measurement function (cosine in this paper).

<u>M</u>asked <u>A</u>ugmentation. This component augments the model training using masked variations. Specifically, let q' represent the **perturbed** query (a variation to q). Existing work [15, 19, 20] leverages the triple of $\{q', p^+, p_j^-\}$ ($\forall j \in [1, n]$) as the augmented data to fine-tune the model, with the same manner of $\{q, p^+, p_j^-\}$; by contrast, we further propose to inject [MASK] tokens to q' before the augmented training. That is, let b_M be a pre-defined masking budget (or the fraction of masked tokens). Then, MIRS injects M [MASK] tokens randomly within q', following a uniform sampling until the masking budget b_M is met. The formed **masked variation** is labeled as q'_M, and $M = \lceil |x| * b_M \rceil$. Then, similar to Eq. (1), the masked augmentation component enables the retriever to identify positive passages given q'_M:

$$\mathcal{L}_{MA} \triangleq -\log \frac{e^{\text{sim}(\mathbf{F}(q'_M), \mathbf{F}(p^+))}}{e^{\text{sim}(\mathbf{F}(q'_M), \mathbf{F}(p^+))} + \sum_{j=1}^{n} e^{\text{sim}(\mathbf{F}(q'_M), \mathbf{F}(p_j^-))}}. \tag{2}$$

Furthermore, to mitigate the impact of positional encoding, position IDs of inserted [MASK] tokens are set to 0, while those of previous query tokens retains unchanged.

<u>S</u>emantic <u>A</u>lignment. This component is to suppress noisy signals conveyed in masked variations, such that the semantic of the original query is preserved with the presence of [MASK] tokens and other variations. Toward this end, the alignment comportment is to explicitly encourage the encoder $\mathbf{F}$ producing semantic-similar representation between the original query ($\mathbf{F}(q)$) and that of masked variation ($\mathbf{F}(q'_M)$). The similarity measurement, (*i.e.*, the cosine function as Eq. (1)), is employed for this purpose leading to the following objective:

$$\mathcal{L}_{SA} \triangleq \text{sim}(\mathbf{F}(q), \mathbf{F}(q'_M)). \tag{3}$$

Note that this alignment loss is different from that of the masked augmentation, in the sense that it does not involve any supervision signal (*i.e.*, the matching score with p^+), but to reduce the representation dissimilarity obtained by q and q'_M.

Overall Objective Function. To summarize, MIRS involves simultaneously training on masked samples, which are query variations with injected [MASK] tokens, and original queries. Additionally, the semantic alignment ensures masked samples remain semantically close to original ones. The model is fine-tuned using the following joint loss:

$$\mathcal{L} = \frac{1}{2}\mathcal{L}_{OR} + \frac{1}{2}\mathcal{L}_{MA} + \mathcal{L}_{SA}. \tag{4}$$

During inference, the masked augmentation and semantic alignment components are discarded; testing queries are inserted with same amount of [MASK] tokens (as training), and their latent representations are extracted by **F** before estimating the similarity with candidate passages for ranking.

Mainstream approaches for robustifying the retrieval model, mentioned in Sect. 2, include **RoDR** [1], **DACL** [15], **DRTA** [19], and **CBST** [20], while their connection and difference with our algorithm is discussed in the following: (i) all methods employ query variations for the model training, which can be cast as a data augmentation in addition to original queries; (ii) the majority existing work (except DACL) focus on their `score` similarity (the matching results between queries and candidate passages), instead of maximizing the `representation` similarity between clean and perturbed queries. In other words, for other approaches, the model scores, obtained from original/clean queries, provide the supervision signals to their perturbations; yet, ours directly unifies their latent representations. On the other hand, DACL also encourages clean-and-perturbed queries stay together in the latent space while being apart from other distinct queries. It, however, involves the extra effort to identify those distinct queries, as they could both be inquired from the same positive passages. By contrast, the proposed MIRS only targets on grouping the pair of clean-and-perturbed queries; (iii) more importantly, MIRS manipulates variations via inserting [MASK] tokens as a "wild card", while existing methods keep variations unchanged. The benefit of the [MASK] injection is discussed in the next *Analysis* section.

3.2 Analysis

The impact of the query variation w.r.t. the original query is under-explored from existing methods, while this section provides the theoretical analysis from the geometry perspective to bridge this gap. We hypothesize that the latent representations from both the original query and its variation come from two separated convex hulls; yet the variation with [MASK] increases the intersection between its own hull with the original one, thereby preserving more semantic from the original query.

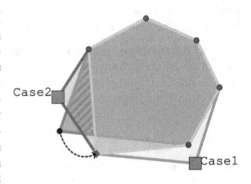

Fig. 2. An illustration of the convex hulls intersections in 2-D (best viewed in color). Blue circles are normal tokens, and black are perturbed tokens, which are replaced later by tokens shown as the red diamond (this process is indicated by the dotted curve with an arrow). Gray squares are inserted [MASK] tokens, showing two situations, Case1 and Case2, corresponding to two geometric relations that may occur, denoted as [MASK1] and [MASK2] respectively. Colored polygons show the convex hulls. The dark gray region indicates that, in Case1 we have $\mathcal{C}(q) \cap \mathcal{C}([q' \cup [\text{MASK1}]])$ which is also $\mathcal{C}(q) \cap \mathcal{C}(q')$. As for Case2, the red region with red hatching lines is the extra intersected area from $\mathcal{C}(q) \cap \mathcal{C}([q' \cup [\text{MASK2}]])$.

To begin with, given the original query q, its latent representation $\mathbf{F}(q)$, estimated using self-attention (from PLM encoders), is mapped from

$$\tilde{\mathbf{Q}} = \text{softmax}(\mathbf{Q}\mathbf{W}_1\mathbf{W}_2^\top\mathbf{Q}^\top)\mathbf{Q}\mathbf{W}_3,$$

where $\mathbf{Q} \in \mathbb{R}^{|q| \times d}$ is the initial embedding for q, d is the hidden dimension, and $\mathbf{W}_k$ ($\forall k \in [1, 3]$) are projection matrices with compatible dimensions. It is well known that $\mathbf{Q}$ is the sum of three parts: token ($\mathbf{E}_t$), segment ($\mathbf{E}_s$), and position ($\mathbf{E}_p$) embedding, or $\mathbf{Q} = \mathbf{E}_t + \mathbf{E}_s + \mathbf{E}_p$. In the context of the query encoding, $\mathbf{E}_s$ is negligible due to the same query and no segmentation required (as a constant). Similar to $\mathbf{E}_p$, as we retain positional encoding for query tokens and set those of inserted [MASK] to 0, so that $\mathbf{E}_p$ become input-independent as it only reflects the absolute/relative positions of tokens. To this end, in following analysis, we simply take $\mathbf{Q} \approx \mathbf{E}_t$. Furthermore, the property of softmax indicates that each row of $\tilde{\mathbf{Q}}$ (say $\tilde{\mathbf{Q}}_i$) is a convex construction of $\mathbf{Q}\mathbf{W}_3$, i.e., $\forall i$, $\tilde{\mathbf{Q}}_i \in \mathcal{C}(q)$ where $\mathcal{C}(q)$ stands for the convex hull of q (see Fig. 2 for the gray area enclosed by black and blue circles). The same process happens in multi-head/layer attention modules. They operate in different projected spaces but the observation of the convex construction still holds. In summary, the convex hull $\mathcal{C}(q)$ is a subspace or more precisely the solution space that determines the latent representation of q, and it is calculated based on the initial embedding $\mathbf{Q}$ with self-attention.

With the variation(s) occurring, original tokens from q are modified, so as $\mathbf{Q}$ and further $\mathcal{C}(q)$ (i.e., relevant vertices in $\mathcal{C}(q)$ deviate from their original positions or even are completely removed). Furthermore, let $\mathcal{C}(q')$ and $\mathcal{C}(q'_M)$ be the convex hull created by q' and q'_M, respectively. We hypothesize that, to preserve the semantic information from q (for stabilizing the retrieval performance) is equivalent to enforcing the encoder to produce invariant representation, *i.e.*, $\mathbf{F}(q) \approx \mathbf{F}(q')$. Since the latent representation is mapped from the original convex hull $\mathcal{C}(\mathbf{q})$ as its solution space, the robustification task is then reformulated as maximizing the intersection of the newly-formed convex hull (either $\mathcal{C}(q')$ or $\mathcal{C}(q'_M)$) with the original convex hull ($\mathcal{C}(q)$); furthermore, the larger intersection results in a better semantic preservation with a higher probability. That is, given that $\mathrm{Vol}(\cdot)$ is a function to estimate the volume of a geometric object, then the probability of the semantic preservation ($\mathbb{P}$) w.r.t. q is determined by $\mathbb{P} \propto \mathrm{Vol}(\mathcal{C}(q) \cap \mathcal{C}(q*))$, where $\mathcal{C}(q*)$ (either $\mathcal{C}(q')$ or $\mathcal{C}(q'_M)$) is the newly-formed convex hull. Notably, MIRS employs a simple [MASK] insertion without removing existing (original and perturbed) tokens from q', which leads to the fact that $C(q'_M)$ always contains $C(q')$ as guaranteed by the following lemma.

Lemma 1. *Given a finite set* $\mathbf{X} = \{x_i\}_{i=1}^{n}$ *in the vector space, we have* $\mathcal{C}(\mathbf{S}) \subseteq \mathcal{C}(\mathbf{X})$ *for any subset* $\mathbf{S} \subseteq \mathbf{X}$. *The equality holds when* $\mathbf{S} = \mathbf{X}$ *trivially or otherwise* $\mathbf{S}$ *contains all the anchor points of* $\mathcal{C}(\mathbf{X})$, *i.e., the convex hull vertices.*

Proof. Let $\mathcal{X}$ be the index set for $\mathbf{X}$ and a subset $\mathcal{S} \subseteq \mathcal{X}$ gives the indices for $\mathbf{S}$. For any point $p \in \mathcal{C}(\mathbf{S})$, $p = \sum_{i \in \mathcal{S}} \lambda_i \mathbf{x}_i$ such that $\lambda_i \geq 0$ and $\sum \lambda_i = 1$, *i.e.*, the convex condition. Apparently $p \in \mathcal{C}(\mathbf{X})$ as well by setting $\lambda_j = 0$ for $j \in \mathcal{X} \backslash \mathcal{S}$. For any point $x_i \in \mathbf{X}$, it is either an anchor point or an internal point referring to $\mathcal{C}(\mathbf{X})$. If $\mathbf{S}$ contains only anchor points, $\mathcal{C}(\mathbf{S}) = \mathcal{C}(\mathbf{X})$ as the internal points can be "absorbed". To see this, assume $\mathbf{x}_1$ is an internal point, then $\mathbf{x}_1 = \sum_{i>1} \beta_i \mathbf{x}_i$ and all β_is for $i > 1$ satisfying convex condition. Then $p = \sum_{i=1} \lambda_i \mathbf{x}_i = \sum_{i>1} (\lambda_i + \lambda_1 \beta_i) \mathbf{x}_i$. Therefore, $\mathcal{C}(\mathbf{X}) = \mathcal{C}(\mathbf{X}_{-1})$ where $\mathbf{X}_{-1}$ is the set of vectors after removing $\mathbf{x}_1$. After eliminating internal points, the convex hull will still be the same.

The immediate result from above lemma is the following corollary stating the relations between $C(q'_M)$ and $C(q')$.

Corollary 1. *Injected* [MASK] *tokens leads to* $C(q') \subseteq C(q'_M)$ *(a 2-D illustration of the convex-hull intersection is also shown in Fig. 2).*

Proof. Let $T_{q'}$ and $T_{q'_M}$ be the token set of q' and q'_M. Clearly, $T_{q'_M} = \{ T_{q'} \cup [\text{MASK}] \}$ and $T_{q'} \subseteq T_{q'_M}$. Similarly, for their initial embeddings, we have $\mathbf{Q}' \subseteq \mathbf{Q}'_M$, where $\mathbf{Q}'$ and $\mathbf{Q}'_M$ is related to q' and q'_M, respectively. According to **Lemma** 1, it is easy to induce that $\mathcal{C}(q') \subseteq \mathcal{C}(q'_M)$. Equality holds only when $\mathcal{C}(q')$ contains all anchor points set in $\mathcal{C}(q'_M)$.

Remark 1. *We point out that* $C(q')$ *may contain subsets that are not part of* $C(q)$, *for example, the areas shaded in red with no hatching lines shown in Fig. 2. The semantic alignment component is then to enforce the model to learn a stable representation that is from* $C(q) \cap C(q'_M)$. *Again, a larger intersection of* $C(q'_M)$ *with* $C(q)$ *provides a larger solution space for that purpose and hence leads to better performance.*

Furthermore, the involvement of the original retrieval target $\mathcal{L}_{OR}$ *in the model training is not essential as the backbone models such as BERT have been pre-trained with* [MASK] *(or alike tokens), so that they can help recover the original convex hull to some extent even under the query variation. Notably, this reflects the **ColBERT*** *model [6]. However, there will still be deviations from the original convex hull depending on the severity of the distortion caused by the query variation, which are responsible for the performance degradation. Therefore, including q in the model provides a true anchor for q'_M to pursuit and achieve robustness.*

4 Experiments

4.1 Setup

Datasets. To make a fair comparison, we follow settings from [15, 19, 20]. Three benchmarking datasets are employed, including TREC Deep Learning Track Passage Retrieval Task 2019 (**DL2019**) and 2020 (**DL2020**) [2] and **MS MARCO (dev/v1)** [9]. Candidate passages are from **MS MARCO (passage/dev)** (approximately 8.8 million).

Query Variation Generation. Four generation strategies are considered [12] to perturb input queries, including: **Misspelling** for substituting existing characters with randomly chosen ASCII ones; **Naturality** for removing all stop words; **Order** for randomly exchanging positions of two words; and **Paraphrasing** for replacing non-stop words with alternatives, according to the similarity of counter fitted-Glove word embeddings.

Implementation Details. All experiments are performed with random seeds. For each run, the batch size is set to 18, the max query length is 32, and the max passage length is 128. For training, the strategy of in-batch negative samples is applied, and seven hard negatives (from the top 200 passages retrieved by BM25) with one positive passage are adopted with each individual training query. Furthermore, the AdamW optimizer with a $5e^{-6}$ learning rate is initialized with a linear learning rate scheduled for 200 thousand updates. At last, our model is trained with a single Tesla A100 40G GPU that takes approximately 40 h to complete. To measure the retrieval performance, the official metric, *i.e.*, Mean Reciprocal Rank for the top 10 retrieved documents (MRR@10), is employed to evaluate the model, and higher MRR means the better retrieval outcome.

4.2 Main Results

The following state-of-the-arts are employed to compare with MIRS, including **RoDR** [1], **DACL** [15], **DRTA** [19], and **CBST** [20]. In addition, the vanilla DPR model is also employed as the **Base** [5]. All models are trained using the BERT-Base as the encoder. Results from contender methods are either directly from original papers or re-implemented using their released codes (if results not available). Additionally, for MIRS the masking budget b_M (or the number of inserted [MAKS]) is set as 20%.

Table 1 shows averaged results across different methods over ten runs. To begin with, MIRS achieves the competitive retrieval accuracy in terms of MRR@10, in comparison to existing contenders with three employed datasets and four query variations on average. Notably, except the Naturality case with MSMARCO (ours scores the second best), MIRS yields the best retrieval performance. On the other hand, the Misspelling type (compared to others) leads to the worst performance across all four variations, which demonstrates the difficulty of querying with misspellings (or typos). Yet,

Table 1. Averaged retrieval performance (over ten runs) obtained by MIRS and current SOTAs with four query variation generators. The number with **bold** represents the best result. Statistical significance testing at p-value < 0.01 (using T-test) are marked with †.

Datasets	Methods	Misspelling	Naturality	Order	Paraphrasing
DL2019	Base	42.7±5.6	66.7±3.1	62.3±2.2	57.4±4.6
	RoDR	51.8±6.0	67.1±2.5	68.5±1.9	64.9±2.5
	DACL	64.6±5.1	68.8±3.3	**72.4±1.5**	68.8±2.0
	DRTA	58.2±6.6	68.1±2.9	72.3±1.1	67.5±2.3
	CBST	61.5±4.6	67.8±2.3	71.1±1.1	68.0±1.7
	MIRS	**67.9±4.5**†	**69.2±2.3**†	**72.4±1.0**†	**71.2±2.0**†
DL2020	Base	45.9±4.6	73.0±2.1	74.8±2.1	63.9±3.3
	RoDR	51.3±6.5	80.3±3.1	75.1±1.8	65.5±2.9
	DACL	63.1±3.1	80.8±2.9	77.1±1.4	69.6±2.0
	DRTA	63.0±3.4	79.3±2.5	76.0±1.1	69.9±2.2
	CBST	62.9±3.2	76.4±1.8	76.5±0.4	68.2±2.4
	MIRS	**64.4±3.2**†	**83.8±2.0**†	**77.8±1.0**†	**70.7±2.2**†
MSMARCO	Base	15.5±0.5	29.0±0.2	30.9±0.1	22.0±0.1
	RoDR	25.0±0.6	**32.9±0.5**	33.9±0.3	26.5±0.2
	DACL	22.8±0.3	31.6±0.2	32.0±0.8	25.9±0.1
	DRTA	21.5±0.4	30.8±0.3	31.5±0.1	24.5±0.3
	CBST	22.8±0.2	31.6±0.3	32.1±0.1	27.1±0.1
	MIRS	**26.2±0.4**†	32.2±0.2†	**34.5±0.1**†	**29.0±0.1**†

MIRS still achieves the marginal improvement (in comparison to the strongest baseline DACL) with an averaged increase of 1.5, 1.5, and 2.4 absolute MRR@10 points for DL2019, DL2020, and MSMARCO with four different variations, respectively. By contrast, the Naturality and Order variations seemingly have less impact on the retrieval performance, compared to the types of Misspelling and Paraphrasing. Note that Naturality is to remove all stop words while Order is to swap two random words' positions. As such, these two variations mainly differs from the original query in terms of the position embedding. The results clearly show that varying position embeddings (the case of Naturality and Order) has limit distortion and even the **Base** model achieves much better performance than those from Misspelling and Paraphrasing. This supports the choice in our analysis, *i.e.* token embeddings have dominant influence on the latent representation (and further the resultant convex hull) compared to position ones. On the other hand, our analysis holds in the sense that adding [MASK] increases the chance of the new formed convex hull (of Naturality and Order) overlapping with the original one. Empirically, the proposed MIRS shows superior performance compared to state-of-the-arts via achieving a considerable margin. We also implement the significance test for ten runs via randomly selecting various seeds and performing the Student's T-test on each dataset. The averaged p-values obtained from DL2019, DL2020, and MSMARCO are $2.41e^{-15}$, $3.78e^{-7}$, and $4.98e^{-5}$, respectively, which verifies the MIRS effectiveness.

4.3 Ablation Study

To better understand the effectiveness of the proposed method, a series of careful studies are carried out. As the most-effective variation, the Misspelling is taken into account hereafter. The following experiments are considered using the DL2019 as the dataset, and results are again reported as the averaged accuracy (MRR@10) over ten runs.

On the Backbone Encoder. To evaluate the flexibility of the proposed MIRS towards different encoders, the Character-BERT(-base) [3] encoder is employed in this experiment, while all other configurations remain the same. Note that CBST [20] also adopts CharacterBERT to deal with the Misspelling case. The Character-BERT enhances the model tolerance (towards typos) via splitting one word into a sequence of characters, before aggregat-

Table 2. Model flexibility comparison of utilizing CharacterBERT as the backbone encoder, with different numbers of misspelling characters.

Datasets	Methods	1-typo	3-typo	5-typo
DL2019	CBST	70.6±4.6	53.5±3.8	45.1±2.6
	MIRS	**75.6±4.5**	**61.3±4.1**	**51.9±3.2**
DL2020	CBST	72.2±3.3	55.6±3.9	42.6±4.6
	MIRS	**74.1±2.9**	**58.5±4.6**	**48.5±4.6**
MSMARCO	CBST	26.3±0.2	20.2±0.2	15.3±0.5
	MIRS	**27.7±0.3**	**22.8±0.1**	**18.5±0.4**

ing representations from individual characters to form one single vector representation for each word directly. Following this setting, we also explicitly employ Character-BERT as the backbone encoder. The comparison between ours and CBST, as a function of the number of misspelling characters (ranging from $[1, 3, 5]$), is illustrated in Table 2.

First, compared to results from Table 1, models using BERT as the backbone encoder clearly underperform those of CharacterBERT. This finding is consistent with [20], as BERT's (WordPiece) tokenizer could tokenize typos into a dramatically-different token series (compared to original words), and further perturb the input embeddings. As such, simply replacing BERT with CharacterBERT improves the retrieval performance. Second, with the increasing numbers of misspelling characters, not surprisingly, the accuracies of both models decrease. Yet, the performance degradation from CBST is much worse than ours. On average, the performance of CBST drops 36.1%, 41.0%, and 41.8% in terms of MRR@10 on the DL2019, DL2020, and MSMARCO, respectively, comparing to that of 31.3%, 34.5%, and 33.2% from MIRS. Overall, MIRS obtains consistently higher performance than CBST, regardless of the backbone encoder and the number of misspelling characters. Hereafter, without explicitly mentioning, the CharacterBERT (instead of BERT) is adopted as the backbone encoder, to minimize the impact from the encoder side and focus on the model design.

On the Breakdown. In the following, we investigate individual components of MIRS and the comparison is considered with the following variants: BA represents the model trained by only pairs of original queries and passages; MA only employs masked query variations and related passages for the model training; and SA maximizes the semantic similarity between the original-masked queries.

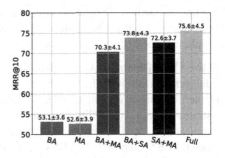

Fig. 3. Comparison of the retrieval performance from individual components of training with masked variations (MA) and aligning original-and-masked semantics (SA).

Figure 3 shows contributions from individual components with $b_m = 20\%$. To begin with, the worst performance is observed by training with only masked query variations (*i.e.*, MA), statistically not significant though compared to that of BA. This reflects *Remark* 1 in the *Analysis*: [MASK] tokens can only recover the convex hull of the original query to some extent; the lack of the true anchor q for the masked query variation q'_M is responsible for the performance degradation. Furthermore, results also reveal that both the proposed masked augmentation and semantic alignment components stably improve the performance of the base model (BA). For instance, BA+MA achieves the 70.3 accuracy as a benefit of augmenting training data with additional (masked) variation samples. For two proposed components, SA seemingly helps in a larger performance boost, which is evidenced by the score of 73.8 from BA+SA compared to 70.3 of BA+MA. Notably, the BA+SA model is trained without fine-tuning the supervised loss of masked samples but encouraging the semantic alignment. The result indicates that the key contributor to MIRS is to restrict the representation similarity of original queries and their variations.

On the Masking Budget. This experiment is to evaluate the impact of the masking budget (b_M) on the proposed method. Obviously, with a higher value of b_M, more [MASK] tokens will be inserted in the variation

Table 3. Comparison of the retrieval performance as a function of the masking budget (b_M).

$b_M=0\%$	$b_M=10\%$	$b_M=20\%$	$b_M=30\%$
68.6±4.9	75.4±6.0	75.6±4.5	75.7±4.0

and lead to more perturbed samples. Specifically, experiments are conduced by varying b_M from the range of 0%, 10%, 20%, and 30%, with the random injection. Note that, when $b_M=0\%$, there is no [MASK] involved, which is equivalent to the method of DACL. On the other hand, with a 10% masking budget, approximately 1 [MASK] is injected in the variation (given the length of the tokenized query). The comparison from Table 3 reflects a notable improvement in the retrieval performance with/out [MASK] (approximately 6.8 absolute MRR@10 point), that demonstrates the advantage of injecting [MASK] tokens to preserve the original query semantic with a high probability. In addition, MIRS also observes a stable accuracy regardless varying masking budgets (with $b_M > 0$) as their difference is relatively moderate (*e.g.*, ±0.25 absolute point). As such, the result clearly demonstrates the effectiveness and stability of the proposed method. This observational experiment once again confirms our inference in the *Analysis*, that is, the span of the convex hull is utterly important rather than its multiplicity (due to the injection location).

On the Out-of-Domain Generalizability. MIRS is then evaluated by the resultant model generalizability. That is, the model is first trained with one query variation, and then evaluated on other unseen variations without further fine-tuning. Figure 4 illustrates the generalizability comparison between ours and CBST, with the DL2019 dataset, and $b_m = 20\%$. In terms of the absolute MRR@10 point (results within the bracket), MIRS outperforms CBST in all 16 cases. For the relative model adaptation (%), only two cases (with Misspelling → Naturality and Naturality → Order) CBST is observed with higher scores. The results clearly demonstrate the strong generalizability of MIRS. Note that, during the model fine-tuning, CBST is trained with the *same* type of query variation as that for inference, which makes it difficult to adopt another variation type. By contrast, MIRS involves the additional [MASK] tokens as the "wild card", that increases the model flexibility and further improves the model transferability to unseen variations. We also observe for MIRS, the higher retrieval accuracy is associated with the transferred model of Paraphrasing → Misspelling and Order → Naturality. Note that those variations share some common aspects, such as the modification of existing word (Paraphrasing → Misspelling) or the swap of word position (Order → Naturality). The results indicate the benefit of leveraging the relationship among different variations to robustify the model, which we leave for the future study.

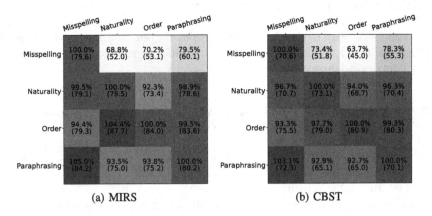

(a) MIRS   (b) CBST

Fig. 4. Comparison of the model generalizability, where a pre-trained retriever (from one variation) is applied to other unseen variations without fine-tuning.

4.4 Discussion

In this section, we investigate different strategies of inserting [MASK] tokens, and further seek for a reasonable explanation for the result. Again, following experiments are conducted with the random injection, and $b_M = 20\%$.

To begin with, we analyze the insertion objects or where to add [MASK] tokens. Notably, the [MASK] tokens were inserted to q' in all previous experiments, whereas other options exist. That is, to justify our choice, three variants are

Table 4. Impact analysis of the [MASK] location.

Original	Variation	Both
72.8±3.6	75.6±4.5	70.8±4.5

tested, including the insertion of [MASK] tokens to: (1) the Original query q only, (2) the query Variation q' only, and (3) Both q and q'. All aforementioned variants employ three proposed components in their cost functions with corresponding entities, *i.e.*, q and q', replaced by the masked ones.

The comparison is shown in Table 4. The Both variant is observed with the worst performance, while the Variation one, *i.e.* our primary option, stands out as the best. This is a solid confirmation of our *remark* regarding to the anchor in the *Analysis*. The ideal retrieval should be based on the clean query q so that the semantics between query and retrieved document can be maximally aligned. When q is masked to be q_M, its convex hull is altered (or "lifted" as a pictorial illustration) and the right solution to the retrieval may be drawn from outside $\mathcal{C}(q)$ although still within $\mathcal{C}(q_M)$. This increases the robustness of the model when combined with semantic alignment as the model has to accommodate the variation brought by this inserted "wild card" for the retrieval (referring to the Original variant). However, our strategy in MIRS, *i.e.* the Variation case where q provides the clear supervision information for q_M, stabilizes the representations of query and document tokens. When both are lifted, there will be too much flexibility and hence the performance could be further reduced. Nevertheless even the Both variant, the worst case in this batch, still produces the slightly better performance than that of CBST (from Table 2).

Additionally, we consider to inject another token ([HOLDER] in this context, instead of [MASK]) in MIRS. That is, the entire process maintains the explicitly same except a different token to be injected (with the same setting of and $b_M = 20\%$). The comparison is shown in Table 5 and the [HOLDER] injection achieves a similar performance as that of [MASK]. The result highlights that including [MASK] (or equivalent) in the model fine-turning plays a role as a

Table 5. Comparison of the retrieval performance in terms of different inserted tokens.

	Misspelling	Naturality
[MASK]	75.6	79.5
[HOLDER]	75.8	81.9
	Order	Paraphrasing
[MASK]	84.0	80.2
[HOLDER]	81.2	78.8

placeholder or a "wild card". Accordingly, the capacity of spanning the convex hull to more likely intersect with original one is further enhanced, which leads to the robust retrieval performance.

5   Conclusion

This paper presents the [MASK] Insertion based Retrieval Stabilizer (MIRS) to robustify the retriever with the presence of query variations. Specifically, the proposed method randomly injects [MASK] tokens within variations, and further encourages the representation similarity between the original query and masked variation. The theoretical analysis shows that the injection of [MASK] tokens helps in forming a convex hull with a large overlapping with that of the original query, so to increase the probability of preserving its original semantics. Intensive experimental results demonstrate the superiority of the proposed algorithm, in comparison to state-of-the-arts, using three benchmark datasets with four types of query variations.

We need to point out that, our analysis is constructed on a crucial assumption asserting that the probability of preserving semantic (or latent representation) is determined by the volume intersection of two convex hulls. Although empirical results confirmed this assumption repeatedly, we will still seek for direct dynamics of the convex hull with regard to the input query for the future study. On the other hand, we will also consider to improve the model robustness from the perspective of masking passages/documents.

References

1. Chen, X., Luo, J., He, B., Sun, L., Sun, Y.: Towards robust dense retrieval via local ranking alignment. In: Raedt, L.D. (ed.) Proceedings of the Thirty-First International Joint Conference on Artificial Intelligence, IJCAI-22, pp. 1980–1986. International Joint Conferences on Artificial Intelligence Organization (7 2022)
2. Craswell, N., Mitra, B., Yilmaz, E., Campos, D., Voorhees, E.M.: Overview of the TREC 2019 deep learning track. In: Proceedings of the Twenty-Ninth Text REtrieval Conference (NIST Special Publication). National Institute of Standards and Technology (NIST) (2020)
3. El Boukkouri, H., Ferret, O., Lavergne, T., Noji, H., Zweigenbaum, P., Tsujii, J.: Character-BERT: reconciling ELMo and BERT for word-level open-vocabulary representations from characters. In: Proceedings of the 28th International Conference on Computational Linguistics, pp. 6903–6915. Barcelona, Spain (Online) (2020)
4. Jeong, S., Baek, J., Cho, S., Hwang, S.J., Park, J.: Augmenting document representations for dense retrieval with interpolation and perturbation. In: Proceedings of the 60th Annual Meeting of the Association for Computational Linguistics (Volume 2: Short Papers), pp. 442–452. Association for Computational Linguistics, Dublin, Ireland (2022)
5. Karpukhin, V., et al.: Dense passage retrieval for open-domain question answering. In: Proceedings of the 2020 Conference on Empirical Methods in Natural Language Processing (EMNLP), pp. 6769–6781. Association for Computational Linguistics, Online (Nov 2020)
6. Khattab, O., Zaharia, M.: ColBERT: efficient and effective passage search via contextualized late interaction over BERT, pp. 39–48. Association for Computing Machinery, New York, NY, USA (2020)
7. Ma, X., Nogueira dos Santos, C., Arnold, A.O.: Contrastive fine-tuning improves robustness for neural rankers. In: Findings of the Association for Computational Linguistics: ACL-IJCNLP 2021, pp. 570–582. Association for Computational Linguistics, Online (Aug 2021)
8. Mao, Y., et al.: Generation-augmented retrieval for open-domain question answering. In: Proceedings of the 59th Annual Meeting of the Association for Computational Linguistics and the 11th International Joint Conference on Natural Language Processing (Volume 1: Long Papers), pp. 4089–4100. Association for Computational Linguistics, Online (2021)
9. Nguyen, T., Rosenberg, M., Song, X., Gao, J., Tiwary, S., Majumder, R., Deng, L.: Ms marco: A human generated machine reading comprehension dataset. In: Proceedings of the Workshop on Cognitive Computation: Integrating neural and symbolic approaches 2016 co-located with the 30th Annual Conference on Neural Information Processing Systems (CEUR Workshop Proceedings, Vol. 1773). CEUR-WS.org (2016)

10. Nogueira, R., Cho, K.: Passage re-ranking with bert. arXiv preprint arXiv:1901.04085 (2019)
11. Parkin, L., Chardin, B., Jean, S., Hadjali, A., Baron, M.: Dealing with plethoric answers of SPARQL queries. In: Strauss, C., Kotsis, G., Tjoa, A.M., Khalil, I. (eds.) Database and Expert Systems Applications, pp. 292–304. Springer, Cham (2021)
12. Penha, G., Câmara, A., Hauff, C.: Evaluating the robustness of retrieval pipelines with query variation generators. In: Hagen, M., et al. (eds.) ECIR 2022. LNCS, vol. 13185, pp. 397–412. Springer, Cham (2022). https://doi.org/10.1007/978-3-030-99736-6_27
13. Raman, N., Shah, S., Veloso, M.: Structure and semantics preserving document representations. In: Proceedings of the 45th International ACM SIGIR Conference on Research and Development in Information Retrieval, pp. 780–790. SIGIR 2022, Association for Computing Machinery, New York, NY, USA (2022)
14. Robertson, S.E., Walker, S.: Some simple effective approximations to the 2-poisson model for probabilistic weighted retrieval. In: Croft, B.W., van Rijsbergen, C.J. (eds.) SIGIR 1994, pp. 232–241. SIGIR '94, Springer, London (1994). https://doi.org/10.1007/978-1-4471-2099-5_24
15. Sidiropoulos, G., Kanoulas, E.: Analysing the robustness of dual encoders for dense retrieval against misspellings. In: Proceedings of the 45th International ACM SIGIR Conference on Research and Development in Information Retrieval, pp. 2132–2136. SIGIR 2022, Association for Computing Machinery, New York, NY, USA (2022)
16. Tang, H., Sun, X., Jin, B., Wang, J., Zhang, F., Wu, W.: Improving document representations by generating pseudo query embeddings for dense retrieval. In: Proceedings of the 59th Annual Meeting of the Association for Computational Linguistics and the 11th International Joint Conference on Natural Language Processing (Volume 1: Long Papers), pp. 5054–5064. Association for Computational Linguistics (2021)
17. Xiong, L., et al.: Approximate nearest neighbor negative contrastive learning for dense text retrieval. In: International Conference on Learning Representations (2021)
18. Zhu, X., Hao, T., Cheng, S., Wang, F.L., Liu, H.: A self-supervised joint training framework for document reranking. In: Findings of the Association for Computational Linguistics: NAACL 2022, pp. 1056–1065. Association for Computational Linguistics, Seattle, United States (2022)
19. Zhuang, S., Zuccon, G.: Dealing with typos for BERT-based passage retrieval and ranking. In: Proceedings of the 2021 Conference on Empirical Methods in Natural Language Processing, pp. 2836–2842. Association for Computational Linguistics, Online and Punta Cana, Dominican Republic (Nov 2021)
20. Zhuang, S., Zuccon, G.: Characterbert and self-teaching for improving the robustness of dense retrievers on queries with typos. In: Proceedings of the 45th International ACM SIGIR Conference on Research and Development in Information Retrieval, pp. 38–45. SIGIR '22, Association for Computing Machinery, New York, NY, USA (2022)

Parallel Pattern Enumeration in Large Graphs

Abir Farouzi[1,3]([✉]), Xiantian Zhou[2], Ladjel Bellatreche[1], Mimoun Malki[3], and Carlos Ordonez[2]

[1] LIAS/ISAE-ENSMA, Chasseneuil-du-Poitou, France
[2] University of Houston, Houston, USA
[3] Ecole Nationale Supérieure en Informatique de Sidi Bel Abbès, Sidi Bel Abbès, Algeria
a.farouzi@esi-sba.dz

Abstract. Graphlet enumeration is a fundamental problem to discover interesting patterns hidden in graphs. It has many applications in science including Biology and Chemistry. In this paper, we present a novel approach to discover these patterns with queries, in a parallel database system. Our solution is based on an efficient partitioning strategy based on randomized vertex coloring, that guarantees perfect load balancing and accurate graphlet enumeration (complete and consistent). To the best of our knowledge, our work is the first to provide an abstract and efficient database solution with queries to enumerate both 3-vertex and 4-vertex patterns on large graphs.

1 Introduction

In graph analytics, finding small structures is crucial to studying the relationships between a set of individuals/objects. These structures are called graphlets or patterns, which are defined as small induced subgraphs. The problem of graphlet enumeration is involved in many fields. In Biology for example, [13] studied the interaction and the function of proteins in the entire proteome, which is based on protein-protein interaction (PPI) networks. For that, [13] developed a graph-based technique that condenses a protein's nearby topology within a PPI network by using a vector of graphlet degrees known as the protein's signature. It then determines the similarity between the signatures of all pairs of proteins. Another relevant example would be in Chemistry, where [6] studied the chemical compounds classification by developing a method based on two steps: (1) substructure discovery process which includes the graphlet enumeration, and (2) the classification process allowing an intelligent chemical compounds classification using the graphlets. Other applications can be found in [4,5].

On the other hand, it is well-known that the graphlets beyond three vertices are complicated to enumerate since *the number of instances* can rapidly grow with $O(n^\alpha)$, where α is the order of the graphlet. Indeed, we enumerate 2 structures for the 3-vertex graphlets, 6 structures for the 4-vertex graphlets,

C. Strauss et al. (Eds.): DEXA 2023, LNCS 14146, pp. 408–423, 2023.
https://doi.org/10.1007/978-3-031-39847-6_32

21 for 5-vertex graphlets, and so on. In practice, it is typical for both the list of graphlets and the time required for their enumeration to increase quickly as the size of the graph increases, resulting in a computationally expensive task [11].

In this context, we presented in a previous work [7] a distributed solution with queries, to enumerate efficiently all the embedded triangles in large graphs. The triangle constitutes a special case of graphlets, where $\alpha = 3$. However, it is not as hard as larger graphlets, particularly those of 4 vertices. Indeed, enumerating 4-vertex graphlets is more challenging, since it requires checking and outputting 6 instances with different structures. Our focus in this work is to present an efficient parallel solution to enumerate all the 4-vertex graphlets. Our solution underlines an efficient partitioning strategy based on vertex-coloring and inspired by our previous work [7] on triangles. Furthermore, 4-vertex graphlet enumeration requires more join operations compared to the triangles. Thus, in the present work, we developed a staged enumeration strategy, where the simple graphlets of 3 and 4 order are used to reveal larger or complex subgraphs dynamically. This reduces memory usage and accelerates the running time.

Our contributions are:

(1) We provide a distributed algorithm for staged graphlet enumeration. Our approach can be implemented on any parallel system including DBMS.
(2) We propose a vertex cut partitioning strategy based on coloring that guarantees a perfect load balancing, and a local enumeration.
(3) We present the computational model optimization for an efficient and optimized execution of our solution.
(4) We study the partitioning strategy, the result correctness, and the isomorphism between the resulting graphlets.
(6) We finally present an experimental validation of our findings and compare our results with a competing graphlet enumeration solution.

Our paper is organized as follows. Preliminaries are described in Sect. 2. In Sect. 3, we explain our data partitioning strategy, the computational model optimization, and our solution for graphlet enumeration. In Sect. 4, we study the partitioning strategy, the graph isomorphism, and the load balancing. We present an experimental validation in Sect. 5. Section 6 explains closely related work and Sect. 7 concludes the paper with general remarks.

2 Preliminaries

2.1 Graph

Let $G = (V, E)$ be an undirected unweighted graph with $n = |V|$ (V is the set of vertices) and $m = |E|$ (E is the set of edges). For our algorithm, we read the input graph G as an edge list $E(u, v)$, where an edge goes from u to v. We define $H = (V_H, E_H)$ as a subgraph of $G = (V, E)$ if $V_H \subseteq V$ and $E_H \subseteq E$, and as an induced subgraph of $G = (V, E)$ if $\forall u, v \in V_H$ and $(u, v) \in E_H$ then $(u, v) \in E$.

2.2   Graphlets (Patterns in Connected Subgraphs)

A graphlet H is a connected induced subgraph of G. We denote two types of 3-vertex graphlets: (1) Wedges (W) that are paths of two edges, and (2) Triangles (TR) which are cycles with three connected edges. Moreover, there are six types of 4-vertex graphlets: (3) 3-Path (P), (4) 3-Star (S), (5) Rectangle (R), (6) Tailed Triangle (T), (7) Diamond (D), (8) 4-vertex Clique (C) (see Fig. 1).

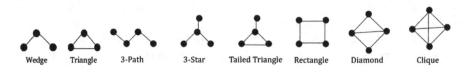

Wedge   Triangle   3-Path   3-Star   Tailed Triangle   Rectangle   Diamond   Clique

Fig. 1. 3 and 4-vertex graphlets.

2.3   Computational Model

In order to run our algorithm, we use the $k - machine$ model (a.k.a *Big Data* model), introduced by [9]. It consists of a set of $k \geq 2$ machines built on a shared-nothing architecture. Each machine can communicate directly with the other machines via message passing (no shared memory) while running an instance of the distributed algorithm. Since there is no shared memory, an efficient data partitioning strategy is mandatory. It aims to minimize the data communication between machine during the distributed algorithm execution. Note that the data communication is a time and space consuming task.

3   Our Approach for Pattern Enumeration

3.1   Data Partitioning

The key idea underlying our partitioning approach is to perform a vertex-cut partitioning based on coloring, so that each vertex and its incident edges are sent to the same machine (see Fig. 2). Our partitioning strategy aims to partition the vertex set V into c subsets of $O(n/c)$ vertex each, where c is the number of color subsets ($c \geq 2$). Then, each edge between two subsets of colors is sent to one random machine from the $k - machine$ model called the proxy machine. Finally, each local machine collects its required edges according to its quadruplet of colors (hard-coded in the algorithm) to perform locally the graphlet enumeration.

In practice, we create table $V\_s(u, u\_color)$ to store each vertex color. Indeed, each entry in the table V_s is a couple of a vertex and its color chosen uniformly and independently at random from the c colors. Then, the table $E\_s\_proxy(u, v, u\_color, v\_color)$ is created to send edges to proxies. It holds the end-vertices color of each edge. Finally, the small table $quadruplet(machine, c_1, c_2, c_3, c_4)$ is created and replicated on each machine of the $k - machine$. This small table is used by each local machine to collect its required edges and stores them in the table $E\_s\_local(machine, u, v, u\_color, v\_color)$. The edges stored in E_s_local table depends on the type of graphlets to generate. Hence, we need to recreate this table according to the required edges to form the graphlets. Notice that each entry in the table *quadruplet* consists of one of the possible permutation of the c colors. Suppose we have two colors: black (b) and white (w). We can generate (b, b, b, b), (b, b, b, w), $\cdots$ etc. We can generate c^4 quadruplets using these c colors.

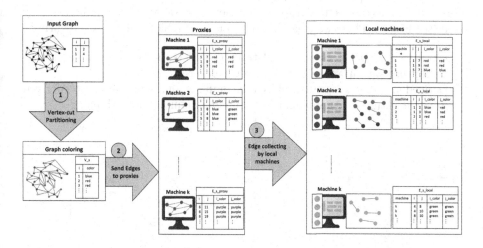

Fig. 2. Data partitioning for graphlet enumeration.

3.2 $k - Machine$ Distributed Model Optimization

The ideal execution of our algorithm requires one machine for each color quadruplet. This means we need a $k - machine$ model of size:

$$k = c^4 \tag{1}$$

However, with larger number of c, the size of the $k - machine$ model can grow rapidly. Hence, we propose $k - machine$ model optimization without impacting our algorithm functions. For that, we need to allow each machine to manage more than one quadruplet. Generally, we can use the following equation:

$$k = c^l \ where \ 1 \le l \le 4 \tag{2}$$

Here, each machine needs to manage c^{4-l} quadruplets of colors. For example, with $c = 2$ and $k = 8$, each machine manages 2 color quadruplets.

3.3 Graphlet Enumeration

Enumerating graphlets of order $O(n^\alpha)$ results in many instances. Suppose we have a graph with n vertices, enumerating all its embedded graphlets requires studying and checking $C_n^{\alpha} = \frac{n!}{\alpha! * (n-\alpha)!}$ possible combinations. Hence, we need to generate 6 instances of 4-vertex graphlets. Moreover, each graphlet of order 4 outputs $4! = 24$ permutations. These permutations represent repetitions, and their elimination is a hard task in most cases. Thus, only some configurations of the graphlets must be considered, and a redundancy elimination process need to be integrated into our algorithm. Figure 4 summarizes the graphlets configuration to consider for each 4-vertex graphlet type. Note that each graphlet is made of 4 vertices $\{u, v, w, z\}$ where $u < v < w < z$.

In order to accelerate the graphlet enumeration, we allow an enumeration by stage (see Fig. 3). Thus, we classify the 4-vertex graphlets into two categories:

(1) Intuitive graphlets: it consists of three graphlets: 3-Path, 3-Star, and rectangle; that we can reuse to generate the other graphlets.
(2) Derived graphlets: represented by the complex graphlets that can be derived from the intuitive graphlets: tailed triangle, diamond, and clique.

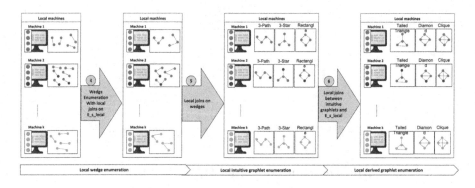

Fig. 3. Graphlet enumeration by stage.

Wedge Enumeration. The main idea behind our approach is to enumerate graphlets by stage. Thus, we enumerate wedges to output intuitive graphlets, that will be used to list derived graphlets. Wedges are graphlets consisting of 3 vertices and 2 edges. We compute them using one self join on the table E_s_local. We enumerate four types of wedges according to color quadruplets. Precisely, the wedges of type 1 ($Wedge\_T1$) correspond to the first three colors ($\{c_1, c_2, c_3\}$) in the table $quadruplet$. This wedge is involved in the enumeration of all the intuitive graphlets. Then, we enumerate wedges of type 2 ($Wedge\_T2$) with ($\{c_2, c_3, c_4\}$) colors to output paths, and wedges of type 3 ($Wedge\_T3$) with ($\{c_1, c_2, c_4\}$) colors to list 3-stars. Finally, we output wedges of type 4 ($Wedge\_T4$) with ($\{c_1, c_3, c_4\}$) colors to enumerate rectangles. For example and without loss of generality, on machine M_1 with (c_1, c_2, c_3, c_4) as quadruplet of colors, we generate $\{Wedge\_T1, Wedge\_T2, Wedge\_T3, Wedge\_T4\}$

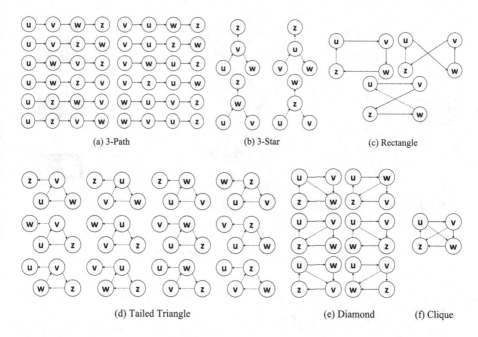

(a) 3-Path   (b) 3-Star   (c) Rectangle

(d) Tailed Triangle   (e) Diamond   (f) Clique

Fig. 4. Graphlets configuration to output.

whose vertex colors are in $\{(c_1, c_2, c_3), (c_2, c_3, c_4), (c_1, c_2, c_4), (c_1, c_3, c_4)\}$ respectively. We create the following tables to store the generated wedges for each machine.

$Wedge\_T1(machine, u, v, w, u\_color, v\_color, w\_color)$
$Wedge\_T2(machine, u, v, w, u\_color, v\_color, w\_color)$
$Wedge\_T3(machine, u, v, w, u\_color, v\_color, w\_color)$
$Wedge\_T4(machine, u, v, w, u\_color, v\_color, w\_color)$

Furthermore, triangles can easily be extracted using the wedges of type 1. Indeed, triangles are a specific case of wedges with an edge connecting each couple of vertices of the wedge (3-vertex clique). The triangle enumeration problem is largely discussed in our previous paper [7].

Intuitive 4-Vertex Graphlet Enumeration. Wedges are the building blocks of intuitive graphlets, wherein we generate the required wedges to enumerate each variant of these graphlets. Subsequently, we carry out joins between them. Figure 5 summarizes this process.

3-Path consist of two wedges (u, v, w) and (v, w, z). It mainly depends on its end-vertices order, in such a way that $u < z$. Hence, we enumerate $\binom{4}{2} = 12$ permutations (by symmetry elimination). For that, we create table $Path(machine, u, v, w, z)$ that holds the 3-Paths resulting of a local join between the table $Wedge\_T1$ and the table $Wedges\_T2$. The query $Q(P)$ summarizes this computation:

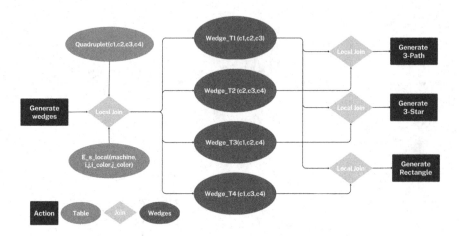

Fig. 5. Intuitive 4-vertex graphlet enumeration process.

```
Q(P):INSERT INTO Path SELECT T1.machine as machine, T1.u as u, T1.v as v,
    T1.w as w, T2.w as z
        FROM Wedge_T1 T1 JOIN Wedge_T2 T2
            ON T1.machine=T2.machine AND T1.v=T2.u AND T1.w=T2.v
        WHERE T1.u<T2.w;
```

3-Star is formed by two wedges (u, v, w) and (u, v, z) connected with the common edge (u, v) of the wedges. The generation of this graphlet depends on the outer vertices of the star. Hence, we need to enumerate $\binom{4}{3} = 4$ permutations. To save the results, we create table *Star(machine, u, v, w, z)*. Then, we join the table *Wedge_T1* with table *Wedge_T3* considering $u < w < z$ to remove all the repetitions. The query $Q(S)$ bellow summarises the 3-Star graphlet enumeration:

```
Q(S):INSERT INTO Star SELECT T1.machine as machine, T1.u as u, T1.v as v,
    T1.w as w, T3.w as z
        FROM Wedge_T1 T1 Join Wedge_T3 T3
            ON T1.machine=T3.machine AND T1.u=T3.u AND T1.v=T3.v
        WHERE T1.u<T1.w AND T1.w<T3.w;
```

Rectangle is represented by two opposite wedges $\{(u, v, w), (u, z, w)\}$ with $v! = z$. For this graphlet, we can have 4 configurations for the cyclic symmetry and 2 configurations for each clockwise counter-clockwise symmetry. As a result, we need to generate one configuration of each of them, so we generate $\frac{24}{4 \times 2} = 3$ permutations. Hence, we create the table *Rectangle(machine, u, v, w, z)* and we perform a local join between the table *Wedge_T1* and the table *Wedge_T4* for each case ($u < v < w < z$, $u < v < z < w$ and $u < w < v < z$) and we union the results together, as presented in the query $Q(R)$:

```
Q(R):INSERT INTO Rectangle SELECT T1.machine as machine, T1.u as u,
    T1.v as v, T1.w as w, T4.v as z
        FROM Wedge_T1 T1 join Wedge_T4 T4
            ON T1.machine=T4.machine AND T1.u=T4.w AND T1.w=T4.u
        WHERE T1.u<T1.v AND T1.v<T1.w AND T1.w<T4.v
```

```
UNION SELECT T1.machine as machine, T1.u as u, T1.v as v, T1.w as w,
T4.v as z
     FROM Wedge_T1 T1 join Wedge_T4 T4
          ON T1.machine=T4.machine AND T1.u=T4.w AND T1.w=T4.u
     WHERE T1.u<T1.v AND T1.v<T4.v AND T4.v<T1.w
UNION SELECT T1.machine as machine, T1.u as u, T1.v as v, T1.w as w,
T4.v as z
     FROM Wedge_T1 T1 join Wedge_T4 T4
          ON T1.machine=T4.machine AND T1.u=T4.w AND T1.w=T4.u
     WHERE T1.u<T1.w AND T1.w<T1.v AND T1.v<T4.v;
```

Derived 4-Vertex Graphlet Enumeration. The graphlets of this class are
generated using the intuitive graphlets class.

Tailed Triangle can be seen as a 3-Star (u, v, w, z) with v at the center, and
an edge (u, w), (w, z) or (u, z) connecting two of its endpoints. Its enumeration relies on the vertex at the center and the vertex at the end of its tail,
for that we consider $\binom{4}{2} = 12$ permutations. To list them, we create table
$Tailed(type, machine, u, v, w, z)$, where *type* distinguish whether the connecting
edge is between $\{u, w\}$, $\{u, z\}$ or $\{w, z\}$. Than we recreate table E_s_local to
have, on each machine all the edges that endpoints are of colors (c_1, c_3), (c_1, c_4),
or (c_3, c_4) of each color quadruplet. Finally, we perform a local join between the
table *Star* and the table E_s_local, as mentioned in the query $Q(T)$:

```
/*1=T(u,v,w),2=T(u,v,z),3=T(v,w,z), T=Triangle*/
Q(T):INSERT INTO Tailed SELECT 1, E1.machine as machine, E1.u as u,
          E1.v as v, w, z
     FROM Star E1 Join E_s_local E2
          ON E1.machine=E2.machine AND E1.u=E2.u AND E1.w=E2.v
UNION SELECT 2, E1.machine as machine, E1.u as u, E1.v as v, w, z
     FROM Star E1 Join E_s_local E2
          ON E1.machine=E2.machine AND E1.u=E2.u AND E1.z=E2.v
UNION SELECT 3, E1.machine as machine, E1.u as u, E1.v as v, w, z
     FROM Star E1 Join E_s_local E2
          ON E1.machine=E2.machine AND E1.w=E2.u AND E1.z=E2.v;
```

Diamond can be recognized as a rectangle with an edge on one of its diagonals. Since only 3 configurations are needed for the rectangles, each rectangle with one of it diagonal can be output as diamond. So, we need to output $3 \times 2 = 6$ permutations for this graphlet. To enumerate them, we create
table $Diamond(machine, u, v, w, z)$. Then, we recreate the table E_s_local to
have edges whose end-vertices are of colors (c_1, c_3) and (c_2, c_4) of each color
quadruplet. We finally join locally the table *Rectangle* with the table E_s_local
according to the query $Q(D)$:

```
Q(D):INSERT INTO Diamond SELECT E1.machine as machine, E1.u as u,
          E1.v as v, w, z
     FROM Rectangle E1 Join E_s_local E2
          ON E1.machine=E2.machine AND E1.u=E2.u AND E1.w=E2.v
UNION SELECT E1.machine as machine, E1.u as u, E1.v as v, w, z
     FROM Rectangle E1 Join E_s_local E2
          ON E1.machine=E2.machine AND E1.v=E2.u AND E1.z=E2.v;
```

Clique is a complete subgraph of four vertices with an edge between each couple of vertices, hence only one configuration should be output, which is made of the lexicographical order between the vertices. This graphlet is a rectangle with its both diagonals. Hence, we create table $Clique(machine, u, v, w, z)$ to hold all the 4-vertex cliques. For that, we recreate the table E_s_local to have the edges whose end-vertices are of colors (c_1, c_3) ans (c_2, c_4) of each color quadruplet. Then, we join locally the table $Rectangle$ with the table E_s_local twice, as presented in the following query $(Q(C))$:

```
Q(C):INSERT INTO Clique SELECT E1.machine as machine, E1.u as u,
      E1.v as v, w, z
         FROM Rectangle E1 JOIN E_s_local E2
            ON E1.machine=E2.machine AND E1.u=E2.u AND E1.w=E2.v
         JOIN E_s_local E3
            ON E1.machine=E3.machine AND E1.v=E3.u AND E1.z=E3.v
         WHERE E1.u<E1.v AND E1.v<E1.w AND E1.w<E1.z;
```

4   Graphlet Enumeration Theoretical Analysis

4.1   Partitioning Strategy Effectiveness

Our partitioning strategy aims to send an induced subgraph to each machine of the $k - machine$ model. It performs a vertex-cut partitioning using a coloring method. First, we partition the vertex set V into c color subsets. After that, each machine receives c^{4-l} quadruplets of colors, which is used to collect its required edges to form an induced subgraph of size $O(m/k)$.

Results Correctness. To study the correctness of our results, we define the following lemmas:

Lemma 1. *Consistency: all the graphlets are output once.*

Proof. The enumeration of each of the 4-vertex graphlets is local without any data communication between machines. Hence, we need to be sure that each graphlet is enumerated once.

1. Wedges: wedges of type 1 are output once on the model following the Eq. 2 with $l \leq 3$. Without loss of generality, suppose we have two quadruplet $q_1 = \{c_1, c_1, c_1, c_1\}$ and $q_2 = \{c_1, c_1, c_1, c_2\}$ on machine M_1. Notice that the first color triplet in q_1 and q_2 is the same. This color triplet is used to output the wedges of type 1 involved in all the intuitive graphlets. In our quadruplet assignment, we ensure that each machine from the $k-machine$ model acquires the same first triplet of color, hence each machine will exclusively output the wedges corresponding to its triplet of color.
2. Intuitive graphlets: Suppose on machine M, we have the quadruplet (c_1, c_2, c_3, c_4). The wedges with colors (c_1, c_2, c_3), (c_2, c_3, c_4), (c_1, c_2, c_4) and (c_1, c_3, c_4) are held by the tables $Wedge\_T1$, $Wedge\_T2$, $Wedge\_T3$ and $Wedge\_T4$ resp. on machine M without local repetitions. When

we generate 3-Path (3-Star or rectangle) graphlets on M, we compute $Wedge\_T1 \bowtie_{\substack{T1.u\,=\,T2.v \\ \&T1.w\,=\,T2.v}} Wedge\_T2$ ($Wedge\_T1 \bowtie_{\substack{T1.u\,=\,T3.v \\ \&T1.w\,=\,T3.v}} Wedge\_T3$ or $Wedge\_T1 \bowtie_{\substack{T1.u\,=\,T4.v \\ \&T1.w\,=\,T4.v}} Wedge\_T4$). Each wedge is unique on M, thus the generated paths (stars or rectangles) are unique on M because they consist of unique wedges on M. Furthermore, depending on the cluster size, each machine defines $4 - l$ quadruplet of colors that are uniquely and specifically assigned to it. Hence, the order of the colors of each quadruplet is unique on each machine (there is no repetition). As a result, the generated intuitive graphlets are output once, since their enumeration depends on the colors and the order of their corresponding quadruplet on each machine.

3. Derived graphlet: we proved above that each intuitive graphlet is output once, and since the output of the derived graphlets is based on the intuitive ones, then each derived graphlet is output once.

Lemma 2. *Completeness: There is no missing graphlet in the output.*

Proof. In our partitioning strategy, each vertex choose independently and uniformly a color from the c colors, in the same time, we create color quadruplets consisting of the different permutations of the c colors. As a result, each edge with its colored end-vertices involved in one or more 3-vertex or 4-vertex graphlets, its end-vertices colors are in one or more color quadruplets. Without loss of generality, suppose the vertex u chose color c_1 and the vertex v chose the color c_2. The edge (u, v) will be sent to the machine M defining (c_1, c_2) in one of its quadruplets. So, if the edge (u, v) is a part of a wedge or derived graphlet, it will be on M and according to the quadruplets on M, the wedge defining the intuitive graphlet or the derived graphlet involving (u, v) will be output. Finally, there will be no missing wedges or graphlets.

Graph Isomorphism. We explained previously that we only consider some permutations when outputting the graphlets. In fact, when we output $P = (u, v, w, z)$ as a 3-path graphlet, we need to eliminate the opposite direction $P' = (z, w, v, u)$, since P et P' are isomorphic. Hence, only the permutation defining $u < z$ is considered. In addition, we need to consider four permutations for 3-star graphlets, depending on the vertex at the center. For $S = (u, v, w, z)$ with v at the center, we need to consider the graphlet where $u < w < z$. All the other graphets with v at the center, are isomorphic to S. The rectangles $R = (u, v, w, z)$ are output in three permutations ($u < v < w < z$, $u < v < z < w$, $u < w < v < z$). All the remaining configurations are isomorphic to R, since the rectangle is a cycle of length 4. The tailed triangles $T = (u, v, w, z)$ accept twelve permutations. For each configuration of 3-star, we can generate three tailed triangles, considering $u < w < z$. The other configurations are the same as T. The diamonds $D = (u, v, w, z)$ need six permutations to be considered. We output two permutations for each rectangle. Hence, all the 18 remaining graphlets are isomorphic to D. Finally, for the cliques $C = (u, v, w, z)$ only one permutation is needed. All the other configurations are isomorphic to C.

4.2   Complexity and Load Balancing

Complexity. 4-vertex graphlet enumeration is more difficult than the triangle enumeration. The complexity of triangle enumeration is $O(n^3)$, whereas it is $O(n^4)$ for 4-vertex graphlets. Our algorithm is bounded by $O(N_\angle^2)$ for the intuitive graphlets, where $N_\angle$ is the wedge count. On the other hand, since the derived graphlets depends on the intuitive graphlets, their complexity is bounded by $O(max\{m \times S, m \times R\})$, with $S(R)$ is the number of 3-star (rectangle) graphlets.

Load Balancing. We mentioned previously that the vertex set V is partitioned into c subsets of $O(n/c)$ vertex each. Each machine then receives an induced sub-graph $G_x = (V_x, E_x)$ of G. The number of edges among the sub-graphs G_x is relatively balanced with high probability. Indeed, each vertex chose a color in a uniform manner, so it has $\frac{1}{c}$ to choose one of the c colors. Then, the edge between a couple of vertices has $\frac{1}{c} \times \frac{1}{c} = \frac{1}{c^2}$ possibility. This balances the load between the proxies. Each machine then collects the required edges to enumerate the graphlets, so it holds $(\frac{1}{c})^4 = \frac{1}{c^4}$ of each type of graphlets. As a result, each machine processes essentially the same number of graphlets, which leads to balance the workload.

5   Experimental Study

5.1   Hardware and Software Setup, and Data Set

Hardware: Experiments are conducted on a cluster with 9 machines. Each machine has 4 cores CPU running at 2.2 Ghz on average, 4 GB of main memory, 500 GB of storage, 32 KB L1 cache, 1 MB L2 cache and Linux Ubuntu server 18.04 as operating system. The machines are connected on 1GB network cards with 128 MB/s as bandwidth. Each machine manages 2 or 3 quadruplets.

Software: We used the columnar DBMS Vertica to execute our queries (the code is available at https://github.com/lias-laboratory/sqlgraphlet), since it is 10× faster than the row DBMSs for graph problems. However, any other parallel system that provides partitioning control can be used, including systems like SparkSQL and TigerGraph. Moreover, we compared our algorithm against D4GE [11]; a tool based on Spark for sub-graph enumeration.

Data Sets: We used four real data sets from Stanford data set collection (https://snap.stanford.edu), summarized in Table 1.

5.2   Graphlet Enumeration

Our experiments are presented in two sets: (1) the first concerns the evaluation of our approach results, including its load balancing and its speed-up, and (2) the second set aims to compare our approach with D4GE [11] which is a distributed graphlets enumeration solution based on Spark. Each experiment is repeated three times and the average time measurement is reported.

Table 1. Data sets: order (n), size (m), triangle count (Δ) and maximum degree (d_max).

Data sets	n	m	Δ	d_max
Facebook	4,039	88k	1,612k	1,045
Pennsylvania	1,088k	1,541k	67k	9
Amazon	334k	925k	75k	549
DBLP	317k	1,049k	13k	343

Graphlet Enumeration Evaluation

Graphlet Counting: Table 2 summarizes the wedges and the 4-vertex graphlets counting for each graph data set. Our partitioning strategy ensures to have each vertex with its incident edges on the same machine according to its color quadruplet. Thus, each graphlet is enumerated once on one machine if it exists.

Table 2. Graphlet counting output.

Data sets	P	S	R	T	D	C
Facebook	1,055,326,189	727,318,426	144,023,053	703,783,680	138,773,046	30,004,668
Pennsylvania	7,384,597	1,707,904	157,802	318,190	5,795	21
Amazon	80,983,900	142,823,893	3,125,323	24,485,894	2,702,808	275,961
DBLP	675,637,762	431,568,151	55,107,655	316,232,255	54,904,261	16,713,192

P: 3-Path, S: 3-Star, R: Rectangle, T: Tailed Triangle, D: Diamond, C: Clique.

Our Solution Speed-Up: Figure 6 depicts the local execution time for each graphlet algorithm on the graph data sets on a cluster of 4 machines, 8 machines and 9 machines. Notice that the partitioning time is negligible and it happens once at the beginning, so we didn't include it. From Fig. 6, we deduce that the most time consuming graphlet is the tailed triangle, because we classify this graphlet in three categories depending on the position of the triangle (at the beginning, in the middle or at the end), then we union the results together. Moreover, we can notice that the execution time of the queries improves as the size of the cluster increase. Hence, two conclusions can be drawn: (1) we can get up to 2× speed-up with larger number of colors, and (2) we obtain better running time on models having less quadruplets managed by each machine. Furthermore, we notice that the derived graphlets require negligible execution time, since they are based on the intuitive graphlets. This opens doors for our stepwise enumeration strategy to list quickly larger graphlets based on smaller ones.

Balanced Workload: To evaluate the load balancing of our partitioning strategy, we present Fig. 7, that presents line charts for the output of the graphlets on each machine, using a cluster with 8 machines. The workload balancing ensured by the partitioning strategy results in outputting almost the same count of

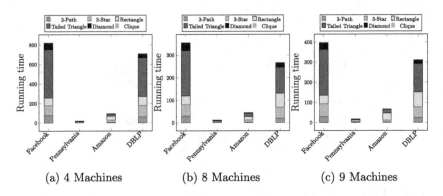

(a) 4 Machines   (b) 8 Machines   (c) 9 Machines

Fig. 6. Local running time for graphlet enumeration (sec) with $k = \{4, 8, 9\}$ and $c = \{2, 2, 3\}$ resp.

graphlets on each machine. The slight variation in counting is due to the randomization in the vertex coloring step. Precisely, the 3-Star query produced a variation between the machines because of the graph structure. Facebook and amazon data sets are particularly skewed graphs, that's why we can enumerate more 3-Star on only some machines.

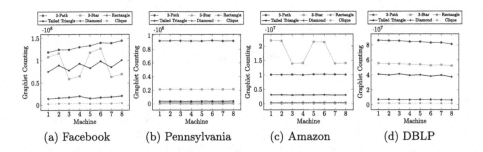

(a) Facebook   (b) Pennsylvania   (c) Amazon   (d) DBLP

Fig. 7. Load balancing on a cluster with 8 machines ($k = 8$ and $c = 2$).

Comparison with D4GE. The distributed Spark solution D4GE [11] requires compressed input graphs, hence the use of graph compression tool like webGraph library [3] is mandatory. As a result, we compressed all the graph data sets using webGraph library before performing the experiments. This compression time is included in our evaluation as the pre-treatment time.

Table 3 provides a comparison between our approach and D4GE on a cluster of 8 machines using 2 colors. Despite its use of compressed data, D4GE shows more running time with sparse graphs as Pennsylvania. This data set is the largest in our chosen data sets, but also the most sparse. On the other hand, our solution showed less efficient with dense graphs compared to D4GE. Note that unlike D4GE that only provide the graphlets count, we list the results. Our ultimate goal is to provide a tool that works by stage and save the results of each step. This has two impacts: (1) speed-up the enumeration process with larger graphlets, and (2) provide a base to develop a recursive strategy to enumerate larger graph structure including the maximum cliques. To sum up, our solution gives an acceptable running time. It scales well, it does not require any graph compression or graph preparation beforehand, and it lists clearly and entirely all the 4-vertex graphlets in the input graph.

Table 3. Comparison of our solution against D4GE (sec).

Data sets	D4GE			Our solution		
	Pre-treatment	Enumeration	Total	Partitioning	Enumeration	Total
Facebook	8	20	28	1	250	251
Pennsylvania	25	18	43	4	10	14
Amazon	18	16	34	1	26	27
DBLP	16	28	44	2	177	179

6 Related Work

Graph Analytics is becoming a first-class challenge in database research [10]. Subgraph enumeration is among the fundamental problems that have received a lot of attention recently. It is based on recursive queries and transitive closure; two main graph problems that are largely and deeply studied in [8,14]. Triangle enumeration and counting is the simplest sub-graph enumeration problem and has been discussed in database perspective in many works such [1,2,7]. These works presented different partitioning strategies to minimize data exchange and to perform triangle enumeration locally. Moreover, many solutions for 4-vertex graphlet enumeration have been developed outside DBMS. [11] is a distributed solution based on Spark to enumerate all the triad and the 4-vertex graphlets in large compressed graphs. [11] was inspired by [15], who presented an efficient partitioning strategy based on coloring, that balances the workload to enumerate subgraphs but with repetitions. Other works for the induced subgraph enumeration, such as FanMod [16] and Rage [12] have emerged. However, they do not perform well on million-scale graphs and are less efficient. On the other hand, we are, to the best of our knowledge, the first to present an integrated solution to discover both 3-vertex and 4-vertex graphlets with queries, that can be reprogrammed using a programming language such as Python and MPI.

7 Conclusion

We present a novel distributed approach to solve 3-vertex and 4-vertex graphlets enumeration problem. Our current solution is programmed using SQL, but we can implement it with a programming language as Python, or a parallel system like Spark. Moreover, we experimentally proved that our partitioning strategy provides a perfect load balancing, and our solution scales well with the graph size. This study is promising and it can be efficiently extended to larger graphlets.

In future work, we will study the impact of the number of colors on query time. We will also study larger graphlet enumeration (of order α), compressing the graph using triangles as super-vertices. We plan to extend our algorithms to multicore CPUs and GPUs. As a longer term goal, we aim to solve clique enumeration, a significantly harder problem.

References

1. Ahmed, A., Enns, K., Thomo, A.: Triangle enumeration for billion-scale graphs in RDBMS. In: Barolli, L., Woungang, I., Enokido, T. (eds.) AINA 2021. LNNS, vol. 226, pp. 160–173. Springer, Cham (2021). https://doi.org/10.1007/978-3-030-75075-6_13
2. Al-Amin, S.T., Ordonez, C., Bellatreche, L.: Big data analytics: exploring graphs with optimized SQL queries. In: Elloumi, M., et al. (eds.) DEXA 2018. CCIS, vol. 903, pp. 88–100. Springer, Cham (2018). https://doi.org/10.1007/978-3-319-99133-7_7
3. Boldi, P., Vigna, S.: The webgraph framework I: compression techniques. In: Proceedings of WWW (2004)
4. Bröcheler, M., Pugliese, A., Subrahmanian, V.: COSI: cloud oriented subgraph identification in massive social networks. In: Proceedings of IEEE ASONAM (2010)
5. Charbey, R., Prieur, C.: Stars, holes, or paths across your Facebook friends: a graphlet-based characterization of many networks. Netw. Sci. 7(4), 476–497 (2019)
6. Deshpande, M., Kuramochi, M., Wale, N., Karypis, G.: Frequent substructure-based approaches for classifying chemical compounds. IEEE TKDE 17(8), 1036–1050 (2005)
7. Farouzi, A., Bellatreche, L., Ordonez, C., Pandurangan, G., Malki, M.: A scalable randomized algorithm for triangle enumeration on graphs based on SQL queries. In: Song, M., Song, I.-Y., Kotsis, G., Tjoa, A.M., Khalil, I. (eds.) DaWaK 2020. LNCS, vol. 12393, pp. 141–156. Springer, Cham (2020). https://doi.org/10.1007/978-3-030-59065-9_12
8. Jachiet, L., Genevès, P., Gesbert, N., Layaida, N.: On the optimization of recursive relational queries: application to graph queries. In: Proceedings of ACM SIGMOD (2020)
9. Klauck, H., Nanongkai, D., Pandurangan, G., Robinson, P.: Distributed computation of large-scale graph problems. In: Proceedings of ACM-SIAM SODA (2015)
10. Lan, M., Wu, X., Theodoratos, D.: Answering graph pattern queries using compact materialized views. In: Proceedings of DOLAP (2022)
11. Liu, X., Santoso, Y., Srinivasan, V., Thomo, A.: Distributed enumeration of four node graphlets at quadrillion-scale. In: Proceedings of SSDBM (2021)
12. Marcus, D., Shavitt, Y.: Rage - a rapid graphlet enumerator for large networks. Comput. Netw. 56(2), 810–819 (2012)
13. Milenković, T., Przulj, N.: Uncovering biological network function via graphlet degree signatures. Cancer Inform. 6, CIN–S680 (2008)

14. Ordonez, C., Cabrera, W., Gurram, A.: Comparing columnar, row and array DBMSs to process recursive queries on graphs. Inf. Syst. **63**, 66–79 (2017)
15. Park, H., Silvestri, F., Pagh, R., Chung, C., Myaeng, S., Kang, U.: Enumerating trillion subgraphs on distributed systems. ACM TKDD **12**(6), 71:1–71:30 (2018)
16. Wernicke, S., Rasche, F.: FANMOD: a tool for fast network motif detection. Bioinformatics **22**(9), 1152–1153 (2006)

S2CTrans: Building a Bridge from SPARQL to Cypher

Zihao Zhao[1,2], Xiaodong Ge[1,2], Zhihong Shen[1(✉)], Chuan Hu[1,2], and Huajin Wang[1]

[1] Computer Network Information Center, Chinese Academy of Sciences, Beijing, China
{zhaozihao,gexiaodong,bluejoe,huchuan,wanghj}@cnic.cn
[2] University of Chinese Academy of Sciences, Beijing, China

Abstract. In graph data applications, data is primarily maintained using two models: RDF (Resource Description Framework) and property graph. The property graph model is widely adopted by industry, leading to property graph databases generally outperforming RDF databases in graph traversal query performance. However, users often prefer SPARQL as their query language, as it is the W3C's recommended standard. Consequently, exploring SPARQL-to-Property-Graph-Query-Language translation is crucial for enhancing graph query language interoperability and enabling effective querying of property graphs using SPARQL. This paper demonstrates the feasibility of translating SPARQL to Cypher for graph traversal queries using graph relational algebra. We present the S2CTrans framework, which achieves SPARQL-to-Cypher translation while preserving the original semantics. Experimental results with the Berlin SPARQL Benchmark (BSBM) datasets show that S2CTrans successfully converts most SELECT queries in the SPARQL 1.1 specification into type-safe Cypher statements, maintaining result consistency and improving the efficiency of data querying using SPARQL.

Keywords: RDF · Property graph · SPARQL · Cypher

1 Introduction

Currently, knowledge graph storage primarily relies on two models: Resource Description Framework (RDF) [5] and property graph [1]. RDF databases, such as Jena, maintain the former, while property graph databases, like Neo4j, manage the latter.

In general, property graph databases outperform RDF databases in graph traversal and pattern matching tasks. However, users tend to favor SPARQL for

This work was supported by the National Key R&D Program of China(Grant No.2021YFF0704200) and Informatization Plan of Chinese Academy of Sciences(Grant No.CAS-WX2022GC-02).
Z. Zhao and X. Ge—Contributed equally to this paper.

C. Strauss et al. (Eds.): DEXA 2023, LNCS 14146, pp. 424–430, 2023.
https://doi.org/10.1007/978-3-031-39847-6_33

data querying, as it is a long-standing W3C recommended standard language. The 2019 W3C Workshop on Web Standardization for Graph Data [11] called for bridging the gap between RDF and property graph query languages, allowing systems to manage data using the property graph data model while enabling users to query data with SPARQL.

Differences in semantic representation and processing logic exist between SPARQL and the property graph query language, represented by Cypher [3], making the standardization process challenging. There are three main challenges of this translation: (a) Proving the semantic equivalence of SPARQL and Cypher in graph traversal query. (b) Resolving the conflict between RDF model and property graph model storage through schema mapping and data mapping. (c) Designing the pattern matching mapping and solution modifier mapping method to translate SPARQL into Cypher.

In this study, we establish a graph relational algebra-based semantics for SPARQL and introduce S2CTrans, a provably semantics-preserving SPARQL-to-Cypher translation method. We then evaluate S2CTrans using comprehensive query features on public datasets. This paper introduces the S2CTrans framework, which offers a mapping method for pattern matching and solution modifiers, enabling the translation from SPARQL to Cypher. We perform a comprehensive query test on large-scale datasets to evaluate the performance improvement of Cypher in graph databases after translating SPARQL using the S2CTrans framework.

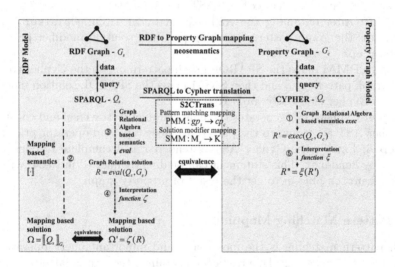

Fig. 1. Overview of SPARQL-to-Cypher translation.

The diagram of our work is illustrated in Fig. 1. At the data level, we implement a syntactic and semantic transformation of RDF graph to property graph using the neosemantincs plug-in [10] developed by Neo4j Labs. This involves storing RDF triples into property graphs as nodes, relationships, and properties. At the query level, the figure illustrates the first two contributions discussed above. The dashed arrow ① represents the graph relational algebra of Cypher, while the dashed arrow ② represents the mapping-based semantics of SPARQL defined in [8]. Our contributions are represented by the dashed arrows ③, ④, and ⑤, which define a graph relational algebra based semantics of SPARQL. Additionally, the solid arrows represent our contributions to the definition of the SPARQL-to-Cypher translation, which includes the pattern matching mapping (**PMM**) and the solution modifier mapping (**SMM**).

2   S2CTrans

2.1   System Architecture

We design and implement S2CTrans, a framework which could equivalently translate SPARQL into Cypher. S2CTrans has been open-sourced[1]. S2CTrans takes SPARQL query as input, and generates Cypher statement with the original semantics by using Jena ARQ [9] parse strategy, graph pattern matching and solution modifiers transformation strategy and Cypher-DSL [7] construction strategy. The S2CTrans works as a five-step execution pipeline:

- **Step 1:** The input SPARQL query is first parsed by the Jena ARQ module. It can check for syntax errors, verify whether it is a valid SPARQL query and generate an abstract syntax tree (AST) representation.
- **Step 2:** After obtaining the AST parsed by SPARQL, `OpWalker` is used to access the graph pattern matching part and solution modifier part from bottom up.
- **Step 3: PMM** maps the SPARQL graph pattern gp_s to the Cypher combining graph pattern cp_c, and then **SMM** maps the SPARQL solution modifiers $\mathbf{M}_s$ to Cypher clause keywords $\mathbf{K}_c$.
- **Step 4:** Cypher-DSL generates the final conjunctive traversal and constructs Cypher AST according to the pattern element type and operator priority.
- **Step 5:** Finally, the Cypher AST is rendered as a complete Cypher statement by `Renderer`. This statement can be directly queried in Neo4j with the `neosemantics` plug-in to get the result of property graph.

2.2   Pattern Matching Mapping

Graph pattern matching is the most basic and important query operation in graph query languages [2]. Due to page constraints, the graph pattern mapping algorithm is introduced in the appendix of S2CTrans-tech-report [12]. The mapping function **PMM** in the algorithm translates SPARQL graph pattern into Cypher graph pattern elements.

---

[1] https://github.com/MaseratiD/S2CTrans.

2.3 Solution Modifiers Mapping

After the graph pattern is obtained by **PMM** algorithm, conditions are usually added to modify the solution of graph pattern matching. Based on the semantic equivalence of SPARQL and Cypher in graph relational algebraic expressions, **SMM** algorithm constructs a mapping table (as shown in Table 1) to implement the mapping of SPARQL solution modifiers M_s to Cypher clause keywords K_c. This table summarizes graph query modification operations and the corresponding graph relational algebra, as well as the forms of SPARQL and Cypher clause construction. The variables and expressions have been mapped to graph pattern elements in **PMM** algorithm.

Table 1. A consolidated list of SPARQL solution modifiers and corresponding Cypher clause keywords.

Operation	Algebra	SPARQL Solution Modifiers - M_s	Cypher Clause Keywords - K_c
Selection	$\sigma_{condition}(r)$	FILTER($Expr_1$ &&($\|\|$) $Expr_2$)	WHERE $Expr_1$ $and(or)$ $Expr_2$
Projection	$\pi_{x_1,x_2,...}(r)$	SELECT ?x_1 ?x_2 ...	RETURN x_1, x_2, ...
De-duplication	$\delta_{x_1,x_2,...}(r)$	SELECT DISTINCT ?x_1 ?x_2 ...	RETURN DISTINCT x_1, x_2, ...
Restriction	$\lambda_s^l(r)$	LIMIT l SKIP s	LIMIT l SKIP s
Sorting	$\varsigma_{\uparrow x_1,\downarrow x_2,...}(r)$	ORDER BY ASC(?x_1) DESC(?x_2)	ORDER BY x_1 ASC, x_2 DESC

Through **PMM** algorithm and **SMM** algorithm, we get the Cypher graph pattern and clause keywords. Cypher-DSL constructs Cypher AST according to graph pattern elements and operator precedence. Finally, we use **Renderer** to construct a complete Cypher statement.

3 Experiments

3.1 Evaluation Criteria

We execute SPARQL queries on several top-of-the-line RDF databases, and execute translated Cypher queries on graph database Neo4j. We evaluate S2CTrans by the translation speed, query execution time and result consistency.

3.2 Experimental Setup

Dataset: This experiment uses the Berlin SPARQL Benchmark(BSBM) dataset recommended by W3C, which consists of synthetic data describing e-commerce use cases, involving categories such as products, producers, etc. We generated 10M triples respectively by BSBM-Tools, and the corresponding property graph version is mapped using the neosemantics plug-in. The details of dataset are introduced in the appendix of S2CTrans-tech-report [12].

Query Statements: We created a total of 40 SPARQL queries, covering 30 different query features. These queries were selected after systematically studying the semantics of SPARQL queries [8]. The queries are detailed in the appendix of S2CTrans-tech-report [12].

System Setup: We execute the query statements on the following databases to evaluate the performance improvement of S2CTrans: **Property Graph Database:** Neo4j v4.2.3 **RDF Databases:** Virtuoso v7.2.5, Stardog v7.6.3, RDF4J v3.6.3, Jena TDB v4.0.0 All experiments were performed on the following machine configurations: CPU: Intel Core Processor (Haswell) 2.1GHz; RAM: 16 GB DDR4; HDD: 512 GB SSD; OS: CentOS 7. In order to ensure the reproducibility of the experimental results, we provide the experimental script, dataset and query statement[2].

3.3   Result Evaluation

According to the evaluation criteria described above, we perform SPARQL query on RDF databases and the translated Cypher query on property graph database Neo4j on the dataset. Finally, we compare and analyze the query results, make sure the consistency. Among them, each query runs an average of 10 times to get the average value. Due to the limited space of the paper, the statements translations and query results are shown in the appendix of S2C-tech-report [12]. The average translation time of S2CTrans of 40 queries on BSBM-10M is **23.7 ms**. Compared with the query time, it accounts for a small proportion. We meticulously conducted tests on datasets of various scales under both cold-start and warm-start scenarios, and all tests yielded similar results. Figure 2 presents the query execution time during the system's cold-start phase. Among most query statements, Neo4j performs better than the RDF databases. Moreover, in the queries with multi-hop paths and long relationships, the performance of Neo4j is 1 to 2 orders of magnitude higher than RDF database. The main reason is that RDF databases spend a lot of time in executing join operations and forming execution plans, while Neo4j uses index-free adjacency, which greatly improves the query efficiency.

The experiment results prove that the proposed S2CTrans is successful in equivalent translating and executing SPARQL queries. S2CTrans enables the users to query property graph by SPARQL.

---

[2] https://github.com/MaseratiD/S2CTrans.

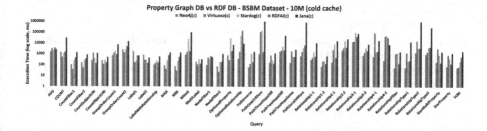

Fig. 2. Property graph database V.S. RDF database - BSBM Dataset 10M

4 Conclusion

In this paper, we introduce S2CTrans, a novel approach that supports SPARQL-to-Cypher translation. This method can convert most SPARQL statements into type-safe Cypher statements. Moreover, we employ property graph databases and RDF databases to conduct experimental evaluations on large-scale datasets, validating the effectiveness and applicability of our approach. The evaluation highlights the substantial performance gains achieved by translating SPARQL queries to Cypher queries, particularly for multiple relationship and star-shaped queries. Although S2CTrans currently has several limitations, it represents an important step toward promoting the standardization of graph query languages and enhancing the interoperability of data and queries between the Semantic Web and graph database communities. In the future, we plan to further refine S2CTrans to support more SPARQL translations and investigate the translation from Cypher to SPARQL.

References

1. Angles, R.: The property graph database model. In: Proceedings of the 12th Alberto Mendelzon International Workshop on Foundations of Data Management, vol. 2100 (2018)
2. Angles, R., Arenas, M., Barceló, P., Hogan, A., Reutter, J., Vrgoc, D.: Foundations of modern query languages for graph databases. ACM Comput. Surv. **50**(5), 68:1-68:40 (2017)
3. Francis, N., et al.: Cypher: an evolving query language for property graphs. In: Proceedings of the 2018 International Conference on Management of Data (2018)
4. Hölsch, J., Grossniklaus, M.: An algebra and equivalences to transform graph patterns in neo4j. In: Proceedings of the Workshops of the EDBT/ICDT 2016 Joint Conference, EDBT/ICDT Workshops 2016, volume 1558 of CEUR Workshop Proceedings (2016)
5. Klyne, G., Carroll, J.J., McBride, B.: RDF 1.1 concepts and abstract syntax, W3C Recommendation (2018)
6. Marton, J., Szárnyas, G., Varró, D.: Formalising openCypher graph queries in relational algebra. In: Kirikova, M., Nørvåg, K., Papadopoulos, G.A. (eds.) ADBIS 2017. LNCS, vol. 10509, pp. 182–196. Springer, Cham (2017). https://doi.org/10.1007/978-3-319-66917-5_13

7. Meier,G., Simons, M.: The neo4j Cypher-dsl (2021). https://neo4j-contrib.github.io/Cypher-dsl/current/
8. Pérez, J., Arenas, M., Gutiérrez, C.: Semantics and complexity of SPARQL. ACM Trans. Database Syst. **34**(3), 16:1-16:45 (2009)
9. Wilkinson, K.: Jena property table implementation. In: Smart, P.R., (ed.) Proceedings of the 2nd International Workshop on Scalable Semantic Web Knowledge Base Systems, pp. 35–46 (2006)
10. Neo4j Labs: neosemantics (n10s): Neo4j RDF & Semantics toolkit (2021). https://neo4j.com/labs/neosemantics/
11. Taelman, R., Vander Sande, M., Verborgh, R.: Bridges between GraphQL and RDF. In: W3C Workshop on Web Standardization for Graph Data. W3C (2019)
12. Zhao, Z., Ge, X., Shen, Z.: S2CTrans: Building a Bridge from SPARQL to Cypher. arxiv

Rewriting Graph-DB Queries to Enforce Attribute-Based Access Control

Daniel Hofer[1,2(✉)] , Aya Mohamed[1,2] , Dagmar Auer[1,2] ,
Stefan Nadschläger[1], and Josef Küng[1,2]

[1] Institute for Application-oriented Knowledge Processing (FAW),
Johannes Kepler University (JKU) Linz, Linz, Austria
{daniel.hofer,aya.mohamed,dagmar.auer,
stefan.nadschlaeger,josef.kueng}@jku.at
[2] LIT Secure and Correct Systems Lab, Linz Institute of Technology (LIT),
Johannes Kepler University (JKU) Linz, Linz, Austria

Abstract. To provide Attribute-Based Access Control (ABAC) in a
data-store, we can either rely on built-in features or, especially if they
are not present, implement access control as a service (ACaaS) on top of
the database. We address the latter, in particular for graph databases,
by rewriting queries which are violating access control conditions. We
intercept the insecure queries right before sending them to the database
to add additional filters. Thus, the database returns only authorized
data and implicitly enforces ABAC beyond its own access control features. Our contributions are an authorization policy model influenced by
XACML and a query rewriting algorithm for enforcing the defined autho-
rizations with respect to this model. Our concept is application- and
database-independent and operates on simple freely formulated queries,
i.e. the queries do not have to follow a predefined structure. A proof-of-
concept prototype has been implemented for Neo4j and its query language Cypher.

Keywords: query rewriting · attribute-based access control (ABAC) ·
graph databases · database security · Cypher

1 Introduction

To enforce access control on a database with limited or even no access control
features, like the community version of Neo4j, we have various options. The
approach we chose is rewriting *insecure* queries before they are handed over to
the database as *secure* queries [1], including authorization-specific filters. Our
approach even supports attribute-based access control (ABAC) [9] by operating
on data stored in the database. We already motivated this approach in our
previous work [6]. However, our current work does not rely on a predefined
query structure, but can handle freely formulated queries.

Authorization requirements are expressed in terms of rules in the authorization policy, referencing *filter templates* to be used in the query rewriting. A *filter*

template defines authorization-specific constraints to be added to the insecure query. Graph database query languages like Cypher distinguish between nodes and relationships. For this work, we refer to both using the term *element*. To rewrite the insecure query applying authorization-specific constraints, we consider the following research questions:

RQ1 Which elements of a query influence the result?
RQ2 What information must be provided in the authorization policy?
RQ3 How can we find mappings between a policy and a query?
RQ4 How do we apply access control filter templates on queries?

Our contributions are (1) identifying the influencing elements and how they impact the query result in Sect. 3, (2) a policy model influenced by XACML as a policy having a set of rules with conditions and references to filter templates in Sect. 4, (3) a query rewriting approach to extend the insecure query with filters encoding authorization requirements in Sect. 5, and (4) a proof-of-concept prototype[1] using Cypher and a preliminary evaluation in Sect. 6. Related works and a summary including an outlook on future work are provided in Sects. 2 and 7 respectively.

2   Related Work

The idea of protecting data by query rewriting is influenced by Browder et. al. and their work about per-user views in Oracle databases [3]. Another influence comes from Bogaerts et. al. [2] as they propose entity-based access control, taking not only attributes but also the relations between entities into account. While their focus is on relational databases, we primarily consider graph databases and thus attributes on nodes and edges. The dynamic rewriting approach was already proposed by Jarman et. al. [7], however, on relational databases and role-based access control. Our policy model is highly influenced by XACML (Ramli et. al. [10]), although we reduced the features to a subset suitable for our requirements. Colombo et. al. proposed an approach similar to ours in [4], as they generate authorized views to replace the original collection in the query. However, their focus is on document-oriented stores with focus on IoT data analysis. Access control by query rewriting for RDF and SPARQL was also proposed by Kirrane in [8]. A slightly different approach is presented by Shay et. al. [11] which checks queries against a policy and blocks them altogether if necessary. The current work is also based on our previous work, especially [5] for query parsing and modification and [9] for XACML policies for graphs.

3   Relevant Information in the Insecure Query

To answer RQ1, we start with checking the elements of the query pattern and identify the relevant elements influencing the query result. The pattern for example in the Cypher query "`MATCH (a:L1)-[c]->(b:L2) WHERE a.id=8 RETURN`

---

[1] https://github.com/jku-lit-scsl/CypherRewritingCore

b" is "(a:L1)-[c]->(b:L2)". While a confidential node in the *RETURN* clause of a query clearly reveals information, other cases are less obvious. For example, we have a graph database with information about students and their grades. A node stores all student data and links to a node with the student's grade for a certain exam. To protect grades, we block returning the grade-nodes. However, a malicious user could return a student's node and include a *WHERE* clause filtering only for a name (which is not confidential) and a specific grade. By only returning the data-node, no access violation is detected, but it implicitly confirms the *guessed* grade. Therefore, an element which has a filter applied might still lead to information leaks although it is not directly returned. On the other hand aggregating functions (e.g. average) prevent access to individual elements (e.g., a student's grade). Thus, we check the combination of filter and return status for each element (see example in Table 1). The filter status is (1) filtered or (2) unfiltered and the return status is (1) aggregated, (2) direct value or (3) not included in the return clause.

Table 1. Influencing factors in MATCH (a:L1)-[c]->(b:L2) WHERE a.id=8 RETURN b.

Element in pattern	Filter	Return	Influencing
a	yes	no	yes
b	no	yes (direct value)	yes
c	no	no	no

4 Policy Model

The purpose of the policy is to specify all authorization-relevant information. A policy P describes a pattern of elements E (i.e., nodes and relationships) and a set of rules R. The function $\Phi(e_{policy}, e_{query}) \rightarrow \{true, false\}$ decides whether an element of the policy pattern (e_{policy}) can be mapped to an element of the insecure query pattern (e_{query}). Let one policy be:

$$P = (E, R, \Phi)$$
$$E = \langle e_1, e_2, ..., e_n \rangle$$
$$R = (e, C, f)$$
$$C = \{c_1, c_2, ..., c_n\}$$

Each rule $r \in R$ references a single element of the policy pattern ($e \in E$) and specifies one or more boolean combined conditions C on the pattern elements and references a filter template f to be applied to e. A condition $c \in C$ checks whether filter and return properties (cp. Sect. 3) are satisfied by any element of the policy pattern $e' \in E$.

Filter templates F are used to exclude unauthorized results in the secure query. They define authorization-relevant constraints to be added to the insecure query. We define a filter template f with placeholders for runtime-specific information as follows:

$$F = \{f_1, f_2, ..., f_n\}$$
$$f = (t, A)$$
$$A = \langle a_1, a_2, ..., a_n \rangle$$

Every filter template $f \in F$ includes a query fragment t containing placeholders. For each placeholder in t, its kind a (e.g., `ruleElement` or `username`) is given in A. The kind `ruleElement` indicates that the placeholder stands for the element in the rule which references this filter template.

5   Query Processing

For a policy and its rules to be applicable, each element defined in the policy e_{policy} is mapped to an equivalent one in the query e_{query} based on its labels and pattern structure. To find a mapping (cp. RQ3), we define a function `getPaths` returning a set of paths from the pattern. Each path consists of a start node, a relationship and an end node $(e_{start}, e_{relationship}, e_{end})$. The relationship and end node can be empty if the start node is isolated.

$$getPaths(E) \rightarrow E^{\star}$$
$$E^{\star} = \{(e_{start}, e_{relationship}, e_{end}), ...\}$$

This step converts the patterns of policy and query into a common and comparable structure. We search for mappings using the function $map(e_{policy}) \rightarrow e_{query}$:

$$map(e_{policy}) \rightarrow e_{query} \Leftrightarrow \forall\, (a, b, c) \in E^{\star}_{policy}\; \exists\, (x, y, z) \in E^{\star}_{query} :$$
$$\Phi(a, x) \wedge \Phi(b, y) \wedge \Phi(c, z) \wedge (a = e_{policy} \wedge x = e_{query}) \wedge$$
$$((b = \emptyset \wedge c = \emptyset) \vee (b = e_{policy} \wedge y = e_{query} \wedge c = e_{policy} \wedge z = e_{query}))$$

The overall mapping is valid if (1) the path elements of the policy and the insecure query are successfully mapped using the function Φ (e.g., $\Phi(a, x)$), (2) the start nodes are matched, and (3) the relationships and end nodes are either empty or matched. Accordingly, we evaluate all conditions C in all rules R for a policy P. We generate a set S of 2-tuples (e_i, f_i) denoting an element from the query e_i and a filter template f_i to be applied on the insecure query q.

$$S = \{(e_i, f_i) \mid \exists(E, R, \Phi) \in P, (e, C, f) \in R,\, e_i \in q,\, f_i \in F\; \forall c \in C :$$
$$\Gamma(q, c) \wedge map(e) = e_i \wedge f_i = f) \}$$

The function $\Gamma(q, c) \rightarrow \{true, false\}$ checks whether a condition c in the rule's conditions C is satisfied by the insecure query q. Further, the element from the rule e must map to the element in the insecure query e_i and the applied filter template f_i is the same as the one f in the rule. To apply the filters of the matched rules on the insecure query (RQ4), we use the following function:

$$\Xi(q, S) \rightarrow q'$$

It takes an insecure query q and for each assignment $(e_i, f_i) \in S$, it instantiates $f_i \rightarrow f_{i_q}$ according to Section 4. This f_{i_q} can then be added to e_i or it extends existing filters using boolean *AND*. With all filters in place, we have rewritten an insecure query q to a secure version q'.

6 Evaluation

We evaluate our query rewriting approach by implementing a proof-of-concept prototype[2] using Cypher, ANTLR, Spring Boot and Kotlin. We rewrite the insecure Cypher query based on the specified policy. The secure query and information about the applied rules are returned. In our prototypical implementation, we only support reading queries with one MATCH clause. In experiments with a set of queries, we tested all currently supported features and visually confirmed that the filters were applied correctly. However, in our prototype we did not consider potential vulnerabilities or attack vectors not addressed by ABAC.

When measuring the performance of the query rewriting (no database access), we noticed the standard deviation to be higher than the average rewriting time ($\approx$ 0.2 ms on a HP ELITEBOOK 850 G6 with 32 GB, CPU i7-8665U). Therefore, we assume the performance overhead to be negligible.

7 Conclusion

In this paper, we proposed a runtime rewriting approach for freely formulated graph-DB queries to enforce ABAC independent of the underlying database and application. First, we defined how various elements of a query contribute to its result (RQ1). We introduced the strategy of categorizing the elements based on whether they have a filter applied and how they are used for returning data. Next, we introduced a policy model encoding our authorization requirements. Then, we formally defined a policy model including a filter template. The policy consists of a pattern and rules to decide if access control constraints apply to an element of the insecure query and which filter template to use (RQ2). The policy pattern is a sequence of elements, which is used in the query processing.

The policy and the insecure query are processed by first splitting their patterns into tuples representing either paths or isolated nodes. Accordingly, we mapped the policy elements with their respective ones in the insecure query

---

[2] https://github.com/jku-lit-scsl/CypherRewritingCore.

(RQ3). A mapping is valid if each path tuple of the insecure query matches one of the policy. In this case, if all conditions of a rule are successfully evaluated, its filter template is instantiated replacing its placeholders with runtime values from the insecure query and/or user information. The last step of query processing is enhancing the insecure query with these access control filters (RQ4).

As we only consider one policy, we plan to support policy sets according to the XACML policy language model in the future. This further demands for combining algorithms. Additionally, we currently support reading queries only with one `MATCH` clause. Thus, we not only need to increase the supported number of `MATCH` clauses, but also the types of supported queries. This could be added using additional conditions or dedicated rules for reading and writing access. Above all, intensive evaluation is needed especially with complex authorization policies and large graph models. Finally, we need to identify possible potential security vulnerabilities.

Acknowledgements. This research has been partly supported by the LIT Secure and Correct Systems Lab funded by the State of Upper Austria and by the COMET-K2 Center of the Linz Center of Mechatronics (LCM) funded by the Austrian federal government and the federal state of Upper Austria.

References

1. Bao, H.N.P., Clavel, M.: A model-driven approach for enforcing fine-grained access control for SQL queries. SN Comput. Sci. **2**(5), 370 (2021)
2. Bogaerts, J., Decat, M., Lagaisse, B., Joosen, W.: Entity-based access control: supporting more expressive access control policies. In: Proceedings of the 31st Annual Computer Security Applications Conference, pp. 291–300 (2015)
3. Browder, K., Davidson, M.A.: The virtual private database in oracle9ir2. Oracle Tech. White Paper, Oracle Corporat. **500**(280) (2002)
4. Colombo, P., Ferrari, E.: Fine-grained access control within NoSQL document-oriented datastores. Data Sci. Eng. **1**(3), 127–138 (2016)
5. Hofer, D., Mohamed, A., Nadschläger, S., Auer, D.: An intermediate representation for rewriting cypher queries. In: Submitted to Workshop (2023)
6. Hofer, D., Nadschläger, S., Mohamed, A., Küng, J.: Extending authorization capabilities of object relational/graph mappers by request manipulation. In: Database and Expert Systems Applications: 33rd International Conference, DEXA 2022, Vienna, Austria, 22–24 August 2022, Proceedings, Part II, vol. 13427, pp. 71–83. Springer, Cham (2022). https://doi.org/10.1007/978-3-031-12426-6_6
7. Jarman, J., McCart, J.A., Berndt, D., Ligatti, J., et al.: A dynamic query-rewriting mechanism for role-based access control in databases (2008)
8. Kirrane, S.: Linked data with access control. Diss. National University of Ireland, Galway (2015)
9. Mohamed, A., Auer, D., Hofer, D., Küng, J.: Extended authorization policy for graph-structured data. SN Comput. Sci. **2**(5), 351 (2021)
10. Ramli, C.D.P.K., Nielson, H.R., Nielson, F.: The logic of XACML. Sci. Comput. Program. **83**, 80–105 (2014)
11. Shay, R., Blumenthal, U., Gadepally, V., Hamlin, A., Mitchell, J.D., Cunningham, R.K.: Don't even ask: database access control through query control. ACM SIGMOD Rec. **47**(3), 17–22 (2019)

A Polystore Querying System Applied to Heterogeneous and Horizontally Distributed Data

Lea El Ahdab[✉], Olivier Teste, Imen Megdiche, and Andre Peninou

Université de Toulouse, IRIT, Toulouse, France
{lea.el-ahdab,olivier.teste,imen.megdiche,andre.peninou}@irit.fr

Abstract. Data storage in various systems such as SQL and NoSQL leads to important problems when trying to unify data querying. Multiple storage systems conduct to heterogeneous data structures and to multiple query languages. In the context of horizontally and disjointed distributed data, this paper proposes a system that allows the user to natively query a polystore system without taking care of data distribution and heterogeneity. Our approach relies on two mechanisms: (i) mapping dictionaries to define the navigation between systems, (ii) operator rewriting mechanisms from native query operators (selection, projection, aggregation and join) to execute queries on any polystore system. Using a dataset from TPC-H benchmark and a horizontally distributed between document and relational database management system, we conduct experiments showing that the rewriting process has a minimum impact when compared to executing queries in both systems.

Keywords: Polystore · Heterogeneity · Data distribution

1 Introduction

Nowadays, data is more likely to be found distributed in classical (SQL) or multiple heterogeneous and flexible data sources (NoSQL), which forms polystores [2]. It complexifies querying in multiple languages based on non-standardized data modeling paradigms and data querying operators. New solutions are based on new languages [7], operators [8,10], models [6], and sometimes flexible schemas [3,4], which depend on data manipulation. This paper deals with horizontal data distribution in which one entity class is stored in different datastores (relational and document-oriented). We introduce a solution for querying polystore systems, based on automatic rewriting and decomposition of queries, facilitating access to horizontally distributed and heterogeneous data using a mapping dictionary able: i) to link each attribute of any dataset to the corresponding attributes in other datasets and, ii), to integrate possible heterogeneity of data inside any dataset. Working on a TPC-H dataset, we experiment our solution with a query rewriting process without impacting the initial query execution time on relational databases and document-oriented datastores. The remainder of the paper is structured as follows. In Sect. 2, we discuss existing solutions and present their limits. Section 3 defines our polystore data model. Section 4 presents and illustrates the proposed rewriting process and the mapping dictionary with data distribution. Section 5 shows our solution results on real data. Finally, in Sect. 6 we give some perspectives about future work.

2   Related Work

With the complexity of data storage systems in polystores (distribution and heterogeneity), query and accessibility should stay as simple as it is in a mono system type store. Some works focus on inferring schemas to access data: graph representation [4,5] or a u-schema model [3] showing structural variations. Existing works mainly focus on vertical distribution where each entity class is found in one database. Some systems [3,8] introduce an external function to manipulate several entities for binary operators. The join operator is not always executed inside DBMS which requires to have an external algorithm joining the sub results [8,9]. However, horizontal data distribution is possible, where every entity class is divided inside multiple databases: user query gets complex and should be expressed by taking into account data location, query formulation according to polystore systems languages. Another aspect is data heterogeneity: semantic [2,9] or syntactic [4] issues. They use synonyms in mapping solutions to build their queries [2,9]. Surprisingly, they do not deal with structural heterogeneity which is induced by the schema-less principle of NoSQL stores. A specific query is translated and parsed into languages of the considered datastores [2,9]. To support our comparison with existing works, table 1 illustrates the differences we can find between our works and others working on query rewriting and mapping. It shows their position about data distribution (H: horizontal, V: vertical), the supported systems (Relational, Document, Column, Graphs) and the query expression with operators. It also provides the query of relational and document-oriented systems in their native languages using algebraic equivalences and presents to the user results in their native form without transforming data. Our solution consists in the rewriting of the combination of SPAJ operators (selection, projection, aggregation and join).

3   Algebraic Definition of Polystores for Horizontally Distributed Data

In SQL approaches, data is represented according to the relational model [1], where data is structured according to relation schemas. NoSQL (documents) approaches are "schema-less" - each record has its own structure that may be different from those of other records in the same dataset. A **polystore** system is defined as $PL = \{DB_1, ..., DB_B\}$ where each **database** is $DB_i = \{DS_1, ..., DS_{S_i}\}$. $\forall j \in [1 ... S_i]$, DS_j is a dataset. Our model gives a universal representation of these different databases. Each **dataset**, DS_j, is

Table 1. A comparison of existing solutions on polystores

Authors	Data Distrib	R	D	C	G	Heterogeneity	Query	σ	π	γ	$\bowtie$
El Ahdab et al.	H	✓	✓			Structural Semantic Syntactic	SQL MongoDB	✓	✓	✓	✓
Barret et al. [4]	V	✓	✓		✓	Syntactic	SparkQL				
Candel et al. [3]	V	✓	✓		✓		SQL	✓	✓		
Ben Hamadou et al. [5]	V	✓	✓	✓		Structural Semantic	SQL MongoDB	✓	✓	✓	✓
Hai et al. [9]	V	✓	✓		✓	Semantic	SQL JSONiq	✓	✓	✓	✓
Duggan et al. [2]	V	✓	✓	✓		Semantic	Declarative	✓	✓		✓
Curino et al. [10]	V	✓				Structural	SQL	✓	✓		

defined by an extension and an intention $DS_j = (\,Int_j\,,\,Ext_j\,)$. An **extension** is a set of instances $i_k = (\chi_k, v_k)$. χ_k is its *key*, internal identifier in database systems, and v_k is the *instance value* which can be atomic or recursively an instance value or an array of values. The **intention** inferred from the extension is the set of all absolute paths deduced from all instance structures existing in the extension $Int = \bigcup_{k=1}^{N_j} S_k$. We focus on polystores where $\forall i_1 \in [1...B]$, $DS_{j_1} \in DB_{i_1}$, $\exists i_2 \in [1...B]$ such as $DS_{j_2} \in DB_{i_2}$ and DS_{j_1}, DS_{j_2} contain different instances of the same class of an entity. A horizontal distribution is **strict** when each attribute of a dataset has at least one equivalent designation in all equivalent datasets. A data distribution is *disjointed* when $\forall i_{k_1} \in Ext_{j_1}, \nexists i_{k_2} \in Ext_{j_2}, j1 \neq j2 \mid v_{k_1} = v_{k_2}$ where v_{k_1} and v_{k_2} are values corresponding to the same entity in the real world. The **mapping dictionary** map_{DS_j} matches each path of a dataset to all its corresponding paths (including itself) in all equivalent datasets dealing with *structural, syntactic* and *semantic* heterogeneity. Due to space limitation, we do not detail in this paper how the mapping dictionaries are built; they are maintained with the definition and using data alignment and schema-matching algorithms [4]. For example, a path A in DS_i is mapped with every corresponding paths in the equivalent dataset DS_j as: $\{(A, DS_j), (X.A, DS_j), (A'.DS_j)\}$.

4 Rewriting Process Definition

We introduce a closed set of operators to formalize a universal algebra: $K = \{\sigma, \pi, \gamma, \bowtie\}$ where σ is a selection operator (restriction), π is a projection operator, γ is an aggregate operator and $\bowtie$ is a binary operator used to join two datasets. Their combination formulates a query $Q = q_1 \circ ... \circ q_r$ where $\forall k \in [1...r], q_k$ is a simple operator or a composition of operators as a sub-query itself. Each q_k of Q is rewritten according to the mapping dictionaries of the queried datasets $map_{DS_{in}}$. A list of mappings for one field f_i is inferred from the set identified in its respective dictionary.

Selection. $\sigma_P(DS_{in})$ is rewritten as $\sigma_{P_{new}}(DS_j)$ where DS_j is a targeted dataset during query rewriting process and the rewriting of P is $P_{new} = \wedge(\vee(\vee p_{k_l} \ \omega_k \ v_k))$ where p_{k_l} are the paths obtained from the rewriting dictionary associated to DS_j and that corresponds to f_i.

Projection. $\pi_E(DS_{in})$ is rewritten as $\pi_{E_{new}}(DS_j)$ where DS_j is a targeted dataset and E_{new} is the rewriting of $E = e_1, ..., e_n$. If $e_i = f_i$ (projection): f_i is replaced by the combination of its corresponding absolute paths according to the mapping dictionary: $p_{k_1}|\ldots|p_{k_m} \forall p_{k_l}$ for DS_j. The "|" operator leads to the projection of the existing path p_{k_l} in any instance value of Ext_{DS_j}. If $e_i = f_i' : f_i$ (projection and renaming): f_i is replaced by $f_i' : p_{k_1}|\ldots|p_{k_m}$ for all DS identified.

Aggregation. $_G\gamma_F(DS_{in})$ is rewritten to a dataset DS_j as $_G\gamma_F(\pi_{E_{new}}(DS_j))$ where E_{new} is a projection rewriting of fields of $G \cup \{f_i\}$. The projection on f_i of the function F is rewritten to $f_i : p_{k_1}|\ldots|p_{k_m} \forall p_{k_l} \in \Delta_{f_i}^{DS_j}$. The same process is applied to all fields of G.

Join. $DS_{in1} \bowtie_J DS_{in2}$ is rewritten for a database DB_j using the corresponding datasets of DS_{in1} and DS_{in2} in DB_j as: $DS_{j_{in1}} \bowtie_{J_{new}} DS_{j_{in2}}$. J_{new} corresponds to the join condition containing the mapped fields.

The user queries one dataset in one language and the query is translated in its algebraic form. Query rewriting rules are used to produce B queries, one for each DB_i of PL. Rewritten queries are then translated into their specific language (SQL or MongoDB) before being executed on DB_i. The sub-results are presented to the user in their original form (using JSON notations). In some complex queries, they may represent only intermediate results and may need more computation to give the target result; in case of aggregation using sum function, some cases may require an additional aggregation to sum intermediate results.

5   Experiments

TPC-H Benchmark. Considering TPC-H data (https://www.tpc.org/) and queries, we have stored data in one SQL database (MySQL) and one Document oriented database (MongoDB). Tables and collections were created in each respective systems. We have considered two volumes of data v1 as 1 Mo (3600 tuples) and v2 as 10 Mo (30000 tuples).

Query Rewriting Evaluation. TPC-H queries are classified according to the number of queried datasets and to their operators composition. Almost half of them are an association of selection (σ), projection (π), aggregation (γ) and joins ($\bowtie$). Our evaluation focuses on (1) analyzing query execution time over an equal data distribution inside both system and when this distribution is unbalanced, and (2) on the impact of operations on the query execution time in the same context of data distribution. We considered a condition of fragmentation on *nation name* and which respects a disjointed repartition of 50% of instances in

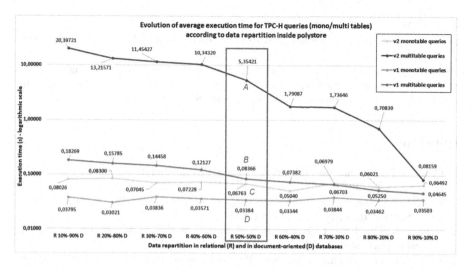

Fig. 1. Evolution of average execution time for TPC-H queries according to data distribution inside polystore with different data volumes v1 (1Mo) and v2 (10Mo)

the relational database and 50% in the document-oriented one. We have evolved this distribution to consider other situations (10%-90%, 20%-80%...).

As illustrated in Fig. 1, the relational system shows a lower execution time than the document system. Focusing on the 50%-50% distribution, SPA operations have no impact on execution time but the join operation presents a higher difference between systems: execution time is 80% times higher for multitable queries than monotable queries (value A, value C of Fig. 1). When data is distributed 90% in documents, query rewriting time is maximize in comparison of 90% of data distribution inside relations. Since each query is executed in each database, it results in a set of separate pieces of data, with possible different structures presented to the user (tuples and documents).

6   Conclusion

In this paper we focus on polystore systems with relational and document-oriented datasets, where data is distributed horizontally. A mapping dictionary represents links between fields and their correspondences in every data source and in their heterogeneous forms. A universal query algebra composed of SPAJ operators is defined for querying both considered systems supporting query rewriting rules and bringing transparency for users. Data remain in native form and only dynamic rewriting of queries and the mapping dictionary are impacted by eventually new data structures. Experiments on a TPC-H dataset show the effectiveness of the proposed solution without significantly impacting the query execution time on top of relational databases (MySQL) and document-oriented databases (MongoDB). In the future, we will focus on the extension of the existing algebra to other systems (column, graph). Another direction is to consider operators more specific to storage systems in order to find their rewriting forms.

Acknowledgements. This work was supported by the French Gov. through the Territoire d'Innovation program, an action of the *Grand Plan d'Investissement* backed by France 2030, Toulouse Métropole and the GIS neOCampus.

References

1. Codd, E.F.: Further normalization of the data base relational model. Data Base Syst. **6**, 33–64 (1972)
2. Duggan, J., Elmore, A.J., Stonebraker, M., et al.: The BigDAWG polystore system. ACM Sigmod Record **44**(2), 11–16 (2015)
3. Candel, C.J.F., Ruiz, D.S., García-Molina, J.J.: A unified metamodel for NoSQL and relational databases. Inf. Syst. **104**, 101898 (2022)
4. Barret, N., Manolescu, I., Upadhyay, P.: Toward generic abstractions for data of any model. In: Proceedings of the 31st ACM International Conference on Information & Knowledge Management, pp. 4803–4807 (2022)
5. Ben Hamadou, H., Gallinucci, E., Golfarelli, M.: Answering GPSJ queries in a polystore: a dataspace-based approach. In: Laender, A.H.F., Pernici, B., Lim, E.-P., de Oliveira, J.P.M. (eds.) ER 2019. LNCS, vol. 11788, pp. 189–203. Springer, Cham (2019). https://doi.org/10.1007/978-3-030-33223-5_16
6. Daniel, G., Gómez, A., Cabot, J.:UMLto [No] SQL: mapping conceptual schemas to heterogeneous datastores. In: 2019 13th International Conference on Research Challenges in Information Science (RCIS), pp. 1–13. IEEE (2019)
7. Misargopoulos, A., Papavassiliou, G., Gizelis, C.A., Nikolopoulos-Gkamatsis, F.: TYPHON: hybrid data lakes for real-time big data analytics – an evaluation framework in the telecom industry. In: Maglogiannis, I., Macintyre, J., Iliadis, L. (eds.) AIAI 2021. IAICT, vol. 628, pp. 128–137. Springer, Cham (2021). https://doi.org/10.1007/978-3-030-79157-5_12
8. Kolev, B., Valduriez, P., Bondiombouy, C., Jiménez-Peris, R., Pau, R., Pereira, J.: CloudMdsQL: querying heterogeneous cloud data stores with a common language. Distributed and Parallel Databases **34**(4), 463–503 (2015). https://doi.org/10.1007/s10619-015-7185-y
9. Hai, R., Quix, C., Zhou, C.: Query rewriting for heterogeneous data lakes. In: Benczúr, A., Thalheim, B., Horváth, T. (eds.) ADBIS 2018. LNCS, vol. 11019, pp. 35–49. Springer, Cham (2018). https://doi.org/10.1007/978-3-319-98398-1_3
10. Curino, C.A., Moon, H.J., Deutsch, A., Zaniolo, C.: Update rewriting and integrity constraint maintenance in a schema evolution support system: PRISM++. Proc. VLDB Endowment **4**(2), 117–128 (2010)

Knowledge Representation

Semantically Constitutive Entities
in Knowledge Graphs

Chong Cher Chia$^{(\boxtimes)}$ (ID), Maksim Tkachenko (ID), and Hady W. Lauw (ID)

School of Computing and Information Systems, Singapore Management University,
Singapore, Singapore
{ccchia.2018,hadywlauw}@smu.edu.sg, maksim.tkatchenko@gmail.com

Abstract. Knowledge graphs are repositories of facts about a world. In this work, we seek to distill the set of entities or nodes in a knowledge graph into a specified number of constitutive nodes, whose embeddings would be retained. Intuitively, the remaining accessory nodes could have their original embeddings "forgotten", and yet reconstitutable from those of the retained constitutive nodes. The constitutive nodes thus represent the semantically constitutive entities, which retain the core semantics of the knowledge graph. We propose a formulation as well as algorithmic solutions to minimize the reconstitution errors. The derived constitutive nodes are validated empirically both in quantitative and qualitative means on three well-known publicly accessible knowledge graphs. Experiments show that the selected semantically constitutive entities outperform those selected based on structural properties alone.

Keywords: semantically constitutive · knowledge graph · embeddings

1 Introduction

Graphs are predominantly used to represent real world data, including social networks, citation network, hyperlink network, etc. One important analysis deals with determining which vertices are the most 'important' in a graph. Because the essential nature of graphs is the very connectivity among its vertices, this notion of 'importance' is frequently formulated in terms of how well a vertex is connected to others in the graph, giving rise to notions such as centrality [4] and influence maximization [25] that would be further explored in related work.

In this work, we are interested in *knowledge graphs*, a machine-friendly way of representing real world facts. These facts are extracted from various sources such as encyclopedic Wikipedia [33], lexical WordNet [14], or even the open Web [34]. The use of knowledge graphs have been extended to applications including question answering [21], recommendations [52], fact-checking [8], etc.

Given its pertinence and myriad applicability, we explore notions of what make a vertex 'important' in a knowledge graph. In addition to the graph-theoretic sense of connectivity, another essential nature of a knowledge graph is its *semantics*. Every triplet instance involving a *head entity*, *relation*, and *tail*

© The Author(s), under exclusive license to Springer Nature Switzerland AG 2023
C. Strauss et al. (Eds.): DEXA 2023, LNCS 14146, pp. 445–461, 2023.
https://doi.org/10.1007/978-3-031-39847-6_36

entity represents a fact, the totality of which collectively represents our semantic understanding of an underlying 'world'. Suppose we retain only a subset of the entities; which subset best preserves our semantic understanding of the 'world'?

For a concrete representation of semantics, we allude to knowledge graph embeddings [51], which embeds entities and relations into continuous vector spaces. The plausibility of facts (triplets) can then be assessed from the embeddings of the corresponding entities and relations. In this work, we assume that such embeddings have been derived and specified as input to our problem.

As output, we seek to identify a relatively small subset of (*"constitutive"*) entities, whose embeddings would be used to reconstitute the remaining (*"accessory"*) entities. To remain true to the raison d'être of a knowledge graph, this reconstitution is faithful to a known fact (triplet) within the graph.

Fig. 1. Subgraph Using 1 or 2 Constitutive Nodes

This concept is illustrated by the knowledge graph subset in Fig. 1, where constitutive nodes (top) are connected to accessory nodes (bottom) by relational edges (in this case `Genre`). Embeddings of accessory nodes can potentially be "forgotten", and "reconstituted" by constitutive nodes. For example, we could use a single constitutive node (`Michael Jackson`) to reconstitute all the accessory nodes (solid edges only). While compact, it is not sufficient for distinguishing different musical genres, and using two constitutive nodes produces more informative reconstitutions (`John Mayer`, `Michael Jackson` both produce `Soul Music`).

Contributions. In this work, we make several contributions.

- The problem of identifying semantically constitutive entities in a knowledge graph is novel, and distinct from existing work solely focused on structural connectivity. We propose a reconstitution function consistent with translational embeddings, and produces interpretable reconstitutions by virtue of being supported by actual triplets within the knowledge graph.
- We propose a new algorithmic formulation to identify constitutive nodes, as well as the selection of triplets for each reconstitution. While related to matching or assignment problems, our formulation is novel in allowing up to k constitutive nodes per accessory node. We describe algorithmic solutions based on Integer Linear Programming (ILP), and propose heuristics that speed up the computation particularly for larger graphs.
- We experiment on 3 well-established knowledge graphs, outperforming baselines both quantitatively (downstream tasks), and qualitatively (user study).
- We make our code publicly available[1] for reproducibility.

---

[1] https://github.com/PreferredAI/semantically-constitutive-entities.

2   Related Work

Node Centrality. Finding the "important" nodes in a graph had previously been approached from structural connectivity. One class of techniques referred to as *point centrality* looks at the quality of an individual node that makes it most central. There are primarily three categories of point centrality measures: local centrality, iterative centrality, or global centrality [48]. *Local centrality* measures centrality by local network topology. A common metric is Degree Centrality, which ranks each node based on the number of edges in the graph. For directed graphs, the in-degrees or out-degrees of a node may be used. Another well-known local centrality metric is *h-index* [19,26], where a node has index h if it has at least h neighbouring nodes of h degree. *Iterative centrality* metrics measure a node's centrality through some (possibly fixed) number of iterative calculations. One such metric is Eigenvector Centrality [1,2], which repeatedly updates the centrality for each node based on the centrality of its neighbours. PageRank [40] builds on Eigenvector Centrality by dampening the influence of further neighbours on the centrality of a given node. *Global centrality* measures a node's centrality in the context of the entire network topology, such as Betweenness Centrality [16], derived from the number of shortest paths passing through it.

In the experiments, we compare against representative point centrality metrics, such as degree centrality and PageRank. Such point centrality measures select nodes based on its individual quality. In our problem, we seek to select a *group* of constitutive nodes. Hence, we compare against the group version of these metrics in the experiments. For example, compared to PageRank that selects nodes individually, another formulation of influence maximization seeks to identify a group of "influential" nodes based on their ability to affect other nodes within the graph, in order to maximize social influence [25]. Although NP-hard, algorithms such as SSA guarantee a $(1 - 1/e - \epsilon)$-approximate solution [20,37,38].

Knowledge Graph. A core concept in our work is the representation of semantic information within a knowledge graph. Such representations commonly take the form of Knowledge Graph Embeddings, as discussed in [24]. One class of Knowledge Graph Embeddings are linear/bilinear models, as exemplified by *TransE* [3], which represents relations between two entities as the translation of one point within the embedding space to another. Given a triplet (h, r, t) representing the head entity, relationship, and tail entity respectively, *TransE* minimizes the L_1/L_2 distance between $h + r$ and t. Other Translational Knowledge Graph Embeddings have since been proposed, such as *TransH* [53] which extends the translation operation onto a hyperplane, and *TransD* [23] which uses separate mapping matrices for the head and tail entities, and each projection is defined by both the entity and relation embeddings.

Other classes of embeddings include factorization models (e.g., RESCAL [39], LFM [22]), neural network models (e.g., ConvE [12], ConKB [36]), and transformer-based models (e.g., CoKE [50], KG-BERT [54]).

Inductive Knowledge Graph Completion [11,18,30,49] generates embeddings for unseen entities. This is done from combining embeddings of known entities, and is therefore not comparable with our work.

Another widely studied aspect of knowledge graphs is the summarization of such graphs, typically through the addition and/or removal of nodes (as discussed in [31]). Summaries typically take the form of either a supergraph or a sparsified graph. Supergraphs refer to graphs where the (super)nodes and (super)edges are a collection of nodes/edges from the original graph, and may be obtained by grouping nodes [13,41,42,56] or identifying patterns within the original knowledge graph [6,9,55]. Supergraphs do not retain the entities and edges of the original graph, and are therefore not comparable with our work. Sparsified graphs are subsets of the original knowledge graph, and reduce the number of nodes and/or edges as compared to the original knowledge graph. This may be accomplished by the introduction of "compressor nodes" [32] or "virtual nodes" [5] to the graph for (edge) dedensification. Other techniques may require a query to base the summary, such as Ontovis [44] or Egocentric Abstraction [28].

3   Semantically Constitutive Entities

Our goal, as stated in Sect. 1, is to select a (user-specified) number of constitutive nodes from a given knowledge graph. Graph embeddings of constitutive nodes can be used to reconstitute non-selected (i.e., accessory) node embeddings.

Problem Definition. A knowledge graph $G = (E, R, T)$ consists of a set of entities E, relations R, and relational triples $T \subset E \times R \times E$. Triple $(h, r, t) \in T$ indicates that relation r is present between head h and tail t entities. Let $\text{H}(\cdot)$ return the corresponding embeddings for entity or relation. For a given target size $\mathcal{P}$, we seek to select a semantically constitutive graph $\hat{G} = (\hat{E}, \hat{R}, \hat{T})$ that ties subset of constitutive entities $\hat{E} \subset E$ with the accessory entities $\hat{E}'$ via relations $\hat{R} \subseteq R$: $\hat{T} \subset \hat{E} \times R \times \hat{E}'$. Formally, we seek to solve the minimization problem:

$$\underset{\hat{G}:|\hat{E}|=\mathcal{P}}{\arg\min} \sum_{e \in \hat{E}'} \text{d}(\text{H}(e), f(e|\hat{G})), \tag{1}$$

where d is a distance function on embeddings (L2 in this work) and $f(e|\hat{G})$ reconstitutes the accessory node e from entities and relations in $\hat{G}$.

To define a particular reconstitution function $f(\cdot|\hat{G})$, we draw on the knowledge graph embedding training procedures: a family of related models (Trans*) learn embeddings by treating the relations between entities as translations between two points in a high-dimensional space, which effectively turns into the following equation: $\text{H}(h) + \text{H}(r) \approx \text{H}(t)$. A target entity e in principle can be reconstituted using multiple head entities and relations as long as we have an appropriate relation between them:

$$f(e|\hat{G}) = \sum_{(h,r,t)\in\hat{T}} \left[\mathbb{1}_{t=e} \cdot (\text{H}(h) + \text{H}(r)) \right] \Bigg/ \sum_{(h,r,t)\in\hat{T}} \left[\mathbb{1}_{t=e} \right], \tag{2}$$

where $\mathbb{1}_{t=e}$ is 1 when t and e refer to the same node and 0 otherwise.

We also experimented with the use of deep neural networks, such as Multi-Headed Attention [47] encoders, as the basis for an alternate reconstitution function, in order to allow varying levels of reconstitution importance for each constitutive node. However, such networks are challenging to train as modelling reconstruction from an unordered set of constituent node/relation pairs is complex. Furthermore, it is not clear how we can retain the translational embedding relationships in such approaches. As such, we opt to use Eq. 2, which is simple and effective for accessory node reconstitution in our experiments, and leave the exploration of alternative reconstruction functions for future work.

Optimization. The optimization problem above is similar to the well-known P-Median Problem (PMP) [10], which selects $\mathcal{P}$ facilities such that the total cost of serving all locations is minimized. However, PMP covers only a basic scenario where each accessory entity must be reconstituted with only a single semantically constitutive node, which is too limiting (as noted in Sect. 1). It is also not feasible to use a fixed number of constitutive nodes, simply because there may not be sufficient triplets in G to reconstitute each accessory node. Thus, we introduce "phantom" nodes, which "reconstitutes" any accessory node at a higher cost. These phantom nodes serve as padding nodes for the entities with low in-degree and are discarded after selection process is completed.

Given that all nodes in G are both facilities and locations, we allow facilities to serve themselves without cost, mirroring the memorization of retained constitutive node embeddings. We update the constraint on location assignment to allow exactly $\mathcal{G}$ facilities to serve the same location, mirroring accessory node reconstitution with multiple constitutive nodes, and introduce "free" facilities which serve locations that are also facilities at no cost.

Let $\mathcal{P}$ be the desired number of facilities, and $\mathcal{G}$ be the maximum number of constitutive nodes used to reconstitute a given accessory node. Given a set of locations $I = E$, the set of facilities J is defined as $J = I \cup P \cup F$, where $P = \{p_1, p_2, \ldots, p_{\mathcal{P}}\}$ and $F = \{f_1, f_2, \ldots, f_{\mathcal{G}-1}\}$ are the set of phantom and free nodes respectively with I, P, and F being mutually disjoint.

Let X be the facility assignment matrix, such that $X_{ij} = 1$ if location i is served by facility j, and 0 otherwise. Y is the facility opening matrix, such that $Y_j = 1$ if facility j is open, and 0 otherwise. C_{ij} denotes the cost of serving location i from facility j. For the nodes i and j from the knowledge graph G, such that $(i, r, j) \in T$ for some r, we define the cost consistently with the reconstitution function: $d(H(i) + H(r), H(j))$. If an entity pair has multiple relations, we select the relation that minimizes distance, and denote it as R_{ij}. Free facilities serve locations at no cost (i.e., $C_{ij} = 0$) for any $j \in F$. We arbitrarily set a high cost ($\alpha \geq 1$) for phantom nodes to encourage the preferential selection of real entities, and discard both free and phantom nodes post-selection.

$$C_{ij} = \begin{cases} \min_{(j,r,i) \in T} d(H(j) + H(r), H(i)) & \text{if } \exists r \in R : (i, r, j) \in T, \\ \alpha \max_{(j,r,i) \in T} d(H(j) + H(r), H(i)) & \text{if } j \in P, \\ 0 & \text{if } i = j \text{ or } j \in F, \\ +\infty & \text{otherwise,} \end{cases} \tag{3}$$

Since PMP is known to be NP-hard, we use an Integer Linear Programming (ILP) solver (i.e., Gurobi [17]) to find an approximate solution:

$$\min \sum_{i \in I} \sum_{j \in J} C_{ij} X_{ij} \qquad \text{subject to} \qquad (4)$$

$$\sum_{j \in J} X_{ij} = \mathcal{G} \qquad\qquad \forall i \in I \qquad (5)$$

$$\sum_{j \in I} Y_j = \mathcal{P} \qquad\qquad (6)$$

$$X_{ij} \leq Y_j \qquad\qquad \forall i \in I, j \in J \qquad (7)$$

$$X_{i\hat{j}} \geq Y_i \qquad\qquad \forall i \in I, \forall \hat{j} \in F \qquad (8)$$

$$Y_j \in \{0,1\} \text{ and } X_{ij} \in \{0,1\} \qquad\qquad \forall i \in I, j \in J \qquad (9)$$

Having a solution to the program above, we can generate the semantically constitutive graph $\hat{G} = (\hat{E}, \hat{R}, \hat{T})$ from X, Y, and R, where $\hat{E} = \{e \in I | Y_e = 1\}$, $\hat{R} = \{R_{ij} | i, j \in I\}$ and $\hat{T} = \{(h, R_{ht}, t) | h \in \hat{E}, t \in \hat{E}'\}$.

Approximation. While it is possible to obtain an integer solution directly, we observed that a 2-step procedure achieves slightly better performance at the cost of marginally higher computational costs. We first solve a relaxed version of the problem where the Eq. (9) is removed. This results in a partial solution $\bar{Y}$ containing fractional assignment of facilities. We replace the facility set J in the original program with a restricted set $\bar{J} = \{i : \bar{Y}_i \geq \epsilon\}$, and solve the new program directly. In our experiments, we default to the 2-step procedure, and set $\epsilon = 0.01$ to discard non-significant facilities.

Discussion. We note that our problem definition is distinct from the Capacitated P Median Problem [15,35,45], which limits the number of locations allowed in each cluster. Our work, conversely, increases the number of clusters each location can belong to, and is therefore not comparable. We also note that phantom (P) and free nodes (F) are artificial constraints, and are removed in Y.

4   Experiments

Our experimental objective is to validate whether paying attention to the semantics in the selection of semantically constitutive entities within a knowledge graph would outperform baselines that focus primarily on structural centrality.

4.1   Experimental Setup

Datasets. We experiment on publicly-available datasets (Table 1) which are common benchmarks for evaluating Knowledge Graph Embeddings.

Table 1. Dataset Summary

Dataset	# Entities	# Relations	# Training Triples	# Validation Triples	# Testing Triples
FB15k-237	14,541	237	272,115	17,535	20,466
WN18RR	40,943	11	86,835	3,034	3,134
CoDEx-L	77,951	69	551,193	30,662	30,662

FB15k-237. FreeBase is a knowledge base containing general facts, and contains reversible (i.e., symmetric) relations. The FB15k-237 dataset [3,46] is a collection of FreeBase triples which retains only a single copy of reversible relation pairs, preventing information leakage during downstream evaluation.

WN18RR. WordNet is a knowledge base consisting of different usages of a given word ("senses"), as well as the lexical relations between these "senses". The WN18RR dataset is selected from a collection of WordNet triples [3], where reversible relations have been removed in the same manner as FB15k-237 [12].

CoDEx. Wikipedia is a crowdsourced encyclopedia that is openly edited. The CoDEx dataset is sampled from Wikipedia using a selection of seed entities and relations [43]. We use CoDEx-L, the largest version of CoDEx.

Baselines. We compare our semantically constitutive nodes to nodes selected by the graph centrality approaches that focus on structural connectivity:

Point Centrality. We expect that highly connected nodes are better suited for accessory node reconstitution as compared to low degree nodes, due to the larger number of possible reconstitutions. We calculate the degrees for all nodes in each dataset, and select the top k nodes as a baseline. We experimented with using in-degrees (Point-In-Centrality) and out-degrees (Point-Out-Centrality) for selection, and observed that the latter generally performs better.

Group Centrality. Point Centrality approaches prioritizes nodes within a dense subgraph at the expense of sparser nodes, as they are selected based on local network topology. We attempt to address this by selecting the nodes iteratively in a greedy fashion; after a node e_i is selected, we remove edges to/from e_i from the degree counts of the remaining nodes, stopping after we have selected k nodes or after all edges have been removed. In the latter case, we then randomly select nodes to ensure that there are k facilities. We note that this is similar to the SingleDiscount heuristic [7]. We report the results when using only in-degrees (Group-In-Centrality) and out-degrees (Group-Out-Centrality), as above.

Eigenvector Centrality. We observe that the above baselines only consider the centrality of each node (i.e., degree), and places no weight on the influence of their neighbours. We therefore also compare to PageRank[2], which considers both the centrality as well as neighbouring influence when ranking node importance.

Influence Maximization. We note that PageRank ranks nodes individually, and may therefore not return the best group of nodes. Our last baseline selects a group of nodes which maximizes the social influence of the group. As this is an NP-Hard problem [25], we use the SSA algorithm (Linear Threshold, $\epsilon = 0.03$, $\delta = 0.01$), which guarantees a $(1 - 1/e - \epsilon)$-approximate solution [20,37,38].

---

[2] Adapted from https://github.com/louridas/pagerank, $a = 0.85, c = 1 \times 10^{-32}$.

Embedding Models. As our focus is on reconstruction, we obtain embeddings from the OpenKE implementation and suggested parameters for TransE, TransH and TransD. We target $\mathcal{P}$ to be a similar proportion (30%) of entities (7K for CoDEx-L, 4K for FB15k-237, 9K for WN18RR) in all following experiments.

4.2   Quantitative Comparisons

A measure of quality is the ability of the selected nodes to retain the semantic meaning of accessory nodes. As we use knowledge graph embeddings to represent node semantics, we turn to knowledge graph embedding evaluation tasks.

Link Prediction. Knowledge graph embedding quality is commonly compared via downstream task such as the well-known Link Prediction task. We form embeddings for each node selection by replacing the embedding for discarded entities with the reconstituted embedding. We use the (filtered) Link Prediction Task [3,53]. Given a true testing triple $(\hat{h}, \hat{r}, \hat{t})$, we wish to rank $\hat{t}$ given $(\hat{h}, \hat{r})$ amongst the set of testing entities $\tilde{E}$ (or $\hat{h}$ given $(\hat{r}, \hat{t})$).

Table 2 shows the experimental results for each dataset, where Hit@10% (of entities in the dataset; similar results observed for Hit@5%) is used as metric to facilitate comparison between differently-sized datasets. The first line is the performance of the original (i.e., full-sized) TransE embeddings. Subsequent lines are performances (relative to the original, in percentage) of each selection method. Semantically-Constitutive consistently achieves a higher Hit@10% as compared to the baselines in all cases for FB15k-237 and WN18RR. For CoDEx-L, Semantically-Constitutive outperforms most baselines, tying with one.

Table 2. Link Prediction Task Hit@10%, Relative % to Original (TransE Embeddings, Higher is Better)

Model	FB15k-237	WN18RR	CoDEx-L
Original	0.968	0.755	0.989
Point-In-Centrality	80.5	35.0	7.3
Point-Out-Centrality	86.1	37.3	24.9
Group-In-Centrality	78.7	40.0	7.2
Group-Out-Centrality	86.6	41.9	**25.1**
SSA	71.1	25.4	14.3
PageRank	77.1	39.3	6.1
Semantically-Constitutive	**87.9**	**43.2**	**25.1**

Multiple Node Reconstitution. We now study the effect of multiple nodes for reconstitution, which is controlled by the parameter $\mathcal{G}$. We expect that larger $\mathcal{G}$ allows reconstitutions to better capture the semantic meaning of the accessory entity, as shown in Sect. 1. We conduct an ablation study for each dataset, by reducing the number of reconstitution nodes allowed (from $\mathcal{G} = 10$) from the same partial solution $\bar{Y}$ (as described in Sect. 3), and repeat the Link Prediction task with the resulting reconstitutions (Table 3).

We observe that while the Hit@10% generally remains fairly consistent as $\mathcal{G}$ is reduced for all models, small but noticeable differences in performance can be observed. For example, in FB15k-237, Semantically-Constitutive outperforms all baselines at every $\mathcal{G}$ level. Next, we observe that the best performance for Semantically-Constitutive is at $\mathcal{G}-4$ (88.29%). This suggests that the number of constitutive nodes $\mathcal{G}$ can be tuned to best utilize the selected semantically constitutive nodes, improving downstream performance. WN18RR shows similar improvements ($\mathcal{G}-2$, 43.29%), but CoDEx-L performance is flat across $\mathcal{G}$.

Embedding Models. Lastly, although not the focus of our work, we study the generalizability of our approach. We replace the entity embeddings with the encoding representations from translational knowledge graph embedding models such as TransH and TransD, and show the Hit@10% for WN18RR in Table 4 (consistent results are observed for other datasets).

First, we observe that among the original embeddings, TransE performs the best (0.755), while TransD (0.738) is able to outperform TransH (0.723). We speculate that this is related to the choice of encoding representation function f, which does not fully capture the translation operation in the TransH and TransD training processes, and leave the selection of a suitable f as future work.

Turning to the baseline approaches, we observe that all models generally perform at similar relative levels across the embedding models. Group-Out-Centrality, the best performing baseline, achieves the best baseline performance on TransE (41.90), similar to the performance of the full embeddings.

Lastly, we observe that while Semantically-Constitutive shows a similar drop in relative performance on TransH (42.86) as compared to TransE (43.21), it was able to achieve a minor improvement on TransD (43.47). We note that the absolute performance of Semantically-Constitutive is still higher on TransE.

Effect of Graph Size on Runtime. We study the effects of knowledge graph size on runtime between Semantically-Constitutive and the "Direct" one-step program. We sample (from 23,616 unique) CoDEx-L tail entities to between 23,000 and 14,000 (with intervals of 1,000), and retain only triples containing those sampled tail entities. We then run both "Direct" and Semantically-Constitutive on these sub-graphs, and set $\mathcal{P}$ to 30% of the number of sampled entities to ensure consistent difficulty. We report the mean runtimes on 5 samples in Fig. 2 (labelled with the initial sample sizes, from 23K to 14K). As discussed in Sect. 3, we observe that Semantically-Constitutive achieves a general improvement in model performance at the expense of slightly longer runtimes, particularly at smaller sampled sizes.

454   C. C. Chia et al.

Table 3. $\mathcal{G}$ Reduction Hit@10%, Relative % to Original (TransE Embeddings, Higher is Better)

Model	$\mathcal{G}=10$	$\mathcal{G}-1$	$\mathcal{G}-2$	$\mathcal{G}-3$	$\mathcal{G}-4$
FB15k-237					
Original	0.968				
Point-In-Centrality	80.5	80.5	80.5	80.4	80.4
Point-Out-Centrality	86.1	86.1	86.1	86.1	86.1
Group-In-Centrality	78.7	78.7	78.7	78.7	78.7
Group-Out-Centrality	86.6	86.7	86.7	86.7	86.7
SSA	71.1	70.9	70.9	70.0	70.9
PageRank	77.1	77.1	77.2	77.1	77.1
Semantically-Constitutive	**87.9**	**88.0**	**88.1**	**88.2**	**88.3**
WN18RR					
Original	0.968				
Point-In-Centrality	35.0	35.0	35.0	35.0	35.0
Point-Out-Centrality	37.3	37.3	37.3	37.0	37.0
Group-In-Centrality	40.0	40.0	40.0	40.0	40.0
Group-Out-Centrality	41.9	41.9	41.9	41.9	42.0
SSA	25.4	25.4	25.4	25.5	25.4
PageRank	39.3	39.3	39.3	39.3	39.3
Semantically-Constitutive	**43.2**	**43.2**	**43.3**	**43.2**	**43.1**
CoDEx-L					
Original	0.968				
Point-In-Centrality	7.3	7.3	7.3	7.3	7.3
Point-Out-Centrality	24.9	24.9	24.9	24.9	24.9
Group-In-Centrality	7.2	7.2	7.2	7.2	7.2
Group-Out-Centrality	**25.1**	**25.1**	**25.1**	**25.1**	**25.1**
SSA	14.3	14.3	14.3	14.3	14.3
PageRank	6.1	6.1	6.1	6.1	6.1
Semantically-Constitutive	**25.1**	**25.1**	**25.1**	**25.1**	**25.1**

4.3 User Study

We conducted a user study to investigate the real-world informativeness of Semantically-Constitutive, and expect Semantically-Constitutive to provide reconstructions (i.e., relation between accessory and constitutive nodes) with higher relevance due to the semantic reconstruction process. We first filter the CoDEx-L dataset to retain only entities that have at least 10 unique edges. We then compare Group-Out-Centrality (best performing baseline) to Semantically-Constitutive (TransE embeddings, $\mathcal{G}=10, \hat{\mathcal{G}}=3$), and select accessory nodes where

Table 4. Translational Knowledge Graph Embedding Hit@10%, Relative % to Original (WN18RR, Higher is Better)

Model	TransE	TransH	TransD
Original	0.755	0.723	0.738
Point-In-Centrality	35.0	35.8	33.9
Point-Out-Centrality	37.3	38.7	38.2
Group-In-Centrality	40.0	39.2	40.9
Group-Out-Centrality	41.9	41.3	41.5
SSA	25.4	25.0	25.1
PageRank	39.3	42.3	41.2
Semantically-Constitutive	**43.2**	**42.9**	**43.5**

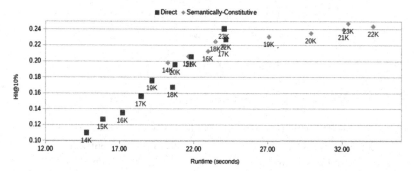

Fig. 2. Model Runtime and Hit@10% on Sampled CODEXL

the triplet relation is "occupation"[3] and all 3 constitutive nodes differ. We randomly select 20 (from 45 total) such accessory nodes for the user study.

Each user was presented a accessory node (e.g.,"Film Actor" in Table 5a) in each question[4], and asked to rank the relevance of all 6 constitutive nodes (supplemented with their Codex-L description) on a five-level Likert Scale [29]. We compare the collected responses by assigning a score between −1 and 1 to each level (Table 5b). Figure 3 shows the average score by 13 users[5] for Group-Out-Centrality (mean = 0.103) and Semantically-Constitutive (mean = 0.458) on each question. We observe that Semantically-Constitutive (in red) generally achieves a higher average score on all questions as compared to Group-Out-Centrality (in blue), showing that the nodes selected by Semantically-Constitutive is better related to the query occupation, and are therefore more informative.

---

[3] Selected in order to limit the obscurity of triplets in the user study.

[4] The order of questions and Likert items were randomized for every user.

[5] This was the number of study participants who agreed to take part in the study. They were neither co-authors, nor aware of the subject of this paper.

Inter-rater Reliability. Next, we wish to study the agreement between different raters. Fleiss' Kappa [27] is commonly used for understanding the inter-rater reliability of ordinal rating data, and range from -1 to 1, with values above 0 indicate agreement (beyond chance) between the raters.

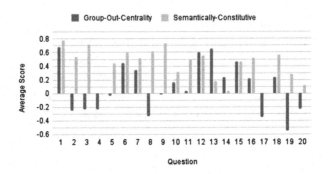

Fig. 3. User Study Scores

Table 5. Example User Study Question ("Film Actor")

(a) Likert Items (Constitutive Node + Codex-L description)

Semantically-Constitutive	Group-Out-Centrality
Robin Williams (American actor and stand up comedian (1951-2014))	John Cale (Welsh composer, singer-songwriter and record producer)
Justin Timberlake (American singer, record producer, and actor)	John Lennon (English singer and songwriter, founding member of The Beatles)
Nicolas Cage (American actor)	A. R. Rahman (Indian singer and composer)

(b) Ranking Options and Associated Score

Option	Relevant	Somewhat relevant	Neither relevant or irrelevant	Somewhat irrelevant	Irrelevant
Score	1	0.5	0	-0.5	-1

We first combine all "Relevant" and "Somewhat relevant" responses, and do the same for "Neither relevant or irrelevant", "Somewhat irrelevant" and "Irrelevant" classes. Next, we calculate the (2-Rater 2-Class) Fleiss' Kappa for each pair of raters, and average them. The expected Fleiss' Kappa in this setting is 0.245, which suggests that a random pair of raters would likely show fair agreement (as suggested by [27]) on the (binary) relevance of each choice.

Next, we investigate the overall reliability of the User Study. We report that the Multiple-Rater 5-Class Fleiss' Kappa (0.119 > 0), indicates that there is likely to be agreement amongst the raters.

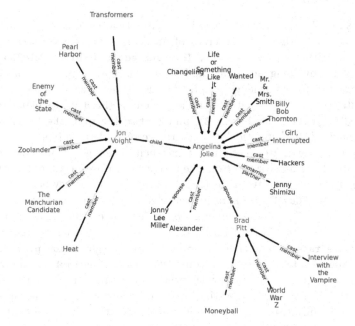

(a) Subgraph Centered on Accessory Node **Angelina Jolie**

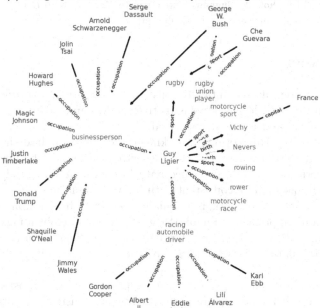

(b) Subgraph Centered on Constitutive Node **Guy Ligier**

Fig. 4. Codex-L Case Studies

4.4 Case Studies

Next, we show 2 subgraphs generated by Semantically-Constitutive on Codex-L. Figure 4a shows the reconstruction of an accessory node {Angelina Jolie}, from two constitutive nodes, ({Girl, Interrupted} and {Billy Bob Thornton}). Other triplets involving accessory nodes such as {Mr. & Mrs. Smith} are discarded from the full knowledge graph. We also show accessory nodes that are reconstituted by other constitutive nodes, such as {Brad Pitt} being reconstituted by {Moneyball}, {World War Z}, and {Interview with the Vampire}. This subgraph shows how Semantically-Constitutive reconstitutes (specific) nodes by combining multiple more general relations, such as being cast in a movie.

Figure 4b shows a subgraph from the CoDEx-L dataset, centered on the node representing the constitutive node {Guy Ligier}. We also show nodes which are reconstituted by {Guy Ligier}, other retained nodes, and reconstituted nodes from these retained nodes. From Fig. 4b, we can infer that {Guy Ligier} was probably involved in rowing ({occupation}→{rowing}), racing ({occupation}→{motorcycle racer}), and rugby ({occupation}→{rugby union player}), Note that while {Guy Ligier} is used to reconstitute {racing automobile driver}, this reconstitution is in conjunction with other constitutive nodes such as {Karl Ebb} and {Eddie Jordan}, suggesting that the concept of {racing automobile driver} is not fully captured by a single node. Next, we observe that while a relation exists between {Guy Ligier} and the accessory node {businessperson}, it is reconstituted by {Howard Hughes} and {Donald Trump}, who may be relatively better recognized as businesspersons, instead.

5 Conclusion

In this work, we identify semantically constitutive entities in a knowledge graph (KG). Intuitively, embeddings of "constitutive" nodes can be used to reconstitute "accessory" nodes, and is based on actual KG triples, providing credence and interpretability. Experiments on three knowledge bases validate the proposed methodology in several ways. On the downstream Link Prediction task, our method outperforms structural connectivity baselines. A user study validates our reconstitutions as more consistent with human evaluation.

One limitation of our work is a reliance on pretrained graph embedding as input, as our approach is unable to generate these embeddings from the knowledge graph directly. Next, reconstructions are only as accurate as the KG provided; problematic reconstructions can be avoided by auditing the underlying KG.

One future direction is to adapt our approach to non-translational KG embeddings. Another is to explore how semantically constitutive entities could enhance related tasks such as KG summarization.

References

1. Bonacich, P.: Factoring and weighting approaches to status scores and clique identification. J. Math. Sociol. **2**(1), 113–120 (1972)
2. Bonacich, P.: Power and centrality: a family of measures. Am. J. Sociol. **92**(5), 1170–1182 (1987)
3. Bordes, A., Usunier, N., Garcia-Duran, A., Weston, J., Yakhnenko, O.: Translating embeddings for modeling multi-relational data. In: NeurIPS, vol. 26 (2013)
4. Borgatti, S.P., Everett, M.G.: A graph-theoretic perspective on centrality. Soc. Netw. **28**(4), 466–484 (2006)
5. Buehrer, G., Chellapilla, K.: A scalable pattern mining approach to web graph compression with communities. In: WSDM, pp. 95–106 (2008)
6. Chen, C., Lin, C.X., Fredrikson, M., Christodorescu, M., Yan, X., Han, J.: Mining graph patterns efficiently via randomized summaries. PVLDB **2**(1), 742–753 (2009)
7. Chen, W., Wang, Y., Yang, S.: Efficient influence maximization in social networks. In: KDD, pp. 199–208 (2009)
8. Ciampaglia, G.L., Shiralkar, P., Rocha, L.M., Bollen, J., Menczer, F., Flammini, A.: Computational fact checking from knowledge networks. PLoS ONE **10**(6), e0128193 (2015)
9. Cook, D.J., Holder, L.B.: Substructure discovery using minimum description length and background knowledge. JAIR **1**, 231–255 (1993)
10. Cornuéjols, G., Nemhauser, G., Wolsey, L.: The uncapacitated facility location problem. Cornell University Operations Research and Industrial Engineering, Technical report (1983)
11. Dai, D., Zheng, H., Luo, F., Yang, P., Chang, B., Sui, Z.: Inductively representing out-of-knowledge-graph entities by optimal estimation under translational assumptions. arXiv preprint arXiv:2009.12765 (2020)
12. Dettmers, T., Minervini, P., Stenetorp, P., Riedel, S.: Convolutional 2D knowledge graph embeddings. In: AAAI. No. 1 (2018)
13. Dunne, C., Shneiderman, B.: Motif simplification: improving network visualization readability with fan, connector, and clique glyphs. In: CHI, pp. 3247–3256 (2013)
14. Fellbaum, C.: WordNet. In: Poli, R., Healy, M., Kameas, A. (eds.) Theory and applications of ontology: Computer Applications, pp. 231–243. Springer, Dordrecht (2010). https://doi.org/10.1007/978-90-481-8847-5_10
15. Fleszar, K., Hindi, K.S.: An effective VNS for the capacitated p-median problem. Eur. J. Oper. Res. **191**(3), 612–622 (2008)
16. Freeman, L.C.: A set of measures of centrality based on betweenness. Sociometry **40**, 35–41 (1977)
17. Gurobi Optimization, LLC: Gurobi Optimizer Reference Manual (2022)
18. Hamann, F., Ulges, A., Krechel, D., Bergmann, R.: Open-world knowledge graph completion benchmarks for knowledge discovery. In: Fujita, H., Selamat, A., Lin, J.C.-W., Ali, M. (eds.) IEA/AIE 2021. LNCS (LNAI), vol. 12799, pp. 252–264. Springer, Cham (2021). https://doi.org/10.1007/978-3-030-79463-7_21
19. Hirsch, J.E.: An index to quantify an individual's scientific research output. PNAS **102**(46), 16569–16572 (2005)
20. Huang, K., Wang, S., Bevilacqua, G., Xiao, X., Lakshmanan, L.V.: Revisiting the stop-and-stare algorithms for influence maximization. PVLDB **10**(9), 913–924 (2017)
21. Huang, X., Zhang, J., Li, D., Li, P.: Knowledge graph embedding based question answering. In: WSDM, pp. 105–113 (2019)

22. vol. Jenatton, R., Roux, N., Bordes, A., Obozinski, G.R.: A latent factor model for highly multi-relational data. In: NeurIPS, vol. 25 (2012)
23. Ji, G., He, S., Xu, L., Liu, K., Zhao, J.: Knowledge graph embedding via dynamic mapping matrix. In: COLING-IJCNLP, pp. 687–696 (2015)
24. Ji, S., Pan, S., Cambria, E., Marttinen, P., Philip, S.Y.: A survey on knowledge graphs: representation, acquisition, and applications. TNNLS **33**(2), 494–514 (2021)
25. Kempe, D., Kleinberg, J., Tardos, É.: Maximizing the spread of influence through a social network. In: KDD, pp. 137–146 (2003)
26. Korn, A., Schubert, A., Telcs, A.: Lobby index in networks. Physica A **388**(11), 2221–2226 (2009)
27. Landis, J.R., Koch, G.G.: The measurement of observer agreement for categorical data. Biometrics **33**, 159–174 (1977)
28. Li, C.T., Lin, S.D.: Egocentric information abstraction for heterogeneous social networks. In: ASONAM, pp. 255–260. IEEE (2009)
29. Likert, R.: A technique for the measurement of attitudes. Arch. Psychol (1932)
30. Liu, S., Grau, B., Horrocks, I., Kostylev, E.: Indigo: GNN-based inductive knowledge graph completion using pair-wise encoding. In: NeurIPS, vol. 34 (2021)
31. Liu, Y., Safavi, T., Dighe, A., Koutra, D.: Graph summarization methods and applications: a survey. CSUR **51**(3), 1–34 (2018)
32. Maccioni, A., Abadi, D.J.: Scalable pattern matching over compressed graphs via dedensification. In: KDD, pp. 1755–1764 (2016)
33. Mahdisoltani, F., Biega, J., Suchanek, F.: Yago3: a knowledge base from multilingual wikipedias. In: CIDR (2014)
34. Mitchell, T., et al.: Never-ending learning. Commun. ACM **61**(5), 103–115 (2018)
35. Mulvey, J.M., Beck, M.P.: Solving capacitated clustering problems. Eur. J. Oper. Res. **18**(3), 339–348 (1984)
36. Nguyen, D.Q., Nguyen, T.D., Nguyen, D.Q., Phung, D.: A novel embedding model for knowledge base completion based on convolutional neural network. arXiv preprint arXiv:1712.02121 (2017)
37. Nguyen, H.T., Dinh, T.N., Thai, M.T.: Revisiting of revisiting the stop-and-stare algorithms for influence maximization. In: Chen, X., Sen, A., Li, W.W., Thai, M.T. (eds.) CSoNet 2018. LNCS, vol. 11280, pp. 273–285. Springer, Cham (2018). https://doi.org/10.1007/978-3-030-04648-4_23
38. Nguyen, H.T., Thai, M.T., Dinh, T.N.: Stop-and-stare: optimal sampling algorithms for viral marketing in billion-scale networks. In: SIGMOD, pp. 695–710 (2016)
39. Nickel, M., Tresp, V., Kriegel, H.P.: A three-way model for collective learning on multi-relational data. In: ICML (2011)
40. Page, L., Brin, S., Motwani, R., Winograd, T.: The pagerank citation ranking: Bringing order to the web. Technical report, Stanford InfoLab (1999)
41. Purohit, M., Prakash, B.A., Kang, C., Zhang, Y., Subrahmanian, V.: Fast influence-based coarsening for large networks. In: KDD, pp. 1296–1305 (2014)
42. Riondato, M., García-Soriano, D., Bonchi, F.: Graph summarization with quality guarantees. DMKD **31**(2), 314–349 (2017)
43. Safavi, T., Koutra, D.: CoDEx: a comprehensive knowledge graph completion benchmark. In: EMNLP, pp. 8328–8350 (2020)
44. Shen, Z., Ma, K.L., Eliassi-Rad, T.: Visual analysis of large heterogeneous social networks by semantic and structural abstraction. IEEE TVCG **12**(6), 1427–1439 (2006)

45. Stefanello, F., de Araújo, O.C., Müller, F.M.: Matheuristics for the capacitated p-median problem. ITOR **22**(1), 149–167 (2015)
46. Toutanova, K., Chen, D.: Observed versus latent features for knowledge base and text inference. In: CVSC, pp. 57–66 (2015)
47. Vaswani, A., et al.: Attention is all you need. In: NIPS, vol. 30 (2017)
48. Wan, Z., Mahajan, Y., Kang, B.W., Moore, T.J., Cho, J.H.: A survey on centrality metrics and their network resilience analysis. IEEE Access **9**, 104773–104819 (2021)
49. Wang, P., Han, J., Li, C., Pan, R.: Logic attention based neighborhood aggregation for inductive knowledge graph embedding. In: AAAI, vol. 33, pp. 7152–7159 (2019)
50. Wang, Q.,et al.: Coke: contextualized knowledge graph embedding. arXiv preprint arXiv:1911.02168 (2019)
51. Wang, Q., Mao, Z., Wang, B., Guo, L.: Knowledge graph embedding: a survey of approaches and applications. TKDD **29**(12), 2724–2743 (2017)
52. Wang, X., He, X., Cao, Y., Liu, M., Chua, T.S.: KGAT: knowledge graph attention network for recommendation. In: KDD, pp. 950–958 (2019)
53. Wang, Z., Zhang, J., Feng, J., Chen, Z.: Knowledge graph embedding by translating on hyperplanes. In: AAAI, vol. 28 (2014)
54. Yao, L., Mao, C., Luo, Y.: KG-BERT: bert for knowledge graph completion. arXiv preprint arXiv:1909.03193 (2019)
55. Zhang, N., Tian, Y., Patel, J.M.: Discovery-driven graph summarization. In: ICDE, pp. 880–891. IEEE (2010)
56. Zhu, L., Ghasemi-Gol, M., Szekely, P., Galstyan, A., Knoblock, C.A.: Unsupervised entity resolution on multi-type graphs. In: Groth, P., et al. (eds.) ISWC 2016. LNCS, vol. 9981, pp. 649–667. Springer, Cham (2016). https://doi.org/10.1007/978-3-319-46523-4_39

KBQA: Accelerate Fuzzy Path Query
on Knowledge Graph

Li Zeng[✉], Qiheng You, Jincheng Lu, Shizheng Liu, Weijian Sun, Rongqian Zhao,
and Xin Chen

Huawei Technologies Co., Ltd., Shenzhen, China
{zengli43,youqiheng,lujincheng7,liushizheng1,sunweijian,
zhaorongqian,chenxin}@huawei.com

Abstract. Fuzzy path query is widely used to find the deep association of entities
in many real-world applications such as knowledge graph answering and social
network analysis. However, existing engines fail to support fuzzy path queries on
large property graphs due to the imprecise string matching and indefinite search
space. In this paper, we propose an extremely fast graph query engine KBQA,
which can perform semantic matching in both entities and properties, and search
arbitrarily long paths efficiently. Facing the performance problem, KBQA designs
two-phase filtering strategy to accelerate candidate selection. Also, bitwise operations are adopted for fast graph exploration. Furthermore, KBQA adaptively
prunes unpromising search paths based on path similarity. Extensive experiments
show that KBQA outperforms all state-of-the-art graph databases by $2\times \sim 10\times$
and searches all 6-hop paths within ten seconds. Our system has been applied in
the ICT field and has achieved remarkable results.

Keywords: Knowledge Graph · Path Query · Fuzzy Matching · Graph
Database

1 Introduction

With the development of information technology, the Internet has evolved towards the
semantic network [3, 17, 25] (i.e., from the link between web pages to the link between
data). Semantic network is a network of knowledge, e.g., social graph as shown in
Fig. 1. Users can query the information on the semantic network like "Who is the wife
of Obama", which is translated into a precise path query $(Obama, Wife, ?)$ by existing
knowledge engine, then the precise answer are displayed in graphs (Fig. 2).

Though existing engines can support precise path queries well, in practical applications the entity names as well as the word sequence are usually inaccurate and incomplete, which is called *fuzzy path query*. Figure 3 gives an example of fuzzy path queries,
i.e., "AAU heaviness". Note that the *AAU* node does not have the property *heaviness* or
its synonym *weight*. Besides, the words *AAU* and *heaviness* may not be precise, thus the
similarity of string matching needs to be computed. The engine needs to start the search
from *AAU* and checks the existence of the required property in each successive node
until all possible results are found. All valid search paths are required to be displayed,

C. Strauss et al. (Eds.): DEXA 2023, LNCS 14146, pp. 462–477, 2023.
https://doi.org/10.1007/978-3-031-39847-6_37

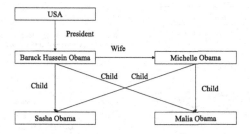

Fig. 1. Example of social graph

which can improve the interpretability. In precise path queries, the query should be the form of an exact path (e.g., "AAU AAU3910 Type02312 Specification Weight"), otherwise existing engines can not find any answer. But in fuzzy path queries, the path length as well as the intermediate nodes can be arbitrary between *AAU* and *heaviness*. Obviously, the problem of fuzzy path query is more difficult due to the similar entity/property matching, synonym replacement and variable path completion.

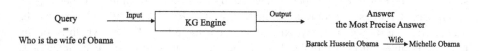

Fig. 2. Example of precise path query on social graph

In the ICT (Information and Communication Technology) field, engineers frequently use the knowledge engine during the planning, construction, maintenance, optimization, and marketing of Telecom network. The ICT domain features massive knowledge and diverse topology, while queries tend to be vague and short. Existing engines [7,9,10,12,18,29] can not process such kinds of queries efficiently because the indefinite search space as well as the imprecise string matching is rather costly. For example, the lengths of different paths between *AAU* and the required property can vary from 1 to 6, leading to a large search space. Considering the query "AAU heaviness" in Fig. 3, if there is > 100 candidates in each hop, the number of 4-hop paths should be larger than $100^4 = 10^8$. On modern large-scale graphs, the out degree of a super node is usually larger than 1000, thus the search space of 6-hop paths is higher than $1000^6 = 10^{18}$. Things can be worse if synonym and string matching are considered in each hop. This is not feasible in practical applications such as social network analysis and the engineering operations. Therefore, an efficient fuzzy query engine is required to accelerate fuzzy path matching on large graphs.

To answer a fuzzy path query q, the routine includes the selection of candidate nodes for the first word (e.g., *AAU*) and the exploration of valid paths that are similar to q. Let C, d and n be the size of candidate set, the average degree in each hop, and the path length, respectively. The search space can be formulated as $C \times d^n$, which is proportional to the size of candidate set. Thus, the candidate selection needs to be

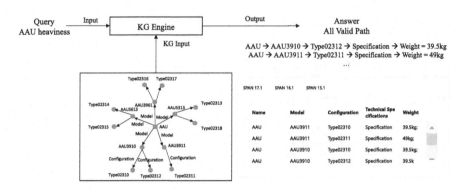

Fig. 3. System Execution Instance

high-precision and lightweight. However, traditional engines adopt naive one-to-one mapping [23], whose computational complexity is $|V| \times z$ (let $|V|$ and z be the number of nodes and the average string matching cost respectively). This is not feasible on large graphs with tens of thousands of nodes, thus we design 2-phase filtering to strike a good balance between performance and precision. Path exploration consists of two phases: DFS-based medium node matching and BFS-based valid endpoint search. The brute-force strategy used by existing solutions has a $C \times d^n$ cost, which is longer than ten minutes on large graphs. Due to the large search space of fuzzy matching, pruning of dissimilar paths is rather prominent in improving the search efficiency of DFS (Depth-first Search) [13]. In addition, a frontier queue is maintained and used for exploration during BFS (Breadth-first Search). Existing solutions generate the frontier queue by ordinary *STL queue* container with sequential *push* operations, which can be costly in both time and memory when the frontier queue is very large. Instead, we elaborate bitmap-based graph storage and generate the frontier queue by innovative bitwise operations, which is extremely fast due to the fitness to modern CPU architecture.

In this paper, we propose a novel system *KBQA* (Knowledge-base Query Answering), which can answer vague queries by supporting semantic string matching and arbitrary path exploration. To the best of our knowledge, KBQA is the first query engine that can process fuzzy path queries in reasonable time even on knowledge graphs with hundreds of millions of edges. Our primary contributions are the following:

- We design an advanced framework of fuzzy path query processing, which can support the matching based on both string similarity and path similarity.
- During the candidate selection, novel 2-phase filtering are devised to accelerate the string matching.
- To optimize the frontier queue generation during path exploration, bitwise primitives are elaborated and carefully implemented.
- Unpromising search paths are adaptively pruned based on dynamically computed path similarity.
- Experiments on both synthetic and real-world graph datasets show that KBQA outperforms the state-of-the-art approaches by up to $10\times$.

The rest of the paper is organized as follows. Section 2 reviews the problem definition and the related work. Then, Sect. 3 presents the framework and accelerative techniques of KBQA. Section 4 shows all experimental results and their analysis. Finally, Sect. 5 concludes the paper.

2 Background

2.1 Problem Definition

Definition 1 *(Graph). A graph is denoted as $G = \{V, E, L, Y\}$, where V is the set of vertices; $E \subseteq V \times V$ is the set of undirected edges; L is a labeling function that maps a vertex (of $V(G)$) to a dictionary of properties (each property is a pair of key and value); Y is the dictionary of each node/property and its synonym list. The labeling function of G can also be specified as L_G. $V(G)$ and $E(G)$ are used to denote vertices and edges of graph G, respectively.*

Definition 2 *(Subgraph). Given a graph $G = \{V, E, L\}$, a subgraph of G is denoted as $G' = \{V', E', L'\}$, where vertex sets V' and edge sets E' in G' are subsets of V and E, respectively, denoted as $V' \subseteq V$ and $E' \subseteq E$. Furthermore, for vertex labeling functions, $L' \subseteq L$.*

Definition 3 *(Path Query). The path query q consists of a beginning word s, k medium words $\{m_1, ..., m_k\}$, and the final property key prop.*

Definition 4 **(Path Match).** *Given a graph g and a path query $q = \{s, m_1, ..., m_k, prop\}$, a path match of q in g is the subgraph $\{v_0, v_1, ..., v_j, val\}$ of g such that:*
(1) $\overline{v_i v_{i+1}} \in E(g)$, $\forall 0 \leq i < j$
(2) s or one of $Y[s]$ is similar to v_0, $prop \in L(v_j)$ or one of $Y[prop] \in L(v_j)$ and $val = L(v_j)[prop]$
(3) $\exists$ a subsequence $\{u_1, ..., u_k\}$ of $\{v_1, ..., v_j\}$ such that u_z or one of $Y[u_z]$ is similar to m_z, $\forall 1 \leq z \leq k$

Definition 5 *(Problem Statement). Given a graph g and a path query q, the fuzzy path query problem is to find out all matching paths of q in g.*

A running example is given in Fig. 3. According to Definition 3, the beginning word and the final property key is s and *heaviness* respectively, while no medium word exists in this case. Many valid path matches exist and one of them is $AAU \rightarrow AAU3910 \rightarrow Type02312 \rightarrow Specification \rightarrow 39.5kg$ (note that *heaviness* and *weight* are synonyms). This paper aims to accelerate fuzzy path query processing on large-scale knowledge graph. Note that a node is also called an entity, and different nodes may have the same entity names. Without loss of generality, we assume the graph g is connected and the result set of path query is not empty. Though our solution can be easily extended to process directed graphs, vertex/edge labels or label sets, that is not our focus. Unless otherwise specified, we use u, $N(u)$, $deg(u)$, $num(L)$, and $|A|$ to denote a vertex, the neighbor set of u, degree of u, the number of currently valid elements in set L, and the size of set A, respectively.

2.2 Related Work

Existing work related to path queries can be mainly divided into three categories: relational table join, subgraph matching, graph exploration.

Relational Table Join. Some earlier solutions (Jena [18], Virtuoso [10]) store the graph data as relational tables and answer path queries by joining these tables. The structure of relational table and the algorithm of join are different in various systems, e.g., binary join [19] and worst-case optimal join [21,22]. However, these solutions conduct too many self-joins and the total computational cost is exponential to the path length, which marks them inefficient.

Subgraph Matching. Neo4j [12] and gStore [29] treat each query as a query graph and find out all its matches in the data graph. The subgraph matching is usually done by a filter-and-verification framework, as detailed in [27,30,31]. The query languages of Neo4j and gStore are Cypher [11] and SPARQL [1], which are not Turing-complete [8,14] and can not support fuzzy path queries. As gStore is open-source and has better performance than Neo4j, we can modify the query processing procedure of gStore to support fuzzy path queries.

Graph Exploration. Other graph databases (ArangoDB [9] and TigerGraph [7]) adopt the paradigm of graph exploration such as breadth-first search (BFS) and depth-first search (DFS). The query language GSQL of TigerGraph is Turing-complete, thus it naturally support fuzzy path queries. As for the language AQL of ArangoDB, it can be enhanced by implementing *javascript* plug-ins. Thus, both ArangoDB and TigerGraph are used in our experiments.

KBQA adopts the graph exploration paradigm, which is obviously more flexible and efficient than relational table join. ArangoDB is inefficient as it lacks optimizations of the graph exploration. Besides, ArangoDB needs to be equipped with *javascript* to support fuzzy path matching, but the performance of *javascript* is not competitive when compared with *C++/Java*. In contrast, gStore is implemented in *C++* and has done sophisticated optimizations of the matching process. However, it targets at more general subgraph matches, thus neglecting special techniques for accelerating fuzzy path matching, e.g., pruning of unpromising paths based on path similarity. As for Tiger-Graph, it is rather powerful and has thoroughly optimized the graph exploration with techniques such as fine-grain parallelism and query compilation. The main shortcoming of TigerGraph is the missing of optimization in the candidate selection and pruning of path similarities. In addition, these solutions do not support special graph storage and bitwise-based frontier queue generation for super nodes. In contrast, KBQA proposes lightweight semantic filtering strategies and advanced bitwise-based graph structures, and leverages powerful pruning techniques to early terminate dissimilar search paths.

3 The Proposed System

3.1 Architecture

The architecture of our KBQA system is shown in Fig. 4. The knowledge graph is built offline, while the user's query are processed online. For each raw query, a series of words (i.e., a path query $q = \{s, m_1, ..., m_k, prop\}$ defined in Definition 3) are obtained by the entity recognition algorithm [5]. Our main innovations include *starting node search algorithm* and *variable path search algorithm*. The starting node search algorithm calculates the similarity score between s and each nodes in the knowledge graph and chooses the highest n ones as candidates. Later, the variable path search algorithm searches from each candidate and checks whether the similarity of current path and q satisfies the semantic constraints (Definition 4).

Fig. 4. The architecture of KBQA system

In KBQA system, the starting candidates C and final results (all valid paths in g) can be displayed on the graph visualization interface. Users can directly observe the effects, as shown in Fig. 3.

3.2 Starting Node Search

Since words of q may not be precise entities, we need to calculate the similarity between each word and all nodes of g. When loading a knowledge graph, all entities are segmented and the weight of each word is calculated according to *Inverse Document Frequency* (IDF [24]). Then, when processing online queries, KBQA adopts a 2-phase filtering strategy which consists of *rough filtering* and *semantic matching*. Instead of matching all nodes directly, our strategy filter out most candidates by a coarse-grain method and only performs fine-grain filtering on a small node set. Therefore, our 2-phase strategy achieves much better performance than the naive method.

Rough Filtering. The similarity score of the first query word s to an entity l is computed by the number of occurrences of s in l and the average number of words in all the entities. If the IDF weight (idf) of s does not exist, the matching ends. Otherwise, the score is computed as below:

$$score = idf \times value \times \frac{1 + k}{value + k \times (1 - b \times \frac{1-d}{avg_d})} \tag{1}$$

where k and b are hyper-parameters ($0 \leq b \leq 1$ and $k \geq 0$). $value$ indicates the number of times that s appears in the entity l. d is the number of words of l, while avg_d is the average number of words for all entities.

Then, all calculated scores are sorted in descending order, and the top n names (e.g.,, five names) with their corresponding nodes are selected. For these nodes, semantic matching is performed with the original starting query word. The nodes that are successfully matched form a set, which is marked as candidate set C.

Semantic Matching. The entity l and the first query word s are viewed as semantically similar if one of the conditions below is satisfied:

- The *levenshtein* distance [20] cannot exceed the parameter x.
- The levenshtein distance between a synonym s' of s and l cannot be greater than the parameter y.
- The entity name consists of the original word s plus a string of digits.
- Other reasonable user-defined conditions approved by business.

For example, if the levenshtein distance is 0, the corresponding entity is retained.

Optimization by Inverted Indexing. To process large graphs, we further propose an efficient accurate score calculation based on lexical inverted tables [15]. When calculating the scores of all entities during preprocessing, indexes are created for each word and the corresponding entities. During query answering, only the names of all entities that contain the word are obtained and calculated. This greatly reduces the number of useless calculation.

3.3 Variable Path Search

After the starting candidate set C is found, we need to find out all feasible paths with variable lengths. The framework of our method is shown in Fig. 5: DFS is performed on g starting from each candidate, and all medium words of q are matched one by one to check the similarity of current path. For each path P that completes medium node matching, BFS is conducted from the endpoint of P for the nearest valid end node (i.e., the node with attributes).

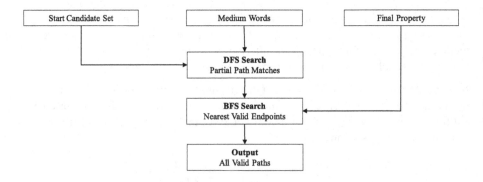

Fig. 5. The framework of variable path search

Medium Node Matching. Taking each node of the starting candidate set C as the source, DFS is carried out to extend the partial path and check whether the similarity of current path and q satisfies the constraint. The similarity calculation of medium words is similar to that of starting node search in Sect. 3.2. For each node v, if v successfully

---

Algorithm 1: The algorithm of medium node matching

Input: candidate set C, query $q = \{s, m_1, ..., m_k, prop\}$, graph g
Output: partial path matches $\{P_i\}$
1 **procedure** $EXPAND(C, q, g)$
2    $R \leftarrow \emptyset$
3 **foreach** $v \in C$ **do**
4 | $path \leftarrow \{v\}$
5 | $DFS(v, 1, path)$
6 **return** R
7 **procedure** $DFS(u, pos, path)$
8 **if** $pos = k$ **then**
9 | **if** $checkSimilarity(path, q)$ **then**
10 | | $R \leftarrow R \cup \{path\}$
11 | **return**
12 **foreach** $v \in N(u)$ **do**
13 | $path \leftarrow path \cup \{v\}$
14 | $DFS(v, pos, path)$
15 | recover the original $path$
16 | **if** v or one of $Y[v]$ is similar to m_{pos} **then**
17 | | $path \leftarrow path \cup \{v\}$
18 | | $DFS(v, pos + 1, path)$
19 | | recover the original $path$

---

matches current medium word, the path P_i is extended by v. Otherwise, though v and m_{pos} are not similar, P_i can be extended by v but m_{pos} remains unmatched in this case. There may be several nodes between medium words, thus the path length is variable, which is the inherent hardness of fuzzy path matching. The pseudo code of medium node matching is given in Algorithm 1.

Valid Endpoint Search. After matching all medium words in q, the processing of fuzzy path queries does not end because the property that named $prop$ needs to be found out. For each path P_i, BFS starts the search from the final node of P_i to match a valid endpoint which contains the given property $prop$ or contains one synonym of $prop$. As long as there is an endpoint v that is successfully matched, the path from the source to v is recorded, which will be returned as one of the final answers. Algorithm 2 lists the pseudo code of valid endpoint search. BFS is selected as the search mechanism to provide better performance because it facilitates parallelism, has better cache locality, and can detect endpoints within the shortest hop count.

Optimization of Graph Storage. Nodes in the original graph g are distinguished by string IDs, which is costly and limits the optimization techniques. We utilize the state-of-the-art preprocessing method [26] that assigns consecutive integer IDs to nodes of g, meanwhile enhancing the locality. Later, efficient data structures can be designed by leveraging the integer IDs.

1. Multi-form neighbor structure. There are some nodes in g that have many neighbors, which is called *supernode*. The common adjacency list occupies too much memory

---

Algorithm 2: The algorithm of valid endpoint search

Input: partial matches $\{P_i\}$, query $q = \{s, m_1, ..., m_k, prop\}$, graph g
Output: entire path matches $\{R_i\}$
1 **procedure** $DETECT(\{P_i\}, q, g)$
2    $R \leftarrow \emptyset$
3 **foreach** $path \in \{P_i\}$ **do**
4 | let u be the final node of $path$
5 | $BFS(u, path)$
6 **return** R
7 **procedure** $BFS(u, path)$
8 **if** $prop \in L(u)$ or one of $Y[prop] \in L(u)$ **then**
9 | $R \leftarrow R \cup \{path \cup L(u)[prop]\}$
10 | **return**
11 **foreach** $v \in N(u)$ **do**
12 | $BFS(v, path \cup \{v\})$

---

in this case, as shown in Fig. 6 (left-top). A tightly compressed bit array [28] is used to store the neighbors of the supernode, which greatly reduces memory usage, as shown in Fig. 6 (right-top). In the bit array, if bits 2~37 are 1, the *AAU3910* has neighbors numbered from 2 to 37.

2. Inversion table of node properties. When matching the final property, each traversed node needs to be matched. However, when the graph search space is large, the number of matching times may reach hundreds of millions, which seriously slows down the query answering. Thus, an inverted index from the property name to the corresponding node set is constructed, as shown in Fig. 6 (bottom). Before matching the endpoint, we filter out nodes that do not contain the required property, and then match only the remaining nodes. This greatly reduces the number of matching times and shortens the matching time. In the bit array, bits 791~880 are 1 indicating that nodes numbered from 791 to 880 have the *Weight* property.

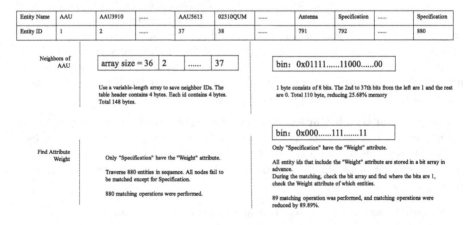

Fig. 6. Inverted table index of the property

Optimization of Pruning. During variable path search, semantic matching is required for each medium word, and hybrid DFS+BFS search (called ADBFS [28]) is used to traverse the search space. In each search path, KBQA combines the graph diameter gd, the number of unmatched medium words (uk), and the current path length pl to make judgment to prune unnecessary search space, thus speeding up the search and reducing the memory occupation.

1. When $uk + pl$ exceeds the diameter of the graph, it is impossible to obtain a valid result, thus the current search path can be terminated.
2. If a perfect path p match exists (i.e., the similarity of p and q is 1.0), all other imperfect search paths can be terminated as the path similarity never increases.

The matching process of the example question "AAU 02310 heaviness" is shown in Fig. 7. Note that *heaviness* is equivalent to *Weight* in *g*. All neighbors of a node are grouped in pairs, while each group is considered as a whole for ADBFS traversal. Each group traversal is equivalent to traversing all nodes of the group in parallel, that is, BFS expansion of nodes in the group. After medium words are successfully matched and the partial path $AAU \rightarrow AAU3910 \rightarrow Type02310$ is obtained, BFS searches for the required property *Weight* using $Type02310$ as the new starting point. In the figure, the blue box indicates nodes that have the property *Weight*, and the black box indicates nodes that do not have the property *Weight*. Whether the node has the required property can be determined by the inverted table of node properties. During exploration, the index is used to filter out invalid nodes to obtain the required property *Weight* at the endpoint and yield the valid path match $AAU \rightarrow AAU3910 \rightarrow Type02310 \rightarrow specifications \rightarrow weight = 39.5kg$.

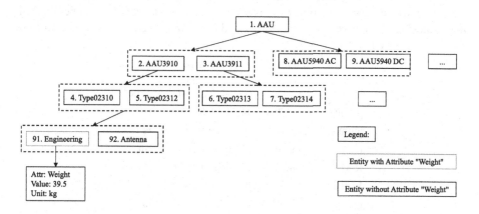

Fig. 7. Matching process of the question"AAU 02310 heaviness"

4   Experiments

In this section, we evaluate our method (KBQA) against state-of-the-art graph databases (ArangoDB [9], gStore [29] and TigerGraph [7]). ArangoDB and gStore need to be enhanced to support fuzzy path queries, as discussed in Sect. 2.2. All experiments are carried out on a workstation running CentOS 7 and equipped with Intel Xeon E5-2620 2.40 GHz CPU, 64 GB host memory and 256 GB disk.

4.1 Datasets and Queries

The experiments are conducted on both real and synthetic datasets. The statistics are listed in Table 1. The patent citation network (*patent*), the online social network (*journal*), as well as the mesh-like *road* [16] are downloaded from SNAP [16]. *gsc* is the telecom graph in ICT field, representing the properties and relationship of equipments. Each node of gsc has one non-unique label, and there is 173 kinds of properties in gsc. Figure 3 is an instance of gsc. Synthetic graphs include R-MAT and WatDiv. R-MAT is generated by [4] under Graph500 settings [6]. WatDiv is a RDF graph with semantic node/edge labels, which is generated by the Waterloo Benchmark [2].

Due to the lack of string node IDs and node properties in all graphs except for gsc, we perform some data augmentation. For node properties, we generate 100 kinds of properties and randomly assign them to each node following the 20/80 rule. The string node IDs are also generated and assigned in a similar way. The value corresponding to each property name can be arbitrary as it does not affect the query performance.

For each dataset, 100 queries are generated by randomly exploring paths from the graph. The length of all paths can not exceed six because in most real-life graphs all relationships can be found within six hops. In each path, string edit and synonym replacement is randomly applied on labels or properties to test fuzzy cases. Besides, medium words are also randomly eliminated from each path. For example, a fuzzy query on gsc is shown in Fig. 3, where heaviness is a *synonym* of *weight* and all medium words (e.g., *AAU3910*) are removed. The time limit is set to 10 min, i.e., a query can not be selected if it is not responded within 600 s. For each dataset, the average running time of all queries is reported.

Table 1. Statistics of Datasets

Name	$|V|$	$|E|$	MD[1]	Type[2]
patent	3,774,768	16,518,948	793	rs
journal	4,847,570	33,099,465	20,290	rs
road	1,965,206	2,766,607	8	rm
R-MAT	1,048,576	15,680,861	67,086	s
WatDiv	10,899,920	109,959,180	671K	s
gsc	5,169,893	62,207,935	486	rs

[1] Maximum degree of the graph.
[2] Graph type: r:real-world, s:scale-free, and m:mesh-like.

4.2 Evaluation of Techniques of Starting Node Search

We evaluate the effect of the 2-phase filtering technique, as shown in in Table 2. Since the length of queried paths is as high as six, all algorithms need several seconds to answer fuzzy queries due to the explosion of search space as 6-hop subgraph touches nearly all nodes in most real graphs. Compared with the naive implementation that performs one-to-one mapping between the first word and all nodes, our technique achieves >1.24× speedup on all graphs except for road. The improvement on road is only 1.12× due to its small graph scale and low degree. Generally, the gain is more obvious on graphs with larger size and higher degree. For example, on journal and WatDiv the performance gain is much more prominent (>1.3×).

The reduction in time comes from the optimization of start node search. Though the final candidate set is the same, the naive method compares the first word with all nodes one by one, and checks whether the string similarity is enough. Before our optimization, nearly 10%~20% time (several seconds) is consumed in the start node search. In contrast, with our 2-phase filtering strategy the filtering time can be reduced to several milliseconds, which can be marginally omitted in the total time cost. Most invalid candidate nodes can be filtered out by the rough filter and inverted table lookup, which reduces the burden of the more precise filter (semantic matching).

Table 2. The performance of start node search (ms)

System	patent	journal	road	R-MAT	WatDiv	gsc
KBQA-[1]	5.56K	8.03K	1.75K	6.1K	57K	21K
KBQA*[2]	4.36K	6.37K	1.56K	4.8K	44K	17K
speedup	1.28×	1.71×	1.12×	1.27×	1.3×	1.24×

[1] KBQA- represents the implementation without the optimization of start node search and variable path matching.
[2] KBQA* represents the implementation without the optimization of variable path matching.

4.3 Evaluation of Techniques of Variable Path Search

In this section, we evaluate the effect of our optimization of variable path search, which consists of medium node matching and valid endpoint search. As stated in Table 3, the overall minimum speedup is 1.22× compared with naive search mechanism, which is a simple DFS process without the optimization of graph storage and pruning. The minimum speedup is achieved only on the smallest dataset road, which is a mesh-like and low-degree graph. Its small and regular adjacency list has no need for bitwise-based graph storage and operations. Except for road, our technique achieves > 2.29× speedup on all other graphs.

Obviously, the improvement is more prominent on larger and more skewed graphs like WatDiv and gsc, showing 4.49× speedup and 2.83× speedup respectively. Generally, large skewed graphs have several super nodes, whose degree is much larger than others. Inherently, the search space grows exponentially on super nodes, especially on the fuzzy matching case, incurring both time and memory pressure on the query processing. On the one hand, the valid endpoint search is costly if the last medium node does not contain the required property. Before our optimization, nearly 30%~40% time (as high as dozens of seconds) is consumed in the valid endpoint search, which is a BFS process and its main cost is the frontier queue generation. But with the multi-form graph storage and bitwise-based frontier queue generation, the proportion is reduced to 5%~10% (no more than 2 s). On the other hand, the explosion of temporary paths is prohibitive on large skewed graphs, implying more potential for our dynamical pruning of dissimilar paths. Based on the monotonicity of path similarity as well as the restriction of graph diameter and current matching process, KBQA adaptively eliminates most unpromising search paths as early as possible. Thanks to our multi-form graph structure and adaptive pruning techniques, the explosion of search space on super nodes is greatly alleviated.

Table 3. The performance of variable path matching (ms)

System	patent	journal	road	R-MAT	WatDiv	gsc
KBQA*	4.36K	6.37K	1.56K	4.8K	44K	17K
KBQA[1]	1.82K	2.37K	1.28K	2.1K	9.8K	6K
speedup	2.4×	2.69×	1.22×	2.29×	4.49×	2.83×

[1] KBQA represents the implementation with the optimization of start node search and variable path matching.

4.4 Overall Performance

In this section, we compare KBQA with Neo4j, Virtuoso, ArangoDB, gStore and TigerGraph, which are the state-of-the-art graph databases. The overall performance is reported in Table 4. Obviously, our optimized KBQA is the single winner, showing $2 \sim 10\times$ speedup compared with all counterparts. Among other systems, the rank is TigerGraph > gStore > ArangoDB > Virtuoso > Neo4j.

Compared with TigerGraph, the minimum speedup is $2.03\times$ on the mesh-like graph road. As the degree of road is rather low and its degree distribution is balanced, the search space is not very large and the effects of our techniques are limited. On the power-law graph WatDiv and gsc, the improvement is much more prominent: $> 3.3\times$ and $4\times$ respectively. They have rather skewed degree distribution and much larger search space, which is a disaster for other engines but can be handled well by KBQA. For example, our elaborate graph structure (bitmap and adjacency list) is very suitable for the storage and traversal of supernodes in scale-free graphs.

The performance of ArangoDB is strictly limited by its inefficient *javascript* plugins. In contrast, gStore is implemented in $C++$, but it lacks special techniques for path matching rather than general subgraph matching. Though optimized by fine-grain parallelism and query compilation, TigerGraph does not have any optimization in the computation and pruning of path similarities. Furthermore, existing engines do not support multi-form graph storage and bitwise-based frontier queue generation for super nodes. As a result, all these solutions encounter the explosion of response time on large skewed graphs, just like our KBQA- implementation. Comparatively, KBQA proposes advanced graph structures for storing irregular neighbors and inverted tables, and leverages powerful pruning techniques for early termination of infeasible search paths, which contribute to its extraordinary performance.

To sum up, on all datasets only KBQA can answer all fuzzy path queries within 10 s, which means KBQA can be practically used in online applications.

Table 4. The elapsed time of different algorithms (ms)

System	patent	journal	road	R-MAT	WatDiv	gsc
Neo4j	9.56K	12K	5.2K	11K	115K	76K
Virtuoso	4.98K	6.6K	3.6K	5K	62K	39K
ArangoDB	4.74K	5.41K	3.1K	5.4K	55K	31K
gStore	4.71K	5.28K	3K	4.8K	40K	28K
TigerGraph	4.12K	5.30K	2.6K	4.5K	32K	20K
KBQA	1.82K	2.37K	1.28K	2.1K	9.6K	5K

5 Conclusions

We introduce an extremely fast query engine KBQA, which supports fuzzy path query answering efficiently on knowledge graphs. Extensive experiments show that KBQA outperforms the state-of-the-art graph databases by $2\times \sim 10\times$. In future, KBQA can be enhanced by analyzing the impact of query result complexity and adding distributed computing ability.

References

1. SPARQL 1.1. In: Alhajj, R., Rokne, J.G. (eds.) Encyclopedia of Social Network Analysis and Mining, 2nd Edition. Springer (2018)
2. Aluç, G., Hartig, O., Özsu, M.T., Daudjee, K.: Diversified stress testing of RDF data management systems. In: Mika, P., et al. (eds.) ISWC 2014. LNCS, vol. 8796, pp. 197–212. Springer, Cham (2014). https://doi.org/10.1007/978-3-319-11964-9_13
3. Borge-Holthoefer, J., Arenas, A.: Semantic networks: structure and dynamics. Entropy **12**(5), 1264–1302 (2010). https://doi.org/10.3390/e12051264
4. Chakrabarti, D., Zhan, Y., Faloutsos, C.: R-MAT: a recursive model for graph mining. In: Berry, M.W., Dayal, U., Kamath, C., Skillicorn, D.B. (eds.) Proceedings of the Fourth SIAM International Conference on Data Mining, Lake Buena Vista, Florida, USA, April 22–24, 2004, pp. 442–446. SIAM (2004)
5. Chang, Y., Kong, L., Jia, K., Meng, Q.: Chinese named entity recognition method based on BERT. In: 2021 IEEE International Conference on Data Science and Computer Application (ICDSCA), pp. 294–299. IEEE (2021)
6. D'Azevedo, E.F., Imam, N.: Graph 500 in OpenSHMEM. In: Gorentla Venkata, M., Shamis, P., Imam, N., Lopez, M.G. (eds.) OpenSHMEM 2014. LNCS, vol. 9397, pp. 154–163. Springer, Cham (2015). https://doi.org/10.1007/978-3-319-26428-8_10

7. Deutsch, A., Yu, X., Wu, M., Lee, V.: Tigergraph: A native MPP graph database. arXiv (2019)

8. Deutsch, A., et al.: Graph pattern matching in GQL and SQL/PGQ. In: SIGMOD, pp. 2246–2258. ACM (2022)

9. Dohmen, L.: Algorithms for large networks in the NoSQL database Arangodb. Bachelor Thesis of RWTH Aachen University (2012)

10. Erling, O.: Virtuoso, a hybrid RDBMS/graph column store. IEEE Data Eng. Bull. **35**(1), 3–8 (2012)

11. Francis, N., et al.: Cypher: An evolving query language for property graphs. In: SIGMOD, pp. 1433–1445. ACM (2018)

12. Guia, J., Soares, V.G., Bernardino, J.: Graph databases: Neo4j analysis. In: ICEIS 2017 - Proceedings of the 19th International Conference on Enterprise Information Systems, Volume 1, Porto, Portugal, 26–29 April 2017, pp. 351–356 (2017)

13. Han, M., Kim, H., Gu, G., Park, K., Han, W.: Efficient subgraph matching: harmonizing dynamic programming, adaptive matching order, and failing set together. In: Proceedings of the 2019 International Conference on Management of Data, SIGMOD Conference 2019, Amsterdam, The Netherlands, 30 June–5 July 2019, pp. 1429–1446. ACM (2019)

14. Hogan, A., Reutter, J.L., Soto, A.: Recursive SPARQL for graph analytics. arXiv abs/2004.01816 (2020)

15. Knuth, D.E.: Retrieval on secondary keys. Art Comput. Program. Sorting Searching **3**, 550–567 (1997)

16. Leskovec, J., Krevl, A.: SNAP Datasets: Stanford large network dataset collection (2014). http://snap.stanford.edu/data

17. Lytras, M., Downes, S.: Semantic networks and social networks. Learn. Organ. **12**(5), 411–417 (2005)

18. McBride, B.: Jena: a semantic web toolkit. IEEE Internet Comput. **6**(6), 55–59 (2002)

19. Mhedhbi, A., Salihoglu, S.: Optimizing subgraph queries by combining binary and worst-case optimal joins. VLDB (2019)

20. Navarro, G.: A guided tour to approximate string matching. ACM Comput. Surv. **33**(1), 31–88 (2001)

21. Ngo, H.Q.: Worst-case optimal join algorithms: techniques, results, and open problems. In: PODS (2018)

22. Ngo, H.Q., Porat, E., Ré, C., Rudra, A.: Worst-case optimal join algorithms: [extended abstract]. In: PODS (2012)

23. Nolé, M., Sartiani, C.: Regular path queries on massive graphs. In: Proceedings of the 28th International Conference on Scientific and Statistical Database Management, SSDBM 2016, Budapest, Hungary, 18–20 July 2016, pp. 13:1–13:12. ACM (2016)

24. Rajaraman, A., Ullman, J.D.: Data Mining, pp. 1–17. Cambridge University Press, Cambridge (2011). https://doi.org/10.1017/CBO9781139058452.002

25. Sowa, J.F.: Semantic networks. Encycl. Cogn. Sci. (2012)

26. Wei, H., Yu, J.X., Lu, C., Lin, X.: Speedup graph processing by graph ordering. In: Özcan, F., Koutrika, G., Madden, S. (eds.) Proceedings of the 2016 International Conference on Management of Data, SIGMOD Conference 2016, San Francisco, CA, USA, 26 June 01 July 2016, pp. 1813–1828. ACM (2016)

27. Zeng, L., Jiang, Y., Lu, W., Zou, L.: Deep analysis on subgraph isomorphism. arXiv (2020)

28. Zeng, L., Zhou, J., Qin, S., Cai, H., Zhao, R., Chen, X.: SQLG+: efficient-hop query processing on RDBMS. In: Bhattacharya, A., et al. (eds.) DASFAA 2022. LNCS, pp. 430–442. Springer, Cham (2022)

29. Zeng, L., Zou, L.: Redesign of the gStore system. Front. Comput. Sci. **12**, 623–641 (2018)

30. Zeng, L., Zou, L., Özsu, M.T., Hu, L., Zhang, F.: GSI: GPU-friendly subgraph isomorphism. In: 36th IEEE International Conference on Data Engineering, ICDE 2020, Dallas, TX, USA, April 20–24, 2020, pp. 1249–1260. IEEE (2020)
31. Zeng, L., Zou, L., Özsu, M.T.: SGSI - a scalable GPU-friendly subgraph isomorphism algorithm. IEEE Trans. Knowl. Data Eng. 1–17 (2022)

Tour Route Generation Considering Spot Congestion

Takeyuki Maekawa[1], Hidekazu Kasahara[2], and Qiang Ma[3]($\boxtimes$) ⓘ

[1] Kyoto University, Yoshida-Hommachi, Sakyo-ku, Kyoto 606-8501, Japan
maekawa.takeyuki.28s@st.kyoto-u.ac.jp
[2] Osaka Seikei University, Aikawa 3-5-9, Higasiyodogawa-ku, Osaka, Japan
kasahara.hidekazu.k13@kyoto-u.jp
[3] Kyoto Institute of Technology, Matasugasaki, Sakyo-ku, Kyoto 606-8585, Japan
qiang@kit.ac.jp

Abstract. Crowding in tourism has been gaining attention in recent years, and its effects include tourism pollution and a lower-quality travel experience. Tourists often have rough plans and take action to avoid crowding. Therefore, we can help them by planning tourist routes that consider crowding. However, conventional route planning and recommendation methods do not consider dynamic factors such as congestion. In this study, we introduce two novel concepts, "dynamic stay duration" and "environmental tax metaphor", into tour route generation methods to plan routes while considering congestion. We implement the proposed method based on a pointer network and the REINFORCE algorithm. The results of an experiment and a user study confirm that the proposed method is superior to the conventional methods.

Keywords: Sightseeing · Congestion · Route Planning · Reinforcement Learning

1 Introduction

The problem of congestion in tourism harms both the destination city and tourists and has attracted attention in recent years. Overtourism occurs when tourists are concentrated in a few tourist spots in a tourist city, causing problems such as traffic congestion, garbage, and noise. The negative effects on tourists include a decrease in the quality of the travel experience and a decline in satisfaction. For example, destination arrival times may be delayed, itineraries may change, and planned sightseeing may not be possible.

The impact of such congestion should be reduced. One way to reduce the impact of congestion and achieve sustainable sightseeing [14] is to support tourists by generating routes that consider congestion. Tourists are likely to make rough travel plans, such as lists of places they want to visit. Therefore,

This work was partly supported by JSPS KAKENHI (23H03404,21K12140) and MIC SCOPE (201607008).

while sightseeing in the destination area, tourists often use navigation apps to search for a way to travel from their current location to their destination on the spot and follow the search results.

Most conventional sightseeing route planning methods use heuristics to solve the issue as an optimization problem. Therefore, it may be difficult to deal with complex problems such as having a large number of spots that need to be traveled to. Gama et al. [2] proposed a method using reinforcement learning and a pointer network [15] to reduce the computational cost of large-scale planning. Their method achieved superior results in a static environment. However, in the real world, certain factors change from moment to moment, such as the congestion and sightseeing value of spots [17], and the method proposed by Gama et al. cannot cope with such dynamic factors. Among these dynamic factors, crowding is time-dependent. As shown in our user study (Sect. 5), this is an important factor for tourists. Additionally, crowded spots may require more time to visit. Because tourists have limited time, generating sightseeing routes that consider congestion is necessary.

In this study, we propose two novel concepts and introduce them as dynamic factors into the model proposed by Gama et al. One is the "dynamic stay duration," which estimates the necessary visiting (stay) duration at a spot based on its congestion. The other is the "environmental tax metaphor", an additional reward mechanism dependent on congestion. A negative reward is assigned to routes consisting of more crowded spots. A brief overview of the proposed method is shown in Fig. 1. We define two congestion indices: the "occupancy rate" and "relative congestion level". The occupancy rate is the ratio of the number of people staying at a tourist spot at a given time to the capacity (maximum number

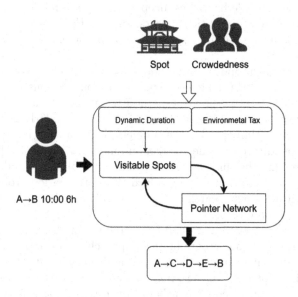

Fig. 1. Overview of Proposed Method

of people) of that spot. Relative congestion is the normalized occupancy rate, assuming that the distribution of the occupancy rates of spots in a tourist city follows a normal distribution.

- Dynamic stay durations change the time spent at a spot based on the occupancy rate. The higher the occupancy rate, the more time the tourists spend in a spot.
- We consider two rewards in our environmental tax metaphor: an environmental tax and a subsidy. The environmental tax is a negative additional reward for discouraging visits to spots with high occupancy rates. It is calculated based on relative congestion. Conversely, subsidies are positive additional rewards for visiting places with low occupancy rates.

Inspired by [2], we implement the proposed method based on a pointer network and the REINFORCE algorithm. The proposed method provides users with routes that allow them to enjoy sightseeing while avoiding congestion. Travelers can enjoy their trips without being bothered by crowds. In this way, destinations can disperse tourists and reduce congestion.

The remainder of this paper is organized as follows. First, the related work is introduced in Sect. 2, and then the problem is defined in Sect. 3. The proposed method is described in Sect. 4, and experiments to verify the method's usefulness are described in Sect. 5. Finally, the conclusion is given in Sect. 6.

2   Related Work

Many studies have been conducted regarding sightseeing route planning. Optimization methods include heuristics, deep learning, and reinforcement learning. There are also many optimization targets, such as the shortest route, shortest travel time, maximization of user preferences, and maximization of environmental rewards.

Gavalas et al. [3] proposed a method for constructing a route by dynamically setting the travel times between spots. The route is constructed by inserting spots. Wu et al. [16] proposed a method to maximize satisfaction, which is determined by the time and financial cost of the trip. Similarly, Hirano et al. [6] used a genetic algorithm to solve the problem of optimizing multiple functions to minimize expenditures of time, energy, and money. Zhang et al. [20] proposed a method for setting the time-varying attractiveness of a spot and constructing a route to maximize this value. They solved this optimization problem by using a genetic algorithm. Route searches using methods based on genetic algorithms have also been proposed [10, 21].

Isoda et al. [8] proposed a method for searching for the route with the highest score using a dynamic programming approach. They introduced dynamic features that vary with time and season, in addition to static features such as the value of the spot itself. Congestion was included in the dynamic features, and the values were obtained from Yahoo! Congestion Radar. Muccini et al. [11] developed a mobile application that eliminates the waiting time before a reserved museum visit by stopping at surrounding museums. The application uses real-time museum waiting times as congestion information. It then builds a tree of spots to visit and searches until a route that satisfies certain time constraints is generated. Xu et al. [18] proposed a method that represents spots as a graph network, calculates the value of spots based on their congestion and popularity, and uses heuristics to find routes to visit spots in order of increasing popularity. Mahdis et al. [1] created data on spots and travel between spots based on big data regarding traffic, weather, and tourist spots. They proposed a planning method using metaheuristics that reflect user preferences. This method focused on road congestion rather than spot congestion. Cristina et al. [13] solved for the shortest route through a specified number of spots on a road network using a backtracking method. Their method calculates a route that avoided congestion by setting the congestion level and eliminating the most congested spots.

Geng et al. [4] proposed a reinforcement method for training agents on a road network without prior information. The optimization target was the shortest travel time. By employing deep reinforcement learning, the training time became approximately half that of conventional shortest-path algorithms, even for unknown road networks. Kong et al. [9] proposed a method for generating sightseeing routes that are not overcrowded by tourists, using a multi-agent reinforcement learning approach.

Many studies have proposed methods for solving optimization problems using heuristic approaches. In addition, deep learning and reinforcement learning methods have emerged in recent years. For example, Gama et al. [2] utilized a pointer network and the REINFORCE algorithm to facilitate computation in complex environments, which previous research found to be difficult. In contrast, this study aims to reproduce a complex environment by introducing dynamic factors that change with the time of day and to perform route planning.

3 Preliminaries

The method proposed by Gama et al. [2] does not consider dynamic factors such as crowding, although it facilitates the introduction of reinforcement learning. It tackles the problems in existing research and complicates computations when large-scale searches are involved. However, their method also refers to fixed information, such as travel time depending on the coordinates and the static reward for visiting a spot. As mentioned before, such information should be dynamic. In the real world, the transit time depends on the geographic coordinates and means of transportation. The quality of the tourist experience when visiting a spot may vary depending on the time, season, and crowds [14,17].

Among the types of dynamic information, congestion is considered to have the greatest impact on visiting time. The extra time spent waiting in line during a sightseeing trip often places pressure on the itinerary. In some cases, this can lead to changes in the itinerary, forcing visitors to give up on spots they had hoped to visit. Failure to visit a spot reduces the quality of the travel experience. Therefore, it is necessary to generate routes that consider congestion. To this

483 T. Maekawa et al.

end, in this study, we propose two concepts for tour trip generation that consider congestion. The first is the "dynamic stay duration", which changes the time spent at a spot. The other is the "environmental tax metaphor", which provides additional rewards according to the relative degree of congestion.

Here, we explain the terms used in this paper.

Spot P Generally, this term refers to tourist and scenic spots. For example, there are 91 spots in and around Kyoto City [17].

Instance This denotes tourist queries and information regarding tourist spots that exist in a certain tourist city.

Query A tuple of a tourist's starting point, destination, departure time, and time budget, which are the inputs for route generation in the Gama method and the proposed method.

Time Budget T The time a tourist plans to spend from the start to the end of the route.

Occupancy Rate C The ratio of people staying in a spot to its capacity. The higher the ratio, the more people are concentrated in the spot.

Relative Congestion Level Z_c This indicates how crowded a spot is compared to other spots in the instance, and is calculated based on the occupancy rate. This value is a standardized measure of a spot's occupancy rate, assuming that the occupancy rate of the spots in an instance follows a normal distribution. That is, assuming that the spot filling rate C follows $N(\mu, \sigma^2)$, the relative congestion Z_c is $Z_c = \frac{C-\mu}{\sigma}$.

Reward This study uses three rewards corresponding to spots: visiting rewards, environmental taxes, and subsidies.

Visiting Reward R_v Positive reward for visiting a spot. Conventional methods typically use popularity as the visiting reward. Our study uses the scenery score [17] as the visiting reward.

Environmental Tax R_p Negative reward based on relative congestion.

Subsidy R_b Positive reward based on relative congestion.

Route S A chronological sequence of spots to be visited is an output element in the Gama and proposed methods.

Value V The value of a route is the sum of all the visiting rewards, environmental taxes, and subsidies of the spots included in the route.

Stay Duration D This value indicates how long a tourist stays at a spot. The standard stay duration D corresponds to the case when a spot is deserted. The dynamic stay duration d assumes that the duration varies depending on the spot's occupancy rate.

Transit Time $m_{P_i P_j}$ Travel time from spot P_i to P_j.

4 Methodology

We propose two concepts, "dynamic stay duration" and the "environmental tax metaphor", and introduce them into Gama's method [2] to generate tour routes by considering crowdedness. The proposed method can be represented as an optimization problem as in Eq. (1).

$$\max \mathcal{V}_{route} = \sum_{P_i \in route} (1-\alpha)R_v(P_i, t) + \alpha \text{ penalty}(P_i, t)$$

$$\text{subject to} \sum_{P_i, P_j \in route} d_{P_i}(t) + m_{P_i P_j} \leq T \tag{1}$$

$$\alpha \in [0, 1]$$

where, α is a hyper-parameter. t denotes the time a tourist visits the spot P_i. $R_v(P_i, t)$ denotes the dynamic visiting reward of P_i at time t. $\text{penalty}(P_i, t)$ denotes the environmental tax (and subsidy) for visiting P_i at t. $d_{P_i}(t)$ denotes the dynamic stay duration of P_i at t.

4.1 Dynamic Stay Duration

Dynamic stay duration represents a mechanism in which the time spent at a spot varies depending on congestion. In conventional tour route planning that does not consider congestion, the time spent at a spot is often static and fixed. However, in the real world, crowding occurs due to the presence of other tourists. The delay caused by congestion can result in spending more time than expected at a sightseeing spot.

Therefore, it is assumed that the time spent at a spot is extended because of congestion, which is considered as a factor in route generation. We assume that the congestion here is related to the spot's own congestion, that is, the occupancy rate. The following assumptions are made regarding the relationship between the occupancy rate and the time spent in a spot. First, two thresholds L, H are set for the occupancy rate C. Let $0 \leq L < H \leq 100$.

- If $0 \leq C < L$, we call this situation low-degree congestion and assume that no congestion-related stay time is incurred.
- If $L \leq C < H$, we call this moderate congestion, and the congestion-induced stay time increases linearly with the number of people staying.
- If $H \leq C$, we call this high-degree congestion, and the stay time due to congestion is assigned the maximum value because the spot is very crowded.

In this study, the stay durations during low-degree congestion are used as the base durations, and the stay durations during high-degree congestion are set to twice the base durations. The relationship between the occupancy rate and the dynamic stay duration is shown in Fig. 2. The stay duration $d_{P_i}(t)$ at time t for spot i can be expressed as Eq. (2):

$$d_{P_i}(t) = \begin{cases} D_{P_i,t} & (C_{P_i,t} \leq L) \\ (1 + \frac{C_{P_i,t}-L}{H-L})D_{P_i,t} & (L \leq C_{P_i,t} \leq H) \\ 2D_{P_i,t} & (H \leq C_{P_i,t}) \end{cases} \tag{2}$$

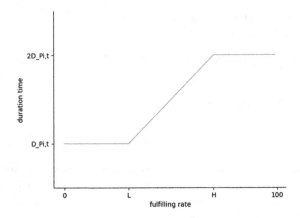

Fig. 2. Assumed Relationship between Occupancy Rate and Stay Duration

4.2   Environmental Tax Metaphor

One way to avoid tourist concentration and reduce pollution is to decentralize tourism [14]. This study aims to contribute to the decentralization of tourism by generating routes that discourage visits to crowded spots. To avoid visiting crowded spots, we introduce a dynamic reward mechanism called the "environmental tax metaphor". Two types of rewards are considered: environmental taxes and subsidies.

Environmental Tax. If a spot is more crowded than other spots in the city, a negative reward is given for visiting that spot. This reward is an environmental tax independent of the visiting reward.

The value of the environmental tax depends on the relative congestion. The environmental tax at time t for spot i is expressed as in Eq. (3). Two threshold values are set: $T_B, T_M (0 \leq T_B < T_M)$. T_B is the allowable relative congestion, and an environmental tax is imposed if it exceeds T_B. The environmental tax will be the minimum value if the relative congestion is above T_M. The minimum value is the negative value of the average visiting rewards $(\overline{R_v(P_*, t)})$ of all the spots at time t .

$$\text{penalty}(P_i, t) = -\min(1, \frac{Z_{cP_i,t} - T_B}{T_M - T_B}) * \overline{R_v(P_*, t)} \tag{3}$$

Subsidy. A possible counterpart to an environmental tax would be a subsidy to direct tourists to uncrowded places. The environmental tax penalizes visits to crowded places, whereas the subsidy rewards visits to uncrowded places.

Similar to the environmental tax, the subsidy depends on relative congestion to determine its value. The subsidy at time t for spot i is expressed as in Eq. (4). Two threshold values are set: $T_B, T_M (T_M < T_B \leq 0)$. T_B is the allowable relative congestion. A bonus subsidy will be given when relative congestion is below T_B. When relative congestion is below T_M, the maximum value of the subsidy will be given. The maximum value is the mean value of the visiting rewards of all the spots at time t.

$$\text{bonus}(P_i, t) = \min(1, \frac{Z_{cP_i,t} - T_B}{T_M - T_B}) * \overline{R_v(P_*, t)} \tag{4}$$

4.3   Dataset

We used the dataset of Kyoto Sightseeing Map2.0 [17] and trajectory data provided by Yahoo Japan Corporation [19] to prepare the following data used in our work:

- Visiting reward matrix
- Base stay duration
- Population matrix
- Spot business hours
- Transit Time Matrix

The visiting reward matrix and base stay duration were generated using Kyoto Sightseeing Map2.0. This is a dataset of 91 tourist spots in Kyoto extracted from social images posted on Flickr. The data include the name of the spot, its geographic coordinates, and the hourly aesthetic score. The aesthetic score [7] is an estimated beauty score of the photos taken around the spot. This study used the aesthetic score as the visiting reward for a spot. Independent of the spot data, GPS trajectory data from Kyoto Sightseeing Map2.0 were also used. From the coordinates, we determined which spots the visitor stayed at, according to the trajectory data, and calculated the number of minutes the visitor stayed at each spot, assuming that log updates within an hour were continuous. If the time spent was extremely short (20 min or less) or extremely long (80 min or more), values of 40 and 70 min were used, respectively. These values were used as the base stay durations.

The population matrix was constructed using GPS trajectory data from Yahoo Japan Corporation [19]. It includes 6667031 trajectories from 757878 users. The number of visitors per hour was calculated by comparison with the latitude and longitude of each spot. The population matrix was used to calculate the occupancy rates. The occupancy rate is the ratio of the number of people staying at each spot to the capacity of the spot. The capacity was set as follows. For spots that are particularly famous and are expected to be crowded as tourist attractions, the occupancy rate was set such that it reached 100 percent two hours before the maximum number of visitors were present. Therefore, there were periods when the fill rate exceeded 100%. The other spots were set such that the occupancy rate reached 80% when the number of visitors reached its maximum. Using this capacity, we created the occupancy matrix.

The business hours of the spots were obtained using the Place Details API [5] of Google Maps. Spots that could not be obtained using this API were assumed to be open 24 h a day if they were considered to be outdoors, and assumed to be open from 9:00 to 17:00 if they were considered to be indoors. The travel times were calculated using the OpenRouteService matrix API [12]. Because it is impossible to calculate the travel time when using public transportation, the travel time matrix was created by empirically setting the travel time to 2.5 times the travel time by car.

4.4 Training and Inference

The model of the proposed method and its training process were based on those of Gama et al. [2] The input consisted of two types of information: spot information and queries. The former is described in Sect. 4.3. The latter refers to queries provided by the user, consisting of the spot where the trip starts (starting point), the spot where the trip ends (destination), departure time, and time budget. The output is the most valuable route.

We trained our model using reinforcement learning with a revised REINFORCE algorithm [2]. Please note that our reward mechanism differed from the original mechanism used in [2]. The training process is shown in Fig. 3 and Algorithm 1.

The inference generates a route for a given query using our model trained in the target city. We applied a beam search in our inference process to maximize the total route probability using our model. For further details, please refer to [2].

The proposed method was designed to generate routes for a single user. Because the model is trained to maximize the value of the trip, it may return the same route for the same query, and congestion may occur. However, it is theoretically possible to generate routes that contribute to decentralization while maintaining the value of the tourist experience. This goal could be realized by modifying the system to acquire real-time information on the number of visitors and congestion and to learn from these data on a case-by-case basis.

5 Experimental Evaluation

5.1 Parameter Tuning

Tuning was performed to determine the most effective parameters (α in Eq. (1), L, M in Eq. (2), and T_M in Eq. (3)) for further processing. Two variations of the proposed method were compared. In Model A, dynamic stay duration and environmental tax are applied, and in Model B, the subsidy mechanism is additionally applied.

The parameters to be tuned and their candidate values are as follows:

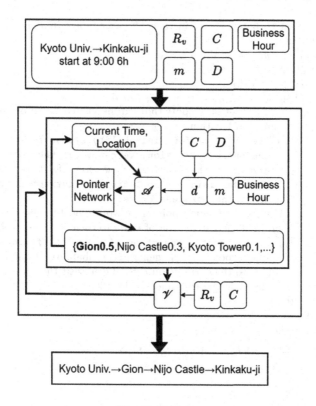

Fig. 3. Training

- $\alpha = 1/3, 1/2, 2/3$ (Eq. (1))
- $L, M = (20, 80), (30, 70), (40, 60)$ (Eq. (2))
- $T_M = 0.5, 1.0, 1.5, T_B = 0$ (Eq. (3))

27 models were compared for three different parameters under three different settings. The models were trained 30,000 times each. The time budget for the queries was 8 h, and the starting point and destination were randomly selected so that they were not the same. Thirty queries were prepared.

Model A with $(L, M) = (30, 70), T_M = 1.5, \alpha = 1/3$ had the lowest average relative congestion level and was used for further comparison. Model B with $(L, M) = (40, 60), T_M = 1.5, \alpha = 2/3$ had the lowest average relative congestion level and was used for further comparison.

5.2 Comparison Experiment

The evaluation experiment compared the proposed method with the original Gama method to determine whether it could generate a route that avoids congestion, contributes to decentralization, and does not impair the tourist experience.

Data: Query set Q, batch size B, instance Φ (set of spots in a city)
Initialize θ
while *Training is not finished* **do**
    for $q \in Q$ **do**
        for $b \in 1, ..., B$ **do**
            while *Destination spot not reached* **do**
                Find the set $\mathcal{A}$ of spots that can be visited based on the current
                time and location.
                /* The criteria for visitability are as follows: From
                   the current location, we can visit the spot and
                   arrive at the destination within the time budget.
                   When we visit the spot from the current location, it
                   is during business hours.   */
                $\mathcal{A}$ is inputted to the pointer network model to generate a
                probability distribution of visits to each spot ($p_\theta(S|\Phi)$).
                // S is a route consisting of a series of spots.
                Sample a spot from the generated probability distribution and
                add it to the current route S_b. Then, visit the spots and
                update the current time and location.
            end
        end
        $V' = \frac{1}{B}\Sigma_{b=1}^{B}V(S_b)$
        $g_\theta = -\frac{1}{B}\Sigma_{b=1}^{B}(V(S_b) - V'\nabla_\theta logp_\theta(S_b|\Phi)$
        Update θ by using g_θ
    end
end

Algorithm 1: Training with REINFORCE Algorithm

We compared models A and B with the Gama model. We modified the original Gama model so that the conditions were the same as those of the proposed model: the starting point and destination could differ, and the training was performed by referring to data from Kyoto City. The difference between the Gama model and the other models is whether the dynamic stay duration is applied. This allowed the Gama model to generate routes with fewer time constraints, resulting in higher visiting rewards. Therefore, we introduced a Gama-D model with dynamic stay duration and included it in the comparison.

The models were trained for 30,000 iterations each. The time budget for the queries was 8 h, and the starting and destination locations were randomly selected such that they were not identical. 30 queries were used.

The experimental results are listed in Table 1. All items in Table 1 were averaged over 30 queries.

Dynamic Stay Duration. The effect of the dynamic stay duration was confirmed by comparing Models Gama and Gama-D. Table 1 shows that Model Gama-D is superior to Gama in all indices of occupancy rate and relative congestion level. Note that the visiting rewards cannot be compared because the number of spots

Table 1. Experimental Results. Gama-D is an extension of the original Gama model with dynamic stay duration. Model A is the proposed method with only the environmental tax. Model B is the proposed method with both the environmental tax and subsidy.

Method	Occupancy Rate	Relative Congestion Level	Value
Gama	0.797	0.0121	*42.478*
Gama-D	0.722	−0.587	**24.126**
A	0.639	−1.197	23.847
B	**0.538**	**−1.900**	22.906

that can be visited within the same time budget varies depending on whether the dynamic stay duration is applied.

Environmental Tax Metaphor. We confirmed the effect of environmental taxes and subsidies by comparing models that use the dynamic stay duration, that is, models A and B. Table 1 shows that Model B is more effective in terms of the average occupancy rate and average relative congestion level, whereas Model Gama is more effective for the route value. The superiority of Model B was demonstrated in terms of the occupancy rate and relative congestion level.

A comparison of the criteria for the occupancy rate shows that Model B recorded superior scores. Model B has a high occupancy rate, almost 20% points lower than that of Model Gama-D. This may be because the introduction of environmental taxes and subsidies led tourists to visit places with low relative congestion. Because the relative congestion is calculated based on the occupancy rate, small relative congestion means that the occupancy rate is relatively low and the route contains many spots that do not meet the criteria for high occupancy. Because of the decrease in the high occupancy rate, the average occupancy rate also decreased.

The relative congestion levels were the smallest for Model B, followed by Model B and then Model Gama. The average relative congestion level of -1.900 in Model B refers to a lower range of approximately 3.8 percent, which means that users can take action to avoid congestion.

For these three models, the route value was highest for Model Gama-D, followed by model A and then model B. This is related to the degrees of freedom in spot selection. Model Gama-D is affected only by the dynamic stay duration and chooses the spot with the highest reward based on the occupancy rate. Model A applies an environmental tax and avoids spots with high relative congestion. Model B applies an environmental tax and a subsidy, so it tends to primarily select spots with low relative congestion. Thus, the more choices a model has, the higher the visiting reward. However, the largest difference is 8 percent between Models Gama-D and B. Therefore, it is unlikely to have a significant impact when the visiting reward is considered to the quality of the travel experience.

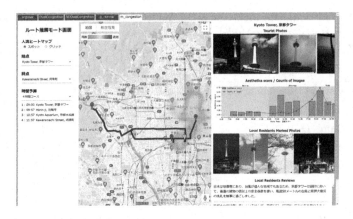

Fig. 4. Example of User Interface for User Study

5.3 User Study

Three subjects were asked to evaluate the routes generated by Models Gama and B.

The routes generated for the queries in Models B and Gama are presented to the user who was asked to answer a questionnaire.

We generated start and destination spots with a large number of tourists, such as (1) Kyoto Tower→ Kawaramachi; (2) Kawaramachi→ Arashiyama; (3) Kawaramachi→ Kinkakuji;

Three time budgets of 4, 6, and 8 h were used. Because there were two models, 18 routes were generated. To distinguish these queries, we denote them as {Method (G for Gama, B for Model B)}{Query number}_{Time budget}. For example, the query for a time budget of 6 h for Kawahara→Arashiyama with Model Gama is G2_6.

An example of the presentation screen for a subject is shown in Fig. 4. The central part of the screen is divided into three parts from left to right: query information, map display, and spot information. The query information section shows the query for the route, spots to be visited on the generated route, and the time of the visit. The map display section displays the generated route on a map, indicating the route taken during the trip. In addition, the number of visitors is based on GPS statistical data from Yahoo JAPAN. The spot information section displays photos of the spot, congestion information, and tourist impressions. This information is based on the Kyoto Sightseeing Map 2.0 data.

The questionnaire items for the participants were as follows. All items were rated on a five-point scale, with 1 indicating no agreement at all (bad, low, do not want to), and 5 indicating strong agreement (good, high, want to). The smaller the value of the degree of congestion, the better, and the larger the values for the other evaluation criteria, the better. The respondents were also asked to provide optional reasons for their evaluations.

1. Do you care how crowded a spot is regarding sightseeing?

Table 2. Average Scores of User Study (Question 4)

	4 h		6 h		8 h	
	Gama Model	B Model	Gama Model	B Model	Gama Model	B Model
Order	4.00	4.44	2.78	3.55	3.22	3.66
Distance	3.89	3.44	3.44	1.33	3.44	1.89
Crowdedness	3.56	2.33	3.56	2.33	3.78	2.67
Satisfaction	3.78	3.11	3.56	2.89	3.78	3.22
Acceptance	3.88	3.22	3.56	2.33	3.78	3.11

2. Do you care about congestion when traveling for sightseeing?
3. Are you aware of overtourism in tourism?
4. Evaluate for each query: order of visit, distance traveled, crowdedness, satisfaction, acceptance (choose the route for your tour).

The respondents tended to be relatively concerned about crowding during sightseeing trips and in crowded spots, with average response values of 4.333. On the other hand, awareness of overtourism was not very high, with an average response value of 2.667. The average responses to question 4 are shown in Table 2. The crowdedness ratings were lower in Model B. The proposed method could provide routes to avoid congestion. Model B was not observed to be superior in terms of travel distance or satisfaction. One possible reason for this is that although the subjects care about crowding during trips, travel distance is more important to them. Therefore, further study on this subject is necessary.

6   Conclusion

To generate tour routes that consider crowdedness, we propose the two novel concepts of "dynamic stay duration" and the "environmental tax metaphor". We realized these concepts by extending the Gama method using a pointer network and the REINFORCE algorithm.

The results of comparative experiments and user studies confirmed the usefulness of these two concepts and the route generation method. We confirmed that the proposed model can generate routes that maintain the value of the tourist experience and avoid congestion. Future work includes the development of a model that can be trained by reading real-time information. Further user studies that consider user preferences are also planned.

References

1. Dezfouli, M.B., Shahraki, M.H.N., Zamani, H.: A novel tour planning model using big data. In: Proceedings of the IDAP 2018, pp. 1–6 (2018)
2. Gama, R., Fernandes, H.L.: A reinforcement learning approach to the orienteering problem with time windows. Comput. Oper. Res. **133**, 105357 (2021)

3. Gavalas, D., Kasapakis, V., Konstantopoulos, C., Pantziou, G., Vathis, N., Zaroliagis, C.: The eCOMPASS multimodal tourist tour planner. Expert Syst. Appl. **42**(21), 7303–7316 (2015)
4. Geng, Y., et al.: Deep reinforcement learning based dynamic route planning for minimizing travel time. In: Proceedings of the ICC Workshops 2021, pp. 1–6 (2021)
5. Google: https://developers.google.com/maps/documentation/javascript/places? hl=ja#place_details
6. Hirano, Y., Suwa, H., Yasumoto, K.: A method for generating multiple tour routes balancing user satisfaction and resource consumption. In: Proceedings of the Intelligent Environments 2019. Ambient Intelligence and Smart Environments, vol. 26, pp. 180–189. IOS Press (2019)
7. Hosu, V., Goldlucke, B., Saupe, D.: Effective aesthetics prediction with multilevel spatially pooled features. In: proceedings of the IEEE/CVF Conference on Computer Vision and Pattern Recognition, pp. 9375–9383 (2019)
8. Isoda, S., Hidaka, M., Matsuda, Y., Suwa, H., Yasumoto, K.: Timeliness-aware on-site planning method for tour navigation. Smart Cities **3**(4), 1383–1404 (2020). https://doi.org/10.3390/smartcities3040066
9. Kong, W.K., Zheng, S., Nguyen, M.L., Ma, Q.: Diversity-oriented route planning for tourists. In: Strauss, C., Cuzzocrea, A., Kotsis, G., Tjoa, A.M., Khalil, I. (eds.) Database and Expert Systems Applications, DEXA 2022. Lecture Notes in Computer Science, vol. 13427, pp. 243–255. Springer, Cham (2022). https://doi.org/10. 1007/978-3-031-12426-6_20
10. Kurata, Y., Shinagawa, Y., Hara, T.: CT-Planner5: a computer-aided tour planning service which profits both tourists and destinations. In: Workshop on Tourism Recommender Systems, RecSys, vol. 15, pp. 35–42 (2015)
11. Muccini, H., Rossi, F., Traini, L.: A smart city run-time planner for multi-site congestion management. In: 2017 International Conference on Smart Systems and Technologies (SST), pp. 175–179 (2017)
12. openrouteservice: Retrieved January 25th (2023). https://openrouteservice.org/ dev/#/api-docs/v2/matrix/profile/post
13. Păcurar, C.M., Albu, R.G., Păcurar, V.D.: Tourist route optimization in the context of COVID-19 pandemic. Sustainability **13**(10), 5492 (2021)
14. Qiang, M.: Tourism informatics - smart tourism toward tourism informatics - : Discovery and recommendation of tourism resources by using user generated content (in Japanese). IPSJ Magaz. **62**(11), e12–e17 (2021)
15. Vinyals, O., Fortunato, M., Jaitly, N.: Pointer networks. In: Cortes, C., Lawrence, N., Lee, D., Sugiyama, M., Garnett, R. (eds.) Advances in Neural Information Processing Systems, vol. 28. Curran Associates, Inc. (2015)
16. Wu, X., Guan, H., Han, Y., Ma, J.: A tour route planning model for tourism experience utility maximization. Adv. Mech. Eng. **9**(10), 1687814017732309 (2017)
17. Xu, J., Sun, J., Li, T., Ma, Q.: Kyoto sightseeing map 2.0 for user-experience oriented tourism. In: Proceedings of the MIPR 2021, pp. 239–242 (2021)
18. Xu, Y., Hu, T., Li, Y.: A travel route recommendation algorithm with personal preference. In: 2016 12th International Conference on Natural Computation, Fuzzy Systems and Knowledge Discovery (ICNC-FSKD), pp. 390–396 (2016)
19. Yahoo Japan Corporation: Yahoo! JAPAN
20. Zhang, Y., Jiao, L., Yu, Z., Lin, Z., Gan, M.: A tourism route-planning approach based on comprehensive attractiveness. IEEE Access **8**, 39536–39547 (2020)
21. Zheng, W., Liao, Z., Qin, J.: Using a four-step heuristic algorithm to design personalized day tour route within a tourist attraction. Tour. Manage. **62**, 335–349 (2017)

A Knowledge-Based Approach to Business Process Analysis: From Informal to Formal

Antonio De Nicola[1]($\boxtimes$) , Anna Formica[2] , Ida Mele[2] , Michele Missikoff[2] ,
and Francesco Taglino[2]

[1] Agenzia Nazionale per Le Nuove Tecnologie, l'Energia e lo Sviluppo Economico
Sostenibile (ENEA), Via Anguillarese 301, 00123 Rome, Italy
antonio.denicola@enea.it
[2] Istituto di Analisi dei Sistemi ed Informatica (IASI) "Antonio Ruberti"
National Research Council, Via dei Taurini 19, 00185 Rome, Italy
{anna.formica,ida.mele,michele.missikoff,
francesco.taglino}@iasi.cnr.it

Abstract. Business Process Analysis (BPA) is a strategic activity, necessary for enterprises to model their business operations. It is a central activity in information system development, but also for business process design and reengineering. Despite several decades of research, the effectiveness of available BPA methods is still questionable. The majority of methodologies adopted by enterprises are rather qualitative and lack a formal basis, often yielding inadequate specifications. On the other hand, there are methodologies with a solid theoretical background, but they appear too cumbersome for the majority of enterprises. This paper proposes a knowledge framework, referred to as BPA Canvas, conceived to be easily mastered by business people and, at the same time, based on a sound formal theory. The methodology starts with the construction of natural language knowledge artifacts and, then, progressively guides the user toward more rigorous structures. The formal approach of the methodology allows us to prove the correctness of the resulting knowledge base while maintaining the centrality of business people in the whole knowledge construction process.

Keywords: Business Process Analysis · Business Model Canvas · Knowledge Representation · Formal Methods

1 Introduction

Business Process Analysis (BPA) [16] is a strategic activity for an enterprise, used for instance for organizational changes, Business Process (BP) reengineering, and information system development. BPA, positioned in the preliminary phase of a software project, represents a fundamental part of the Requirement Engineering task.

Software projects are among the most challenging engineering undertakings. According to the Standish Group's Annual Chaos Report of 2020, based on the

© The Author(s), under exclusive license to Springer Nature Switzerland AG 2023
C. Strauss et al. (Eds.): DEXA 2023, LNCS 14146, pp. 493–507, 2023.
https://doi.org/10.1007/978-3-031-39847-6_39

analysis of 50K projects, 69% of software projects end in a partial or total failure. It is well known that one of the major causes of software project failure is represented by the problem of business/IT misalignment [15], i.e., the fact that the services of the information system do not fully correspond to the business needs. Such a problem is mainly caused by difficulties in the communications between business people and IT specialists, yielding poor requirement specifications [4].

Traditionally, BPA is a territory of business experts, who adopt methodologies that are mainly descriptive, without a rigorous approach for carrying out the analysis and drawing up related documents [1]. The informal nature of the produced documents, often containing imprecise statements or missing information, is one of the primary causes of poor requirement specification and, then, of the failures in the development of enterprise information systems. There are proposals of formally grounded methodologies but they appear too cumbersome and generally are not adopted by business people.

In this paper, we propose an evolution of the knowledge-driven BPA methodology, referred to as *BPA Canvas*, presented in its preliminary version in [9]. The novel contribution of this work is represented by the formalization of the methodology, necessary to check the correctness of the Business Process Knowledge Base (BPKB). Such a formalization has been introduced without losing the user-friendly characteristics of the methodology, aimed at being easily adopted by business people. The BPA Canvas grants business experts a central role in gathering and modeling business process knowledge. To this end, the methodology proposes a progressive construction of the knowledge artifacts based on a visual layout, inspired by the Business Model Canvas [12], organized into eight sections. The sections are conceived in a sequence that evolves from simple, narrative models, to semantically richer ones. An approach that facilitates knowledge management activities for business experts (substantially reducing the role of IT specialists). At the same time, the formal grounding of the methodology guarantees the achievement of a formally founded knowledge base, easy to be inspected, queried, and automatically verified.

The rest of the paper is organized as follows. Section 2 provides a review of the literature in the area of knowledge management for BPA. Section 3 describes the BPA Canvas methodology, then Sect. 4 shows the application of the methodology by means of a running example in a home delivery pizza shop, called *PizzaPazza*. Section 5 presents the formal grounding of BPA Canvas and, finally, in Sect. 6 the conclusions are given.

2   Related Work

The area of BPA is very active, both at the scientific and industrial levels, however, knowledge-based BPA research supported by a solid formal background presents only a few results and none of them tries to conjugate the rigor of a formal approach with ease of use. Here, we briefly review some of the key results in the area, with a focus on knowledge management aimed at BPA [3]. In the quest for a formal method for BPA, the large majority of the literature proposes

ontology-based solutions. We recall COBRA, a Core Ontology for Business pRocess Analysis [11], that is based on a Time Ontology. Another research line,
with a wider scope, is represented by the adoption of ontologies and semantic
web services for BP management, such as Semantic Business Process Management (SBPM) [6]. Such proposals appear to be more inclined towards the formal aspects than the ease of use for business experts. A different research line,
rooted in the business domain, starts from the international business standard
UBL (Universal Business Language) [19].

An interesting proposal [14] is based on the association of a business ontology
to UBL (Universal Business Language), introducing some formal models of the
UBL components and templates, including the UBL process flows. The formal
implementation of the UBL ontology has been achieved by using OWL (the
W3C Web Ontology Language). Probably due to an excess of formalization, the
proposal was not largely adopted by the business community.

Another interesting proposal is represented by the Business Process Modeling
Ontology (BPMO) [17] that besides UBL also considers other business modeling
standards, including ebXML. BPMO has been mainly conceived for interoperability, i.e., to allow the exchange of information among cooperating enterprises,
rather than to support BPA. Then, SemPrAnn is a tool to semantically enrich
BPMN (Business Process Model and Notation) processes [5].

Finally, it is worth to coming back to the mentioned Business Model Canvas
[12] that inspired the BPA Canvas layout. The former addresses a high-level
enterprise space related to business strategies with respect to our proposal that
is focused on business processes. Furthermore, it remains at an informal level
and lacks a systematic approach for modeling activities and business objects.
Along this line, another work to be mentioned is the Business Process Canvas
(BP Canvas) [7] which has a similar scope to ours, since it aims at supporting the
BPA. However, in the mentioned proposal, the only similarity is the adoption of
a canvas layout to analyze business processes. The BP Canvas is not based on a
formal theory and, thus, does not produce a formally grounded knowledge base.
In fact, the gathered domain knowledge is represented by informal descriptions.
Finally, the BP Canvas still requires validation on the field. On the contrary,
our BPA Canvas has been experimented in two real-world cases, in particular
in an SME (a fashion atelier) and a Public Administration department (Italian
Ministry of Economy and Finance), and the feedback is very encouraging.

In conclusion, in the literature, there is a growing awareness of the importance of a solid, systematic, formally grounded knowledge-driven approach to
the BPA and of the need for the coexistence of formal and informal knowledge
management practices [18] [20]. However, the existing proposals have a limited
practical impact, failing in the objective of convincing business experts to adopt
more rigorous and formal business process modeling methods.

3   The Business Process Analysis Canvas

In this section, we present the main ideas of the BPA Canvas and the related
methodology. It includes a set of knowledge artifacts and a procedure aimed at

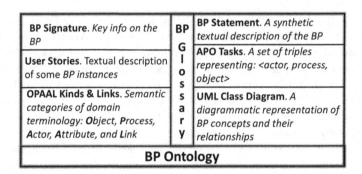

BP Signature. *Key info on the BP*	**BP** **G** **I** **o** **s** **s** **a** **r** **y**	BP Statement. *A synthetic textual description of the BP*
User Stories. Textual description of some *BP instances*		APO Tasks. *A set of triples representing: <actor, process, object>*
OPAAL Kinds & Links. *Semantic categories of domain terminology: Object, Process, Actor, Attribute, and Link*		UML Class Diagram. *A diagrammatic representation of BP concepts and their relationships*
BP Ontology		

Fig. 1. BPA Canvas layout

guiding business experts in collecting and organizing the knowledge of a business process. With respect to the business process modeling methods available in the literature, the BPA Canvas has not the objective of drawing process diagrams, an activity that is postponed to the BP design phase. BPA Canvas is aimed at the careful collection of the knowledge necessary to build a first static model of the business process. The idea is that a rigorous and detailed knowledge base about the BP will substantially support the subsequent design task and improve the quality of the process flow diagrams, improving therefore the quality of the produced information system.

3.1   The BPA Canvas Layout

The BPA Canvas is organized into eight knowledge sections that hold different kinds of knowledge artifacts, i.e., models of the given business process. The models can assume various forms, with different levels of details and formality. In particular, we have (i) plain text, a narrative form of knowledge representation; (ii) structured text, e.g., itemized lists (bullet points) that collect and organize short statements; (iii) tables, typically providing a systematic visualization of knowledge items; (iv) diagrams, where the knowledge is graphically represented, according to a given standard; (v) formal representation of the business domain by means of a BP Ontology.

Figure 1 shows the layout of the eight sections of the BPA Canvas that are listed below.

- **BP Signature.** The first knowledge artifact, in the form of a list, aimed at providing a synthetic profile of the business process.
- **BP Statement.** This is a preliminary plain text description of the business process and its business scenario, described in general terms (i.e., at an intentional level).
- **User Stories.** One or more plain text descriptions of exemplar executions of the BP (i.e., at an extensional level). In essence, it represents one or more instances of the BP Statement.

- **APO Tasks.** This is a set of triples representing a first operational account of the business process, abstracting the actual sequencing of the tasks.
- **BP Glossary.** A collection of terms that characterize the BP domain, together with their descriptions.
- **OPAAL Kinds & Links.** This section is composed of two parts. The first part, *Kinds*, provides a semantic tagging of the terms (concept names) used in the construction of the knowledge artefacts, according to the following categories: *Object, Process, Actor*, and *Attribute*.
  The second part, *Links*, represents semantic relations among concept names, i.e., *ISA* for subsumption relation, *PartOf* for composition relation, and *HasA* to relate an entity with an attribute.
- **UML Class Diagram.** A set of diagrams representing the relationships among the concepts that provide a static view of the BP. The Class Diagram is built by using tasks and links in *APO Tasks* and *OPAAL* sections, respectively.
- **BP Ontology.** It is an encompassing representation of the knowledge collected in the previous sections, encoded in formal terms by using an ontology language (e.g., OWL).

3.2   The BPA Canvas Methodology

The proposed methodology, which will be applied to an example in the next section, suggests starting with the *BP Signature*, and, then, continuing with the *BP Statement* and a number of *User Stories*. The latter represents the instances of the BP, therefore for sake of completeness, we need to report one story for each distinct process execution (i.e., one for each source-sink direct path in the BP graph). These models are built by using plain text descriptions, easily provided by business experts.

The *APO Task* section requires a first linguistic analysis of the two above sections, extracting simple triples in the form: *(subject, verb, direct or indirect object)*. Please note that the tasks are collected in a set, without considering their actual sequencing. Their sequencing, to create the BP diagram, is postponed to the subsequent phase of BP design, not addressed in this paper. This choice is motivated by the progressive approach of the methodology, aimed at lightening the cognitive overload for people not trained in Knowledge Engineering.

Another important section of the BPA Canvas is the *BP Glossary*. It is constructed by collecting all the terms used in the analyzed business domain, together with their descriptions. The Glossary represents a solid reference point for end-users and stakeholders, useful when the picture gets large and complicated.

The *OPAAL Kinds & Links* section requires that the analyst classifies the terminology according to the four anticipated kinds: *Object, Process, Actor*, and *Attribute*. In particular:

i. **Object:** refers to the terms denoting any passive entity with a lifecycle that follows the CRUDA paradigm composed by *Create, Read, Update, Delete* [8], plus *Archive* that is relevant in business processes;

ii. **Process:** terms denoting any form of activity, function, or operation aimed at enacting CRUDA operations on one or more business objects;
iii. **Actor:** terms denoting any active entity involved in one or more processes;
iv. **Attribute:** a property (simple or complex) associated with one or more concepts of the above kinds;

Then, in the second part of the *OPAAL* section, it is necessary to model the structural links among the above terms (i.e., concept names). As anticipated, we consider the *ISA, PartOf, HasA* (i.e., Attribution) relations.

The next BPA Canvas section consists of a *UML Class Diagram*. It is built starting from the triples reported in the *BP Task* and the *Link* part of the *OPAAL* section. In particular, given a task, we use the second element of the triple to represent the label of an arc connecting the first and third elements that name the boxes of the Class Diagram. Then, given a triple in *Links*, the second element labels the arch established between the boxes labeled by the first and the third elements.

In the final step, we have all the knowledge necessary to formalize the whole picture, building the *BP Ontology*.

We presented the eight canvas sections in a sequence, but the methodology adopts the Agile philosophy [13], therefore the sections can be populated iteratively, with a spiral approach, with frequent releases, validation with end-users and stakeholders, and successive improvements.

The rest of the paper illustrates, by means of an example, how the BPA Canvas can be used to build a *BP Knowledge Base (BPKB)*. We know that such an endeavor is a challenging one, requiring time, energy, and constant attention to keep the BPKB aligned with the (ever-changing) business reality. But another key challenge is to keep the BPKB self-consistent, i.e., without contradictions, dangling references, missing or disconnected concepts, etc.

To address the problem of "certifying" a BPKB (with respect to the BPA Canvas) in a systematic way, we need to develop a formal grounding. This is presented in Sect. 5, where we propose a formalization of a BPKB.

In the next section, we present a running example that illustrates a practical use the BPA Canvas methodology.

4   A Running Example

The example illustrates the construction of the BPKB for a home delivery pizza shop, called *PizzaPazza*, achieved following the BPA Canvas methodology. We show how the knowledge artifacts are first built in a step-wise fashion, omitting, for sake of space, the successive refinement cycles.

BP Signature. Table 1 represents the first knowledge artifact of the pizza shop, called BP Signature. This is a structure of seven labeled elements with a meaning clearly explained by the labels.

Table 1. The BP Signature

BP Name	HomeDeliveryPizza
Trigger	OrderArrived
Key Actors	Customer, Cook, Delivery Boy
Key Objects	Order, Dough, Pizza, Delivery Vehicle
Input	Purchase Order
Objective	Cook and deliver pizzas to customers
Output	Pizza Delivered, Customer Happy

BP Statement. The text of the BP Statement is the synthesis of an interview with a (fictitious) pizza shop owner, who describes how a customer order is handled by the shop.

> *My business, PizzaPazza, is a home-delivery pizza shop. The customer fills in the order, by using our Web site, and then submits it to the shop, together with the payment. Making good pizzas requires good quality dough, produced in-house, and careful baking of the pizza. To make clients happy, we need to quickly fulfill the order and the delivery boy needs to know the streets and how to speedily reach the customer's address.*

BP User story. Here the text reports a specific execution of the BP, i.e., it represents an instance of the BP. If necessary, more user stories are reported, to represent various use cases and the corresponding process instances.

> *Mary connects to the PizzaPazza Web site and places her order of two Napoli pizzas, providing also the payment. Upon the arrival of Mary's order at PizzaPazza, John, the cook, puts the order on the worklist. When Mary's turn arrives, John prepares the ordered pizzas, bakes them, and then alerts the delivery boy Ed to come and pick up the pizzas. Thus, Ed collects the pizzas and starts his delivery trip, eventually achieving the delivery to Mary's home.*

The first three knowledge artifacts, *Signature*, *Statement*, and *User story*, represent an important, but informal, starting point easily managed by a business expert. The following BPA Canvas sections are built starting from the textual artifacts, moving toward the semantic analysis of the business scenario.

4.1 Analysis of the BP Statement and User Stories

The analysis starts from the above free-form texts to extract a first structured knowledge artifact, **APO Tasks** (see Table 2). Such an artifact collects the knowledge extracted from the text that concerns who (Actor) is doing what (Process/Actions) yielding what results (Object/Outcome). Then, each task assumes the form of a triple representing an atomic action in the form of *(actor, process, object)*.

In essence, according to linguistic theory, the text is analyzed to extract triples formed by a subject noun phrase, a verb phrase, and a direct or indirect object noun phrase. In the triples, actions are represented using the gerund, which is more readable compared to the more 'technical' stemming form. Furthermore, tasks are represented in active verbal form, therefore if in the text we have a passive form (e.g., the order is issued by the customer), when building the triple we need to turn it into an active form (e.g., customer issuing an order).

Table 2. Some of the BP Tasks

Actor	Action	Object
Customer	Filling	Order
Customer	Submitting	Order
PizzaShop	Receiving	Order
Cook	Preparing	Pizza
Cook	Producing	Dough
Cook	Baking	Pizza
DeliveryBoy	Collecting	Pizza
DeliveryBoy	Delivering	Pizza
Customer	Receiving	Pizza
Customer	Appraising	Service

4.2   OPAAL Kinds and Links

In this step, we start creating a lexicon of terms organized according to four categories: Object, Process, Actor, and Attribute. Then, we introduce a set of triples representing the static relationships (*Links*) among terms, as explained in Sect. 3.2.

Table 3 shows an excerpt of the Lexicon organized according to the OPAAL kinds. Please note that here we do not mean to be complete, the reported structures have mainly an illustrative purpose.

Table 3. The OPAAL Kinds of the BP

Categories	Business terminology
Object	Order, Pizza, Margherita, Dough, Topping, ...
Process	Baking, Submitting, Receiving, Delivering, ...
Actor	PizzaShop, Customer, Cook, DeliveryBoy, ...
Attribute	Price, Quantity, PizzaKind, Address, ...

Then, in Table 4, we report the set of triples representing the structural links of the BP.

Table 4. OPAAL Links of the BP

Structural Links		
Dough	*PartOf*	Pizza
Customer	*HasA*	Address
Margherita	*ISA*	Pizza
...		

Once more, in the Link structure we only represent static relationships. For instance, the operational relation between Customer and Order is reported in the APO Tasks, e.g., (Customer, Submitting, Order).

4.3 Building the Remaining Knowledge Artifacts

The BP Glossary is built starting from the knowledge artifacts that have been produced so far, plus additional terms that the analyst deems necessary. But primarily, it is created by extracting from the text the relevant terminology, i.e., the terms that represent objects, actors, attributes, and activities (processes) characterizing the analyzed business domain. For each term, a short description is provided and, if the case, one or more synonyms. In Table 5, an excerpt from the BP Glossary (the descriptions have been derived from The Free Dictionary[1]) is shown.

Table 5. The BP Glossary

Term	Description
Customer	One who buys goods or services from a store or business
Baking	To cook food with dry heat, especially in an oven
DeliveryBoy	One that performs the act of conveying or delivering
Order	A request made by a customer at a pizza shop for food
Pizza	A baked pie consisting of a shallow bread-like crust covered with toppings
...	...

The two final sections, the *UML Class Diagram* and the *BP Ontology* can be derived from the three central sections of the BPA Canvas. For the sake of space, we will not elaborate on them.

---

[1] https://www.thefreedictionary.com/.

5   A Formal Account of a Business Process Knowledge Base

In this section, we present the formal grounding of the BPA Canvas methodology. Such a formalization aims at guaranteeing the quality of the released knowledge base, avoiding missing information, redundancy, and contradictions. Furthermore, a well-defined BPKB is easier to be inspected, queried, and maintained. This is particularly important since our Agile methodology allows for an iterative construction of the BPKB with frequent, partial releases. At each cycle, the BPKB, which is refined and possibly enriched with the introduction of new knowledge items, needs to be checked to verify its quality.

5.1   Formalising the BPKB Core Components

We first propose a formal account of the BPKB components according to the BPA Canvas methodology. In particular, the formalization focuses on the knowledge base *core*, represented by the *APO Task, Glossary,* and the *OPAAL Kind and Link* sections. Then, we present a number of correctness rules aimed at checking that a BPKB instance has been correctly built[2].

Definition 1. *$\mathcal{BPKB}$. Given a terminology N (i.e., a set of terms), a Business Process Knowledge Base ($\mathcal{BPKB}$) is a complex structure organized according to the layout of the BPA Canvas, where the OPAAL section has been decomposed into two parts: Kind and Link, yielding a 9-tuple defined as follows:*

$$\mathcal{BPKB} = (\mathcal{P}, \mathcal{S}, \mathcal{U}, \mathcal{K}, \mathcal{L}, \mathcal{T}, \mathcal{G}, \mathcal{D}, \mathcal{O})$$

where:

- *$\mathcal{P}$ is the* BP Signature (or Profile)
- *$\mathcal{S}$ is the* BP Statement
- *$\mathcal{U}$ is one or more* User stories
- *$\mathcal{K}$ is a set of pairs representing the categorization of terms* (Kinds)
- *$\mathcal{L}$ is a set of structural* Links
- *$\mathcal{T}$ is the set of triples representing the* Tasks *belonging to the BP*
- *$\mathcal{G}$ is the* BP Glossary *in the form of a set of pairs (conceptName, description)*
- *$\mathcal{D}$ is the* BP UML Diagram
- *$\mathcal{O}$ is the* BP Ontology

The following formalization focuses on the *core* of the $\mathcal{BPKB}$ represented by the four central components, i.e., $\mathcal{K}$, $\mathcal{L}$, $\mathcal{T}$, and $\mathcal{G}$. In the formalization, we omit the first three sections, which are unstructured and expressed in natural language, and the last two sections, the UML Diagram and the Ontology, which can be derived from the core.

---

[2] For sake of precision, each BPKB that is built is an interpretation of the axiomatic definitions that, together with the correctness rules, allows for checking if the BPKB at hand is actually a correct interpretation, i.e., a model of our theory.

Definition 2. Kind. *Given a terminology N, $\mathcal{K}$ is a set of pairs:*

$$\mathcal{K} \subseteq \{(n, k) \mid n \in N, k \in K\}$$

where $K = \{O, P, Ac, At\}$ represents the categories a term can belong to, and:

- *O stands for Object*
- *P stands for Process (or activity)*
- *Ac stands for Actor*
- *At stands for Attribute*

In our running example, for instance, the pairs $(Cook, Ac)$ and $(Pizza, O)$ state that the terms $Cook$ and $Pizza$ represent an $Actor$ and an $Object$, respectively.

Definition 3. Structural Link. *Given a terminology N, $\mathcal{L}$ is a set of triples:*

$$\mathcal{L} \subseteq \{(n_1, r, n_2) \mid n_1, n_2 \in N, r \in R, n_1 \neq n_2\}$$

where $R = \{ISA, PartOf, HasA\}$ defines the structural relations (links) used in the $\mathcal{BPKB}$. A triple (n_1, r, n_2) is in $\mathcal{L}$ if n_1 and n_2 are related according to r.

For example, $(Cook, ISA, Person)$, $(Dough, PartOf, Pizza)$ are triples in $\mathcal{L}$.

Definition 4. Tasks. *This component of the $\mathcal{BPKB}$ represents the tasks of the BP as a set $\mathcal{T}$ of 3-tuple, defined as follows:*

$$\mathcal{T} = \{(ac, p, o) \mid \{(ac, Ac), (p, P), (o, O)\} \subseteq \mathcal{K}, (ac, p) \in Inv, (p, o) \in Ach\}$$

Then, Inv and Ach are two sets of term pairs defined as follows:

$$Inv = \{(ac, p) \mid \{(ac, Ac), (p, P)\} \subseteq \mathcal{K} \text{ and } ac \text{ is involved in } p\}$$

$$Ach = \{(p, o) \mid \{(p, P), (o, O)\} \subseteq \mathcal{K} \text{ and } p \text{ achieves } o\}$$

i.e., Inv contains all the ordered pairs of terms formed by an actor, ac, involved in an activity, p, and Ach includes all the pairs whose first element is an activity, p, achieving or producing the second element that is an object, o.

Note that the set of concept names in $\mathcal{L}$ represents a superset of those involved in $\mathcal{T}$ as there could be pairs of concepts in the ISA relation for which the most general concepts do not necessarily participate in a triple in $\mathcal{T}$. For instance, in our business domain $(Cook, Preparing, Pizza)$ is a possible task that implies $(Person, Preparing, Pizza)$.

Definition 5. BP Glossary. *The glossary $\mathcal{G}$ of the $\mathcal{BPKB}$ is a set of ordered pairs defined as follows:*

$$\mathcal{G} = \{(n,d) \mid n \in N, d \in D\}$$

where D is the set of all possible strings, standing for natural language descriptions.

In our running example, the pair (*Pizza, "Italian open pie made of thin bread dough spread with a spiced mixture of e.g. tomato sauce and cheese"*) is a possible element belonging to the glossary.

Definition 6. Correct concept. *Given a $\mathcal{BPKB}$, a well-formed concept c associated with it is a 5-tuple defined as follows:*

$$c = (n, d, k, \mathcal{T}_n, \mathcal{L}_n)$$

where:

- $n \in N$ *represents the concept name*
- d *is a textual description such that the pair $(n,d) \in \mathcal{G}$*
- k *is a kind such that the pair $(n,k) \in \mathcal{K}$*
- $\mathcal{T}_n = \{(ac,p,o) \mid (ac,p,o) \in \mathcal{T}, ac = n \text{ or } p = n \text{ or } o = n\}$ *is the set of tasks in which n participates*
- $\mathcal{L}_n = \{(n_1, r, n_2) \mid (n_1, r, n_2) \in \mathcal{L}, n_1 = n \text{ or } n_2 = n\}$, *that is the set of Structural Links involving n.*

and the concept c is also correct (under Closed World Assumption (CWA) [2]) if it satisfies the rules in Sect. 5.2.

Although in this paper we do not elaborate on the *UML Class Diagram* and the *Ontology* details, we anticipate that the *UML Class Diagram* can be built starting from the *APO Tasks* and the structural *Links*. In particular, the built *UML Class Diagram* will consist of boxes (i.e., classes), named with *object* or *actor* names, connected by two types of arcs: functional and structural. The functional arcs (i.e., associations) will be labeled with *process* names connecting the actors with the objects, as reported in the *APO Tasks* triples. The structural arcs will be created from the triples in the structural *Links* where the label of the arc is the second element. For the *ISA* and *PartOf* relations, the arc will connect two boxes labeled with the first and third elements. In the case of the *HasA* relation, the first element will be a box name, and the third element one of its attributes that will be listed within the box (according to the *UML Class Diagram* syntax).

At this point, the *Ontology* can be derived from the knowledge so far collected. Note that the construction of the knowledge base does not follow a 'waterfall' approach, but the Agile philosophy [13]. Therefore, its construction is achieved in a spiral fashion, and, at each cycle, it is possible to check and correct it, while enriching the overall content.

Definition 7. Correct $\mathcal{BPKB}$. *A $\mathcal{BPKB}$ is correct if all the concepts associated with it are correct.*

Definition 8. Enterprise Knowledge Base. *Consider a set $\mathcal{BPKB}_i$, $i = 1, ..., k$, where k is the total number of Business Processes of an enterprise. The Enterprise Knowledge Base $\mathcal{EKB}$ is defined as:*

$$\mathcal{EKB} \supseteq \bigsqcup_i \mathcal{BPKB}_i$$

where $\bigsqcup_i$ stands for a merge operation.

We expect that in an enterprise there are several BP, each having its own $\mathcal{BPKB}$ that needs to be merged to form the $\mathcal{EKB}$. We are aware that such a merging is a complex endeavor, and it falls outside the scope of the paper.

5.2   Consistency Rules for Concept Correctness

In this section, we present the consistency rules that need to be satisfied for the correctness of the business concepts, and consequently of the $\mathcal{BPKB}$.

Given a concept:

$$c = (n, d, k, \mathcal{T}_n, \mathcal{L}_n)$$

associated with a $\mathcal{BPKB}$, the following rules must hold:

R1 – Definedness. The concept name n needs to have a description in $\mathcal{G}$.

R2 – Uniqueness. The concept name n needs to be present only once in $\mathcal{G}$.

R3 – Categorization. The concept name n needs to be categorized according to the set of categories $K = \{O, P, Ac, At\}$.

R4 – Disjointness. The concept name n needs to be associated with only one kind in $\mathcal{K}$.

R5 – Structural completeness. All the concept names in G need to participate in at least one triple in $\mathcal{L}$.

R6 – Functional completeness. All the actor, object, and process names need to participate in at least one task, i.e., a triple in $\mathcal{T}$. If a concept does not appear in a task, at least one of its subsumees or components or attributes (as declared in $\mathcal{L}$) needs to participate.

R7 – Pragmatics. For all triples in $\mathcal{T}$, the concept names need to belong to their respective categories, i.e., ac in the first place, p in the second place, and o in the third place.

The correctness rules have been informally presented with descriptive texts, for the sake of space, the formal account is omitted.

6   Conclusions and Discussion

In this paper, we presented the BPA Canvas, a methodology for the acquisition, modeling, and management of business process knowledge. The proposed methodology has been conceived to be easily adopted by business people being, at the same time, based on a solid formal grounding. The knowledge organization is guided by a canvas layout, structured in eight sections representing a sort of knowledge dashboard and providing a synoptic view of the BPKB. As seen, concerning previous proposals in the area of BPA, this methodology presents three key characteristics: (i) it starts with informal, intuitive models to grant business experts a central role; (ii) it adopts an Agile approach, with a cyclic progression of model building, with continuous releases and validity checks; (iii) it is characterized by a theoretical foundation for the *core* of the BPKB that represents its backbone.

Currently, we are working on a platform that, based on the formal part of the methodology, supports the knowledge acquisition task and checks the consistency as well as the completeness of the BPKB (under the CWA). In the most popular BPA methodologies, such properties need to be achieved manually. Then, our work will continue along two main lines. The first consists of the development of several services to support the BPKB construction. We will start with NLP services that analyze the first three canvas sections (BP Signature, Statement, and User Stories) to start populating the core of the BPKB. Then, we will provide semantic services aimed at enriching the BPKB by exploring existing terminological resources, such as DBpedia, Wikidata, and WordNet, available on the Internet.

The work presented in this paper is the continuation of the work carried out in the context of the European Project BIVEE (Business Innovation in Virtual Enterprise Environment) where a first proposal of knowledge-based enterprise analysis has been proposed in [10].

Acknowledgement. We gratefully acknowledge the partial support of the PNRR MUR project PE0000013-FAIR.

References

1. Aguilar-Savén, R.S.: Business process modelling: review and framework. Int. J. Prod. Econ. **90**(2), 129–149 (2004). https://doi.org/10.1016/S0925-5273(03)00102-6, production Planning and Control
2. Álvez, J., Gonzalez-Dios, I., Rigau, G.: Applying the closed world assumption to SUMO-based FOL ontologies for effective commonsense reasoning. In: ECAI 2020. Frontiers in Artificial Intelligence and Applications, vol. 325, pp. 585–592. IOS Press (2020). https://doi.org/10.3233/FAIA200142
3. Andersson, B., et al.: Towards a reference ontology for business models. In: Embley, D.W., Olivé, A., Ram, S. (eds.) ER 2006. LNCS, vol. 4215, pp. 482–496. Springer, Heidelberg (2006). https://doi.org/10.1007/11901181_36

4. Aversano, L., Grasso, C., Tortorella, M.: A Literature review of business/IT alignment strategies. Procedia Technol. **5**, 462–474 (2012). https://doi.org/10.1016/j.protcy.2012.09.051, 4th Conference of ENTERprise Information Systems - aligning technology, organizations and people (CENTERIS 2012)
5. Di Martino, B., Colucci Cante, L., Esposito, A., Graziano, M.: A tool for the semantic annotation, validation and optimization of business process models. Softw. Pract. Exp. (2023). https://doi.org/10.1002/spe.3184
6. Hepp, M., Leymann, F., Domingue, J., Wahler, A., Fensel, D.: Semantic business process management: a vision towards using semantic web services for business process management, vol. 2005, pp. 535–540 (2005). https://doi.org/10.1109/ICEBE.2005.110
7. Koutsopoulos, G., Bider, I.: Business process canvas as a process model in a nutshell. In: Gulden, J., Reinhartz-Berger, I., Schmidt, R., Guerreiro, S., Guédria, W., Bera, P. (eds.) BPMDS/EMMSAD -2018. LNBIP, vol. 318, pp. 49–63. Springer, Cham (2018). https://doi.org/10.1007/978-3-319-91704-7_4
8. Martin, J.: Managing the Data Base Environment, 1st edn. Prentice Hall PTR, USA (1983)
9. Missikoff, M.: A knowledge-driven business process analysis methodology. In: Strauss, C., Cuzzocrea, A., Kotsis, G., Tjoa, A.M., Khalil, I. (eds.) Database and Expert Systems Applications. DEXA 2022. LNCS, vol. 13427, pp. 62–67. Springer, Cham (2022). https://doi.org/10.1007/978-3-031-12426-6_5
10. Missikoff, M., Assogna, P.: The BIVEE Project: An Overview of Methodology and Tools (2015). https://doi.org/10.1002/9781119145622.ch3
11. Pedrinaci, C., Domingue, J., Alves de Medeiros, A.K.: A core ontology for business process analysis. In: Bechhofer, S., Hauswirth, M., Hoffmann, J., Koubarakis, M. (eds.) ESWC 2008. LNCS, vol. 5021, pp. 49–64. Springer, Heidelberg (2008). https://doi.org/10.1007/978-3-540-68234-9_7
12. Pigneur, Y., Osterwalder, A.: Business Model Generation: A Handbook for Visionaries, Game Changers and Challengers. John Wiley and Sons, Hoboken, New Jersey (2010)
13. Prakash, R., Agarwal, N.: Managing business analysis for agile development. Int. J. Mod. Eng. Res. (IJMER) **3**(3), 1393–1395 (2013)
14. Roy, S., Sawant, K.P., Ghose, A.K.: Ontology modeling of UBL process diagrams using owl, pp. 535–540 (2010). https://doi.org/10.1109/CISIM.2010.5643509
15. van Grembergen, W., De Haes, S.: A Research Journey into Enterprise Governance of IT, Business/IT Alignment and Value Creation, pp. 1–13, January 2010. https://doi.org/10.4018/jitbag.2010120401
16. Vergidis, K., Tiwari, A., Maieed, B.: Business process analysis and optimization: beyond reengineering. IEEE Trans. Syst. Man Cybern. Part C Appl. Rev. **38**(1), 69–82 (2008). https://doi.org/10.1109/TSMCC.2007.905812
17. Von Rosing, M., Laurier, W.P., Polovina, S.M.: The BPM ontology, vol. 1 (2015). https://doi.org/10.1016/B978-0-12-799959-3.00007-0
18. Värk, A., Reino, A.: Practice ecology of knowledge management-connecting the formal, informal and personal. J. Doc. **77**(1), 163–180 (2021). https://doi.org/10.1108/JD-03-2020-0043
19. W3C Recommendation: Universal business language v2.0. Technical report, W3C (2006). http://docs.oasis-open.org/ubl/cs-UBL-2.0/UBL-2.0.htm
20. Wen, P., Wang, R.: Does knowledge structure matter? Key factors influencing formal and informal knowledge sharing in manufacturing. J. Knowl. Manag. **26**(9), 2275–2305 (2022). https://doi.org/10.1108/JKM-06-2021-0478

Evaluating Prompt-Based Question Answering for Object Prediction in the Open Research Knowledge Graph

Jennifer D'Souza[1]([✉]) [ID], Moussab Hrou[2], and Sören Auer[1,3] [ID]

[1] TIB Leibniz Information Centre for Science and Technology, Hannover, Germany
{jennifer.dsouza,auer}@tib.eu
[2] Gottfried Wilhelm Leibniz Universität Hannover, Hannover, Germany
[3] L3S Research Center, Leibniz University of Hannover, Hannover, Germany

Abstract. Recent investigations have explored prompt-based training of transformer language models for new text genres in low-resource settings. This approach has proven effective in transferring pre-trained or fine-tuned models to resource-scarce environments. This work presents the first results on applying prompt-based training to transformers for *scholarly knowledge graph object prediction*. Methodologically, it stands out in two main ways: 1) it deviates from previous studies that propose entity and relation extraction pipelines, and 2) it tests the method in a significantly different domain, scholarly knowledge, evaluating linguistic, probabilistic, and factual generalizability of large-scale transformer models. Our findings demonstrate that: i) out-of-the-box transformer models underperform on the new scholarly domain, ii) prompt-based training improves performance by up to 40% in relaxed evaluation, and iii) tests of the models in a distinct domain reveals a gap in capturing domain knowledge, highlighting the need for increased attention and resources in the scholarly domain for transformer models.

Keywords: Question Answering · Prompt-based Question Answering · Natural Language Processing · Knowledge Graph Completion · Open Research Knowledge Graph

1 Introduction

The seminal work by Petroni et al. [7] introduced testing transformer-based [13] language models for their vasts store of linguistic and factual knowledge with explicit relational cloze objectives [12] for extracting new facts from them for knowledge base (KB) population. KBs are effective solutions for accessing relational data such as (Michael Jordan, born-in, x). The traditional method for populating such KBs with additional facts would otherwise leverage complex NLP

Supported by TIB Leibniz Information Centre for Science and Technology, the EU H2020 ERC project ScienceGraph (GA ID: 819536) and the BMBF project SCINEXT (GA ID: 01lS22070).

C. Strauss et al. (Eds.): DEXA 2023, LNCS 14146, pp. 508–515, 2023.
https://doi.org/10.1007/978-3-031-39847-6_40

pipelines involving entity extraction, co-reference resolution, entity linking, and relation extraction components [11] that are known to be plagued by the error propagation problem from earlier to later components in the pipeline. Instead, the powerful transformer language models as rich stores of linguistic and factual information having been pre-trained on billion-word corpus from encyclopedic sources were probed for additional facts, showing to outperform the traditional NLP pipeline method for generating relational knowledge to populate KBs as a downstream task [7].

Pretraining language models on large-scale corpora provides a task-agnostic foundation, which can be enhanced through fine-tuning with task-specific objectives for better performance in downstream tasks. To probe transformer language models for relational facts, the original models' knowledge is accessed by conditioning on their latent context representations. For the Question Answering (QA) task, the approach involves initializing a task-agnostic model with the pretrained models' weights, followed by fine-tuning to create a QA task-specific model using instances from, say, the Stanford Question Answering Dataset (SQuAD) [8,9], which this work follows. This work specifically focuses on fine-tuning language models for the QA task over *scholarly knowledge* by aiming to effectively transfer the models' learned structural representation from SQuAD QA. Next, the *why* and *how* of our work are introduced.

Why Focus on the Scholarly Domain? In the face of rapid publication rates at an alarming rate of millions of articles per year [4], researchers are immensely challenged in keeping up with the latest findings in scholarly publications. The Open Research Knowledge Graph(ORKG) [1] was created to help researchers overcome this challenge. By leveraging semantic scholarly knowledge publishing tools, ORKG offers smart information access methods such as Comparisons, Visualizations, and Benchmarks. This enables researchers to comprehend knowledge in a matter of minutes instead of months or days. The structured science-wide scholarly contributions within ORKG can serve as a valuable resource for discovering new facts using transformer language models, potentially leading to an NLP service that assists with scholarly knowledge curation and completion. Inspired by Petroni et al's fact probing method [7], this service could utilize fine-tuned versions of language models to discover additional objects for new relations. Table 1 shows some example instances of the proposed task. The task involves extracting objects as answers from scientific paper abstracts using a question-answer format, similar to the SQuAD QA task. This approach is chosen for its intuitive transferability to the scholarly domain and the potential to leverage state-of-the-art language models fine-tuned on the QA task. Our study examines the capacity of linguistically-rich, fine-tuned SQuAD QA models to transfer to the scholarly domain, which has been previously unexplored.

How to Obtain an Optimal Model for a New Domain? To pre-train or fine-tune a language model for QA on a new domain, the traditional method is to use expensive human-labeled data. Instead, inspired from prior work [3,10,16], this study uses two strategies: 1) template-based unsupervised generation of structured data similar to SQuAD QA data from the ORKG KB, and 2) structural

Table 1. Example instances depicting our Cloze-style adaptation of SQuAD Question Answering task as an extractive objective from the context given a question prompt. The "Context" is a paper abstract, the "Cloze task" shows the four question variants created by appending a "Wh" question word given an input predicate as: no label, What, Which, and How questions. Text in blue show the answer extraction object from the context; text in red show the appended prompt elements including "Wh" question words and the "?" symbol.

Context	Cloze task	Answer
... In the following process oriented knowledge management as it was defined in the EU-project PROMOTE (IST-1999-11658) is presented and the KM-Service approach to realise process oriented ...	Approach name? _ _ _ What approach name? _ _ _ Which approach name? _ _ _ How approach name? _ _ _	PROMOTE
... to investigate processes of community assembly contributing to biotic resistance to an introduced lineage of Phragmites australis, a model invasive species in North America. ...	Continent? _ _ _ What continent? _ _ _ Which continent? _ _ _ How continent? _ _ _	North America
Solid lipid nanoparticles (SLNs) are nanocarriers developed as substitute colloidal drug delivery systems parallel to liposomes, lipid emulsions, polymeric nanoparticles, and so forth. ...	Type of nanocarrier? _ _ _ What type of nanocarrier? _ _ _ Which type of nanocarrier? _ _ _ How type of nanocarrier? _ _ _	Solid lipid nanoparticles

prompt-based learning over state-of-the-art SQuAD-specific fine-tuned transformer models for the scholarly domain. Similar structured QA data prompts knowledge generalization or optimal knowledge stimulation in fine-tuned language models, optimizing training in a resource-scarce setting.

Summarily, our contributions are: 1) we introduce a template-based unsupervised question generation framework for the scholarly domain, similar in format to the SQuAD dataset; 2) we report, for the first time, a detailed empirical analysis of the scholarly domain object prediction task using prompt-based QA task and state-of-the-art transformer models as rich stores of linguistic, probabilistic, and factual parameters, thereby testing the transferability of these pre-trained models on a novel domain. Our dataset, code, and models are publicly released.

2   Task Definition

The article introduces a new task of object extraction, for RDF knowledge graph statements, which is structurally formulated based on the SQuAD QA task dataset [8,9]. The task is an extractive task in a machine reading comprehension setting, where the model is expected to extract the answer as a contiguous span to a given question. Our task is: given an ORKG *predicate* formulated as a question using an unsupervised template-based generation prompting function $f_{prompt}(x)$, to extract the *object* answer from the corresponding paper *Abstract* context. For unsupervised question generation as $f_{prompt}(x)$, a static template pattern is defined. Inspired by the prior seminal work on this theme [3], the $f_{prompt}(x)$ template is: "Wh"+predicate+"?", where "Wh" ∈ {What, Which, How} as the most common question types. Each "Wh" question results in a question-type-specific homogeneous dataset where the same prompt template considering the specific question-type word was applied to all predicates. Finally, a note on the expected object answer granularity considered in this task: They include three different granularities: tokens, span as a short multi-token phrase, and sentences.

3   PROMPT-ORKG: Our Scholarly Knowledge Question Answering Corpus

The ORKG semantic web model uses contribution triples to describe papers in a structured RDF format. We adapt the model for *scholarly knowledge object prediction*, employing extractive QA structured per SQuAD. By querying specific portions of the ORKG KB using the ORKG Rest API, we extract triples related to the Contribution node. Our challenge is to gather (context, question, answer) data instances. Initially, we create a preliminary dataset by selecting predicates as question candidates and corresponding object nodes as answer units. In the preliminary raw dataset, displayed in the 'Before' column of Table 2, the number of contributions exceeds the number of papers since a paper can have more than one contribution; and the (predicate, object) pairs are sparsely distributed across more than 200 research fields in the ORKG taxonomy). Four dataset variants are generated based on the predicate: 1) three variants using the most common "Wh" question words (What, Which, and How) generated unsupervisedly with the template "Wh+predicate+?," and 2) one variant using the cloze-style question template "predicate+?," replacing the [MASK] token with the "?" symbol. We define our experimental settings based on these four variants, and also evaluate an "unchanged" dataset where the predicate remains as the question unit without any modifications, allowing us to demonstrate the effectiveness of the SQuAD-format prompt-based QA setting. The context in each data instance comprises paper Abstracts. However, not all papers in the ORKG get abstracts from Crossref or SemanticScholar. Specifically, we were able to obtain abstracts for 5,486 (58,5%) out of the 9,379 papers in the raw data. This results in pruning all the rows in our triples dataset whose abstracts could not be fetched. We further narrow down the selection for the answer unit by choosing object candidates that can be found in the paper's abstract, resulting in a reduction from 116,421 to 14,499 (predicate, object) pairs in the raw dataset. Finally, after deduplication and heuristics-based filtering of unsuitable object candidates, our dataset comprised 5,909 (predicate, object) pairs coupled with a context (see row 3 in Table 2) which we name the PROMPT-ORKG corpus.

4   Models

With the PROMPT-ORKG dataset for the scholarly domain in place, we relegated attention to selecting three optimal transformer model variants to test as machine learners on our newly introduced, previously unexplored problem setting. *The machine learning test specifically sought out empirical evidence for the transferability of the probabilistic parameters of the existing large-scale transformer models.* In this respect, we were interested in two main strengths of the

Table 2. Our scholarly knowledge question answering for object prediction task dataset statistics on the raw data ('Before' column) and after data cleaning ('After' column).

Statistic parameter	Before	After
number of unique papers	9,379	2,710
number of unique contributions	14,499	3,059
number of (predicate,object) pairs	116,421	5,909
number of unique predicate labels	3,436	853
number of unique object labels	38,234	3,524
avg. number of tokens per predicate label	–	2.01
avg. number of tokens per object label	–	2.43
number of unique abstracts	–	2649
avg. number of tokens per paper abstract	–	196.97
number of abstracts with more than 510 tokens	–	37
number of unique abstracts with more than 510 tokens	–	14

transformer models: to query over an open class of relations [7] and the ability to train them on structurally similar data [3,10]. Thus selecting optimal SQuAD transformer model variants were a natural choice since the structural patterns in the PROMPT-ORKG dataset emulate SQuAD.

The selected language model variants are based on BERT [2] which seminally introduced the "masked language model" (MLM) cloze-based pre-training objective based on a bidirectional self-attention architecture [13]. They are: BERT pretrained, SQuAD2.0 finetuned (deepset/bert-base-cased-squad2); RoBERTa pretrained [6], SQuAD2.0 finetuned (deepset/roberta-base-squad2); and MiniLM pretraining distillation [14], SQuAD2.0 finetuned (deepset/minilm-uncased-squad2).

5   Results

5.1   Experimental Setup

Hyper-parameter Tuning. All models are tuned for 4 epochs, learning rate $\in \{0.0001, 0.00005\}$, train batch size 8, eval batch size 8, and weight decay 0.01. Only the best model is saved and used in the evaluation phase.

Metrics. Evaluations are considered in two main settings: 1. strict, i.e. exact match; and 2. relaxed, i.e. containment match where the gold answer is checked to be contained in the predicted answer. In both settings, the main metric is F1 score and secondarily accuracy is also applied. Note that after prediction, the answers undergo minimal post-processing to be suitable for evaluation. This entailed trimming the trailing and leading white spaces, converting all answers to lower case, and removing the following special characters: ., comma, ;, :, -,), (, _ and +, if they are at the end of the answers.

Table 3. F1-score (and parenthesized Accuracy) results in the exact-match setting over the 4 dataset variants from the 3 models with cell values as "vanilla models"/"after prompt-based training."

Dataset variant	bert-base-cased -squad2	roberta-base -squad2	minilm-uncased -squad2	*row avg*
unchanged	0.5/11.2 (1.0/29.4)	2.5/22.7 (1.8/35.5)	0.5/16.2 (0.4/35.5)	1.2/16.7 (1.1/33.5)
none	1.4/23.5 (1.4/31.7)	4.0/23.0 (2.5/37.5)	3.6/18.0 (3.5/35.4)	1.8/21.5 (2.5/34.9)
what	1.5/25.7 (1.0/33.8)	5.1/21.2 (4.3/35.9)	6.3/19.8 (5.1/36.1)	4.3/22.2 (3.5/35.3)
how	0.3/19.8 (0.5/34.0)	3.0/20.9 (2.2/36.4)	4.6/22.3 (4.0/33.8)	2.6/21.0 (2.2/34.7)
which	1.9/17.8 (1.6/33.5)	5.5/24.0 (4.5/36.6)	5.9/25.9 (5.5/36.8)	4.4/22.6 (3.9/35.6)
column avg	1.1/19.6 (1.1/32.5)	4.0/22.4 (3.1/36.4)	4.2/20.4 (3.7/35.5)	-

Table 4. F1-score (and parenthesized Accuracy) results in the relaxed setting over the 4 dataset variants from the 3 models with cell values as "vanilla models"/"after prompt-based training."

Dataset variant	bert-base-cased -squad2	roberta-base -squad2	minilm-uncased -squad2	*row avg*
unchanged	6.7/17.1 (9.2/42.9)	7.6/23.7 (5.7/49.7)	18.6/22.9 (18.2/47.5)	11.0/21.2 (11.0/46.7)
none	6.2/31.7 (5.9/43.4)	20.3/32.3 (14.8/50.6)	21.3/26.1 (16.0/48.3)	15.9/30.0 (12.2/47.4)
what	7.3/34.6 (5.6/46.0)	22.8/31.0 (17.0/49.5)	24.4/26.2 (16.5/46.0)	18.2/30.6 (13.0/47.2)
how	6.8/27.6 (8.2/45.2)	23.7/30.4 (16.5/51.2)	22.7/28.5 (16.3/47.0)	17.7/28.8 (13.7/47.8)
which	8.3/24.0 (7.4/43.9)	25.1/35.9 (18.2/48.6)	23.0/36.4 (18.0/47.9)	18.8/32.1 (14.5/46.8)
column avg	7.1/27.0 (7.2/44.3)	19.9/30.7 (14.4/49.9)	22.0/28.0 (17.0/47.3)	-

5.2   Evaluations

In this section, we present the results and discuss observations from each of our 30 total experiments considering 5 dataset variants, 3 optimal SQuAD QA models, and 2 experimental settings i.e. vanilla and after prompt-based learning. The core experimental results are depicted in Tables 3 and 4 in terms of F1-scores, and parenthesized accuracies, in the exact-match vs. relaxed settings, respectively. We discuss the experimental results with respect to two main research questions.

RQ1: What is the Impact of the PROMPT-ORKG SQuAD-format Structural QA Task Formulation on the Transferability of the Large-scale SQuAD Fine-tuned Language Models? This question subsumes two sub-questions. RQ1.1: How do the BERT model performances contrast when queried out-of-the-box, i.e. as vanilla models, versus after being trained on our corpus variants? RQ1.2: Is it effective to adopt the SQuAD-format structural representation to optimally query the large-scale SQuAD fine-tuned language models? For RQ1.1, examining both F1 scores and accuracies, in the exact match setting (Table 3), the results from the trained models on our corpus variants are approximately 25% (and 35%) higher, respectively, than when testing the untrained "vanilla" models on data from a new domain. For RQ1.2, examining the *row avg* exact match F1 scores in Table 3, we see the results from the SQuAD-format dataset variants are 5% or 6% higher than the "unchanged" dataset. Note the SQuAD-format dataset variants report results in the range 21% to 22% F1 scores, while the "unchanged" dataset evaluations report only 16% F1 score. Thus we can conclude that indeed the SQuAD-format task formulation as the four variants in the PROMPT-ORKG corpus is an effective strategy to optimally stimulate the probabilistic parameters of the SQuAD fine-tuned language models towards their transferability on a new domain, i.e. the scholarly domain.

Thus our findings cumulatively are as follows. Training the models on our corpora produce more effective predictors compared to the vanilla models with performances as low as 1% reflecting the domain gap of the language models between the generic domain and the scholarly domain. To obtain optimal versions of the SQuAD fine-tuned language models, formulating the task dataset in SQuAD format is an effective strategy to obtain optimally trained models.

RQ2: Which Dataset Variant Produced the Most Optimal Models? Given that we have 2 evaluation tables, the F1 scores in the exact match setting shown in Table 3 were considered the reference evaluations on which to base conclusions. Examining the *row avg* results in the Table, we observed that the results from the *which* variant of the PROMPT-ORKG corpus was statistically insignificantly better than the *what* variant with 22.6% F1 score versus 22.2% F1 score. This indicates directions for future work to examine the automated generation of the questions by selecting and applying the most suitable question type *which* or *what* for the predicates. Across all four tables, examining the *column avg* results, the RoBERTa SQuAD model proved optimal for transferability to a new domain.

6   Conclusions

The prompt-based learning paradigm [5,15] is increasingly used in NLP for fine-tuning BERT-based QA models on domains with less training data. In this work, we demonstrated the applicability of SQuAD-based prompt format training of BERT models on scholarly data for object prediction in the ORKG. Our experiments showed promising results in terms of domain transferability of the models with the right training strategy, however, the model performances reflect the need for further improvement to be realized as practical solutions.

References

1. Auer, S., et al.: Improving access to scientific literature with knowledge graphs. Bibliothek Forschung und Praxis **44**(3), 516–529 (2020)
2. Devlin, J., Chang, M.W., Lee, K., Toutanova, K.: Bert: pre-training of deep bidirectional transformers for language understanding. arXiv preprint arXiv:1810.04805 (2018)

3. Fabbri, A.R., Ng, P., Wang, Z., Nallapati, R., Xiang, B.: Template-based question generation from retrieved sentences for improved unsupervised question answering. In: Proceedings of the 58th Annual Meeting of the Association for Computational Linguistics, pp. 4508–4513 (2020)
4. Johnson, R., Watkinson, A., Mabe, M.: The STM report. An overview of scientific and scholarly publishing, 5th edn, October 2018
5. Liu, P., Yuan, W., Fu, J., Jiang, Z., Hayashi, H., Neubig, G.: Pre-train, prompt, and predict: a systematic survey of prompting methods in natural language processing. arXiv preprint arXiv:2107.13586 (2021)
6. Liu, Y., et al.: Roberta: a robustly optimized Bert pretraining approach (2019). https://arxiv.org/abs/1907.11692
7. Petroni, F., et al.: Language models as knowledge bases? In: Proceedings of the 2019 Conference on Empirical Methods in Natural Language Processing and the 9th International Joint Conference on Natural Language Processing (EMNLP-IJCNLP), pp. 2463–2473 (2019)
8. Rajpurkar, P., Jia, R., Liang, P.: Know what you don't know: unanswerable questions for squad. In: Proceedings of the 56th Annual Meeting of the Association for Computational Linguistics (Volume 2: Short Papers), pp. 784–789 (2018)
9. Rajpurkar, P., Zhang, J., Lopyrev, K., Liang, P.: Squad: 100,000+ questions for machine comprehension of text. In: Proceedings of the 2016 Conference on Empirical Methods in Natural Language Processing, pp. 2383–2392 (2016)
10. Schick, T., Schütze, H.: Exploiting cloze-questions for few-shot text classification and natural language inference. In: Proceedings of the 16th Conference of the European Chapter of the Association for Computational Linguistics: Main Volume, pp. 255–269 (2021)
11. Surdeanu, M., Ji, H.: Overview of the English slot filling track at the tac2014 knowledge base population evaluation. In: Proceedings of the Text Analysis Conference (TAC2014) (2014)
12. Taylor, W.L.: "Cloze procedure": a new tool for measuring readability. Journal. Q. **30**(4), 415–433 (1953)
13. Vaswani, A., et al.: Attention is all you need. Adv. Neural Inf. Process. Syst. **30** (2017)
14. Wang, W., et al.: Minilm: deep self-attention distillation for task-agnostic compression of pre-trained transformers (2020)
15. Wei, J., et al.: Chain of thought prompting elicits reasoning in large language models. arXiv preprint arXiv:2201.11903 (2022)
16. Zhong, W., et al.: ProQA: structural prompt-based pre-training for unified question answering. arXiv preprint arXiv:2205.04040 (2022)

Variables are a Curse in Software Vulnerability Prediction

Jinghua Groppe$^{(\boxtimes)}$, Sven Groppe, and Ralf Möller

Institute of Information Systems (IFIS), University of Lübeck, Ratzeburger Allee 160,
23562 Lübeck, Germany
`jinghua.groppe@uni-luebeck.de`

Abstract. Deep learning-based approaches for software vulnerability prediction currently mainly rely on the original text of software code as the feature of nodes in the graph of code and thus could learn a representation that is only specific to the code text, rather than the representation that depicts the 'intrinsic' functionality of a program hidden in the text representation. One curse that causes this problem is an infinite number of possibilities to name a variable. In order to lift the curse, in this work we introduce a new type of edge called *name dependence*, a type of *abstract syntax graph* based on the name dependence, and an efficient node representation method named *3-property encoding* scheme. These techniques will allow us to remove the concrete variable names from code, and facilitate deep learning models to learn the functionality of software hidden in diverse code expressions. The experimental results show that the deep learning models built on these techniques outperform the ones based on existing approaches not only in the prediction of vulnerabilities but also in the memory need. The factor of memory usage reductions of our techniques can be up to the order of 30,000 in comparison to existing approaches.

Keywords: deep learning · software security · software vulnerability ·
abstract syntax graph · 3-property encoding · name dependence

1 Introduction

A number of efforts have been dedicated to applying deep learning (DL) to predict the vulnerabilities of software code. However, DL-based approaches have not achieved significant breakthroughs in this field and still have a limited capability to distinguish vulnerable code from non-vulnerable one [1]. Currently, DL approaches, both unstructured [2,5,8,11] or structure-based [1,6,7,9,10], borrowed the method used in the natural language processing to define the semantics of the full code or nodes in a code graph. The full code or a piece of the code is considered plain text like a natural language and it is first split into tokens, and each token is represented by a real-valued vector called embedding.

This work is part of the BMBF project with the contract number 16KIS1337.

C. Strauss et al. (Eds.): DEXA 2023, LNCS 14146, pp. 516–521, 2023.
https://doi.org/10.1007/978-3-031-39847-6_41

Unstructured approaches learn the representation of the code only based on the sequence of the tokens. The sophisticated graph-based approaches learn a presentation based on the tokens appearing in each node and the relations between nodes.

A functionality can be programmed using an infinite number of text representations and one main reason for the infinity is the arbitrariness in naming variables. For example, the functionality of the summation of two variables can be coded as $a + b$, $x1 + x2$ or using any other names. Since different names have different embeddings, a DL model, which learns based on the raw code text, could only find a representation, which is specific to the code text with the used variable names, and would not be able to capture the intrinsic functionality beyond the diversity of code expression using different variable names.

We could not obtain a well-generalized model in the presence of an infinity of text code of a functionality. Therefore, we need solutions to transform an infinite number of text representations of variable names into a finite number and we are suggesting such a solution in this work. Concretely, we suggest a new edge type of name dependence and an abstract syntax graph (ASG) that extends a standard abstract syntax tree (AST) with the edges of name dependence and develop a 3-property node encoding scheme based on the ASG. These techniques can be used to remove variable names from code, and greatly mitigate the semantic uncertainty of variables and the infinity of text code of a functionality. The empirical evidence presented later shows that our techniques do help DL models to learn the intrinsic functionality of the software and improve their prediction performance.

2   Breaking the Curse of Variables

In order to help DL models of software vulnerability prediction to improve their generalization ability, in this section, we suggest techniques of how to transform an infinite number of text representations of varable names into a finite number.

2.1   Name Dependence and Abstract Syntax Graph

In programming languages, a variable is related to its declaration (which is either explicitly given or implied). We can determine this relation by the name of the variable. Software engineering uses the term 'dependence' to describe the relations between two components, like data dependence, and control dependence. To align with it, we define a new kind of dependence called *name dependence* to express the relation between a variable and its declaration. In an AST with full information, the name dependence between two nodes can be inferred by the names of variables and identifiers. When we remove the names of variables and identifiers from the AST, we lose the information on name dependence. Without the information, we will not be able to restore the semantics of the original code. So, we need a way to express the name dependence when names are absent. A solution is to add an edge of name dependence between two related nodes. After

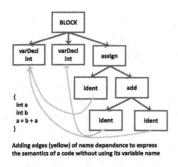

Adding edges (yellow) of name dependence to express
the semantics of a code without using its variable name

Fig. 1. ASG example

Table 1. 3-property encoding scheme

construct	3-prop. encode with variable names			3-prop. encode without variable		
	class	name	type	class	name	type
int a	varDecl	a	int	varDecl	–	int
If (a ≥ 0)	control	IF	–	control	IF	–
a*0.01	mathOp	mul	–	mathOp	MUL	–
f(a)	call	f	–	call	f	–
a	ident	a	int	ident	VAR	int
stdout	ident	stdout	–	ident	stdout	–
{...}	block	–	–	block	–	–
10.01	literal	10.01	float	literal	–	float
'Hi'	literal	'Hi'	str	literal	–	str
int[8] b	varDecl	b	int[8]	varDecl	–	int[N]

adding such edges, the tree structure turns into a graph structure as illustrated in Fig. 1, which we call *abstract syntax graph* (ASG). From the graph, we can construct a fragment of code with the exact semantics as the original code, but perhaps with a different text representation, which would not be a problem at all for the task of vulnerability prediction.

2.2   3-Property Encoding Scheme

Apart from the ASG, we further suggest a method to efficiently represent the nodes in a code graph, *3-property encoding*, which provides a consistent description of the feature of nodes and allows DL models to infer the commons and differences between nodes easily. This 3-property encoding is developed in the context of our ASG but it can be applied to other code graphs and it is also programming languages agnostic.

In a code graph, every node represents an executable syntactic construct in code, which can be an expression, a statement or its constituent parts, like variables and constants (which are of course also executable). Currently, the piece of code that consists of the construct (with or without a notation to the construct like 'varDecl' and 'add') is used as the feature of the node. The feature is encoded by first splitting the piece of code into tokens and then averaging the embeddings of all the tokens. The code-based encoding uses the original piece of code to present the feature of a node, and at the same time, the result of encoding blurs the semantics of the original code since the averaging operation. Our 3-property encoding avoids these two issues by introducing additional information related to the language constructs.

Each language construct has its properties, which may not explicitly appear in the raw code text. Independent of specific programming languages, we found that it is enough to use three properties to describe different constructs: *class*, *name* and *type* of data if any, and each value of the properties will be represented by a unique token. Table 1 demonstrates several common language constructs and their representation with the three properties. With this property-based approach, we can encode all nodes in a consistent way, and this is a very valuable characteristic for many applications. So far, this 3-property encoding has not removed the diversity of text representations and we will further normalize this encoding scheme to mitigate the diversity as much as possible based on the name dependence and ASG.

Besides the variable names which we can remove thanks to the edges of name dependence, there are also other constructs in code, which can have any values. One of them is literals, e.g. 0.01 and 'Hello', which will cause similar issues as variable names, so we will also remove the concrete value of a literal. Another construct is array declarations with size, e.g. $char[8]$, $char[1024]$. We will normalize them as e.g. $char[N]$. A more refined solution could be to create several normalized data types e.g., $char[int8]$, $char[int16]$ and so on, and normalize the data type of arrays according to their sizes. For instance, any char arrays with sizes between 0 and 256 could be normalized to $char[int8]$. Table 1 also provides examples of normalized representations. The definition of the classes of language constructs and the normalized tokens could vary depending on the implementation of applications and the tool for generating code graphs.

3   Evaluation

In order to evaluate our techniques, we build four types of code graphs, AST, AST+, ASG and ASG+, for training DL models of software vulnerability prediction. AST+ is an AST extended with flow and data dependencies and control flow. ASG is an AST with the edges of name dependence and variable names removed, and ASG+ is ASG with flow and data dependencies and control flow.

Models: We develop two models (3propASG and 3propASG+), which use our graph structures and 3-property node encoding scheme, and two baselines (codeAST and codeAST+), which adopt the common graph structures and the pieces of code as the feature of nodes (i.e., the code-based encoding presented in Sect. 2) that is currently adopted by existing models [1,10]. All models share the following architecture: the input data is delivered to the layer of GGRU with one time step, the least expensive option. The output of GGRU is sent to each of three 1D convolution (Conv1d) layers with 128 filters each and perceptive fields of 1, 2, and 3 respectively, and one 1D max pooling (MaxPool1d) is applied over the output of each Conv1D to perform downsample. The results of the Max-Pool1d layers are concatenated together and sent to the hidden layer with 128 neurons, and a 25% dropout is applied to the output of convolution and the hidden layer. We apply Relu for non-linear transformation and the embeddings of 100 dimensions to encode tokens.

Datasets: We use several real-world datasets from different open-source projects: Chromium+Debian [1], which contains 10,699 samples and 7.05% of which are flawed; FFmpeg+Quemu [10] with 13,428 samples and 43.68% flawed; VDISC [8] with 68,398 samples and 46.38% flawed. The tool Joern[1] is utilized to create the AST and AST+ from source code and our AST+ corresponds the code property graph of Joern.

---

[1] https://github.com/joernio/joern.

Table 2. Performance of models over the datasets

Model	Graph	Encoding	Chromium+Debian		FFmpeg+Quemu		VDISC	
			Acc	F1	Acc	F1	Acc	F1
codeAST	AST	code	92.01	30.20	55.36	57.01	77.82	75.57
3propASG	ASG	3-Prop	**92.34**	**44.97**	**60.35**	62.30	**81.27**	**79.86**
codeAST+	AST+	code	90.89	25.86	58.38	46.66	75.67	74.49
3propASG+	ASG+	3-Prop	**92.34**	44.59	57.04	**62.99**	80.94	79.63

Table 3. Memory need of three samples from Chromium+Debian

Hash (Code ID)	#nodes	#tokens	code-based	3-prop.	code-based /3-prop.
-6552851419396579257	4,409	33,659	59G	5.3M	11,220
2388171415474875762	7,012	54,157	152G	8.4M	18,052
5045872831385413038	12,077	96,805	468G	14.5M	32,268

Performance: We use 80% of the datasets as training data and 20% for validation and evaluation. The models are trained with a batch size of 32 and a learning rate of 0.001, and the Adam optimizer [4] is used to minimize the loss function. Since much empirical evidence (e.g. [3]) has shown that the pre-trained embeddings are not necessarily better than random initializations. Therefore, we use the standard normal distribution $\mathcal{N}(0,1)$ to initialize the embeddings and train models with 10 different initializations. Table 2 presents the performance of models with the best F1 values. The evaluation results show that the DL models based on our graph structures (ASG and ASG+) and 3-property encoding scheme outperform those based on existing graph structures (AST and AST+) and code-based encoding over all the datasets. Among these datasets, Chromium+Debian is extremely imbalanced and contains only 592 (6.92%) samples with vulnerability. Over this dataset, our models perform significantly well with F1. These results are strong evidence that our techniques improve the ability of DL models to infer the functionality of code.

Memory Requirement: A huge advantage of our 3-property encoding is that it has a very low memory footprint and can process very large code graphs in comparison to the existing code-based encoding. In our experiments, an 8G memory is enough to process all the data using the 3-property encoding. In comparison, the code-based encoding requires as much as 560G memory. With the 3-property encoding, the feature of each node is represented by only three tokens. With the code-based encoding, the feature of each node is represented by a piece of raw code. Although different pieces of code will create different number of tokens and the minimal node could contain only one token, all nodes are required to have the same number of tokens. This means that all the nodes in a code graph finally consist of the maximal number of tokens.

Table 3 provides the memory footprint required by our 3-property encoding and the existing code-based encoding for processing these samples. The comparison shows that our encoding scheme can be up to 30,000 times more efficient than the code-based encoding. This explains why existing works [1,10] only use the code samples with a number of nodes less than 500.

4 Conclusions

In order to break the curse of variables, we introduce the edges of name dependence and ASG extending AST with this new type of edges and suggest a 3-property node encoding scheme based on the ASG. These techniques not only allow us to represent the semantics of code without using its variable names but also allow us to encode all nodes in a consistent way. The evaluation shows that our techniques do improve the abilities of DL models to predict software vulnerabilities. Furthermore, we also believe that the 3-property encoding will be also a useful technique for many tasks in software analysis and software engineering.

References

1. Chakraborty, S., Krishna, R., Ding, Y., Ray, B.: Deep learning based vulnerability detection: are we there yet. IEEE Trans. Softw. Eng. (2021)
2. Dam, H.K., Tran, T., Pham, T., Ng, S.W., Grundy, J., Ghose, A.: Automatic feature learning for vulnerability prediction. arXiv preprint arXiv:1708.02368 (2017)
3. Groppe, J., Schlichting, R., Groppe, S., Möller, R.: Deep learning-based classification of customer communications of a German utility company. In: Jain, S., Groppe, S., Bhargava, B.K. (eds.) Semantic Intelligence. LNEE, vol. 964, pp. 1–16. Springer, Singapore (2023). https://doi.org/10.1007/978-981-19-7126-6_16
4. Kingma, D.P., Ba, J.: Adam: a method for stochastic optimization. arXiv preprint arXiv:1412.6980 (2014)
5. Li, Z., et al.: Vuldeepecker: a deep learning-based system for vulnerability detection. arXiv preprint arXiv:1801.01681 (2018)
6. Lin, G., et al.: Cross-project transfer representation learning for vulnerable function discovery. IEEE Trans. Ind. Inf. **14**(7), 3289–3297 (2018)
7. Pradel, M., Sen, K.: Deepbugs: a learning approach to name-based bug detection. Proc. ACM Program. Lang. **2**(OOPSLA), 1–25 (2018)
8. Russell, R., et al.: Automated vulnerability detection in source code using deep representation learning. In: 17th IEEE International Conference on Machine Learning and Applications (ICMLA), pp. 757–762 (2018)
9. Wang, S., Liu, T., Tan, L.: Automatically learning semantic features for defect prediction. In: 38th International Conference on Software Engineering, pp. 297–308 (2016)
10. Zhou, Y., Liu, S., Siow, J., Du, X., Liu, Y.: Devign: effective vulnerability identification by learning comprehensive program semantics via graph neural networks. Adv. Neural Inf. Process. Syst. **32** (2019)
11. Zou, D., Wang, S., Xu, S., Li, Z., Jin, H.: Vuldeepecker: a deep learning-based system for multiclass vulnerability detection. IEEE Trans. Dependable Secure Comput. **18**(5), 2224–2236 (2019)

Feature Selection for Aero-Engine Fault Detection

Amadi Gabriel Udu[1,2] ⓘ, Andrea Lecchini-Visintini[3](✉) ⓘ, and Hongbiao Dong[1] ⓘ

[1] School of Engineering, University of Leicester, University Road, Leicester LE1 7RH, UK
[2] Air Force Institute of Technology, PMB 2014, Kaduna, Nigeria
[3] School of Electronics and Computer Science, University of Southampton, University Road, Southampton SO17 1BJ, UK
alv1e22@soton.ac.uk

Abstract. Timely and accurate detection of aero-engine faults is crucial to preventing loss of lives and equipment. In recent times, there has been a focus on data-driven approaches to fault detection in aero-engines owing to the availability of numerous sensor information which addresses the complexities of model-based techniques. However, the increased use of sensors in aero-engines induces problems relating to multicollinearity and high dimensionality in developing fault detection models. Various feature selection approaches have been proposed for tackling dimensionality problems, with each offering advantages based on the peculiarity of the data. This study, therefore, investigates the use of feature-selection approaches to address the dimensionality problems associated with aero-engine data. Our study also reveals that careful evaluation of feature selection approaches is effective in achieving earlier fault detection in aero-engines with enhanced model performance.

Keywords: Feature Selection · Fault Detection · Aero-engine · Machine Learning

1 Introduction

Within the last decade, engine related faults contributed at least 20% of all fatal aviation accidents [1]. This has necessitated an increased focus on adopting improved maintenance philosophies and development of models for predicting aero-engine faults [2]. Generally, two approaches for fault detection are reported in the literature; these include model-based and data-driven approaches. Of the two approaches, the data-driven approach can be easily applied to fault detection and has proven to offer high versatility in locating and classing faults without the burden of modelling sophisticated systems [3, 4]. Data-driven approach utilises data measured from positioned sensors and detects faults based on the analysis of symptoms and prior knowledge of characteristics of a healthy system.

There exist some inherent problems in building data-driven fault detection models using aero-engine data. These include high dimensionality resulting from the increased use of sensors, class imbalance due to low number of faults on operational flights, and data sparsity resulting from malfunctioning transducers/sensors, among others. One of the available approaches to addressing high dimensionality is feature selection. It involves selecting a subset of relevant features from the original feature set that provides the most significant information for model building. Accordingly, features that do not contribute significantly or add noise to the model are removed. In certain cases this could help in enhancing model performance, reducing computational cost involved in model training, improving interpretability, and mitigating overfitting problems [5].

In this study, feature selection is explored as a means of addressing the problem of high dimensionality inherent in aero-engine data and enhancing model performance.

2 Materials and Methods

2.1 Data Description

This study benefits from an aero-engine dataset taken from 10 x Lycoming IO-540 engines which have logged over 4,000 h in flying operations spanning a 5-year utilisation period. There are 36 time-series sensor data which are presented in Table 1 from the engines, with a total of 1,802 flights, which formed the primary data set. From these data, 120 s of the 4 critical flight segments (namely: take-off, climb, descent, and landing) for each flight were retrieved and this formed the new dataset used in this study. These segments are consistent in all flights and important for fault detection owing to the level of stress exerted on the aero-engines during these periods. Also, most air accidents occur during these segments [6].

Table 1. Some sensor information available from IO-540 aero-engine.

Sensor	Parameters
Attitude	Pressure altitude (ft), air speed (kts), angle of attack (°), pitch (°), vertical speed (ft/min)
Category	Exhaust gas temperature (EGT) mixture, power setting (%)
Electric	Alternator contact (V), current (A), voltage (V), canopy contact (V)
Fuel	Fuel level (gal) x 2, fuel remaining (gal), fuel flow (gal/hr), fuel contact (V), fuel pressure (psi)
Pressure	Manifold pressure (inHg), oil pressure (psi), barometer setting (inHg),
Speed	Engine speed (rpm)
Temperature	Cabin temperature (°C), cylinder head temperature (°C) x. 6, EGT (°C) x6, oil temperature (°C), outside air temperature (°C)

2.2 Feature Extraction

Six features comprising three time-domain features - root mean square, skewness, and kurtosis - and three frequency-domain features - mean frequency, median frequency, and spectral flatness - were extracted from the time-series data. Various studies have established the usefulness of these features in building fault detection models [7, 8]. Also, the coefficients of quadratic fit of the four flight segments were computed and taken as features. Subsequently, 8 features that exhibited infinite variance inflation factor scores were removed. This transformed to 1,288 features for each flight.

As it is an operational aero-engine dataset, there exist a class imbalance of flights where fault occurred on the aero-engine, as against normal flights without fault. Thus, a binary classification problem was considered in the study, where **Class 0**, are flights without fault occurrence (1,756 flights) and **Class 1**, represents flights where a fault occurred (46 flights).

2.3 Feature Selection Methods

Two feature selection approaches are considered in the study, comprising sequential feature selection (SFS) and feature importance (FI). SFS involves the iterative selection of a subset of features from the original feature set, where the selected features are used to train a model, and their performance is evaluated [9]. FI ranks the importance of features in a dataset based on their estimated impact on the performance of an ML model and is embedded in the 4 classifiers considered in this study [10]. FI scores are assigned based on various factors such as the frequency of a feature being used to split the data in decision-tree based models, and the impact of a feature on the model's predictive ability measured by coefficient magnitude, among others.

2.4 ML Algorithms

Four ensemble learners were considered in this study. These include random forest (RF), adaptive boosting (AdaBoost), extreme gradient boosting (XGBoost) and light gradient boost machine (LightGBM). These classifiers are robust to noise and handling of high-dimensional/imbalanced datasets [11].

2.5 Evaluation Metrics

Being a typical class imbalance problem, where there are relatively few samples with faults present, evaluation metrics such as accuracy do not give a reliable indication of the model's performance. Therefore, in this study, the area under curve (AUC) of the receiver operator characteristics (ROC) curve and geometric mean were adopted to guarantee a reliable assessment as they capture the varying aspects of model performance for class imbalance problem [12].

3 Results and Discussion

3.1 Set-Up

Python libraries used for the major part of this study include NumPy, scikit-learn, and SciPy. In computing features based on SFS, a forward selection was initiated with AUC-ROC chosen as the scoring parameter. For FI, the importance scores were retrieved based on decision trees of the four respective ML classifiers.

Only the highest scoring 500 features were selected for both feature selection methods considered in the study, as subsequent features made no significant contributions to model performance. Consequently, the selected features were stacked in rows with the highest-ranking feature on the first row and the next highest-ranking feature appended on the subsequent rows. This formed 500 rows of features with the first row containing only the highest-ranked feature, the second row containing the two highest-ranked features, and the last row containing 500 features.

Starting with the first row, each feature array was trained on the four respective ML algorithms using 5-fold stratified k-fold cross validation. Finally, the array that produced the maximum evaluation scores was retrieved. To determine the enhancement in model performance resulting from the feature selection method considered, a control model was also trained without feature selection.

3.2 Performance

Table 2 presents the performance scores for the fault detection models and the respective flight segments where these features were selected. The highest performance values and enhancement on each metrics are emphasised. The results show that feature selection led to an enhancement in model performance for all cases considered. Furthermore, FI delivered comparatively higher performance scores than SFS across all models. The model built on XGBoost had the highest AUC-ROC score of 0.972 using FI. The peak beneficiaries from FI however were models built on AdaBoost and XGBoost algorithms with corresponding AUC-ROC and G-mean enhancement of 0.332 and 0.299, respectively.

Comparatively, the models based on RF and LightGBM required fewer features to achieve their maximum performance scores than those based on AdaBoost and XGBoost. For the control fault detection model where feature selection was not implemented (No FS), the RF algorithm gave the maximum performance for AUC-ROC and G-mean, respectively. The AUC plots for the fault detection models are presented in Fig. 1. It is also

Table 2. Performance scores and features for fault detection models.

Model	FS method	AUC-ROC	G-mean	Δ AUC-ROC	Δ G-mean	No. of features	Take-off	Climb	Descent	Landing
RF	No FS	0.834	0.770			1288	324	324	324	324
	SFS	0.902	0.874	0.075	0.119	30	30	0	0	0
	FI	0.938	0.890	0.111	0.135	66	19	18	18	11
AdaBoost	No FS	0.646	0.660			1288	324	324	324	324
	SFS	0.777	0.786	0.169	0.160	51	51	0	0	0
	FI	0.967	0.920	**0.332**	0.283	141	31	40	35	35
XGBoost	No FS	0.671	0.658			1288	324	324	324	324
	SFS	0.855	0.809	0.215	0.187	216	65	57	65	29
	FI	**0.972**	**0.938**	0.310	**0.299**	106	29	36	17	24
LightGBM	No FS	0.796	0.740			1288	324	324	324	324
	SFS	0.854	0.770	0.068	0.039	159	65	57	37	0
	FI	0.962	0.922	0.173	0.197	45	10	12	9	14

noted that for the model based on RF, SFS required minimal number of features to deliver AUC-ROC and G-mean scores of about 0.902 and 0.874 respectively. Additionally, the features selected by SFS for RF and AdaBoost were taken from the take-off phase only. This suggests an early fault detection using the SFS.

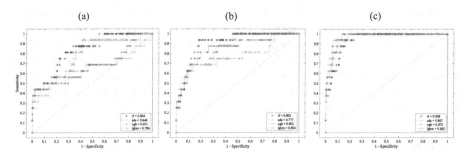

Fig. 1. AUC-ROC scores for fault detection models with (a) no feature selection, (b) sequential feature selection, and (c) feature importance.

4 Conclusion and Future Work

In this study, two feature selection approaches namely sequential feature selection and feature importance were adopted in developing fault detection models for aero-engines. For both approaches, the features were computed using the cascade of features that maximised model performance on each of the four machine learning algorithms considered in the study (*i.e.* RF, AdaBoost, XGBoost and LightGBM). The results indicate that feature selection led to an enhancement in model performance for all cases considered, with feature importance method delivering a comparatively higher performance of the approaches considered. The study suggests that adopting sequential feature selection with the appropriate machine learning algorithms could enable the early detection of faults in aero-engines. Future work could explore the use of feature selection methods in addressing data sparsity resulting from malfunctioning transducers/sensors, during aero-engine operation. This provides an alternative to the regular dropping of "not-a-number" (nan) or missing values which characterises aero-engine data and often alleviates penalties to fault detection model performance.

Declaration of Competing Interest. The authors have no conflict of interest to declare.

References

1. NTSB: Aviation accident database & synopses. Ntsb.Gov (2023). https://www.ntsb.gov/_lay outs/ntsb.aviation/index.aspx
2. Patel, D., Zhou, N., Shrivastava, S., Kalagnanam, J.: Doctor for machines: a failure pattern analysis solution for Industry 4.0. In: Proceedings 2020 IEEE International Conference Big Data, Big Data 2020, pp. 1614–1623 (2020). https://doi.org/10.1109/BigData50022.2020. 9378369

3. Poon, J., Jain, P., Konstantakopoulos, I.C., Spanos, C., Panda, S.K., Sanders, S.R.: Model-based fault detection and identification for switching power converters. IEEE Trans. Power Electron. **32**(2), 1419–1430 (2017). https://doi.org/10.1109/TPEL.2016.2541342

4. Naderi, E., Khorasani, K.: Data-driven fault detection, isolation and estimation of aircraft gas turbine engine actuator and sensors. Mech. Syst. Signal Process. **100**, 415–438 (2018). https://doi.org/10.1016/j.ymssp.2017.07.021

5. Dhal, P., Azad, C.: A comprehensive survey on feature selection in the various fields of machine learning. Appl. Intell. **52**(4), 4543–4581 (2022). https://doi.org/10.1007/s10489-021-02550-9

6. Boyd, D.D., Stolzer, A.: Accident-precipitating factors for crashes in turbine-powered general aviation aircraft. Accid. Anal. Prev. **86**, 209–216 (2016). https://doi.org/10.1016/j.aap.2015.10.024

7. Burns, T., Rajan, R.: A mathematical approach to correlating objective spectro-temporal features of non-linguistic sounds with their subjective perceptions in humans. Front. Neurosci. **13**(Jul), 1–14 (2019). https://doi.org/10.3389/fnins.2019.00794

8. Patel, D., et al.: FLOps: on learning important time series features for real-valued prediction. In: Proceedings 2020 IEEE International Conference Big Data, Big Data 2020, pp. 1624–1633 (2020). https://doi.org/10.1109/BigData50022.2020.9378499

9. Rückstieß, T., Osendorfer, C., van der Smagt, P.: Sequential feature selection for classification. In: Wang, D., Reynolds, M. (eds.) AI 2011. LNCS (LNAI), vol. 7106, pp. 132–141. Springer, Heidelberg (2011). https://doi.org/10.1007/978-3-642-25832-9_14

10. Li, J., et al.: Feature selection: a data perspective. ACM Comput. Surv. **50**(6), 1–45 (2017). https://doi.org/10.1145/3136625

11. Bentéjac, Candice, Csörgő, Anna, Martínez-Muñoz, Gonzalo: A comparative analysis of gradient boosting algorithms. Artif. Intell. Rev. **54**(3), 1937–1967 (2020). https://doi.org/10.1007/s10462-020-09896-5

12. Salim, R., Xizhao, W.: A broad review on class imbalance learning techniques. Appl. Soft Comput. **143**, 110415 (2023). https://doi.org/10.1016/j.asoc.2023.110415

Tracking Clusters of Links in Dynamic Social Networks

Erick Stattner$^{(\boxtimes)}$ ᴵᴰ

University of the French West Indies, Pointe-á-Pitre, France
erick.stattner@univ-antilles.fr

Abstract. In this work, we focus on the tracking of cluster of links, called *conceptual links*, in dynamic networks. We seek to understand how conceptual links appear and evolve during the network development. For this purpose, we propose a set of measures to capture some behaviors characterizing the evolution of these clusters. Our approach is used to understand the evolution of the conceptual links extracted on two real world networks: a scientific co-author network and a mobile communication network. The results obtained highlight significant trends in the evolution of the conceptual links in these two networks.

Keywords: Social network analysis · Network dynamics · Clustering · Cluster tracking

1 Introduction

Several methods of network clustering have been proposed in recent years [3]. It is interesting to note that these different approaches have allowed an evolution of the notion of *"clusters"* in social networks. Indeed, whereas the first contributions exploited only the structure of the networks to extract densely connected groups of nodes, also called *communities* [4], more recent approaches have focused on the extraction of more complex *clusters* defined both by their structure and by the attributes of the nodes [8]. One of the recent network clustering approach is the search for conceptual links [5]. This is a method that performs clusters of links exploiting both network structure and node attributes in order to identify frequent links between groups of nodes sharing common attributes.

However all these approaches make the assumption of static networks while it is now accepted that the temporal dimension is an inherent part of these structures [7]. Indeed, networks are living structures in which nodes and links may appear or disappears [2,6] to undergo significant structural changes.

In this work, we focus on the tracking of conceptual links in dynamic networks. More particularly, we seek to understand how conceptual links appear and evolve during the network development. For this purpose, we propose a set of measures to capture some behaviours characterizing the evolution of these clusters, namely the transitions that occur inside these clusters. Our approach has been used on two real world networks: a scientific co-author network and a mobile communication network. The results obtained highlight significant trends in the evolution of the conceptual links in these two networks.

© The Author(s), under exclusive license to Springer Nature Switzerland AG 2023
C. Strauss et al. (Eds.): DEXA 2023, LNCS 14146, pp. 528–533, 2023.
https://doi.org/10.1007/978-3-031-39847-6_43

2 Clusters of Links in Social Networks

Let us introduce $G = (V, E)$ a network, in which V is the set of nodes (vertexes) and E the set of links (edges) with $E \subseteq V \times V$.

V is defined as a relation $R(A_1, ..., A_p)$ where each A_i is an attribute. Thus, each vertex $v \in V$ is defined by a tuple $(a_1, ..., a_p)$ where $\forall q \in [1..p], v(A_q) = a_q$, the value of the attribute A_q in v.

An item is a logical expression $A = x$ where A is an attribute and x a value. The empty item is denoted $\emptyset$. An itemset is a conjunction of items for instance $A_1 = x$ and $A_2 = y$. An itemset which is a conjunction of k non empty items is called a k-itemsets. We denote I_V the set of all itemsets built from V.

Thus, for any itemset m in I_V, we denote V_m the set of nodes in V that satisfy m and we define:

- the *m-left-hand linkset* LE_m as the set of links in E that start from nodes satisfying m, i.e. nodes in V_m with $LE_m = \{e \in E \ ; \ e = (a, b) \quad a \in V_m\}$
- the *m-right-hand linkset* RE_m as the set of links in E that arrive to nodes satisfying m, i.e. nodes in V_m with $RE_m = \{e \in E \ ; \ e = (a, b) \quad b \in V_m\}$

Definition 1. Conceptual link. For any two itemsets m_1 and m_2 in I_V, the *conceptual link* (m_1, m_2) of G is the cluster of links connecting nodes in V_{m_1} to nodes in V_{m_2} (as shown on Fig. 1).

For instance, if m_1 is the itemset cd and m_2 is the itemset efj, the *conceptual link* $(m_1, m_2) = (cd, efj)$ includes all links in E between nodes in V that satisfy the property cd with nodes in V that satisfy the property efj. Thus, $|(m_1, m_2)|$ gives the number of links connecting nodes in V_{m_1} to nodes in V_{m_2}. Let L_V be the set of conceptual links of $G = (V, E)$ and (m_1, m_2) be any element in L_V.

$$(m_1, m_2) = LE_{m_1} \cap RE_{m_2} = \{e \in E \ ; \ e = (a, b) \quad a \in V_{m_1} \text{ and } b \in V_{m_2}\} \quad (1)$$

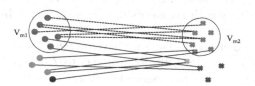

Fig. 1. Example of conceptual link extracted between m_1 and m_2

Definition 2. Support of conceptual link. We call *support* of any element $l = (m_1, m_2)$ in L_V, the proportion of links in E that belong to l.

$$supp(l) = \frac{|(m_1, m_2)|}{|E|} \quad (2)$$

Definition 3. Frequent Conceptual Link. Given a real number $\beta \in [0..1]$, a conceptual link l in L_V is *frequent* if its support is greater than a minimum link support threshold β,

$$supp(l) > \beta \quad (3)$$

3   Methodology for Tracking Clusters of Links

Our objective is to track the evolution of the clusters of links during the evolution of the network. As depicted on Fig. 2 we have adapted the rules used in order to track traditional communities in social networks [1] to conceptual links.

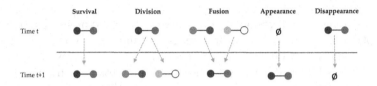

Fig. 2. Transitions that may occur on a conceptual link

More precisely, let $L = E(m_1, m_2)$ and $L' = E(m'_1, m'_2)$ be two clusters of links extracted at successive states of the network. We define the *similarity* between L and L' as being the amount of links shared by these two sets.

$$sim(L, L') = min \left(\frac{|L \cap L'|}{|L|}, \frac{|L \cap L'|}{|L'|} \right) \tag{4}$$

This measurement is between 0 and 1 and provides knowledge on the size of the intersection between L and L'. If it is equal to 0, it indicates that the sets are disjoint, while a value close to 1 indicates that the clusters are very close.

In this way, from this notion of similarity, we define the transitional behaviors that characterize the changes that may occur on a conceptual link between the state G_t of the network and its state G_{t+1}. Typically, let L be a cluster of links extracted from G_t, we introduce $match(L)$ as the set of conceptual links L' in G_{t+1} whose the intersection with L is greater than a certain threshold, that is to say whose the similarity with L exceeds the given threshold. In our experiments the threshold is set to 0.75.

If there is no such cluster in G_{t+1} we have $match(L) = \emptyset$, namely no cluster in G_{t+1} is identified as being sufficient similar to L in G_t. Thus by exploiting $match(L)$, we use a set of rules to identify each of the evolution behaviors depicted in Fig. 2.

- **Fusion:** L in G_t merges with other conceptual links of G_t to form L' in G_{t+1} if $L' \in match(L)$ and $\exists Z \neq L$ in G_t such as $L' \in match(Z)$.
- **Division:** L in G_t divides in several conceptual links $L'_1, L'_2, ...L'_k$ in G_{t+1}, if $\forall i, L'_i \in match(L)$.
- **Survival:** L in G_t becomes L' in G_{t+1}, if $L' \in match(L)$ and $\forall Z \neq L$ in G_t, $L' \notin match(Z)$.
- **Disappearance:** L in G_t disappears, if none of the above cases occur.
- **Appearance:** L in G_t appears, if $\forall L$ in G_t, $L' \notin match(L)$

Finally, note that the fusion and division events are not disjoint. Indeed one part of a cluster may be involved in a merging process while another part may be involved in a division.

4   Experimental Results

We have applied our approach to two dynamic networks having different nature.

(i) A collaboration network, that represents the co-writing links of the scientific articles published at the conference EGC (*Extraction and Knowledge Management*) from 2004 to 2015. The network went from 140 nodes to 1400 nodes from 2004 to 2015, while the links went from 340 to 2500 over the same period.

(ii) A mobile communications network that represents the phone calls made by subscribers of a local mobile phone operator on the 1st of June 2009 from 5am to 3pm. The number of nodes varies from about 7000 to 250000 on the study period, while the number of links varies from about 7000 to 250000.

It should be noted that these networks only grow during their development; no link or node is removed from the network.

In a first step, we have studied the fraction of appearance and disappearance for both networks (see Fig. 3) between the states G_t and G_{t+1} of the network.

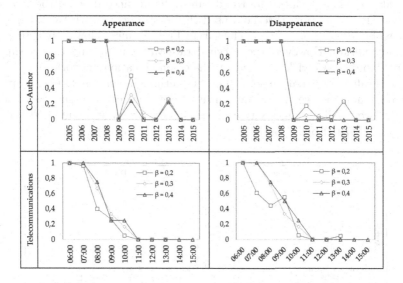

Fig. 3. Appearance and Disappearance rates in both networks

We observe that during the first moments of the study, 100% of the clusters are identified as appearances. This can be explained by the fact that, at the beginning of the study, the clusters are volatile because of the small amount of data that do not reflect strong tendencies. A similar observation can also be made if we focus on the rate disappearance that is 100% during the first moments of the study. However, after a certain period of study, we observe a decrease in the rate of appearances and disappearances.

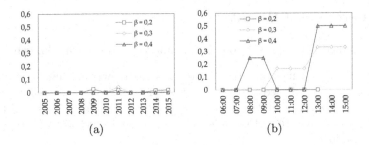

Fig. 4. Survival rate in (a) co-author network and (b) communication network

These results suggest that on these two networks, after a certain period of data accumulation, the tendencies tend to be confirmed from one state to another, which lead to a decrease in the rates of appearance and disappearance.

In order to understand what happens to these clusters that do not to disappear, we focused on the survival rate on both networks. Figure 4(a) shows the results on the co-author network and Fig. 4(b) on the communication network.

Unlike previous results, the trends are different according to the datasets. Whatever is the support threshold used the co-author network has a negligible fraction of clusters that survive. The survival rate is always lower than 10% over the study period. However, the survival rate varies according to the support threshold and grows with the time on the communication network.

The results show that on some datasets, although clusters do not disappear, we also do not observe significant survival rates between two successive states of the network. This suggests that these clusters are only maintained through fusion and division behavior.

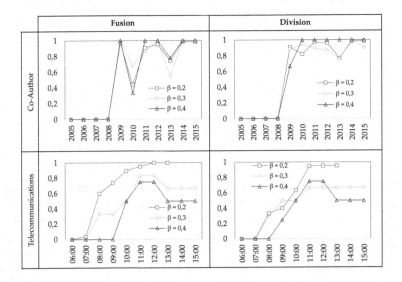

Fig. 5. Fusion and Division rates in both networks

Finally, if we focused on the fusion and division rates (Fig. 5), we can observe that the large majority of transitions that take place in the co-author network appear to be fusions and divisions of clusters (we remind that these two events are not disjoint). On the communication network, the rates of fusion and division also remain important, but vary according to the support threshold β.

Thus this result suggests that, on these two networks, a certain accumulation time is necessary to identify strong and tenacious clusters, which only evolve by mixing with the others in fusion and division processes.

5 Conclusions and Future Directions

In this paper, we have proposed a methodology to track cluster of links in dynamic networks. The contributions of this paper can be summarized as follows. (i) We identified five transitions that could take place on the clusters between two states of the network and introduced the notion of similarity to formally describe these transitions. (ii) Our method has been implemented and tested on two social networks experiencing significant changes in their structure over time. (iii) The results have highlighted significant trends in cluster evolutions.

In perspective, we plan to apply our approach to other datasets in order to confirm the observed trends or identify new ones.

References

1. Aynaud, T., Fleury, E., Guillaume, J.L., Wang, Q.: Communities in evolving networks: definitions, detection, and analysis techniques. In: Mukherjee, A., Choudhury, M., Peruani, F., Ganguly, N., Mitra, B. (eds.) Dynamics on and of Complex Networks. Modeling and Simulation in Science, Engineering and Technology, vol. 2, pp. 159–200. Springer, New York (2013). https://doi.org/10.1007/978-1-4614-6729-8_9
2. Calabrese, F., Smoreda, Z., Blondel, V.D., Ratti, C.: Interplay between telecommunications and face-to-face interactions: a study using mobile phone data. PLoS ONE **6**(7), e20814 (2011)
3. Fortunato, S.: Community detection in graphs. Phys. Rep. **486**, 75–174 (2010)
4. Newman, M.E., Girvan, M.: Finding and evaluating community structure in networks. Phys. Rev. E **69**(2), 026113 (2004)
5. Stattner, E., Collard, M.: Social-based conceptual links: conceptual analysis applied to social networks. In: International Conference on Advances in Social Networks Analysis and Mining (2012)
6. Stehle, J., et al.: High-resolution measurements of face-to-face contact patterns in a primary school. PloS One **6**(8), e23176 (2011)
7. Toivonen, R., Kovanen, L., Kivela, M., Onnela, J., Saramaki, J., Kaski, K.: A comparative study of social network models: network evolution models and nodal attribute models. Social Networks **31**, 240–254 (2009)
8. Yang, J., McAuley, J., Leskovec, J.: Community detection in networks with node attributes. In: 2013 IEEE 13th International Conference on Data Mining (ICDM), pp. 1151–1156. IEEE (2013)

Mind in Action: Cognitive Assessment Using Action Recognition

Sayda Elmi[1,2]([✉]), Sai Karthik Navuluru[2], and Morris Bell[1]

[1] School of Medicine, Yale University, New Haven, USA
{saida.elmi,morris.bell}@yale.edu
[2] University of New Haven, West Haven, USA
snavu3@unh.newhaven.edu

Abstract. Human action recognition aims at extracting features on top of human skeletons and estimating human pose. It has received increasing attention in recent years. However, existing methods capture only the action information while in a real world application such as cognitive assessment, we need to measure the executive functioning that helps psychiatrists to identify some mental disease such as Alzheimer, Schizophrenia and ADHD. In this paper, we propose a skeleton-based action recognition named Mind-In-Action (MIA) for cognitive assessment. MIA integrates a pose estimator to extract the human body joints and then automatically measures the executive functioning employing the distance and elbow angle calculation. Three score functions were designed to measure the executive functioning: the accuracy score, the rhythm score and the functioning score. We evaluate our model on two different datasets and show that our approach significantly outperforms the existing methods.

Keywords: Mental health · Embodied cognition · executive functioning · ADHD · Alzheimer · Action recognition · Skeleton extraction

1 Introduction

Human action recognition, as a central task in video analyses, becomes increasingly crucial and has received an important attention in recent years. Moreover, existing studies have shown promising progress on pose estimation [1,3]. To fully describe the spatial configurations and temporal dynamics in human actions, skeleton-based representations have been used for human action recognition, as human skeletons provide a compact data form to extract dynamic features in human body movements [8]. In practice, human skeletons in a video are mainly represented as a sequence of joint coordinate lists. Many action recognition techniques [12–15] have shown great performance on public benchmarks, where the body joint coordinates are extracted by pose estimators. However, such performance is not necessarily replicated in real-world scenarios, where the data

C. Strauss et al. (Eds.): DEXA 2023, LNCS 14146, pp. 534–539, 2023.
https://doi.org/10.1007/978-3-031-39847-6_44

comes from specific application requirements. The specific real-world application that we are focusing on in this paper is cognitive assessment in adults and children using cognitively demanding physical tasks. In addition, in the existing state-of the-art, only the pose information is extracted and skeleton sequences capture only action information while in a real world application such as cognitive assessment, we need to measure executive functioning in addition to the pose estimation.

Cognitive impairments, represented by inattention, laziness, hyperactivity or acting impulsively, lack of motivation and being forgetful, are commonly found among children and adults. According to [2,16], Attention Deficit/Hyperactivity Disorder (ADHD) is a psychiatric neurodevelopmental disorder that is very hard to diagnose. In fact, symptoms for ADHD arise from a primary deficit in executive functions [10]. Embodied cognition is a well established construct [18], recognizing that mental functioning involves brain and body working together and cognition develops along with and by way of physical movement. Better measures of cognition that closely relate to individual functioning and predict disability are sorely needed to provide proper remedial intervention at the appropriate time [11]. As measures of embodied cognition, two cognitive games can be used: (i) the Cross-Your-Body (CYB) game can provide sufficient psychometric observations and can be used as a measure of behavioral self-regulation. As shown in [11], the game is significantly related to cognitive flexibility, inhibitory control and working memory. The game has four trials with up to four paired behavioral rules: "touch your ears", "touch your shoulders", "touch your hips" and "touch your knees". Subjects first respond naturally, and then are instructed to switch rules by responding in the "opposite" way (e.g., touch their knees when told to touch their ears) and (ii) the Traffic-Lights Game (TLG), which is one of the core tasks with higher cognitive demand, is an attention and response inhibition task. It is similar to computerized continuous performance tests that assess sustained attention and response inhibition but is more complex and requires rhythmic upper body movement in response to commands. The participant is asked to pass a juggling ball (or any other object) from one hand to the other in rhythm to the words "Green Light", to move the ball up and down to the words "Yellow Light" and to not pass the ball when the participant hears "Red Light". The task is subsequently repeated at a faster pace in different trials. The participant is then presented with the same task but using a sequence of pictures of green, red, and yellow traffic lights as visual cues, rather than the spoken cues, thus allowing for comparison between sensory modalities in audio and visual trials. The Traffic-Lights-Game can provide sufficient psychometric observations and can be used as a measure of behavioral self-regulation. The game has four trials and the subjects are expected to perform the sequential movement for every count/beat provided by the therapist. Then, they are evaluated by three scores: (i) Action Score: helps to evaluate the working memory and represents the total number of correct actions, (ii) Rhythm Score: helps to evaluate the coordination and self-regulation and represents the total number of correct rhythms, accurately keeping the beat. Starting late and rushing to catch the beat is not correct and (iii) Functioning Score: helps to evaluate the executive functioning and represents the total number of correct actions in rhythm. Monitoring such a task and scoring it manually is tiresome and requires constant attention from the therapists.

In this paper, inspired by the success of the computer vision tools for action recognition and human skeleton extraction, we investigate how to apply a deep learning model to monitor cognitive abilities, which helps with identifying cognitive impairments.

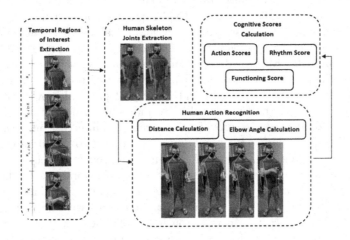

Fig. 1. Overview of the proposed architecture.

2   MIA: Model Overview

2.1   Problem Formulation

In this section, we first introduce the key data structures used in this paper and formally define our problem. Following the convention, we use capital letters (e.g., X) to represent both matrices and graphs, and use squiggle capital letters (e.g., $\mathcal{X}$) to denote sets. Human action recognition requires to temporally segment allframes of a given video. We first extract a set of temporal regions of interest $\mathcal{X}$ where $\mathcal{X} = X_{i,i\in\{1..K\}}$. Then, we predict the action in each temporal region of interest $X_{i,i\in\{1..K\}}$ and calculate a set of cognitive scores $\mathcal{S}$ where $\mathcal{S} = \{\varphi_A, \varphi_R, \varphi_F\}$ and $\varphi_A, \varphi_R, \varphi_F$ are the action, rhythm and functioning scores, respectively. The task can be formulated as follows:

Given K temporal regions of interest X, each $X_{i,i\in K}$ is a sequence of D-dimensional features where $X_{i,i\in K} = (x_1, \ldots, x_{|X_i|})$ and $x_{j,j\in\{1..|X_i|\}} \in \mathbb{R}^D$, the task is to infer the sequence of frame-wise action labels $Y_{i,i\in K} = (y_1, \ldots, y_{|Y_i|})$ and there are C action classes $\mathcal{C}$ where $\mathcal{C} = \{1, \ldots, C\}$ and $y_{j,j\in\{1..|Y_i|\}} \in \mathcal{C}$. A set of cognitive scores $\mathcal{S}$ is calculated for every Y_i for K temporal regions of interest.

2.2 Model Architecture

As shown in Fig. 1, inspired by [4], the architecture of MIA comprises four major modules which are:

- Temporal regions of interest extraction: Given a trial video, a temporal region of interest (TRI) is defined as the trial segment where the subject is told to do an action. TRI is denoted by $X_{i,i\in K}$ where K in the total number of TRIs in the video.
- Human skeleton joints extraction: For the joint localization problem on RGB data, we explored various existing state-of-the-art methods and decided on using the deep-learning architecture proposed in [13]. While some of the other pre-trained models works well with upper-body pose [6] and useful for certain other applications, our problem requires an efficient and accurate pose estimator for full-body human joint. In particular, we build upon the GCN model [13] providing highly accurate results regarding the relative position of human body-joints. We choose to use the pose estimator model stated above as it is the current state-of-the-art pose estimator for multi-person pose estimation.
- Human action recognition: Based on the Traffic-Lights task, there are a total of three classes, i.e., $|\mathcal{C}| = 3$, as represented in Fig. 1 where **class1:** hands are in the initial position at the same level, when the subject is told "Red-Light", **class2:** one hand goes up and down when the subject is told "Yellow-Light", **class3:** one hand is up and ready to pass the ball (or other object) to the other hand.
- Cognitive scoring module: The scoring protocol was created by the psychologist experts and specifies 3 cognitive scores $S = \{\varphi_A, \varphi_R, \varphi_F\}$ where $\varphi_A, \varphi_R, \varphi_F$ represent action, rhythm and function scores, respectively and they are defined as the amount of times that the subject performs the correct action and reflects how the subject follows the instructions in rhythm.

3 Experimental Evaluation

In our experiment, we collected data from a broad range of people who are healthy or clinically well characterized and when possible have had conventional cognitive assessments. The adult data collection covers the life-span from 11 year to 90 years. We collected RGB data from 35 participants that are recruited to follow the instructions provided by the music video and perform the task sequences for 4 trials with a total of around 150 videos. Traffic-Light task was performed twice for a sub-sample, approximately 2 weeks apart. Motion capture data were collected and then converted into cognitive scores.

Our MIA model aims to predict three cognitive scores, i.e., φ_A, φ_R and φ_F. We measure our method by Root Mean Square Error (RMSE) and Mean Absolute Error (MAE) for each of our predicted cognitive scores as follows:

$$\text{RMSE} = \sqrt{\tfrac{1}{n}\Sigma_{t=1}^n\left(s - \varphi(Y)_t\right)^2} \text{ and } MAE = \tfrac{1}{n}\Sigma_{t=1}^n|s - \varphi(Y)_t| \text{ where } \varphi(Y)$$

Table 1. Evaluation of MIA model in terms of RMSE and MAE

Method	Adult Data-set					
	Action Score		Rhythm Score		Function Score	
	RMSE	MAE	RMSE	MAE	RMSE	MAE
DTGRM	1.652	0.791	1.521	1.192	0.101	0.498
MSTCN++	1.532	0.738	1.422	1.132	0.925	0.412
ASRF	1.743	0.818	1.635	1.254	1.132	0.52
MSTCN	2.051	1.215	1.982	1.458	1.532	0.891
MIA	**1.312**	**0.635**	**1.198**	**0.912**	**0.707**	**0.250**

and s are the predicted score value and real score value, respectively; n is the number of all predicted score values.

For the first three trials, the subjects are required to follow the audio instructions, but for the last trial, the challenge becomes cognitively demanding as they are told to follow visual instructions showing green, yellow and red traffic lights. In the following, we evaluate the performance of our MIA model in terms of accuracy of the predicted cognitive scores. The results on four trials were used to evaluate the accuracy, reported in Table 1.

Four deep learning-based methods are used to evaluate our proposed model MIA: (i) MSTCN [5]: introduces an auxilliary self-supervised task to find correct and in-correct temporal relations in videos using smoothing loss to avoid over-segmentation errors, (ii) DTGRM [17]: uses Graph Convolution Networks (GCN) and to model temporal relations in videos. It has the ability for efficient temporal reasoning, (iii) MSTCN++ [9]: is an improvement over MSTCN where the system generates frame level predictions using a dual dilated layer that combines small and large receptive field and (iv) ASRF [7]: alleviates over-segmentation errors by detecting action boundaries. Table 1 shows the comparative performances for MSTCN, DTGRM, MSTCN++, ASRF and MIA for different predicted cognitive scores. On Adult data-sets, MSTCN++ shows a better performance against DTGRM and ASRF while MIA has the best performance comparing to other models, being immune to contextual nuisances, such as background variation and lighting changes.

4 Conclusion

This paper proposes a deep learning based action recognition method for evaluating and monitoring cognitive abilities of human subjects. We deploy a deep learning architecture to analyze human activity and provide informative measures to the experts regarding the performance of the subject, i.e., cognitive scores.

References

1. Chen, Y., Zhang, Z., Yuan, C., Li, B., Deng, Y., Hu, W.: Channel-wise topology refinement graph convolution for skeleton-based action recognition. In: ICCV, pp. 13359–13368 (2021)
2. Cormier, E.: Attention deficit/hyperactivity disorder: a review and update. J. Pediatr. Nurs. **23**(5), 345–357 (2008)
3. Duan, H., Zhao, Y., Chen, K., Lin, D., Dai, B.: Revisiting skeleton-based action recognition. In: CVPR (2022)
4. Elmi, S., Bell, M., Tan, K.L.: Deep-Cogn: skeleton-based human action recognition for cognitive behavior assessment. In: IEEE 34th ICTAI, pp. 692–699 (2022)
5. Farha, Y.A., Gall, J.: MS-TCN: multi-stage temporal convolutional network for action segmentation. In: IEEE CVPR, pp. 3575–3584 (2019)
6. Gattupalli, S., Ghaderi, A., Athitsos, V.: Evaluation of deep learning based pose estimation for sign language recognition. In: PETRA, p. 12. ACM (2016)
7. Ishikawa, Y., Kasai, S., Aoki, Y., Kataoka, H.: Alleviating over-segmentation errors by detecting action boundaries. In: IEEE WACV, pp. 2321–2330. IEEE (2021)
8. Johansson, G.: Visual perception of biological motion and a model for its analysis. Percept. Psychophysics **14**(2), 201–211 (1973)
9. Li, S.J., AbuFarha, Y., Liu, Y., Cheng, M.M., Gall, J.: MS-TCN++: multi-stage temporal convolutional network for action segmentation. CoRR, abs/2006.09220 (2020)
10. McClelland, M.M., Cameron, C.E.: Self-regulation in early childhood: improving conceptual clarity and developing ecologically valid measures. Child Dev. Perspect. **6**(2), 136–142 (2012)
11. McClelland, M.M., et al.: Predictors of early growth in academic achievement: the head-toes-knees-shoulders task. Front. Psychol. **5**(2), 599 (2014)
12. Song, Y.-F., Zhang, Z., Shan, C., Wang, L.: Stronger, faster and more explainable: a graph convolutional baseline for skeleton-based action recognition. In: MM, pp. 1625–1633. ACM (2020)
13. Song, Y.F., Zhang, Z., Shan, C., Wang, L.: Constructing stronger and faster baselines for skeleton-based action recognition. IEEE Trans. Patterns Pattern Anal. Mach. Intell. **45**, 1474–1488 (2021)
14. Sun, K., Xiao, B., Liu, D., Wang, J.: Deep high-resolution representation learning for human pose estimation. In: CVPR, pp. 5693–5703 (2019)
15. Tran, D., Wang, H., Torresani, L., Feiszli, M.: Video classification with channel-separated convolutional networks. In: ICCV, pp. 5552–5561 (2019)
16. Tucha, O.: The history of attention deficit hyperactivity disorder. ADHD Attention Deficit Hyperactivity Disord. **2**(4) (1999)
17. Wang, D., Hu, D., Li, X., Dou, D.: Temporal relational modeling with self-supervision for action segmentation. CoRR, abs/2012.07508 (2020)
18. Wilson, A.D., Golonka, S.: Embodied cognition is not what you think it is. Front. Psychol. **12**(2) (2013)

Author Index

C. Strauss et al. (Eds.): DEXA 2023, LNCS 14146, pp. 541–544, 2023.
https://doi.org/10.1007/978-3-031-39847-6

Printed in the United States
by Baker & Taylor Publisher Services